Bioprospecting of Plant Biodiversity for Industrial Molecules

Bioprospecting of Plant Biodiversity for Industrial Molecules

Edited by

Santosh Kumar Upadhyay
Department of Botany, Panjab University, Chandigarh, India

Sudhir P. Singh
Center of Innovative and Applied Bioprocessing (CIAB), Mohali, India

WILEY

This edition first published 2021

Registered Offices
John Wiley & Sons, Inc., 111 River Street, Hoboken, NJ 07030, USA
John Wiley & Sons Ltd, The Atrium, Southern Gate, Chichester, West Sussex, PO19 8SQ, UK

Editorial Office
The Atrium, Southern Gate, Chichester, West Sussex, PO19 8SQ, UK

For details of our global editorial offices, customer services, and more information about Wiley products visit us at www.wiley.com.

Wiley also publishes its books in a variety of electronic formats and by print-on-demand. Some content that appears in standard print versions of this book may not be available in other formats.

Library of Congress Cataloging-in-Publication Data applied for

Hardback ISBN: 9781119717218

Cover Design: Wiley
Cover Image: © Bernard Radvaner/Corbis/Getty Images

Set in 9.5/12.5pt STIXTwoText by Straive, Pondicherry, India
Printed and bound by CPI Group (UK) Ltd, Croydon, CR0 4YY

C9781119717218_070121

Contents

List of Contributors

Variyata Agrawal
Molecular Biology and Genetic Engineering Laboratory
Department of Botany
University of Allahabad
Prayagraj
India

Surjeet Kumar Arya
Department of Entomology
College of Agriculture, Food and Environment
University of Kentucky
Lexington
KY
USA

Shilpi Bansal
Vegetable Science Division
ICAR-Indian Agricultural Research Institute
Pusa
New Delhi
India

Nikita Bisht
CSIR-National Botanical Research Institute (CSIR-NBRI)
Lucknow
India

Academy of Scientific and Innovative Research (AcSIR)
Ghaziabad
India

Fabrice Fekam Boyom
Antimicrobial and Biocontrol Agents Unit (AmBcAU)
Laboratory for Phytobiochemistry and Medicinal Plants Studies
Department of Biochemistry
Faculty of Science
University of Yaoundé I
Yaoundé
Cameroon

Monica Butnariu
Chemistry & Biochemistry Discipline
Banat's University of Agricultural Sciences and Veterinary Medicine "King Michael I of Romania" from Timisoara
Timis
Romania

Debasis Chakrabarty
Molecular Biology and Biotechnology Division
Tissue Culture and Transformation Lab
CSIR-National Botanical Research Institute
Lucknow
India

Debarati Chakraborty
Department of Molecular Biology and Biotechnology
University of Kalyani
Kalyani
West Bengal
India

Puneet Singh Chauhan
CSIR-National Botanical Research Institute (CSIR-NBRI)
Lucknow
India

Academy of Scientific and Innovative Research (AcSIR)
Ghaziabad
India

Aksar Ali Chowdhary
Department of Botany
Central University of Jammu
Samba
Jammu and Kashmir
India

Sameer Dixit
Department of Biology
University of Western Ontario
London
Ontario
Canada

Achintya Kumar Dolui
Department of Lipid Science
CSIR-Central Food Technological Research Institute
Mysuru
Karnataka
India

Academy of Scientific and Innovative Research
Ghaziabad
Uttar Pradesh
India

Shivanand Suresh Dudhagi
CSIR-National Botanical Research Institute
Lucknow
India

Ren-You Gan
Research Center for Plants and Human Health
Institute of Urban Agriculture
Chinese Academy of Agricultural Sciences
Chengdu
China

Devina Ghai
Department of Botany
Panjab University
Chandigarh
UT
India

Nishita Giri
ICAR-Indian Institute of Soil and Water Conservation (ICAR-IISWC)
Dehradun
Uttarakhand
India

Aradhana Lucky Hans
Department of Biotechnology
Babasaheb Bhimrao Ambedkar University
Lucknow
India

Jyoti Singh Jadaun
Department of Botany
Dayanand Girls Postgraduate College
Kanpur
Uttar Pradesh
India

Subbiah Karuppusamy
Department of Botany
Botanical Research Center
The Madura College
Madurai
Tamil Nadu
India

Arshpreet Kaur
Department of Botany
Panjab University
Chandigarh
UT
India

Ramya Krishnan
CSIR-National Institute of Interdisciplinary Science and Technology
Thiruvananthapuram
Kerala
India

Current: Accubits Technologies
Thiruvananthapuram
Kerala
India

Aakanksha Kumar
Bioclues
Hyderabad
Telangana
India

Skalzang Lhamo
Department of Botany
Central University of Jammu
Samba
Jammu and Kashmir
India

Hua-Bin Li
Guangdong Provincial Key Laboratory of Food Nutrition and Health
Department of Nutrition
School of Public Health
Sun Yat-Sen University
Guangzhou
China

Jia-Hui Li
School of Science
The Hong Kong University of Science and Technology
Hong Kong
China

Min Luo
Guangdong Provincial Key Laboratory of Food Nutrition and Health
Department of Nutrition
School of Public Health
Sun Yat-Sen University
Guangzhou
China

K. Sri Manjari
University College for Women
Osmania University
Hyderabad
Telangana
India

Shakti Mehrotra
Department of Biotechnology
Institute of Engineering and Technology
Lucknow
Uttar Pradesh
India

Sonal Mishra
School of Biotechnology
University of Jammu
Jammu
Jammu and Kashmir
India

Nalini Pandey
Department of Botany
University of Lucknow
Lucknow
Uttar Pradesh
India

Garima Pathak
B.D. College – A Constituent Unit of Patiliputra University
Patna
India

Sandip V. Pawar
University Institute of Pharmaceutical Sciences
Panjab University
Chandigarh
UT
India

Thakku R. Ramkumar
Agronomy Department
IFAS, University of Florida
Gainesville
FL
USA

Sangeeta Saxena
Department of Biotechnology
Babasaheb Bhimrao Ambedkar University
Lucknow
India

Jaspreet K. Sembi
Department of Botany
Panjab University
Chandigarh
UT
India

Ao Shang
Guangdong Provincial Key
Laboratory of Food
Nutrition and Health
Department of Nutrition
School of Public Health
Sun Yat-Sen University
Guangzhou
China

Shatrughan Shiva
Department of Plant Molecular Biology
and Genetic Engineering
CSIR-National Botanical Research
Institute
Council of Scientific and Industrial
Research Rana Pratap Marg
Lucknow
India

Academy of Scientific and Innovative
Research (AcSIR)
Ghaziabad
India

Akanchha Shukla
Department of Biology
University of Western Ontario
London
Ontario
Canada

Ananya Singh
Molecular Biology and Genetic
Engineering Laboratory
Department of Botany
University of Allahabad
Prayagraj
India

Sakshi Singh
Department of Molecular Biology and
Human Genetics
Banaras Hindu University
Varanasi
Uttar Pradesh
India

Sudhir P. Singh
Center of Innovative and Applied
Bioprocessing (CIAB)
Mohali
India

Vinayak Singh
Department of Biology
University of Western Ontario
London
Ontario
Canada

Mrinalini Srivastava
Molecular Biology and Biotechnology
Division
Tissue Culture and Transformation Lab
CSIR-National Botanical Research Institute
Lucknow
India

Pankaj Srivastava
ICAR-Indian Institute of Soil and Water Conservation, (ICAR-IISWC)
Dehradun
Uttarakhand
India

Vikas Srivastava
Department of Botany
Central University of Jammu
Samba
Jammu and Kashmir
India

Kranti Thakur
Department of Botany
Shoolini Institute of Life Sciences and Business Management (SILB)
Solan
Himachal Pradesh
India

Shalini Tiwari
CSIR-National Botanical Research Institute (CSIR-NBRI)
Lucknow
India

Rufin Marie Kouipou Toghueo
Antimicrobial and Biocontrol Agents Unit (AmBcAU)
Laboratory for Phytobiochemistry and Medicinal Plants Studies
Department of Biochemistry
Faculty of Science
University of Yaoundé I
Yaoundé
Cameroon

Santosh Kumar Upadhyay
Department of Botany
Panjab University
Chandigarh
UT
India

Jagdeep Verma
Department of Botany
Government College
Rajgarh
Himachal Pradesh
India

Pankaj Kumar Verma
Department of Botany
University of Lucknow
Lucknow
Uttar Pradesh
India

Shikha Verma
Molecular Biology and Biotechnology Division
CSIR-National Botanical Research Institute
Lucknow
Uttar Pradesh
India

Xiao-Yu Xu
Guangdong Provincial Key Laboratory of Food
Nutrition and Health
Department of Nutrition
School of Public Health
Sun Yat-Sen University
Guangzhou
China

Dinesh Kumar Yadav
Molecular Biology and Genetic Engineering Laboratory
Department of Botany
University of Allahabad
Prayagraj
India

Neelam Yadav
Molecular Biology and Genetic Engineering Laboratory
Department of Botany
University of Allahabad
Prayagraj
India

Preface

Nature has the reservoir for all the desired molecules in the form of biodiversity that includes microbial, animal, and plants. Bioprospection is very well-established method for the identification and isolation of new active molecules of desired activity. Researches are being conducted to exploit the biological resource for obtaining biomolecules of pharmaceutical, bioceutical, agricultural, bioremediation, etc. significance. The exploitation of bioactive significance in the natural compounds in the biosphere is required to be intensified with systematic and sustainable approaches. The expedition and validation of the scientific parameters in the ethnic knowledge, preservation of bioresource, and biotechnological advancement in the generation of efficient biological systems, keeping in mind the approach of societal development exploitation with nature's protection, are the current demand in scientific investigations.

Bioprospection is the exploration of economic potential in the biological resource mostly in terms of nutraceutical value. In recent decades, substantial attention has been given on a variety of bioresources for bioprospecting. For example, macro- and microalgae have been demonstrated to be a biomass value of neutraceutical, pharmaceutical, food, biomedical, bioenergetic importance.

Plants are a crucial biological component of the biosphere in the earth. The plant resource has served the humankind in several ways by providing food, feed, medicine, nutraceuticals, shelter, etc. About 3.9 lakh known plant species make the animals and other organisms' life possible at the earth. Plant bioprospecting is being performed since the existence of human life on the earth. Extensive investigations have been done to explore several phytochemicals, pharmaceuticals, antioxidants, etc. There is a need to develop plant products with prebiotic properties and with high bioavailable mineral micronutrients. A rich cultural knowledge associated with multifarious health beneficial aspects of plants is available in different parts of the world. Many plants of cosmetic and perfumery importance have been shown to be of great economic value. The plants grown for production of spices and condiments have significant societal and medicinal merits. Many lower plants have been demonstrated to exhibit potential in biopesticide development. Furthermore, this is an era of secondary agriculture by the valorization of the abundant residual plant biomass.

We firmly believe that this book will be an essential repository in obtaining the holistic knowledge of plants bioprospecting. The compiled information will be useful to academicians and researchers in augmenting their understandings on the aspects mentioned earlier.

About the Editors

Source: Santosh Kumar Upadhyay

Dr. Santosh Kumar Upadhyay is currently working as an Assistant Professor in the Department of Botany, Panjab University, Chandigarh, India. Prior to this, Dr. Upadhyay was DST-INSPIRE faculty at the National Agri-Food Biotechnology Institute, Mohali, Punjab, India. He did his doctoral work at the CSIR–National Botanical Research Institute, Lucknow, and received his Ph.D. in Biotechnology from UP Technical University, Lucknow, India. He has been working in the field of Plant Biotechnology for more than 14 years. His present research focuses in the area of functional genomics. He is involved in the bioprospecting and characterization of various insect toxic proteins from plant biodiversity and defense and stress signaling genes in bread wheat. His research group at PU has characterized numerous important gene families and long noncoding RNAs related to the abiotic and biotic stress tolerance and signaling in bread wheat. He has also established the method for genome editing in bread wheat using CRISPR–Cas system and developed a tool SSinder for CRISPR target site prediction. His research contribution led the publication of more than 58 research papers in leading journals of international repute. Further, there are more than 5 national and international patents, 22 book chapters, and 6 books in his credit. In recognition of his substantial research record, he has been awarded NAAS Young scientist award (2017–2018) and NAAS-Associate (2018) from the National Academy of Agricultural Sciences, India, INSA Medal for Young Scientist (2013) from the Indian National Science Academy, India, NASI–Young Scientist Platinum Jubilee Award (2012) from the National Academy of Sciences, India, and Altech Young Scientist Award (2011). He has also been the recipient of the prestigious DST-INSPIRE Faculty Fellowship (2012) and SERB-Early Career Research Award (2016) from the Ministry of Science and Technology, Government of India. Dr. Upadhyay also serves as a member of the editorial board and reviewer of several peer-reviewed international journals.

Dr. Sudhir P. Singh is currently Scientist at the Center of Innovative and Applied Bioprocessing (CIAB), Mohali, India. He has been working in the area of molecular biology and biotechnology for more than a decade. Currently, his primary focus of research is gene mining and biocatalyst engineering for the development of approaches for transformation of agro-industrial residues and under or unutilized side-stream biomass into value-added bio-products. He has explored the metagenomic resource from diverse habitats and developed enzyme systems for catalytic biosynthesis of functional sugar molecules such as D-allulose, turanose, fructooligosaccharides, glucooligosaccharides, 4-galactosyl-Kojibiose, xylooligosaccharides, levan, dextran biosynthesis, etc. Dr. Singh has published over 55 research articles, 4 review articles, and 4 books (edited). Further, he has five patents (granted) to his credit as an inventor. He has been conferred International Bioprocessing Association–Young Scientist Award (2017), School of Biosciences–Madurai Kamraj University (SBS-MKU) Genomics Award (2017), Professor Hira Lal Chakravarty Award, ISCA, DST, (2018), and Gandhian Young Technological Innovation Award to team (2019).

Acknowledgments

We are thankful to the Panjab University, Chandigarh, and Centre of Applied and Innovative Bioprocessing (CIAB) for providing facility to complete this book. We are grateful to all the esteemed authors for their exceptional contributions and reviewers for their critical evaluation and suggestions for the quality improvement of the book.

We would like to thank Miss Rebecca Ralf (Commissioning Editor), Miss Kerry Powell (Managing Editor), and Shyamala Venkateswaran (Production Editor) from John Wiley & Sons, Ltd for their excellent management of this project and anonymous reviewers for their positive recommendations about the book.

We also appreciate the support of our friends and research students, whose discussion and comments were beneficial to shape this book. We thank our numerous colleagues for direct or indirect help in shaping this project.

SKU wishes to express gratitude to his parents, wife, and daughter for their endless support, patience, and inspiration. SPS is grateful to his parents and family for consistent moral support. SPS acknowledges the support from CIAB and the Department of Biotechnology, Government of India.

We would like to warmly thank faculties and staffs of the department and university for providing a great working environment.

1

An Introduction to Plant Biodiversity and Bioprospecting

Ramya Krishnan[1,*], *Sudhir P. Singh*[2], *and Santosh Kumar Upadhyay*[3*]

[1] *CSIR-National Institute of Interdisciplinary Science and Technology, Thiruvananthapuram, Kerala, India*
[2] *Center of Innovative and Applied Bioprocessing (CIAB), Mohali, India*
[3] *Department of Botany, Panjab University, Chandigarh, UT, India*

1.1 Introduction

There is an extensive diversity in life that has been disentangled and organized into coherent units called taxa. The five kingdom system of classification has simplified the life forms into five groups. These groups orchestrate information concerning a wide variety of characteristics such as morphological, genetic, metabolomic, ecological, etc. Kingdom Plantae is one of these five kingdoms that consists of all the plant forms on earth and is rich in its metabolomic characteristic. This kingdom is highly diverse and is composed of both seed bearing (Phanerogams) and seedless (Cryptogams) plants forming five broadly classified groups, i.e. algae, bryophytes, pteridophytes, gymnosperms, and angiosperms, which are evolutionarily related. Each of these groups consists of hundreds of thousands of known species, which in turn consist of a variety of chemicals called metabolites or more specifically secondary metabolites. These secondary metabolites or natural products are believed to possess certain biological activities that are used by the producer for their environmental and competitive fitness. Progressively, it became a paradigm that all the plants possess some potent biologically active substance/s that could have great commercial/therapeutic value to humans. It has been often argued that the currently available knowledge regarding the chemical diversity of the plant biome represents only a fraction of that diversity, hence paving way toward further explorations. Thus, their rich metabolomic diversity and its knowledge increase the opportunity for humans to utilize plants as a key resource for bioprospecting.

1.2 What is Bioprospecting

Let us widen our imaginations and visualize an underdeveloped rural village in India, where an old wise man is treating a sick man with his self-made herbal concoction comprising

*Current: Accubits Technologies, Thiruvananthapuram, Kerala, India

Bioprospecting of Plant Biodiversity for Industrial Molecules, First Edition.
Edited by Santosh Kumar Upadhyay and Sudhir P. Singh.

the roots of some wild plants, or, let us visualize him trying to squeeze a stream of juice from a bunch of leaves upon a snake-bitten area of a person's leg. Just after a few days, the concoction healed the patient of his fever and inflammations, and the juice rescued the patient from snake venom. Therefore, it could be assumed that the concoction made from the herbs might consist of metabolites having antimicrobial properties, and the juice of the leaves might consist of chemical/metabolite that had antivenom properties. Now again, let us visualize a group of scientists walking inside the same village, talking to this old wise man, collecting these herbs, and returning to their respective laboratories, where they try to screen these herbs for the presence of active compounds having antimicrobial or antivenom properties using modern technologies. This whole procedure of exploring biologically important/useful compounds from natural resources lays down foundations to the science of "bioprospecting." The former half of our visualization could be considered as "traditional bioprospecting," and the latter half of it could be considered as "modern bioprospecting." Traditional bioprospecting can even be traced back to as old as the bronze age. In 1991, a 5300-year-old corpse of an iceman "Otzi" was discovered in the Tyrolean Alps and was found to have a whipworm (*Trichuris trichiura*) infection. Surprisingly he was already equipped with the corresponding anthelmintic medicine, which is the fruiting body of the fungus *Piptoporus betulinus* [1, 2]. Thus, the utilization of natural resources for the interest of humans is as old as humankind itself, and what we follow today is just a modern and sophisticated version of this science.

The term "bioprospecting" was initially described by Reid et al. [3] as the science, where biological systems are screened for novel components that are of industrial, commercial, or scientific value. It includes the hunt for biological products that possess characteristics interesting to humankind. These characteristics could be considered to have great potentials in the field of therapeutics, agriculture, cosmetics, etc. Although the utilization of the biologically active properties of plant/animal extracts for various purposes was seen even before thousands of years, bioprospecting as a science for commercial and economic gains was introduced and progressed in and around the twentieth century. In 1958, vinblastine and vincristine, two therapeutic agents for cancer, were developed from the rosy periwinkle plant in Madagascar. These therapeutic agents were researched and manufactured by the company Eli Lilly with cues from the local shaman spiritual herbalists [4]. Further, prospecting in the wild has warranted many therapeutic agents, such as antibiotics and several other anticancer drugs. The modern biochemists and pharmacologists have been busy seeking ways to block or enhance the function of a target protein molecule for a cure to a particular disease. The classical combinatorial chemistry has its limits in the synthesis of new compounds when it comes to the exceedingly large and diverse number of the target proteins that are being identified. The diverse and continuously evolving structures of the natural products of Mother Nature may be a possible solution to these problems. Even as the rational drug designing with the help of combinatorial chemistry is becoming more important, natural products have been valuable for pharmaceutical companies owing to their wide structural diversity and their excellent adaptation to biological active structures [4]. A fast exploration of any medicine cabinet or a cosmetic shop directly indicates the share of bioprospecting natural products, by astute businessmen in building the global economy.

Chemicals, genes, and designs are the three major sources of motivation that biodiversity extends to contemporary scientists. Thus, the science of bioprospecting finds its applications with respect to these three domains and are called chemical prospecting, gene prospecting, and bionic prospecting respectively.

1.2.1 Chemical Prospecting

Nature and natural resources are a combination of diverse and repeatedly evolving systems that give rise to varying chemicals. The major defense mechanisms of the herbivores rely mainly on the chemicals synthesized by the plant [5, 6]. Communication, intraspecies and interspecies competition, attraction toward opposite sex, and pollination are also based to a great extent on chemistry and have accorded to the development of diversity [7, 8]. The scan of nature by humans for such useful chemicals has been termed as "Chemical prospecting" by Thomas Eisner [9], who in subsequent years had been tirelessly busy promoting it [10]. Although humans have been busy creating novel and diverse chemicals in the laboratory for different purposes, the contribution of the chemical diversity present in nature toward these creations has been admirable. The extent of chemical diversity found in nature has always found a role in our day-to-day lives either as a lead molecule inspiring the chemists to create certain novel compounds or the lead molecule used as common drugs. There have been many examples of natural compounds used as therapeutic agents, which later have been synthesized commercially and have led to economic gains. Most of these natural compounds were derived from either wild plants, animals, or microorganisms. Snake venom, for example, has also been a source of a number of neurological drugs. The peptides in the venom of snake *Bothrops jararaca* were responsible for the antihypertensive medicines enalapril and lisinopril [11]. This peptide in the snake venom was responsible for the inhibition of an enzyme in the human blood, which converts the enzyme angiotensin I to a hypertensive form angiotensin II. The analysis of the bioactive compounds present in the snake venom finally led to the formulation of antihypertensive drugs captopril and enalapril, etc. The nonsteroidal anti-inflammatory drug, diclofenac, was derived from the lead molecule salicin obtained from the bark of willow tree *Salix* sp. [12]. The antiviral drug vidarabine and anticancer drug cytarabine were obtained from the marine sponge [13]. Similarly, the antiviral drug acyclovir was prepared using prior knowledge of cytosine arabinoside, which was isolated from a Florida sponge [14].

The screening of the natural chemicals can be either random where materials are collected from random plants and animals or is based on the traditional knowledge, where materials are collected from plants and animals with a known function. These materials are subjected to extract preparation and bioassays to discover the bioactivity. The bioactives are further extracted and purified using automated systems. Many of the modern pharmaceutical industries have become huge economic giants, utilizing the ethnobotanical richness and diversity of nature for drug explorations.

1.2.2 Gene Prospecting

The advent of modern gene technology offers many opportunities for the selection and propagation of efficient traits. There have been many products from nature that are in the market or are close to entering the market. One of the examples of the gene prospecting is the protein hirudin from the leeches, which has been used as an anticoagulant of blood, based on the traditional knowledge of the use of leeches in thrombosis and hypertension. Another example of such protein is from the saliva of a bat named *Desmodus rotundus,* which uses this protein to inhibit the coagulation of blood in its preys, preventing thrombolytic blood clots and allowing clot-free drinking. Enzymes have also been used for a number of purposes by humans such as cheese making, meat tenderization, etc. In the present world these enzymes have grown their importance in several industries such as food, textiles, paper, etc. Gene prospecting searches for novel enzymes from natural sources, and the modern gene technology helps in the production and propagation of these enzymes at low cost and in abundance. Several important enzymes encoding protein families have been identified by gene prospecting in recent years which have diverse catalytic potentials. For instance, a number of studies reported the identification of antioxidant enzymes encoding gene families from plants including crop plants [15, 16, 17, 18, 19]. The introduction of novel technologies such as metagenomics, metaproteomics, and metatranscriptomics have allowed the isolation and production of important and useful enzymes even without the conventional cultivation of the microorganisms.

1.2.3 Bionic Prospecting

Bionics can be described as biologically influenced engineering. It relates to the construction and building of materials or systems inspired by natural systems. Most of the bionic prospecting was earlier based on biorational approaches. For example, the waxy coat of the Lotus flower is presumed to have a role in its self-cleaning mechanism, inspired by the flower, similar mechanisms have been used in the buildings and cars to prevent dirt. Novel architectures, bioengineering, sensor technologies, and bio-modeling constitute impressive areas of bionic prospecting.

1.3 Significance of Plants in Bioprospecting

Biodiversity, the heterogeneity of life, is multifariously distributed across the globe. Among all the living beings, plants are the fundamental structural aspects of the terrestrial as well as marine ecosystems and also are the basis of all food webs. With an estimated 300 000 species spread across the world, plants provide key ecosystem organization and primary production structure. High plant diversity is presumably linked to a high biotic heterogeneity, which further leads to a higher probability of specialization potential in different groups [20].

Natural products have been found to be a continuous source of novel pharmaceutical or other commercial products. Especially notable among these are the natural products from plants, which remain an everlasting source of many drugs and cosmetics. It is worth mentioning that a quarter of all the drugs prescribed today by medical practitioners come from plants [21]. This approximation indicates the significance of plant-derived natural products in the pharmaceutical industry. These plant-derived natural products have been termed as botanical therapeutics, which are compounds used to maintain good health and prevent/treat diseases. Botanical therapeutics can be further classified into a range of plant-derived products for several purposes, such as commercial drugs, botanical drugs, dietary supplements, food additives, medicinal and special dietary-use foods, and cosmaceuticals.

Commercial drugs include plant-based single compounds (such as aspirin, paclitaxel and morphine, etc.), that are used for the diagnosis, cure, mitigation, or treatment of diseases. Botanical drugs are plant-based extracts (such as the topical drug polyphenon E, Senokot etc.) that are used for the diagnosis, cure, and mitigation of diseases. Carrageenan and garlic extract are the food additives and dietary supplements, respectively, with health claims. Flavocoxid and Aloe cream are the special dietary-use food and cosmaceutical, respectively, which are also derived from plants.

Botanical therapeutic finds its roots in the indigenous knowledge of certain communities or places. A shrub *Tripterygium wilfordii,* also known as Thundergod vine was initially used in the Chinese medicine for providing relief to certain weather-generated or activity-generated symptoms such as joint pains. Later it was found that these plants consisted of certain inhibitors that prevented the production of certain inflammatory agents and cytokines, thus preventing inflammation or pain [22]. This effect was later found to be arising because of the downregulation of certain transcription factors caused by the action of two diterpenoids [23]. Another plant *Artemisia dracunculus* also known as Russian tarragon, has been used in culinary and as a medicinal herb since a long time. The ethanolic extract showed antidiabetic action in mice with type II diabetes. The extract was reported to enhance insulin-stimulated glucose uptake in case of insulin-resistant mice. Further researches revealed that the extract enhanced the building up of insulin receptor substrate-2 and a protein kinase (specific for serine-threonine) while decreased protein tyrosine phosphatase (PTP-1B) levels, both activities are known to correlate with the increase in insulin sensitivity [24]. Thus, plants have been considered as priceless sources of bioprospecting.

1.4 Pros and Cons of Bioprospecting

Bioprospecting has been a significant method of discovering new drugs and other commercially important compounds since the beginning of the scientific world. A wide range of drugs have been discovered from plants that are prevalent in the medical field [24]. With the advent of novel scientific technologies employed in the screening and extraction of commercially useful compounds from plants and other sources, bioprospecting has progressed a lot. This rapid progress has been responsible for the immense flow of certain life-saving drugs and other commercial or dietary products into the world market. Certain drugs obtained through bioprospecting have proved to be a boon to the drug industry and the medical field as well. Drugs such as vinblastine, vincristine, enalapril, etc. have proved to be significant supports to the respective patients. The recent outbreak of the SARS-CoV-2 has also triggered a massive movement in the scientific community to search for efficient and potential drugs through bioprospecting. Not only does bioprospecting help in obtaining useful products and/or drugs but is also responsible for the economic gains to the host country holding the indigenous knowledge on which the search was based upon. Apart from the economic gains to the host country, it is also responsible for economic gains to the business company and the country hosting the product. The market value of bioprospecting-derived herbal medicines alone has been found to have surpassed US$30 billion in the year 2000 [25]. Along with the dissemination of a specific indigenous knowledge to the entire world, it also provides enormous employment opportunities in different avenues. Thus, on the whole, bioprospecting contributes to the economic status of the world.

Although bioprospecting has a huge list of advantages, it is not immune to certain limitations. Biodiversity and the related indigenous information systems are the actual foundations of highly biodiverse countries such as India. The embezzlement of the indigenous knowledge has increased by the introduction of certain international laws (the Intellectual Property Rights [IPR] etc.). These traditional knowledge are sometimes stolen from indigenous communities, countries, or individuals. The term Biopiracy refers to the violation of the contractual agreement on the control and use of indigenous bioresources and knowledge without the consent of the local community. Although these international laws against stealing of traditional knowledge prevail, it often fails to offer sufficient protection to the traditional knowledge of the underdeveloped and biodiversity-rich countries.

1.5 Recent Trends in Bioprospecting

Bioprospecting finds its foundations in the traditional knowledge about the uses of certain natural products. However, in due course of time, it has progressed a lot involving the current sophisticated technologies in the identification of active compounds, its extraction and purification, etc. The advent of molecular biology techniques and biotechnology has also paved way for the production of these active compounds on a large scale in industries. Although numerous compounds have been identified from plants that have bioprospecting values, the immense wealth of nature still remains unexplored, and with each passing day researchers are busy proving their caliber in discovering these natural compounds.

Plant itself is a vast kingdom, which includes different subkingdoms and families of plants that are much more diverse and rich with respect to bioprospecting. Recent studies from the Angiosperms have revealed certain novel pharmacological aspects of the compounds present in them. Gallic acid, quercetin and catechin from *Psidium guajava* were reported to possess anti-dengue properties in vitro [26]. A bioactive ingredient of black pepper, *piperine*, was reported to have novel pharmacological activities [27]. The antioxidant and anticancer properties of the aqueous extract of *Stryphnodendron adstringens* were also evaluated and reported recently [28]. A genetic and metabolomic approach has also been sought in recent bioprospecting trends. Recently, the potential functions of a predicted biosynthetic gene cluster of plants were evaluated by superimposing their locations on metabolite quantitative trait loci [29]. Apart from medicinal aspects, angiosperms were also bioprospected for their insecticidal activities. Several studies suggested the potential applications of plant lectins as insecticides [30–35]. Plant latex was also reported for its insecticidal or antimicrobial potential [36] and anthelmintic and antifungal actions [37].

Gymnosperms have also been a useful resource for bioprospecting. The genus *Ephedra* from the family *Ephedraceae* has been reported by many researches to have potential medicinal, economic, and ecological aspects [38]. Further, the antioxidant, antiproliferative [39], and antimicrobial effects [40] of *Ephedra* were also reported The male flowers of *Ginkgo biloba* were reported to possess anticancer effects [41]. Inhibitory effects of oligostilbenes in *Gnetum latifolium* upon neuroinflammation are considered to be a boon for medicinal sciences [42].

Even the liverworts have been found to be useful resources in bioprospecting [43]. Several studies have also reported liverworts for their commercial importance. A cannabinoid-like compound in the genus *Radula* has been reported to have a similar structure to that of cannabis and has been termed as legal high [44]. A novel epi-neoverrucosane-type diterpenoid was also reported in the liverwort *Pleurozia subinflata* [45].

1.6 Omics for Bioprospecting and in silico Bioprospecting

"Omics" represent a conjugative strategy to high-throughput analysis of biological entities for multifarious objectives. It mainly involves lipidomics, genomics, transcriptomics, proteomics, and metabolomics. Plants or their endophytic microorganisms are prone to the dynamicity of their immediate environment and thus manifest an immense genetic, metabolic, and accompanying physiological variations. Traditional methods employ plant or microbe collection/isolation, their extractions and performing numerous chemical analyses making the bioprospecting process toilsome, exhaustive, and expensive. Progresses in modern technologies such as next-generation sequencing (NGS) have offered a platform that allows the prediction of the important compounds with the help of the gene sequence. Analysis of the proteome and metabolome could enhance our understanding of the molecular and biochemical attributes of the plant/microbe under study.

It is of great interest in bioprospecting to describe the genes coding for enzymes having a significant role in a biosynthetic pathway. Modern NGS has made gene ontology more compelling, that the molecular interactions between the members of the microbiome and the host can be easily predicted [46]. Elucidation of whole-genome sequences has enabled us to decipher the biosynthetic gene clusters (BCGs) and provide insights into the potentials of the genome. It has also provided opportunities to explore various important gene families in important plants including cereal crops [47, 48, 49]. The metagenome of a decaying wood biomass was helpful in finding and isolating certain novel glycosyl hydrolases in a recent study [50, 51]. Müller et al. [52] summarize the discovery of certain bioactive molecules from plant-associated microorganisms that have been deciphered with the help of plant metagenomics.

It is vital to gather knowledge on the expression patterns of the genes encoding a significant protein involved in a biosynthetic pathway in bioprospecting. Transcriptomic analysis of the biological entity renders information on the genes expressed under a particular condition. The importance of the transcriptomic information was clearly depicted in the closely related strains of a halophilic bacterium *Salinibacter ruber*, where the environmental sensing genes were downregulated in independent cultures and upregulated in coculture conditions [53]. Similarly the expression of an antitumor compound *astin* was found to occur in a fungal endophyte *Cyanodermella asteris* residing in *Aster tataricus* only in symbiotic conditions, where the plant signals were responsible for triggering the biosynthesis [54]. Another study on the actinobacterium *Streptomyces davawensis* reported the involvement of the cre gene homologs (creE and creD) in the production of desferrioxamine only under coculture conditions [55].

Transcriptomics renders information on the expression of a set of genes under certain conditions while whether it is translated to active protein products or not remains doubtful. Further, the transcriptome data enabled the identification of numerous important long-non coding RNAs in important plants, which could not be identified earlier [16, 56, 57, 58, 59]. Another omic approach, i.e. proteomics solves this dilemma and allows to know the active protein products that are translated from the expressed genes. In addition, proteomics using mass spectroscopy (MS) coupled with certain other techniques allows the researcher to know the posttranslational modifications in the active protein products [60, 61].

A promising tool used in molecular biology today is the clustered regularly interspaced short palindromic repeats (CRISPR). A strong tool for manipulation of gene expression and genome editing, CRISPR are a group of bacterial or archaeal sequences. The CRISPR-Cas system consists of a CRISPR RNA (crRNA) that binds to a Cas protein (Cas9 etc.), to form a CRISPR-Cas complex, which directs cleavage of the DNA or RNA target sequences. The number and type of CRISPR-Cas systems are varying in different organisms [62]. Hence an understanding and identification of such systems in microorganisms can support in the prediction as well as modification of gene expressions for secondary metabolite production.

Metabolomics is an omic approach where it can be used independently or as a conjugative technology along with proteomics, lipidomics, or glycomics, where it helps in the elucidation of the respective results obtained. The small molecules known as metabolites present within the cells, tissues, or organisms interact with one another in a biological system and are known as the metabolome of the respective biological entity. Metabolomics aims at deciphering the composition and interactions of the metabolome and its constituents. Unlike other omic approaches such as genomics, transcriptomics, and proteomics, metabolomics faces an analytical challenge due to the varying physical properties of the small molecules [63]. These challenges are overcome with the help of chemistry and other integrated technologies (such as high-resolution mass spectroscopy (HR-MS) and nuclear magnetic resonance (NMR)), which couple metabolomics [64]. A genetic metabolomic approach method for bioprospecting plant biosynthetic gene cluster was evaluated recently by superimposing the BCG location on metabolite quantitative trait loci [29].

On the whole, the bioprospecting territory has seen advances in the technologies involving single cell and single molecule, which has led to the progress and advancement of bioprospecting itself. Recent trends have a bias toward the computational sciences approach as they are easy, reliable, and time-saving. This has led to the revolutionization of the multidisciplinary approach involving all the areas of sciences to decipher the mechanisms of a single biological entity. Due to the large databases of diverse metabolome and proteome available today, there is a higher probability of discovering novel drugs or molecules of commercial importance through the dissemination of this information.

1.7 An Insight into the Book

This book consists of chapters that deal with different aspects of bioprospecting in plants. Our world is highly biodiverse, and this vast biodiversity unfurls an array of opportunities for its exploitation commercially, medically, or scientifically. While on the one hand the higher utilization and exploitation of natural resources for its bioprospecting aspects are considered a progressive step toward the upliftment of both the scientific knowhow and economy, on the other hand it can cause an adverse effect on the environment. Therefore, it is necessary to conserve the resources that are utilized for bioprospecting. The chapter titled “Ecological Restoration and plant biodiversity” deals with the need of conservation and restoration of biodiversity for sustainable development through biochemical and biotechnological approaches. It is very well known that plants are holobionts and not standalone entities, which consists of numerous microorganisms residing on or within them. Among these, the endophytes that reside within the plants have tight associations to them and also consist of several enzymes that may be useful in their maintenance within their host plant. These enzymes have also been useful to the human kind in various aspects and have been a subject of exploration since a long time. Thus, the chapter “Endophyte enzymes and their application in industries” discusses the enzymes characterized from diverse endophytes, their current industrial application, and the strategies applied to increase their yield. Although plants are holobionts and consist of numerous microorganisms residing along with them, not all microorganisms are beneficial to them. Some microorganisms act as plant pathogens and cause harm to them. Plants have their own immune system to recognize and fight against the entry of these pathogens. The fight against the pathogen may trigger in the plants the release of certain secondary metabolites, which have antimicrobial properties and kill the pathogens. The exploration of these antimicrobial compounds may further be useful in combating several diseases in the human population. The chapter “Anti-microbial products from plant diversity” provides recent insights into the possibilities of the important plant-derived antimicrobial compounds useful as an alternative to combat infections. Another chapter “Plant bioprospecting for biopesticides and bioinsecticides” discusses the plants that consist of natural mechanisms to evade pests and insects and further discusses its application in the production of biopesticides and bioinsecticides. The primary immune system and the production of secondary metabolites in plants protect them from biotic stresses such as microbial pathogens or insects and pests. In addition to these primary defenses, plants have mechanisms to overcome abiotic stresses as well. These include active compounds to fight against the reactive oxygen species (ROS) compounds such as antioxidants, etc. These active compounds are present in the plant products and, when included in human diet, perform the same actions as in plants. The chapter titled “Functional plants as natural sources of dietary antioxidants” describes the in vitro antioxidant activity of different kinds of functional plants, including vegetables, fruits, medicinal plants, cereals, flowers, and microalgae. Another interesting and most popular plant-derived products that are prevalent in the dietary intake all over the world are the spices and condiments. The chapter “Plant diversity and ethnobiological knowledge of spices and condiments” discusses the important chemical constituents that are responsible for the value of the spices and the current research prevalent in India with respect to spices. In addition to these, plants and plant-derived compounds are utilized for various other purposes such as in the pharmaceutical industry, food industry, and cosmetic industry. The chapter titled “Biodiversity and plant bioprospecting in cosmetics” outlines various aspects of bioprospecting in the cosmetic industry addressing the need to keep in line with various ethical guidelines and enlists different natural products that are in use in the cosmetic industry. Similarly the chapters “Plants as sources of essential oils and perfumery applications” and “Bioprospection of plants for essential mineral micronutrients” discuss the utilization of plant and plant products in obtaining the essential oils and essential mineral micronutrients, respectively. Plants are a part of not only the terrestrial ecosystem but also the marine ecosystems. In marine systems plants occur in the form of seaweeds and other micro and macroalgae. For a long time marine sources have been considered as the richest in several active compounds or commercially important compounds. The chapter "Marine bioprospecting: seaweeds for industrial molecules" provides a remarkable insight into the high value and multimodal activity profile of seaweed extracts or seaweed-derived molecules, especially industrial molecules. Furthermore, plants are not only used as sources of food, fodder, medicine, industrial molecules, etc. but are also used as an alternate source of energy. The chapters "Bioenergy crops as an alternate energy source" and "Biomass to bioenergy" discuss the utilization of energy crops as an alternate source of energy.

References

1 Dickson, J.H., Oeggl, K.D., Kofler, W. et al. (2019). Seventy-five mosses and liverworts found frozen with the late Neolithic Tyrolean Iceman: origins, taphonomy and the Iceman's last journey. *PLoS One* 14 (10): e0223752.

2 Peintner, U., Pöder, R., and Pümpel, T. (1998). The iceman's fungi. *Mycological Research* 102 (10): 1153–1162.

3 Reid, W.V., Laird, S.A., Meyer, C.A. et al. (1993). Biodiversity Prospecting: Using Genetic Resources for Sustainable Development (No. GTZ-PR 665). Kathmandu: ICIMOD.

4 Onaga, L. (2001). Cashing in on nature's pharmacy. *EMBO Reports* 2 (4): 263–265.

5 Janzen, D.H. (1975). Ecology of Plants in the Tropics. London UK: Edward Arnold Publishers Ltd.

6 Vermeij, G.J. (2016). Plant defences on land and in water: why are they so different? *Annals of Botany* 117 (7): 1099–1109.

7 Dixson, D.L. and Hay, M.E. (2012). Corals chemically cue mutualistic fishes to remove competing seaweeds. *Science* 338 (6108): 804–807.

8 Mateo, N., Nader, W., and Tamayo, G. (2001). Bioprospecting. Encyclopedia of Biodiversity, vol. 1, 471–487. New York: Academic Press.

9 Eisner, T. (1994). Chemical prospecting: a global imperative. *Proceedings of the American Philosophical Society* 138 (3): 385–393.

10 Berenbaum, M.R. (2011). Thomas Eisner: interpreter extraordinaire of nature's chemistry. *Proceedings of the National Academy of Sciences* 108 (49): 19482–19483.

11 Ianzer, D., Santos, R.A.S., Etelvino, G.M. et al. (2007). Do the cardiovascular effects of angiotensin-converting enzyme (ACE) I involve ACE-independent mechanisms? New insights from proline-rich peptides of *Bothrops jararaca*. *Journal of Pharmacology and Experimental Therapeutics* 322 (2): 795–805.

12 Jones, R. (2001). Nonsteroidal anti-inflammatory drug prescribing: past, present, and future. *The American Journal of Medicine* 110 (1): S4–S7.

13 Kijjoa, A. and Sawangwong, P. (2004). Drugs and cosmetics from the sea. *Marine Drugs* 2 (2): 73–82.

14 Newman, D.J. and Cragg, G.M. (2016). Drugs and drug candidates from marine sources: an assessment of the current "state of play". *Planta Medica* 82 (09/10): 775–789.

15 Tyagi, S., Sharma, S., Taneja, M., Kumar, R., Sembi, J.K., and Upadhyay, S.K. (2017). Superoxide dismutases in bread wheat (Triticumaestivum L.): Comprehensive characterization and expression analysis during development and, biotic and abiotic stresses. *Agri Gene* 6, 1–13.

16 Tyagi, S., Sharma, A., and Upadhyay, S.K. (2017). Role of next generation RNA seq data in discovery and characterization of long non-coding RNA in plants. ISBN 978-953-51-5847-9. In book: Next Generation Sequencing, Publisher: InTech - open science.

17 Tyagi, S., Himani, Sembi, J.K., and Upadhyay, S.K. (2018). Gene architecture and expression analyses provide insights into the role of Glutathione peroxidases (GPXs) in bread wheat (Triticumaestivum L.). *Journal of Plant Physiology* 223: 19–31.

18 Tyagi, S, Shumayla, Verma, P.C., Singh, K., and Upadhyay, S.K. (2020). Molecular characterization of ascorbate peroxidase (APX) and APX-related (APX-R)genes in Triticum aestivum L. *Genomics* 112 (6): 4208–4223. https://doi.org.10.1016/j.ygeno.2020.07.023

19 Tyagi, S., Shumayla, Madhu, Singh, K., and Upadhyay, S.K. (2021). Molecular characterization revealed the role of catalases under abiotic and arsenic stress in bread wheat (Triticumaestivum L.). *Journal of Hazardous Materials* 403, 123585. https://doi.org.10.1016/j.jhazmat.2020.123585

20 Wiens, J.A. and Rotenberry, J.T. (1968). Linnut Vcirikuvin, Otava (Helsinki) 15 Hutchinson, GE (1959) Am. Nat. 93. *Ecological Monographs* 50: 308.

21 Newman, D.J. and Cragg, G.M. (2007). Natural products as sources of new drugs over the last 25 years. *Journal of Natural Products* 70 (3): 461–477.

22 Brinker, A.M., Ma, J., Lipsky, P.E., and Raskin, I. (2007). Medicinal chemistry and pharmacology of genus Tripterygium (Celastraceae). *Phytochemistry* 68 (6): 732–766.

23 Kupchan, S.M., Court, W.A., Dailey, R.G. Jr. et al. (1972). Tumor inhibitors. LXXIV. Triptolide and tripdiolide, novel antileukemic diterpenoid triepoxides from *Tripterygium wilfordii*. *Journal of the American Chemical Society* 94 (20): 7194–7195.

24 Schmidt, B.M., Ribnicky, D.M., Lipsky, P.E., and Raskin, I. (2007). Revisiting the ancient concept of botanical therapeutics. *Nature Chemical Biology* 3 (7): 360–366.

25 Pushpangadan, P., George, V., Ijinu, T.P., and Chithra, M.A. (2018). Biodiversity, bioprospecting, traditional knowledge. Sustainable development and value added products: a review. *Journal of Traditional Medicine & Clinical Naturopathy* 7 (1): 1–7.

26 Trujillo-Correa, A.I., Quintero-Gil, D.C., Diaz-Castillo, F. et al. (2019). In vitro and in silico anti-dengue activity of compounds obtained from *Psidium guajava* through bioprospecting. *BMC Complementary and Alternative Medicine* 19 (1): 298.

27 Shityakov, S., Bigdelian, E., Hussein, A.A. et al. (2019). Phytochemical and pharmacological attributes of piperine: a bioactive ingredient of black pepper. *European Journal of Medicinal Chemistry* 176: 149–161.

28 Baldivia, D.D.S., Leite, D.F., Castro, D.T.H.D. et al. (2018). Evaluation of in vitro antioxidant and anticancer properties of the aqueous extract from the stem bark of *Stryphnodendron adstringens*. *International Journal of Molecular Sciences* 19 (8): 2432.

29 Witjes, L., Kooke, R., van der Hooft, J.J. et al. (2019). A genetical metabolomics approach for bioprospecting plant biosynthetic gene clusters. *BMC Research Notes* 12 (1): 1–5.

30 Macedo, M.L.R., Oliveira, C.F., and Oliveira, C.T. (2015). Insecticidal activity of plant lectins and potential application in crop protection. *Molecules* 20 (2): 2014–2033.

31 Upadhyay, S.K., Saurabh, S., Rai, P. et al. (2010). SUMO fusion facilitates expression and purification of garlic lectin but modifies some of its properties. *Journal of Biotechnology* 146: 1–8.

32 Upadhyay, S.K., Mishra, M., Singh, H. et al. (2010). Interaction of *Allium sativum* leaf agglutinin (ASAL) with midgut BBMV proteins and its stability in *Helicoverpa armigera*. *Proteomics* 10: 4431–4440.

33 Upadhyay, S.K., Saurabh, S., Singh, R. et al. (2011). Purification and characterization of a lectin with high hemagglutination property isolated from *Allium altaicum*. *The Protein Journal* 30: 374–383.

34 Upadhyay, S.K., Singh, S., Chandrashekar, K. et al. (2012). Compatibility of garlic (*Allium sativum* L.) leaf agglutinin and Cry1Ac δ-endotoxin for gene pyramiding. *Applied Microbiology and Biotechnology* 93: 2365–2375.

35 Upadhyay, S.K. and Singh, P.K. (2012). Receptors of garlic (*Allium sativum*) lectins and their role in insecticidal action. *The Protein Journal* 31 (6): 439–446.

36 Abarca, L.F.S., Klinkhamer, P.G., and Choi, Y.H. (2019). Plant latex, from ecological interests to bioactive chemical resources. *Planta Medica* 85 (11/12): 856–868.

37 Freitas, C.D., Viana, C.A., Vasconcelos, I.M. et al. (2016). First insights into the diversity and functional properties of chitinases of the latex of *Calotropis procera*. *Plant Physiology and Biochemistry* 108: 361–371.

38 Zhang, B.M., Zhi-Bin, W.A.N.G., Ping, X.I.N. et al. (2018). Phytochemistry and pharmacology of genus Ephedra. *Chinese Journal of Natural Medicines* 16 (11): 811–828.

39 Mellado, M., Soto, M., Madrid, A. et al. (2019). In vitro antioxidant and antiproliferative effect of the extracts of Ephedra chilensis K Presl aerial parts. *BMC Complementary and Alternative Medicine* 19 (1): 1–10.

40 Zang, X., Shang, M., Xu, F. et al. (2013). A-type proanthocyanidins from the stems of *Ephedra sinica* (Ephedraceae) and their antimicrobial activities. *Molecules* 18 (5): 5172–5189.

41 Li, M., Li, B., Xia, Z.M. et al. (2019). Anticancer effects of five biflavonoids from *Ginkgo biloba* L. male flowers in vitro. *Molecules* 24 (8): 1496.

42 Cho, H.M., Ha, T.K.Q., Pham, H.T.T. et al. (2019). Oligostilbenes from the leaves of *Gnetum latifolium* and their biological potential to inhibit neuroinflammation. *Phytochemistry* 165: 112044.

43 Singh, H., Rai, K.M., Upadhyay, S.K., Pant, P., Verma, P.C, Singh, A.P., and Singh, P.K. (2015). Transcriptome sequencing of a thalloid bryophyte; Dumortiera hirsuta (Sw) Nees: assembly, annotation, and marker discovery. *Scientific Reports* 5: 15350.

44 Chicca, A., Schafroth, M.A., Reynoso-Moreno, I. et al. (2018). Uncovering the psychoactivity of a cannabinoid from liverworts associated with a legal high. *Science Advances* 4 (10): eaat2166.

45 Kamada, T., Johanis, M.L., Ng, S.Y. et al. (2020). A new Epi-neoverrucosane-type diterpenoid from the liverwort *Pleurozia subinflata* in Borneo. *Natural Products and Bioprospecting* 10 (1): 51–56.

46 Dominic Mills, J., Kawahara, Y., and Janitz, M. (2013). Strand-specific RNA-seq provides greater resolution of transcriptome profiling. *Current Genomics* 14 (3): 173–181.

47 Kaur, A., Taneja, M.,Tyagi, S., Sharma, A., Singh, K., and Upadhyay, S.K.* (2020). Genome-wide characterization and expression analysis suggested diverse functions of the mechanosensitive channel of small conductance-like (MSL) genes in cereal crops. *Scientific Reports* 10 (1). https://doi.org.10.1038/s41598-020-73627-7

48 Sharma, A., Shumayla, Tyagi, S., Alok, A., Singh, K., and Upadhyay, S.K. (2020). Thaumatin-like protein kinases: Molecular characterization and transcriptional profiling in five cereal crops. *Plant Science*. https://doi.org./10.1016/j.plantsci.2019.110317

49 Shumayla, Tyagi, S., Sharma, A., Singh, K., and Upadhyay, S.K. (2019). Genomic dissection and transcriptional profiling of Cysteine-rich receptor-like kinases in five cereals and functional characterization of TaCRK68-A. *International Journal of Biological Macromolecules* 134: 316ŁŁ–329. https://doi.org./10.1016/j.ijbiomac.2019.05.016

50 Gilbert, J., Li, L.L., Taghavi, S. et al. (2012). Bioprospecting metagenomics for new glycoside hydrolases. In: Biomass Conversion (ed. M. Himmel), 141–151. Totowa, NJ: Humana Press.

51 Li, L.L., Taghavi, S., McCorkle, S.M. et al. (2011). Bioprospecting metagenomics of decaying wood: mining for new glycoside hydrolases. *Biotechnology for Biofuels* 4 (1): 1–13.

52 Müller, C.A., Obermeier, M.M., and Berg, G. (2016). Bioprospecting plant-associated microbiomes. *Journal of Biotechnology* 235: 171–180.

53 González-Torres, P., Pryszcz, L.P., Santos, F. et al. (2015). Interactions between closely related bacterial strains are revealed by deep transcriptome sequencing. *Applied and Environmental Microbiology* 81 (24): 8445–8456.

54 Schafhauser, T., Jahn, L., Kirchner, N. et al. (2019). Antitumor astins originate from the fungal endophyte *Cyanodermella asteris* living within the medicinal plant *Aster tataricus*. *Proceedings of the National Academy of Sciences* 116 (52): 26909–26917.

55 Hagihara, R., Katsuyama, Y., Sugai, Y. et al. (2018). Novel desferrioxamine derivatives synthesized using the secondary metabolism-specific nitrous acid biosynthetic pathway in *Streptomyces davawensis*. *The Journal of Antibiotics* 71 (11): 911–919.

56 Shumayla, Sharma, S., Taneja, M., Tyagi. S., Singh, K., Upadhyay, S.K.*. (2017). Survey of High Throughput RNA-Seq Data Reveals Potential Roles for lnc RNAs during Development and Stress Response in Bread Wheat. Front. *Plant Science* 8, 1019. 3.

57 Bhatia, G., Sharma, S., Upadhyay, S.K., Singh, K. (2019). Long Non-coding RNAs Coordinate Developmental Transitions and other Key Biological Processes in Grapevine. *Scientific Reports*. https://doi.org.10.1038/s41598-019-38989-7

58 Taneja, M., Shumayla, Upadhyay, S.K. (2021). An overview of Long non-coding RNA in plants. Upadhyay SK. (Ed). Long Non-coding RNAs in Plants: Roles in development and stress. *Academic Press* 1–14. https://doi.org./10.1016/B978-0-12-821452-7.00001-5

59 Upadhyay, S.K. (2021). Long Non-coding RNAs in Plants: Roles in development and stress. *Academic Press*, ISBN 9780128214527. https://doi.org./10.1016/C2019-0-03200-1

60 Ctortecka, C., Palve, V., Kuenzi, B.M. et al. (2018). Functional proteomics and deep network interrogation reveal a complex mechanism of action of midostaurin in lung cancer cells. *Molecular & Cellular Proteomics* 17 (12): 2434–2447.

61 Svozil, J. and Bärenfaller, K. (2017). A cautionary tale on the inclusion of variable posttranslational modifications in database-dependent searches of mass spectrometry data. In: Methods in Enzymology, vol. 586 (ed. A.K. Shukla), 433–452. Academic Press.

62 Hou, S., Brenes-Álvarez, M., Reimann, V. et al. (2019). CRISPR-Cas systems in multicellular cyanobacteria. *RNA Biology* 16 (4): 518–529.

63 Kuehnbaum, N.L. and Britz-McKibbin, P. (2013). New advances in separation science for metabolomics: resolving chemical diversity in a post-genomic era. *Chemical Reviews* 113 (4): 2437–2468.

64 Rabinowitz, J.D., Purdy, J.G., Vastag, L. et al. (2011). Metabolomics in drug target discovery. In: Cold Spring Harbor Symposia on Quantitative Biology, vol. 76 (eds. T. Grodzicker, B. Stillman and D. Stewart), 235–246. Cold Spring Harbor Laboratory Press.

2

Entomotoxic Proteins from Plant Biodiversity to Control the Crop Insect Pests

Surjeet Kumar Arya[1], *Shatrughan Shiva*[2,3], *and Santosh Kumar Upadhyay*[4]

[1] *Department of Entomology, College of Agriculture, Food and Environment, University of Kentucky, Lexington, KY, USA*
[2] *Department of Plant Molecular Biology and Genetic Engineering, CSIR-National Botanical Research Institute, Council of Scientific and Industrial Research Rana Pratap Marg, Lucknow, India*
[3] *Academy of Scientific and Innovative Research (AcSIR), Ghaziabad, India*
[4] *Department of Botany, Panjab University, Chandigarh, UT, India*

2.1 Introduction

Plant biodiversity is in danger due to many natural and man-made materials. We need to protect its existence from all these culminating threats against the plant growth. These materials could be predators, pathogens, along with man-made chemicals like pesticides. New strategy should be designed to limit the use of pesticides and protect plant crops against these predators and pathogens. In due course of this, the introduction of *Bacillus thuringiensis* entomotoxic proteins expressing genes is one of those. This strategy found to have positive impact in limiting the spread of lepidopteran pest could also have limited usage due to its antinatural product label by the consumers and biosafety issues in mammals [1]. There was a good review on agricultural land usage by Carlini and Grossi-de-Sá [1] in which they explained continuous reduction in the average cultivated land per capita and demands high protein's production and supplements its need from other animal sources. In this review, they have also explained the decrease in the agricultural exploitation but presents a loss as high as 45%, before and after harvesting, due to the attack of a variety of emerging pests, including nematodes, insects, and virus- and bacteria-induced diseases, and estimated the loss to cost around 100 billion dollars [1, 2]. The major losses are reported from arthropods that cause estimated losses around 17.7 billion dollars every year and methods available to limit its spread are heavily dependent on chemical pesticides [3, 4]. These insect pests cause tremendous losses in agriculture worldwide, which are detrimental to several crops. In a broad way, the major insects that damage various economically important crop plants are lepidoptera, coleoptera, and diptera and homoptera. Pesticide application to control the spread of these pests is costly and has environmental hazardous effects. So, there is a necessity to perform research and development work to identify the alternate approach in combating the pests to reduce crop damages and demands for

Bioprospecting of Plant Biodiversity for Industrial Molecules, First Edition.
Edited by Santosh Kumar Upadhyay and Sudhir P. Singh.

naturally occurring cheaper and more eco-friendly biopesticides. One of the approaches that have gained good attention in recent years is the introduction of genes from one source to the plant system through recombinant DNA techniques. This technique becomes technically feasible in producing the engineered crops having improved nutritional value and production of vaccines against the variety of plant diseases. Introduction of efficient biomolecules into crop plants through this genetic engineering technology made it easier for providing protection against the insect damages.

This landmark was entrenched about 15 years ago where tobacco plant was engineered to express entomotoxic protein from bacterium *Bacillus thuringiensis* [5]. There are about over 180 Bt products registered in U.S. Environmental Protection Agency [6] and about more than 276 Bt microbial formulations registered in China [7]. In spite of the robust usage in some of the countries, it has also faced challenges in acceptance due to biosafety issues in mammals as well as question raised on the ecological impacts of this newly introduced biopesticide. The consumers are considering it as an antinatural products mostly due to the crossing of the species barrier [1]. An alternate planning could be the manipulation of the plant own defense system and introduction of an insect resistance gene from other plant source into desired damaged crop plant. Inadequacy of the proper immune system in the plant system compared to animals made them to produce several structural and chemical defense molecules to combat the attack of the pest. In this chapter, we have discussed about some of the important plant defense arsenals known to cause toxic effects against the pest when administered orally or through transgenic approach. Some of these include, lectins, proteinase inhibitors, ribosome-inactivating proteins (RIPs), arcelins, α-amylase inhibitors, and plant peptides that include defensins, cyclotides, canatoxin-like proteins, ureases and urease-derived encrypted peptides, chitinases, and proteases.

2.2 Lectins

Plants have evolved with different defense strategies to cope up with the threat from phytophagous insects. The morphological and structural features are the part of defense arsenals against the pest. Among these known barriers, the release of chemical compounds also plays major role as protecting mechanism. These chemical products range from low molecular weight compounds called secondary metabolites to peptides and proteins [1, 8]. Such defense proteins include plant lectins. They are considered as nonimmune origin heterogeneous group of proteins with noncatalytic domain that can bind specifically to carbohydrate moiety and RIPs. It is also considered as multivalent proteins that can agglutinate cells. Distribution of these lectins is from various groups ranging from plants, viruses, bacteria, invertebrate, and vertebrates, including mammals. These lectins are highly conserved in the plant kingdom and purified from various plant sources. Seeds are considered as an importance reservoir of lectins. There are about more than 200 three-dimensional structures deposited in the 3D Lectin databank, mostly from legume seeds [8]. Different plant sources lectins exhibited structural similarity in the amino acid homologies that exist from *Canavalia ensiformis* (concanavalin A) and from other leguminous seeds, despite differences in their carbohydrate-binding specificities.

The research on plant lectins was dated 130 years ago. Many biochemist and molecular biologist started taking interest on carbohydrate-binding ability of this molecule. The first discovery of a plant lectin was reported by Stillmark from the seeds of castor bean (*Ricinus communis* L.) and named as "ricin." This ricin is shown to agglutinate red blood cells, and the term hemagglutinin was introduced. This finding gave birth to the word "lectin" which is derived from "legere," the larin verb for "to select" [8].

Previously, efforts were made to organize this very heterogenous group of plant lectins and classified lectins into several specificity group called "natural groups." This classification was based on their ability to recognize and bind-specific sugars [8]. Later this classification was considered irrelevant and artificial with respect to evolutionary relationship between plant lectins. New classification of the lectins was taken under consideration related to sequence information that have become available in the last decades. Based on available genome/transcriptome information, lectins are now classified into twelve distinct families of evolutionary and structurally related lectin domains. These binding domains were arranged in an alphabetical order staring from *Agaricus bisporus* agglutinin homologs, amaranthins, V chitinase homologs, cyanovirin family, *Euonymus europaeus* agglutinin family, *Galanthus nivalis* agglutinin (GNA) family, proteins with hevein domains, jacalins, proteins with a legume lectin domain, LysM domains, agglutinin family, and andricin-B family. Each individual lectin domain bears its own characteristics with one or more binding sites. Moreover, most of these domains are spread all over the plant kingdom [8].

There are many lectins that are present abundantly in seeds or numerous storage tissues which include bulbs, tubers, bark, or rhizomes [8]. Common examples are different isoforms of the *Phaseolus vulgaris* agglutinin (PHA) produced during seed development which can accommodate 10% of total seed protein [8]. Other examples of storage tissue lectins are GNA present in the bulbs of the snowdrop or the Urtica dioicaagglutinin (UDA). These known lectins accumulate in a certain plant tissues or organs and their synthesis is independent of external stimulus. One more lectin named NICTABA is shown to be expressed in tobacco (*N. tabacum*) leaves when treated with plant hormone methyl jasmonate and insect herbivory [9–11].

In the last few decades, many nonstorage tissue lectins have been described whose expression does not involved developmental regulation instead controlled by environmental stimulus like certain biotic and abiotic stress conditions such as insect herbivory, cold or high salt concentration, wounding, and drought. This group of lectins is now referred as "inducible plant lectins," and their non-storage tissues include leaves, roots, or flowers. Mannose-specific jacalin-related lectin is the first example of inducible lectin which accumulates from salt-induced rice seedlings (*Oryza sativa*) called ORYSATA [12, 13].

Basically, plant lectins are classified into four major groups due to the presence of domain architecture and differentiated as merolectins, hololectins, chimerolectins, and superlectins [8]. Hololectins are majorly isolated, and well-characterized lectins that are composed of two or more than two identical carbohydrate-binding domains which agglutinate cells. In chimerolectins, carbohydrate domain fused with each other and shows biological activities irrespective of domain organizations. Sequences analysis of plant genomes reveals abundant presence of chimerolectinsOpposed to other lectins, superlectins recognize structurally nonrelated structures with their more than two carbohydrate domains.

Many lectins reported to be toxic to numerous insect orders belongings to economically important pest groups such as Lepidoptera, Coleoptera, Diptera, or Hemiptera. Since from last two decades, insecticidal properties of the lectins have been exploited to render crop plant resistance to insect attack. One example of these include extraction and expression of the lectin in crop plants in the development of transgenic tobacco expressing pea lectin that showed reduced growth of the *Heliothis virescens* (Fabricius) [1, 14]. Other examples include the mannose specific lectin isolated from pea expressed in nine food crops found to be toxic against *C. maculatus* [15]. Previously, experiment was conducted to test the insecticidal properties of the lectins on artificial diet delivering the purified lectins. Examples of these include screening of the 25 lectins extracted from 15 different plant families and tested against the legume pod borer [16].

The lectins purified from bulbs of Amaryllidaceae species such as snowdrop (*Galanthus*) or daffodils (*Narcissus*) showed protection of bulbs against the pest. The best studied plant lectin, GNA which was purified from snowdrop bulbs shown to have toxicity against wide range of insect orders especially, Hemiptera. This GNA has been engineered successfully in many varieties of crops like as rice, wheat, potatoes, or tobacco to provide resistance against the agricultural important pest. Pea lectin purified from PSA was evaluated on the growth and survival of the pollen beetle (*Meligethes aeneus*) [17]. This pea lectin reported to have caused reduced in mass gain for larvae but had no effect on the adult beetles as was reported by Lehrman et al. [18]. To further confirm the toxicity of several legume lectins for Coleoptera, legume lectin purified from jackbean concanavalin A (ConA) has shown toxicity against Hemipteran pea aphid (*Acyrthosiphon pisum*) [19, 20]. This result clearly indicates that selectivity of a plant lectin against the pest irrespective of binding specificity.

In the past decades, there are many other lectins that have showed entmotoxic properties apart from lectins discussed above. Examples of these plant lectins include leaf lectin, NICTABA expressed after folivory have toxic effect against insect's pest. Ectopical expression studies of NICTABA using transgenic tobacco based on feeding trials clearly explained that NICTABA is toxic against the larvae of two Lepidopteran insect, the tobacco hornworm (*Manduca sexta*) and cotton leafworm (*Spodoptera littoralis*) [11]. Another protein belonging to the NICTABA family named as protein 2 (PP2) reported to show entomotoxic effect [21, 22]. Group of lectins with a ricin domain called cinnamomin expressed in seeds of the camphor tree (*Cinnamomum camphora*) shown to have toxic effect on mosquito (*Culex pipienspallens*) and bollworm (*Helicoverpa armigera*) [23]. There are other lectins that shown to have toxic effect as insect control agents against sap-sucking pest insects are amaranthins and the jacalin-related lectins. There was report of ectopic expression of *Amaranthus caudatus* agglutinin (ACA) expressed under phloem-specific promoter showed enhanced resistance against the nymphs of cotton aphid (*Aphis gossypii*) [24].

Another promising jacalin-related lectin was used in development of transgenic crop plants, called HFR1 shown to produce entomotoxic effect against Hessian fly during infestation on wheat (*Triticum aestivum*) [13, 25]. Various *Allium* species have also been screened for the isolation of mannose-binding lectins having insect toxic activities [26–29, 208]. There are several other important lectins having entomotoxic potential against various important insect pests of agriculture which are listed in Table 2.1.

Table 2.1 List of plant lectins having insecticidal properties.

Lectins	Nature of protein	Plant biological name	Plant common name	Tissue	Plant family	Target insect	Common name-insect	Insect order	Application	Reference
Wheat germ agglutmm (WGA)	GlcNac	*Triticum aestivum*	Bread wheat	Seeds	Poaceae	*Callosobruchus maculatus*	Cowpea seed beetle	Coleoptera	Effects on both developmental time and causes larval mortality	[30]
Recinous lectin	GalNac/Gal	*Ricinus communis*	Castor bean	Seeds	Euphorbiaceae	*Diabrotica undecimpunctata*	Spotted cucumber beetle	Coleoptera	Inhibit larval growth and highly causes mortality	[31]
Eranthis lectin (EHL)		*Eranthis hyemdis*	Winter aconite	Bulbs	Ranunculaceae	*Diabrotica undecimpunctata*	Spotted cucumber beetle	Coleoptera	Inhibits insect larval growth	[32]
Griffonia simplicifolia leaf lectin II (GSII)	Glc/Nac	*Griffonia simplicifolia*	Griffonia	Seeds	Fabaceae	*Callosobruchus maculatus*	Cowpea seed beetle	Coleoptera	Proteolytic degradation by Cowpea bruchid midgut extracts	[33]
Galanthus nivalis agglutinin (GNA)		*Galanthus nivalis*	Snowdrop	Seeds	Amaryllidaceae	*Sitobion avenae*	English grain aphid	Homoptera	Decrease insect fecundity but not affect survival	[34]
Artocarpus lectin		*Artocarpus hirsutus*	Wild jack	Seeds	Moraceae	*Tribolium castaneum*	Red flour beetle	Coleoptera	Grorth inhibition of insect larvae	[35]
ASA (*Allium sativum*)	Mannose-binding lectin (WsMBP1)	*Allium sativum*	Garlic	Bulbs	Amaryllidaceae	*Dysdercus cingulatus*	Red cotton bug	Hemiptera	Highest mortality in insect larvae	[36]
CEA (*Colocasia esculenta*)		*Colocasia esculenta*	Taro	Leaves	Araceae	*Dysdercus cingulatus*	Red cotton stainer	Hemiptera	Affect insect larval growth and development	[36]
DEA (*Differenbachia sequina*)		*Differenbachia sequina*	Dumbcane	Leaves	Araceae	*Dysdercus koenigii*	Red cotton stainer	Hemiptera	Affect insect larval growth and development	[36]

(*Continued*)

Table 2.1 (Continued)

Lectins	Nature of protein	Plant biological name	Plant common name	Tissue	Plant family	Target insect	Common name-insect	Insect order	Application	Reference
Parigidin-br1	Bracelet subfamily of cyclotides	*Palicourea rigida*	Bate-caixa	Leaves, inflorescences, and peduncles	Rubiaceae	*Diatraea saccharalis*	Sugarcane borer	Lepidoptera	Disrupts insect cell membranes	[37]
Wheat germ agglutinin (WGA)		*Triticum aestivum*	Bread wheat	Seeds	Poaceae	*Helicoverpa zea*	Corn earworm	Lepidoptera	WGA causes growth inhibition	[38]
PF2	Lectin	*Olneya tesota*	Desert ironwood	Seeds	Fabaceae	*Zabrotes subfasciatus*	Mexican bean weevil	Coleoptera	PF2 interferes with starch digestion in insect gut	[39]
Tma12	Chitin-binding and chitinase activity	*Tectaria macrodonta*		Fronds and rhizomes	Tectariaceae	*Bemisia tabaci*	Whitefly	Hemiptera	Number of eggs, egg-laying pattern and nymphal development were severely affected	[40]
Withania somnifera lectins	Mannose-binding lectin (WsMBP1)	*Withania somnifera*	Ashwagandha	Leaves	Solanaceae	*Hyblaea puera*	Teak defoliator	Lepidoptera	Delay in growth and metamorphosis, decreased larval body mass and increased mortality	[41]
Withania somnifera lectins		*Withania somnifera*	Ashwagandha	Leaves	Solanaceae	*Probergrothius sanguinolens*	Indian red bug	Hemiptera	Delay in growth and metamorphosis, decreased larval body mass and increased mortality	[41]
Soursop lectin	Carbohydrate-binding proteins	*Annona muricat*	Soursop	Seed kernel	Annonaceae	*Chironomus costatus*	Lake flies	Diptera	Mosquito larvicidal	[42]

2.3 Proteinase Inhibitors

Plant proteinases belongs to family Leguminosae (legumes), Solanaceae (nightshades), and Poaceae (Grasses) and considered as an important components of plants secondary metabolites. These proteinase inhibitors (PIs) target insect gut proteases upon insect herbivory. Plants engineered with PI defensive gene under inducible or constitutive promoter could be a valuable methods in comparison to genes identified from complex pathways which enhances resistance to insect pests [43, 44]. It has been reported that plant PIs target almost all the insect digestive proteases of different classes. During insect herbivory, the expression of plant PIs found to be upregulated even in the presence of endogenous plant proteases [45–47].

The classification of PIs based on reactive amino acid residue in the active site that include (i) serine protease inhibitors or Serpin present in the Bowman-Birk inhibitors (BBIs) family, squash inhibitors, Kunitz family, cereal trypsin/amylase inhibitors, (MSI), potato type I (PI 1), and potato type II protease inhibitors (PI 2), mustard (Sinapis) trypsin inhibitors; (ii) cysteine protease inhibitors (Cystatins); and (iii) aspartyl and metallocarboxypeptidase inhibitors [44, 48].

These PIs played a crucial role of plant defense against insect herbivory [44, 49, 209]. Green and Ryan proposed around four decades ago that during attack of potato beetles, a rapid accumulation of PIs in potato or tomato was observed locally as well as systemically [45]. Few years later, [50] reported transgenic tobacco expressing PIs by using plant genetic transformation methods. In this transgenic crop, plant was transformed with construct having PI-encoding genes from cowpea (Cowpea trypsin inhibitor, CpTI) linked to CaMV 35S promoter. After that, there was lot of research papers coming out that established the concept of overexpression of both native and foreign PI in plants that propelled such studies of transgenic plant development upon insect feeding [51–53].

There is a diverse group of plant PIs that differ mechanistically in their structure against the proteases [54]. Previously, the method of PIs classification was based on their protease specificity. Now a days, it has been added more with sequence information and their 3D-structures [55] In plants, numerous roles of PIs have been described that include their action as storage proteins, endogenous proteolytic activity regulators [43], part of development processes including programmed cell death [56], and as resistance moiety against insect and pathogens to protect plants [1, 57, 58].

Setting up insecticidal bioassay comprising of dietary supplementation of selected PIs which involved either uptake by feeding of purified PIs to insect in combination with artificial diet or by overexpression of the construct in the development of transgenic plants. Increased mortality was observed while delivering the PIs in an artificial diet or transforming plants with overexpressing construct [59, 60] that resulted in retarded growth and development of larvae belonging to various insect orders [44, 54, 61–64].

These plant PIs are known to inhibit the proteolytic enzymes of insect guts resulting in reduced fecundity, extended developmental period, and increased mortality due to the absence of desired amino acid residues. Upon dietary PIs feeding, it is known to start feedback mechanism that leads to compensatory enzymatic hyperproduction of digestive proteases that have caused reduced amount of essential amino acid intake by the insects [43, 44, 65]. Biochemical examination of protease activity reveals substantial molecular adjustments in the midgut cells, to counter dietary stress caused on PI feeding [44, 66, 67]. Other protease inhibitors have also played a critical role against the insect to cause significant mortality and growth hindrance (Table 2.2).

Table 2.2 List of plant protease inhibitors with insect toxic activity.

Plant protease inhibitor	Nature of protein	Plant species	Plants common name	Plant part used	Plant family	Action against insect	Insect common name	Order	Application	Reference
Potato type I inhibitor (StPin1A) + potato type II inhibitor (NaPI)	Serine protease inhibitor	*Solanum tuberosum*	Potato	Leaves	Solanaceae	*Manduca sexta*	Carolina sphinx moth	Lepidoptera	Inhibit insect laeval growth and development	[52]
Potato proteinase inhibitor II (PINII) gene (pin2)	Serine protease inhibitor	*Solanum tuberosum*	Potato	Leaves	Solanaceae	*Sesamia inferens*	Pink stem borer	Lepidoptera	Inhibit insect laeval growth	[51]
Cowpea protease trypsin inhibitor gene (CpTi)	Serine protease inhibitor	*Fragaria × ananassa*	Strawberry	Stem tissue	Rosaceae	*Otiorhynchus sulcatus*	Vine weevi	Coleoptera	Weevils had not developed to the adult stage	[68]
Mustard trypsin inhibitor (MTI-2)	Serine protease inhibitor	*Brassica napus*	Oilseed rape	Seeds	Brassicaceae	*Spodoptera littoralis*	Cotton leafworm	Lepidoptera	Growth inhibition of insect larvae	[69]
Maize proteinase inhibitor MPI	Serine protease inhibitor	*Glycine max*	Soybean	Seeds	Fabaceae	*Chilo suppressalis*	Striped rice stemborer	Lepidoptera	Affect larval growth and insect gut proteinases	[70]
Arabidopsis thaliana Serpin1 (atserpin1)	Serine protease inhibitor	*Nicotiana alata*	Jasmine tobacco	Leaves	Solanaceae	*Epiphyas postvittana*	Light brown apple moth	Lepidoptera	Affects the growth and development of larvae	[71]
Soybean PI (Kunitz and BBPI)	Serine protease inhibitor	*Variety conquista*	Soybean	Seed	Fabaceae	*Scheloribates praeincisus*	Soil mite	Oribatida	Reduce insect survival rate	[72]
N. attenuata trypsin proteinase inhibitors (NaTPIs)	Trypsin proteinase inhibitors	*Nicotiana attenuata*	Coyote tobacco	Seeds	Solanaceae	*Manduca sexta*	Tobacco hornworm	Lepidoptera	Inhibit insect gut proteinase which significantly reduce larval performance	[73]
SaPIN2a	Serine protease inhibitor	*Solanum americanum*	American black nightshade	Leaves and stems	Solanaceae	*Helicoverpa armigera*	Cotton bollworm	Lepidoptera	Reduction in the larval weight of *H. armigera*	[74]
BWI-1a (ISP)	Serine protease inhibitor	*Fagopyrum esculentum*	Buckwheat	Seeds	Polygonaceae	*Phalaena viridana*	Tortrix moth	Lepidoptera	Inhibit growth of larvae	[75]

AtSerpin1	Serine protease inhibitor	*Arabidopsis thaliana*	Arabidopsis	Leaves	Brassicaceae	*Spodoptera littoralis*	Cotton leafworm	Lepidoptera	Inhibit insect digestive proteases	[76]
PA1b (Pea Albumin 1, subunit b)	Inhibitory cysteine-knot (ICK)	*Pisum sativum*	Pea	Seeds	Fabaceae	*Sitophilus granarius*	Grain weevil	Coleoptera	Ability to kill cereal weevils	[77]
Barley cystatins (HvCPI-6)	Cysteine protease inhibitor	*Hordeum vulgare*	Barley	Leaves	Poaceae	*Myzus persicae*	Green peach aphid	Hemiptera	Causes proteolytic digestion in aphids	[78]
L. bogotensis aspartic protease inhibitor (LbAPI)	Cysteine protease inhibitor	*Lupinus bogotensis*	Lupinus	Seeds	Fabaceae	*Meloidogyne incognita*	Southern root-nematode	Tylenchida	Inhibit growth and development	[79]
oryzacystatin II proteinase inhibitor	Serine protease inhibitor	*Oryza sativa*	Rice	Leaves	Poaceae	*Leptinotarsa decemlineata*	Colorado potato beetle	Coleoptera	Reduce larval weights but mortality significantly not affected	[80]
Capsicum annuum proteinase inhibitor (CanPI7)	Serine protease inhibitor	*Capsicum annuum*	Bell peppers	Leaves	Solanaceae	*Helicoverpa armigera*	Cotton bollworm	Lepidoptera	Growth retardation in insect larvae	[81]
Kunitz trypsin inhibitor (AtKTI4 and AtKTI5)	Serine protease inhibitor	*Arabidopsis thaliana*	Arabidopsis	Leaves	Brassicaceae	*Tetranychus urticae*	Red spider mite	Trombidiformes	Increase mite mortality and reduce mite reproduction	[82]
Protease inhibitor from *Allium sativum* "garlic" (ASPI)	Trypsin proteinase inhibitors	*Allium sativum*	Garlic	Peeled garlic bulbs	Amaryllidaceae	*Aedes aegypti*	Yellow fever mosquito	Diptera	Inhibition of gut proteases	[83]
Barley protease inhibitor CI2c	Serine protease inhibitor	*Hordeum vulgare*	Barley	Leaves	Poaceae	*Myzus persicae*	Green peach aphid	Hemiptera	Reduce nymph production per aphid during the lifespan	[84]
Cystatin TaMDC1	Cysteine protease inhibitor	*Triticum aestivum*	Wheat	Seedling	Poaceae	*Leptinotarsa decemlineata*	Colorado potato beetle	Coleoptera	Inhibit insect larval growth	[85]
Cowpea trypsin inhibitor (CpTI)	Serine protease inhibitor	*Vigna unguiculata*	Cowpea	Seeds	Fabaceae	*Callosobruchus maculatus*	Cowpea weevil	Coleoptera	Act on the digestive enzymes of insects	[59]

2.4 α-Amylase Inhibitors

Being expressed in seeds as storage protein plant proteinases inhibitors considered to be part of constitutive and inducible defense mechanism against the attack of various biotic factors like pests and pathogens [86, 87]. These inhibitors work by regulating its action on insect gut proteinases and α-amylases by disrupting the proper digestion of plant proteins and starch. The role of α-amylases from insect and mammals has been well described in the point of biochemical, molecular, and structural view in many reported literatures [1, 88–90]. These enzymes present in the gut of insect when they feed on seed products. It is found to be highly conserved as an important molecule. When the action of these molecules is impeded by any inhibitors could cause decrease in the energy of the insect. These inhibitors are particularly abundant in various plant sources like [91–94] legumes [88, 95, 96]) as a part of plant defense mechanism. In addition to that, plant inhibitors are gaining interest now a days as tool to engineered resistance against pests in transgenic plants [1, 97, 98].

In the last decades focused has been given to lectin-like inhibitors identified in the common bean *P. vulgaris* seeds, which was found to have toxic effect against the insects [1, 99, 100]. aAIs identified in *Phaseoulus* seeds belongs to protein family which already have two other defensive proteins called phytohemagglutinin (PHA) and arcelins (Arc) [1, 101].

Diversity of a-AIs is very vast with different sources and reported to block the digestive enzyme action on mammals and insects [102]. These molecules known to inhibit the action of alpha-amylases are identified from various sources like wheat, Indian finger millets, barley, and *Amaranthus paniculatus* against the insect of particular order Coleoptera [103, 104]. Many transgenic crops expressing a-AIs have been developed that known to inhibit digestion of alpha amylases against three burchids, *Bruchus pisorum* (pea weevils), *Callosobruchus maculatus* (cowpea weevils), and *C. chinensis* (adzuki weevils) [105, 106].

There are many types of α-AIs that have been reported are known to classify as proteinaceous and non-proteinaceous. In the case of proteinase inhibitors, six classes were coming from higher plants and two were reported as Tendamistat from *Streptomyces tendae* and Helianthamide from *Stichodactyla helianthus*. Six different α-Ais classes include lectin-like, knottin-like, cereal-type, Kunitz-like, γ-purothionin-like, and thaumatin-like [93, 107]. These reported classes of inhibitors have diversified action on alpha amylase based on their different structure and mode of action. In the case of non-proteinases inhibitors, they are mostly organic compounds that can act as substrate analogs and inhibit α-amylases, e.g., acarbose.

Based on biochemical studies, these inhibitors reported to possess diverse specificities with various amylases. There are several factors involved to cause differential action and interactions. These differences could be assigned to variations in the sequences and structural modification in α-amylases and α-AIs.

The plant α-AI diversity was also found across different species, most of them are particularly specific to one family. Examples of this is Ragi bifunctional inhibitor (RBI), which is mostly expressed in cereals and could have the ability to inhibit mammalian as well as insect α-amylases. Examples of many more α-AI are present in Table 2.3.

Table 2.3 Amylase inhibitors from various plants having insect toxic potential.

Protein name	Nature of protein	Plant species	Plants common name	Plant part used	Plant family	Action against insect	Insect common name	Order	Application	Reference
AmI1 and AmI2	α-Amylase inhibitor	*Triticum aestivum*	Wheat	Grains	Poaceae	*Tenebrio molitor*	Mealworm	Coleoptera	Affect larval midgut	[108]
Wheat 0.28	Monomeric proteins inhibitor	*Triticum aestivum*	Wheat	Seeds	Poaceae	*Tenebrio molitor*	Mealworm beetle	Coleoptera	Affect growth of insect	[109]
WDAI-3	Dimeric inhibitors	*Triticum turgidum*	Wheat	Grains	Poaceae	*Tenebrio molitor*	Mealworm beetle	Coleoptera	WDAI-3 more active against *Tenebrio molitor* than other insect	[109]
SIα1, SIα2 and SIα3		*Sorghum bicolor*	Sorgham	Seed	Poaceae	*Melanoplus differentialis*	Locust	Orthoptera	Inhibit insect gut α-amylases	[110]
Wheat WRP24	Dimer protein inhibitor	*Triticum aestivum*	Bread wheat	Wheat kernels	Poaceae	*Tribolium castaneum*	Red flour beetle	Coleoptera	WRP24 suppressed larval growth by more than fourfold	[92]
Wheat WRP25		*Triticum aestivum*	Bread wheat	Wheat kernels	Poaceae	*Sitophilus oryzae*	Rice weevil	Coleoptera	Substantial weight loss	[92]
Wheat WRP26		*Triticum aestivum*	Bread wheat	Wheat kernels	Poaceae	*Tenebrio molitor*	Mealworm	Coleoptera	Inhibite α-amylases of Mealworm beetle	[92]
Wheat WRP27	Monomeric proteins inhibitor	*Triticum aestivum*	Bread wheat	Wheat kernels	Poaceae	*Sitophilus oryzae*	Rice weevil	Coleoptera	Strongly inhibiteonly the rice weevil a-amylases	[92]
PAI		*Cajanus cajan*	Pigeonpea	Seeds	Fabaceae	*Helicoverpa armigera*	Cotton bollworm	Lepidoptera	Increased mortality and adverse effects on larval growth and development	[111]
AMY2		*Hordeum vulgare*	Barley	Malt	Poaceae					[112]
Wheat 0.19		*Triticum aestivum*	Wheat	Seeds	Poaceae	*Acanthoscelides obtectus*	Bean weevil	Coleoptera	Inhibit insect α-amylase activity	[93]

(Continued)

Table 2.3 (Continued)

Protein name	Nature of protein	Plant species	Plants common name	Plant part used	Plant family	Action against insect	Insect common name	Order	Application	Reference
Wheat 0.53		*Triticum aestivum*	Wheat	Seeds	Poaceae	*Tenebrio molitor*	Mealworm beetle	Coleoptera	Inhibit *Tenebrio molitor* α-amylase activity	[93]
BIII (rye)	Bifunctional α-amylase/ trypsin inhibitors	*Secale cereale*	Rye	Endosperm	Poaceae	*Zabrotes subfasciatus*	Mexican bean weevil	Coleoptera	More effective against insect α-amylases than against mammalian enzymes	Iulek et al. [113]
Zeamatin	α-amylase and trypsin inhibitor	*Zea mays*	Corn	Poaceae	Seeds	*Tribolium castaneum*	Red flour beetle	Coleoptera	Inhibit α-amylase activity	[114]
α-AI1	Endo-amylases	*Phaseolus vulgaris*	Common bean	Seeds	Fabaceae	*Callosobruchus maculates*	Cowpea seed beetle	Coleoptera	Enzyme inhibitors impede insect digestion through action on insect gut digestive α-amylases	[115]
α-AI2	Endo-amylases	*Phaseolus vulgaris*	Common bean	Seeds	Fabaceae	*Zabrotes subfasciatus*	Mexican bean weevil	Coleoptera	Enzyme inhibitors impede insect digestion through action on insect gut digestive α-amylases	[115]
RBI	Inhibits both α-amylase and trypsin	*Eleusine coracana*	Finger millet	Seeds	Poaceae					[116]
A. hypochondriacus α-amylase inhibitor (AhAI)	α-Amylase inhibitor	*Amaranthus hypochondriacus*	Prince-of-Wales feather	Leaves, inflorescence	Amaranthaceae	*Callosobruchus chinensis*	Adzuki bean weevil	Coleoptera	Affect the growth of *C. chinensis*	[103]

2.5 Ribosome-Inactivating Proteins (RIPs)

RIPs found widely in plant species and within different tissues, and it is considered to have toxic effect due to the presence of *N*-gulcosidases [117]. This *N*-glycosidase depurinate the eukaryotic and prokaryotic rRNAs, thereby arresting the synthesis of protein during translation. RIPs are demonstrated to have antifungal, antibacterial, antiviral, and insecticidal activities in various in vivo; and transgenic plant experiments [117–121]. These studies have provided a vast knowledge to comprehend the biochemical and medicinal properties of RIPs. There were many research articles, and reviews were published that explained its importance [117, 122–124].

RIPs have been divided into three main types on the basis of physical properties such as type I, type II, and type III [117, 125, 126]. There are about various kinds of RIPs that have been reported that cover around 17 plant families and showed its presence in bacteria, fungi, and algae as well [117, 122]. Many identified RIPs belong to numerous small group of plant species, such as Rosaceae, Caryophyllaceae, Euphorbiaceae, Cucurbitaceae, Sambucaceae, Poaceae, Phytolaccaceae, and Rosaceae [117, 122, 127, 128].

Trichosanthrip, was designated as novel RIP, was initially purified from mature seeds *Trichosanthes kirilowii*. This RIP effectively inhibits cell-protein synthesis [129]. RIPs have been identified to be located in various plant tissues like leaves, seeds, roots, and tubers [129]. Certain bacteria produces RIPs have shown to have enzymatic activity similar to plant analogs like type II RIPs Shiga toxin type 1 (Stx1) and Shiga toxin type 2 (Stx2) [117, 130, 131]. Many studies have proven that RIPs possess insecticidal activities upon insect attack that includes Lepidoptera [117, 132–135] and Coleoptera [117, 136]. This RIPs reported to enhance plant resistance against the insects [117, 137]. Artificial diet experiment was conducted to confirm the insecticidal activity of RIPs against the pest, by supplementing the diet with variable concentration of RIPs, such as type II RIP from *Sambucus nigra*, which have caused decreased in the fecundity and survival of *Acyrthosiphon pisum* [138]. There was another experiment in which transgenic plant overexpressing the RIPs SNA-I had caused the retarded development and decreased survival upon insect feeding by *Myzus nicotianae* [138]. In addition to this, when diet was supplemented with different type-1 RIPs had caused decreased survival and fecundity in *Anticarsia gemmatalis* Hübner and *Spodoptera furgiperda* [117, 139]. In 2016, there was a study that showed type-I and type-II RIPs insecticidal properties from the Apple (*Malus domestica* Borkh) against the pea aphids which have caused reduction in nymph survival [140]. RIPs overexpression have also detected to cause resistance against *Helicoverpa zea* [134]. More to this, report from maize resistance to feeding by *Spodoptera frugiperda* and corn earworms (*Helicoverpa zea*) was dedicated to maize ribosome-inactivating protein (MRIP) and wheat germ agglutinin (WGA) [38]. Overexpression of type I and type II RIPs has also produce resistance in tobacco plants against the insect pest, *Spodoptera exigua* [141].

The mechanism of RIPs action is still now clear, and several studies explained that RIPs can also take part in the apoptosis process 117, 142, 143]. Examples of this were reported from feeding experiment on *A. pisum* supplemented with SNA-I-induced apoptosis in the midgut through caspase-3 activation [117, 144]. Other important RIPs are shown in Table 2.4.

Table 2.4 List of ribosome inactivating and other related proteins with insect toxic activity.

Proteins	Nature of protein	Plant species	Plants common name	Plant part used	Plant family	Action against insect	Insect common name	Order	Application	Reference
Ricin	Carbohydrate-binding protein	*Ricinus communis*	Ricinus	Seeds	Euphorbiaceae	*Callosobruchus maculatus*	Cowpea seed beetle	Coleoptera	Highly potent toxin to cowpea seed beatle	[136]
Cinnamomin	rRNA *N*-glycosidase activity	*Cinnamomum camphora*	Camphorwood	Seeds	Lauraceae	*Helicoverpa armigera*	Cotton bollworm	Lepidoptera	Inhibition of protein synthesis in bollworm larvae	[23]
						Culex pipinespallens	Mosquito	Diptera	Inhibit protein synthesis in insects	[23]
Maize RIP	rRNA *N*-glycosidase activity	*Zea mays*	Maize	Seeds	Poaceae	*Helicoverpa zea*	Corn earworm	Lepidoptera	Increase mortality and reduced weights	[134]
Type II RIP	Ribosome-inactivating protein	*Cinnamomum camphora*	Camphorwood	Seeds	Lauraceae	*Bombyx mori*	Silkworm	Lepidoptera	Inhibit protein synthesis in insects	[135]
						Culex pipienspallens	Northern house mosquito	Diptera	Inhibit protein synthesis in insects	[135]
Maize RIP	rRNA *N*-glycosidase activity	*Zea mays*	Maize	Seeds	Poaceae	*Lasioderma serricorne*	Cigarette beetle	Coleoptera	Increase mortality and reduction in feeding	[132]
SNA-I	Ribosome-inactivating protein	*Sambucus nigra* L	Black elder	Bark	Adoxaceae	*Acyrthosiphon pisum*	Pea aphid	Homoptera	Reduced survival and fecundity	[138]
						Myzus nicotianae	Tobacco aphid	Hemiptera	Reduced survival and fecundity	[138]
Saporin	Ribosome inactivating protein	*Saponaria officinalis*	Bouncing Betty	Seeds	Caryophyllaceae	*Anticarsia gemmatalis*	Lelvetbean caterpillar	Lepidoptera	Highly toxic and induced mortality	[139]
						Spodoptera frugiperda	Fall armyworm	Lepidoptera	Highly weight loss	[139]
PAP-S	rRNA *N*-glycosidase activity	*Phytolacca americana*	American pokeweed	Leaves	Phytolaccaceae	*Anticarsia gemmatalis*	Lelvetbean caterpillar	Lepidoptera	Highly toxic and induced mortality	[139]
						Spodoptera frugiperda	Fall armyworm	Lepidoptera	Highly weight loss	[139]

Lychnin	Ribosome-inactivating protein	*Lychnis chalcedonica*	Flower of Bristol	Seeds	Caryophyllaceae	*Anticarsia gemmatalis*	Lelvetbean caterpillar	Lepidoptera	Highly toxic and induced mortality	[139]
						Spodoptera frugiperda	Fall armyworm	Lepidoptera	Highly weight loss	[139]
Gelonin	rRNA *N*-glycosidase activity	*Gelonium multiflorum*	False lime tree	Seeds	Euphorbiaceae	*Anticarsia gemmatalis*	Lelvetbean caterpillar	Lepidoptera	Highly toxic and induced mortality	[139]
						Spodoptera frugiperda	Fall armyworm	Lepidoptera	Highly weight loss	[139]
Momordin	Ribosome inactivating protein	*Momordica charantia*	Bitter melon	Seeds	Cucurbitaceae	*Anticarsia gemmatalis*	Lelvetbean caterpillar	Lepidoptera	Highly toxic and induced mortality	[139]
						Spodoptera frugiperda	Fall armyworm	Lepidoptera	Highly weight loss	[139]
Maize ribosome-inactivating protein (MRIP)	Ribosome-inactivating protein	*Zea mays*	Maize	Seeds	Poaceae	*Spodoptera frugiperda*	Fall armyworm	Lepidoptera	High mortality to spodoptera	[38]
Type-1 RIP	rRNA *N*-glycosidase activity	*Malus domestica*	Apple	Leaves	Rosaceae	*Acyrthosiphon pisum*	Pea aphid	Homoptera	Reduction in fecundity, intrinsic rate of increase, net reproductive rate and doubling time of the insect population	[140]
Type-2 RIP	Catalytic activity and lectin-binding properties	*Malus domestica*	Apple	Leaves	Rosaceae	*Myzus persicae*	Green peach aphid	Hemiptera	Reduction in fecundity, intrinsic rate of increase, net reproductive rate and doubling time of the insect population	[140]
Type-1 RIP and Type-2 RIP		*Malus domestica*	Apple	Leaves	Rosaceae	*Spodoptera exigua*	Beet armyworm	Lepidoptera	Highly entomotoxic activity causes 78% mortality during the larval stage	[140]

2.6 Arcelins

Arcelin (Arc) is a carbohydrate-binding insecticidal proteins mostly found in noncultivated wild type accessions of the common beans. These insecticidal proteins shown to confer resistance against burchid beetles. In some of the tropical and subtropical countries, the common beans suffer post-harvest losses which are primarily caused by the bruchids pests, *Acanthoscelides obtectus* and *Zabrotes subfasciatus*. It is considered as the novel storage protein in the common bean seeds in addition to phaseolin and phytohemagglutinin. Till now, seven arcelin variants have been identified that are designated as arcelin 1 to arcelin 7, among these 1 and 5 shown to provide resistance in leguminous crops [145–148]. Highest resistance to the Mexican bean weevil in wild type, *Phaseolus vulgaris* accessions were observed due to the presence of these two variants [1, 149]. These types of high resistance accession were only maintained in lines generated from crossing between arcelin-1 or arcelin-5 parents.

Arcelin 1 shown to have insecticidal activity observed during feeding experiments with artificial seeds [150]. It was realized that the composition of artificial diets differs to several parameters from that of arcelin-containing beans, the amount of arcelin as a percentage of total protein, and the arcelins by phaseolin ratio. The effectiveness of arcelin-5 was tested through artificial diet against *Z. subfasciatus* larval development or through transgenic seeds in which high expression of arcelin-5 was observed.

In comparison to other arcelin, Arcelin-1 and 5 variant was discovered later and not considered for first series of the breeding experiments [1] and artificial seed assays [150]. Arcelin-5 was to shown be present in the wild-type G02771 accession and consists of three polypeptides (Arc5a, Arc5b, and Arc5c), which are organized as monomers and dimers in their native states [151, 152]. Arcelin-5 is released due to the expression of two genes, namely, arc5-1, that codes for Arc5a and arc5-II which is further encoded by the two genes, namely Arc5b and Arc5c [151, 153]. The insecticidal activity of arcelin-5 was checked in vivo; and feeding assay indicated the presence of arcelin-5. Arcelin considered to be weak lectins despite having sequence homology. The sugar-binding specificity of the arcelin-1 was significantly differs from those of PHA-L and PHA-E. These differences between arcelin-1 and lectin were reported to be come from substitution or deletions of essential amino acid residues during metal and sugar recognition [154].

Arcelin usage as insecticidal has gain attention toward bruchids pests as well as its inhibitory role on larval development of *Z. subfasciatus* was highly appreciated [1, 150, 155]. There are other wild accessions that have showed little resistance against some of important bean burchid, like *Acanthoscelides obtectus* [155, 156]. The mechanism of action of arcelin still contentious, despite having good research on it, was considered toxic by Osborni et al. [150] and could be referred indigestible that caused larval starvation [1, 149]. It was also postulated that Asn-linked glycans of arcelin-1 could be the reason of toxicity through its binding affinity toward lectins [154]. Few research were conducted to understand why arcelin is an insecticidal moiety against *Z. subfasciatus* but not for *A. obtectus*. The experiment explained that arcelin-1 disrupted the midgut epithelial structures in *Z. subfasciatus* which was not possible in midgut of *A. obtectus* [1, 156]. There are many other reported arcelins that are present in Table 2.5.

Table 2.5 Various plants arcelins with insect inhibitory functions.

Arcelins	Nature of protein	Plant species	Plants common name	Plant part used	Plant family	Action against insect	Insect common name	Order	Application	Reference
Bean arcelin		*Phaseolus vulgaris*	Common bean	Seeds	Fabaceae					[147]
Arcelin-1	One of the four electrophoretic variants of arcelin	*Phaseolus vulgaris*	Common bean	Seeds	Fabaceae	*Callosobruchus maculatus*	*Cowpea weevil*	Coleoptera	Inhibited the development of larvae of *Callosobruchus maculatus*	[101]
Arcelin-4		*Phaseolus vulgaris*	Common bean	Seeds	Fabaceae	*Acanthoscelides obtectus*	*Bean weevil*	Coleoptera	Causes larval growth inhibition and mortality	[157]
Arcelin-5	Glycoproteins	*Phaseolus vulgaris*	Common bean	Seeds	Fabaceae	*Zabrotes subfasciatus*	*Mexican bean weevil*	Coleoptera	Inhibit insect growth and development	[151]
Native arcelin-1	Dimeric glycoprotein	*Phaseolus vulgaris*	Common bean	Seeds	Fabaceae	*Zabrotes subfasciatus*	*Mexican bean weevil*	Coleoptera	Inhibit insect growth and development	[158]
Arc5-III		*Phaseolus vulgaris*	Common bean	Seeds	Fabaceae	*Zabrotes subfasciatus*	*Mexican bean weevil*	Coleoptera	Highest inhibitory effect on the development of *Zabrotes subfasciatus* larvae	[159]
L. purpureus Arcelin		*Lablab purpureus*	Hyacinth bean	Seeds	Fabaceae	*Spodoptera litura*	*Asian armyworm*	Lepidoptera	Inhibitory effect on the larval development	Malaikozhundan et al. [160]
Tepary bean Arcelin		*Phaseolus acutifolius*	Tepary bean	Seeds	Fabaceae	*Zabrotes subfasciatus*	*Mexican bean weevil*	Coleoptera	Provides resistance towards the Mexican bean weevil	[161]
L. purpureus Arcelin		*Lablab purpureus*	Hyacinth bean	Seeds	Fabaceae	*Rhyzopertha dominica*	*Lesser grain borer*	Coleoptera	5% Dose of the *L. purpureus* fraction resulted in complete mortality of all larvae	[162]
Arcelin, phytohemagglutinin and α-amylase inhibitor (ABA)		*Phaseolus vulgaris*	Common bean	Seeds	Fabaceae	*Zabrotes subfasciatus*	*Mexican bean weevil*	Coleoptera	Inhibit insect growth and development	[163]
L. purpureus Arcelin		*Lablab purpureus*	Hyacinth bean	Seeds	Fabaceae	*Callosobruchus maculatus*	*Cowpea weevil*	Coleoptera	Inhibit larval growth	[164]
P. lunatus Arcelin		*Phaseolus lunatus*	Lima bean	Seeds	Fabaceae	*Callosobruchus maculatus*	*Cowpea weevil*	Coleoptera	Inhibit larval growth	[165]

2.7 Defensins

These are considered as a small cationic peptide that have 44–54 amino acid residues which are stabilized by three to four disulfide bridges having molecular mass of approximately 5 kDa [166, 167]. Till date, there are several defensins molecules that have been extracted from various plant tissues like leaf, stem, root, and endosperm. These isolated defensins possess numerous antibacterial, antifungal, and insecticidal properties [167]. The role of defensins is primarily reported to inhibit insect enzymes, importantly α-amylases and proteases. The first reported defensin molecules were extracted from plant sorghum (*Sorghum bicolor*) [110, 166]. It showed toxic effects against insect, *Periplaneta americana* alpha amylases but does not cause any effect on the mammalian digestive enzymes [110].

The mung bean isolated defensins have been thoroughly studied and well characterized in terms of its structure and function and referred as VrD1. Many studied reported the expression of VrD1 in the yeast was found to have inhibitory effect against the *Callosobruchus chinensis* in the bioassay [168]. The defensins, VuD1, isolated from cowpea have showed inhibitory effect against *A. obtectus* and *Z. subfasciatus*, insect α-amylases but had not produce effect against weevil *Callosobruchus maculatus* [169]. Later studies have shown inhibitory action of VuD1 against *C. maculatus*, α-amylases at micromolar concentration, without causing any effect on the mammalian enzymes [166, 170]. Many studies in the identification of defensins have been carried out which elaborated the usage of these peptides in developing transgenic plants resistance to insect pests. Examples of some of these include defensins, BrD1, which were isolated from turnip (*Brassica rapa*) expressed in GM rice cultivars. These developed transgenic crops exhibited increased resistance against the insect, *N. lugens* compared to the nontransformed plants. Other defensins molecules found toxic against insect is presented in Table 2.6.

2.8 Cyclotides

These are the peptide molecules that are like as defensins and have low molecular mass cationic peptides that is approximately 30 amino acid residues in length. Unlike defensins, they do not have *N*- and *C*-termi [166, 185]. Presently, most of cyclotides have been extracted and well characterized, these peptides possess many antibacterial, antiviral, insecticidal, and hemolytic properties [185]. First reported cyclotides extracted from African plant, *Oldenlandia affinis*, was kalata B1 and was found to have insecticidal action against lepidopteran *Helicoverpa punctigera* [186]. Kalata B1 was recently expressed in the transgenic crop plant, *Nicotiana benthamiana*, to study the cyclization of this peptide. Three mostly conserved regions that were considered for posttranslational modifications identified to be at C-terminal region of kalata B1 [166, 187]. Apart from kalata B1, the insecticidal activity of kalata B2 was identified that have inhibitory affect against *H. armigera* larvae which was isolated from *O. affinis* [175]. These inhibitory effects were due to less consumption of food by the *H. armigera* larvae rather than due to toxicity [188]. There was another cyclotide identified from blue pea (*Clitoria ternatea*) that has shown inhibitory effect against *Z. subfasciatus* and *A. obtectus* when fed on an artificial diet [189]. After that there was another study that has extracted cyclotides from the Brazilian Savannah Rubiaceae flower plant, *Palicourea rigida* referred as paragidin-BR1, that have caused mortality more than 60% against *Diatraea saccharalis* larvae. In in vitro; assay, the efficacy of paragidin-BR1 was also checked at micromolar concentration against the Sf-9 cell line of *S. furgiperda* [37, 166].

There is tremendous potential in the use of cyclotides in the future application in developing transgenic plants against the insect. However, till date, no transgenic plants expressing cyclotides genes against the insect pests have been reported.

2.9 Canatoxin-Like Proteins

The first protein ever crystallized from the jackbean, *C. ensiformis*, is lectin concanavalin A [1]. Jacbeans also contain a potent neurotoxic protein named canatoxin [1, 171]. This potent toxic protein is a noncovalently linked dimer of a 95 kDa polypeptide chain, which accounts for 0.5% of the dry seed weight. When this toxin ingested by insect is cleaved by cathepsin-like enzymes from gut releasing entomotoxic peptide named pepcanatox. Other types of digestive enzymes like trypsin are not susceptible from the attack of this toxin [190]. Diversified presence of this canatoxin-like proteins in other plant sources was also reported which includes ureases and urease-derived encrypted peptides. The role of this toxin on different insects was tested to predict its role as a plant defense molecule [190]. The insects, *C. maculatus* and *R. prolixus*, that relied on cathepsins B and D as main digestive enzymes were found to be have lethal effects due to canatoxin. The two major pests of agricultural importance, *Dysdercus peruvianus* (cottonsucker bug) and *Nezara viridula* (Southern greensoybean stinkbug) are also susceptible to the toxic effect of canatoxin. There was a report that confirmed the toxicity effect of entomotoxic protein, canatoxin, in the presence of insect cathepsin-like enzymes to produce entomotoxic peptide(s) [191]. Canatoxin is found to be less toxic in comparison to α-amylase inhibitors, proteinase inhibitors, and some lectins but was found to be 40-fold more toxic than arcelins against insect, *Z. subfasciatus* [1]. Other canatoxins-like proteins have also played a critical role against the insect, which is represented in Table 2.6.

2.10 Ureases and Urease-Derived Encrypted Peptides

The ureases are the metalloenzymes that has role to hydrolyze urea into ammonia and carbon dioxide. Their presence was reported in plants, fungi, and bacteria [1, 178]. This is a first enzyme extracted from jack beans to be crystallized that consists of a 90.7 kDa homohexamer chains [1].

The important role of plant ureases is to utilize urea as a nitrogen source internally or externally. These are found to be highly expressed within the seeds. The stored ureases in seeds help in early germination by utilizing nitrogen sources [178]. Apart from this, it also exerts insecticidal and antifungal properties. Importantly, insects that produce cathepsin-like enzymes during digestion are found to be very susceptible to ureases, whereas insect having trypsin-like digestive enzymes does not found to susceptible to ureases [178]. The insects that are susceptible to ureases are *C. maculatus* and *Rhodnius prolixus* (kissing bug) and unsusceptible insects include, such as *Schistocerca americana* (locust), *Manduca sexta*, and *Drosophila melanogaster*

Table 2.6 List of other proteins with insect toxic activities.

Proteins	Nature of protein	Plant species	Plants common name	Plant part used	Plant family	Action against insect	Insect common name	Order	Application	References
Canatoxin-CNTX		*Canavalia ensiformis*	Jack bean	Seeds	Fabaceae					[171]
Soybean urease (SBU)	Urea amidohydrolase	*Glycine max*	Soybean	Seeds	Fabaceae	*Dysdercus peruvianus*	Peruvian larder beetle	Coleoptera	Ubiquitous enzyme	[172]
Potato urease	Ubiquitous enzyme	*Solanum tuberosum*	Potato	Leaves	Solanaceae					[173]
JBURE-II		*Canavalia ensiformis*	Jack bean	Seeds	Fabaceae	*Nezara viridula*	Green stink bug	Hemiptera	100% lethality after 72 h of second instar larvae	[174]
Kalata B2	Subfamily of cyclotides	*Oldenlandia affinis*	*Hedyotis affinis* Roem	Leaves	Rubiaceae	*Helicoverpa armigera*	Cotton bollworm	Lepidoptera	Inhibits the growth and development of *Helicoverpa armigera* larvae	[175]
G. hirsutum seed urease (GHU)		*Gossypium hirsutum*	Cotton	Seeds	Malvaceae	*Dysdercus peruvianus*	Cotton stainer bug	Hemiptera	Degradation of digestive enzymes present in the target insects	[176]
Momordica urease		*Momordica charantia*	Bitter melon	Seeds	Cucurbitaceae					[177]
Jack bean urease – JBURE-I	Nickel-dependent enzymes	*Canavalia ensiformis*	Jack bean	Seeds	Fabaceae	*Rhodnius prolixus*	Triatomid bug	Hemiptera	96% mortality after 24 h and reduced body weight gain	[178]
Jaburetox-2Ec		*Canavalia ensiformis*	Jack bean	Seeds	Fabaceae	*Dermestes peruvianus*	Peruvian larder beetle	Coleoptera	100% mortality after 11 days	[178]
Pigeon pea urease (PPU)		*Cajanus cajan*	Pigeonpea	Seeds	Fabaceae	*Callosobruchus chinensis*	Kdzuki bean weevil	Coleoptera	Affect insect digestive system	[179]

Cyclotides	Binds to phospholipid membranes	*Clitoria ternatea*	Butterfly pea	Leaves	Fabaceae	*Spodoptera frugiperda*	Fall armyworm	Lepidoptera	Cytotoxicity against Sf9 cell	Oguis et al. [180]
Concanavalin A (ConA)	Carbohydrate-binding protein	*Canavalia ensiformis*	Jack bean	Seeds	Fabaceae	*Bactericera cockerelli*	Potato psyllid	Hemiptera	Apoptotic response was induced by ConA in psyllid midgut cells	[181]
Jaburetox (JBTX)	Ureases activity	*Canavalia ensiformis*	Jack bean	Seeds	Fabaceae	*Oncopeltus fasciatus*	Large milkweed bug	Hemiptera	Ureases activity	[182]
Ύ1-hordothionin		*Hordeum vulgare*	Barley	Seeds	Poaceae					[183]
Ύ1-purothioni		*Triticum aestivum*	Wheat	Seeds	Poaceae					[183]
VrCRP	A cysteine-rich protein	*Vigna radiata*	Mung bean	Seeds	Fabaceae	*Callosobruchus chinensis*	Chinese bruchid	Coleoptera	Lethal to larvae of the bruchid	[9]
AtPDF1.1		*Arabidopsis thaliana*	Arabidopsis	Leaves	Brassicaceae					[184]
AtPDF1.2a		*Arabidopsis thaliana*	Arabidopsis	Leaves	Brassicaceae					[184]

(fruit fly) [1, 178]. The differential action of plant ureases by insect digestive enzymes in different developmental stages affects susceptibility status of adult and nymph pest [1, 178]. The toxic effect of JBURE-1 isoform isolated from major jack bean depends on the release of entmotoxic-derived encrypted peptide called pepcanatox and cathepsin-like enzymes [191]. Based on sequence information of pepcanatox, another recombinant peptide was also produced named, Jaburetox [192]. This recombinant peptide has a molecular weight of 11 kDa, which is toxic to various insect pests that are not susceptible to native urease, JBURE-I [178].

The insects *Dysdercus peruvianus* (cotton stainer bug), *Oncopeltus fasciatus* (large milkweedbug), and *R. prolixus*are are susceptible to the toxic effect of JBURE-I [178, 193, 194]. JUBRE-II isoform exerts toxic effect against the insect *R. prolixus* [195]. *S. furgiperda* was susceptible to Jaburetox [178, 192, 196].

2.11 Chitinases

As we know, chitin is the extracellular layer of insect exoskeleton and could be considered as important target for the pesticidal action [166, 197]. Apart from lectins, plant also produces chitinases, a hydrolytic enzyme that can interact with chitin monomers to disrupt the action of insect chitin synthase [197].

These chitinases help in the hydrolysis of chitin that present β-1,4-linked *N*-acetylglucosamine residues [198, 199]. Chitin is considered as monomeric protein having a molecular mass of 25–35 kDa [166, 198]. Plant chitinases were classified as endo- and exo-chitinases based on the specific cleavage site in the target chitin moiety. According to primary structure, plant chitinases are divided into four different groups that include, class I, class II, class III, and class IV. In class I, *N*-terminal cysteine-rich domain was present with having near about 40 amino acid residues in length that is found to be highly conserved in the main structure [166]. In class II, these chitinases do not have cysteine-rich domain at the *N*-terminus. The Class III include enzymes with no sequence similarity with class I or II, but in class IV, sequence similarity was observed with class I that contain cysteine-rich domain [166, 198].

There was report that described the toxic effect of chitinase, when engineered into tomato plants against the pest, Coloradopotato beetle larvae (*Leptinotarsa decemlineata*), named WIN6 was isolated from popular plants (*Populus trichocarpa*). Another report came from *D. melanogaster* where they have shown toxic effect of two chitinases, namely, LA-a and LA-b when administered in an artificial diet [200].

2.12 Proteases

These are classified as peptidases or proteinases that are present in animals, plants, bacteria, archaea, and viruses that hydrolyze the covalent bonds present in the polypeptide chains. Many of them have emerged as a protective moiety against herbivorous pest. However, those proteases that do not evolve as enterotoxins could also have insecticidal effect when administered ectopically [166, 201].

There is a very limited study that described behavior of proteases as toxic against insect pests. One example of these types include Mir1-CP, which is also known as papain-like cysteine isolated from maize lines and have shown resistance against *S. furgiperda* [202–204].

Engineered GM plant calluses expressing Mir1-CP exerts growth inhibition when insect fed on it [205]. In addition to this, purified Mir1-CP recombinant is known to enhance the toxic effect of Bt Cry toxin on peritrophic matrix of *S. furgiperda* and many other insects [206]. Papain, which is another protease, found to be present in the latex papaya (*Carica papaya*) and one more cysteine protease called ficin identified in the wild fig (*Ficus virgata*) caused to have detrimental effect against three important lepidopteran species, namely, *Samia ricini* (Indian erisilkmoth), *Mamestra brassicae* (cabbage moth), and *Spodoptera litura* (tobacco cutworm) [166, 207]. Therefore, these plant proteases could be one of the unexplored promising agents in the development of transgenic plants against insect pests [201].

2.13 Conclusions

Here, we have discussed about the plant defense systems that cause detrimental action against insect growth and phenotype. Different plant systems have variable effect on insect growth established through feeding bioassay and transgenic approach. We could manipulate the selected defense proteins against the insect either by overexpressing the gene of interest or introducing the plant genes into different plant species. These plant proteins have tremendous potential in controlling the insect spread and saving the loss happened in the agricultural field to food productions.

References

1 Carlini, C.R. and Grossi-de-Sá, M.F. (2002). Plant toxic proteins with insecticidal properties. A review on their potentialities as bioinsecticides. *Toxicon* 40: 1515–1539. https://doi.org/10.1016/S0041-0101(02)00240-4.

2 Oerke, E.-C., Dehne, H.-W., Schönbeck, F., and Weber, A. (1999). *Crop Production and Crop Protection*. Elsevier https://doi.org/10.1016/C2009-0-00683-7.

3 Oliveira, C.M., Auad, A.M., Mendes, S.M., and Frizzas, M.R. (2014). Crop losses and the economic impact of insect pests on Brazilian agriculture. *Crop Protection* 56: 50–54. https://doi.org/10.1016/j.cropro.2013.10.022.

4 Paul, S. and Das, S. (2020). Natural insecticidal proteins, the promising bio-control compounds for future crop protection. *Nucleus* https://doi.org/10.1007/s13237-020-00316-1.

5 Vaeck, M., Reynaerts, A., Höfte, H. et al. (1987). Transgenic plants protected from insect attack. *Nature* 328: 33–37. https://doi.org/10.1038/328033a0.

6 RED, EPA (1998). *FACTS: Bacillus thuringiensis. EPA-738-F-98-001*. Washington, DC: United States Environmental Protection Agency.

7 Huang, D.-F., Zhang, J., Song, F.-P., and Lang, Z.-H. (2007). Microbial control and biotechnology research on *Bacillus thuringiensis* in China. *Journal of Invertebrate Pathology* 95: 175–180. https://doi.org/10.1016/j.jip.2007.02.016.

8 Van Damme, E.J.M., Peumans, W.J., Barre, A., and Rougé, P. (1998). Plant lectins: a composite of several distinct families of structurally and evolutionary related proteins with diverse biological roles. *Critical Reviews in Plant Sciences* 17: 575–692. https://doi.org/10.1080/07352689891304276.

9 Chen, Y., Peumans, W.J., Hause, B. et al. (2002). Jasmonate methyl ester induces the synthesis of a cytoplasmic/nuclear chitooligosaccharide binding lectin in tobacco leaves. *The FASEB Journal* 16: 905–907. https://doi.org/10.1096/fj.01-0598fje.
10 Vandenborre, G., Miersch, O., Hause, B. et al. (2009). *Spodoptera littoralis*-induced lectin expression in tobacco. *Plant and Cell Physiology* 50: 1142–1155. https://doi.org/10.1093/pcp/pcp065.
11 Vandenborre, G., Van Damme, E.J.M., and Smagghe, G. (2009). Nicotiana tabacum agglutinin expression in response to different biotic challengers. *Arthropod-Plant Interactions* 3: 193–202. https://doi.org/10.1007/s11829-009-9075-6.
12 Peumans, W.J., Barre, A., HoulesAstoul, C. et al. (2000). Isolation and characterization of a jacalin-related mannose-binding lectin from salt-stressed rice (*Oryza sativa*) plants. *Planta* 210: 970–978. https://doi.org/10.1007/s004250050705.
13 Vandenborre, G., Smagghe, G., and Van Damme, E.J.M. (2011). Plant lectins as defense proteins against phytophagous insects. *Phytochemistry* 72: 1538–1550. https://doi.org/10.1016/j.phytochem.2011.02.024.
14 Boulter, D., Edwards, G.A., Gatehouse, A.M.R. et al. (1990). Additive protective effects of different plant-derived insect resistance genes in transgenic tobacco plants. *Crop Protection* 9: 351–354. https://doi.org/10.1016/0261-2194(90)90005-R.
15 Sauvion, N., Peumans, W.J., Damme, E.J.M.V., and Gatehouse, J.A. (1996). Effects of GNA and other mannose binding lectins on development and fecundity of the peach-potato aphid *Myzus persicae*. *Entomologia Experimentalis et Applicata* 79 (3): 285–293. https://doi.org/10.1007/BF00186287.
16 Machuka, J., Damme, E.J.M., Peumans, W.J., and Jackai, L.E.N. (1999). Effect of plant lectins on larval development of the legume pod borer, *Maruca vitrata*. *Entomologia Experimentalis et Applicata* 93: 179–187. https://doi.org/10.1046/j.1570-7458.1999.00577.x.
17 Melander, M., Åhman, I., Kamnert, I., and Strömdahl, A.-C. (2003). Effects of GNA and other mannose binding lectins on development and fecundity of the peach-potato aphid *Myzus persicae*. *Transgenic Research* 12: 555–567. https://doi.org/10.1023/A:1025813526283.
18 Lehrman, A. (2007). Does pea lectin expressed transgenically in oilseed rape (Brassica napus) influence honey bee (Apis mellifera) larvae? *Environmental Biosafety research* 6 (4): 271–278.
19 Sauvion, N., Charles, H., Febvay, G., and Rahbe, Y. (2004). Effects of jackbean lectin (ConA) on the feeding behaviour and kinetics of intoxication of the pea aphid, *Acyrthosiphon pisum*. *Entomologia Experimentalis et Applicata* 110: 31–44. https://doi.org/10.1111/j.0013-8703.2004.00117.x.
20 Sauvion, N., Nardon, C., Febvay, G. et al. (2004). Binding of the insecticidal lectin Concanavalin A in pea aphid, *Acyrthosiphon pisum* (Harris) and induced effects on the structure of midgut epithelial cells. *Journal of Insect Physiology* 50: 1137–1150. https://doi.org/10.1016/j.jinsphys.2004.10.006.
21 Beneteau, J., Renard, D., Marché, L. et al. (2010). Binding properties of the *N*-acetylglucosamine and high-mannose *N*-glycan PP2-A1 phloem lectin in Arabidopsis. *Plant Physiology* 153: 1345–1361. https://doi.org/10.1104/pp.110.153882.
22 Dinant, S., Clark, A.M., Zhu, Y. et al. (2003). Diversity of the superfamily of phloem lectins (phloem protein 2) in angiosperms. *Plant Physiology* 131: 114–128. https://doi.org/10.1104/pp.013086.

23 Zhou, X., Li, X., Yuan, J. et al. (2000). Toxicity of cinnamomin-a new type II ribosome-inactivating protein to bollworm and mosquito. *Insect Biochemistry and Molecular Biology* 30: 259–264. https://doi.org/10.1016/S0965-1748(99)00126-5.

24 Wu, J., Luo, X., Guo, H. et al. (2006). Transgenic cotton, expressing *Amaranthus caudatus* agglutinin, confers enhanced resistance to aphids. *Plant Breeding* 125: 390–394. https://doi.org/10.1111/j.1439-0523.2006.01247.x.

25 Subramanyam, S., Smith, D.F., Clemens, J.C. et al. (2008). Functional characterization of HFR1, a high-mannose *N*-glycan-specific wheat lectin induced by hessian fly larvae. *Plant Physiology* 147: 1412–1426. https://doi.org/10.1104/pp.108.116145.

26 Upadhyay, S.K., Mishra, M., Singh, H. et al. (2010). Interaction of *Allium sativum* leaf agglutinin (ASAL) with midgut BBMV proteins and its stability in *Helicoverpa armigera*. *Proteomics* 10: 4431–4440. https://doi.org/10.1002/pmic.201000152.

27 Upadhyay, S.K., Saurabh, S., Rai, P. et al. (2010). SUMO fusion facilitates expression and purification of garlic lectin but modifies some of its properties. *Journal of Biotechnology* 146: 1–8. https://doi.org/10.1016/j.jbiotec.2010.01.013.

28 Upadhyay, S.K., Saurabh, S., Singh, R. et al. (2011). Purification and characterization of a lectin with high hemagglutination property isolated from *Allium altaicum*. *The Protein Journal* 30: 374–383. doi: 10.1007/s10930-011-9342-0.

29 Upadhyay, S.K., Singh, S., Chandrashekar, K. et al. (2012). Compatibility of garlic (*Allium sativum* L.) leaf agglutinin and Cry1Ac δ-endotoxin for gene pyramiding. *Applied Microbiology and Biotechnology* 93: 2365–2375. doi: 10.1007/s00253-011-3547-1.

30 Murdock, L.L., Huesing, J.E., Nielsen, S.S. et al. (1990). Biological effects of plant lectins on the cowpea weevil. *Phytochemistry* 29: 85–89. https://doi.org/10.1016/0031-9422(90)89016-3.

31 Czapla, T.H. and Lang, B.A. (1990). Effect of plant lectins on the larval development of European corn borer (Lepidoptera: Pyralidae) and Southern corn rootworm (Coleoptera: Chrysomelidae). *Journal of Economic Entomology* 83: 2480–2485. https://doi.org/10.1093/jee/83.6.2480.

32 Kumar, M.A., Timm, D.E., Neet, K.E. et al. (1993). Characterization of the lectin from the bulbs of *Eranthis hyemalis* (winter aconite) as an inhibitor of protein synthesis. *The Journal of Biological Chemistry* 268: 25176–25183.

33 Zhu-Salzman, K., Shade, R.E., Koiwa, H. et al. (1998). Carbohydrate binding and resistance to proteolysis control insecticidal activity of *Griffonia simplicifolia* lectin II. *Proceedings of the National Academy of Sciences of the United States of America* 95: 15123–15128. https://doi.org/10.1073/pnas.95.25.15123.

34 Stoger, E., Williams, S., Christou, P. et al. (1999). Expression of the insecticidal lectin from snowdrop (*Galanthus nivalis* agglutinin; GNA) in transgenic wheat plants: effects on predation by the grain aphid Sitobionavenae. *Molecular Breeding* 5: 65–73. https://doi.org/10.1023/A:1009616413886.

35 Gurjar, M.M., Gaikwad, S.M., Salokhe, S.G. et al. (2000). Growth inhibition and total loss of reproductive potential in *Tribolium castaneum* by *Artocarpus hirsuta* lectin. *Invertebrate Reproduction and Development* 38: 95–98. https://doi.org/10.1080/07924259.2000.9652443.

36 Roy, A., Banerjee, S., Majumder, P., and Das, S. (2002). Efficiency of mannose-binding plant lectins in controlling a homopteran insect, the red cotton bug. *Journal of Agricultural and Food Chemistry* 50: 6775–6779. https://doi.org/10.1021/jf025660x.

37 Pinto, M.F.S., Fensterseifer, I.C.M., Migliolo, L. et al. (2012). Identification and structural characterization of novel cyclotide with activity against an insect pest of sugar cane. *The Journal of Biological Chemistry* 287: 134–147. https://doi.org/10.1074/jbc.M111.294009.

38 Dowd, P.F., Johnson, E.T., and Price, N.P. (2012). Enhanced pest resistance of maize leaves expressing monocot crop plant-derived ribosome-inactivating protein and agglutinin. *Journal of Agricultural and Food Chemistry* 60: 10768–10775. https://doi.org/10.1021/jf3041337.

39 Lagarda-Diaz, I., Geiser, D., Guzman-Partida, A.M. et al. (2014). Recognition and binding of the PF2 lectin to α-amylase from *Zabrotes subfasciatus* (Coleoptera:Bruchidae) larval midgut. *Journal of Insect Science* 14 https://doi.org/10.1093/jisesa/ieu066.

40 Shukla, A.K., Upadhyay, S.K., Mishra, M. et al. (2016). Expression of an insecticidal fern protein in cotton protects against whitefly. *Nature Biotechnology* 34: 1046–1051. https://doi.org/10.1038/nbt.3665.

41 George, B.S., Silambarasan, S., Senthil, K. et al. (2018). Characterization of an insecticidal protein from *Withania somnifera* against lepidopteran and hemipteran pest. *Molecular Biotechnology* 60: 290–301. https://doi.org/10.1007/s12033-018-0070-y.

42 Parthiban, E., Arokiyaraj, C., Janarthanan, S., and Ramanibai, R. (2020). Purification, characterization of mosquito larvicidal lectin from *Annona muricata* and its eco-toxic effect on non-target organism. *Process Biochemistry* 99: 357–366. https://doi.org/10.1016/j.procbio.2020.09.025.

43 Farmer, E.E. and Ryan, C.A. (1990). Interplant communication: airborne methyl jasmonate induces synthesis of proteinase inhibitors in plant leaves. *Proceedings of the National Academy of Sciences of the United States of America* 87: 7713–7716. https://doi.org/10.1073/pnas.87.19.7713.

44 Singh, S., Singh, A., Kumar, S. et al. (2020). Protease inhibitors: recent advancement in its usage as a potential biocontrol agent for insect pest management. *Insect Science* 27: 186–201. https://doi.org/10.1111/1744-7917.12641.

45 Green, T.R. and Ryan, C.A. (1972). Wound-induced proteinase inhibitor in plant leaves: a possible defense mechanism against insects. *Science* 175: 776–777. https://doi.org/10.1126/science.175.4023.776.

46 O'Donnell, P.J., Calvert, C., Atzorn, R. et al. (1996). Ethylene as a signal mediating the wound response of tomato plants. *Science* 274: 1914–1917. https://doi.org/10.1126/science.274.5294.1914.

47 Singh, A., Singh, I.K., and Verma, P.K. (2008). Differential transcript accumulation in *Cicer arietinum* L. in response to a chewing insect *Helicoverpa armigera* and defence regulators correlate with reduced insect performance. *Journal of Experimental Botany* 59: 2379–2392. https://doi.org/10.1093/jxb/ern111.

48 Akbar, S.M.D., Jaba, J., Regode, V. et al. (2018). *Plant protease inhibitors and their interactions with insect gut proteinases*. In: *The Biology of Plant-Insect Interactions: A Compendium for the Plant Biotechnologist*, 1–47. New York: CRC Press ISBN 9781498709736.

49 Kessler, A. and Baldwin, I.T. (2002). Plant responses to i nsect herbivory: the emerging molecular analysis. *Annual Review of Plant Biology* 53: 299–328. https://doi.org/10.1146/annurev.arplant.53.100301.135207.

50 Hilder, V.A., Gatehouse, A.M., Sheerman, S.E. et al. (1987). A novel mechanism of insect resistance engineered into tobacco. *Nature* 330 (6144): 160–163.

51 Duan, X., Li, X., Xue, Q. et al. (1996). Transgenic rice plants harboring an introduced potato proteinase inhibitor II gene are insect resistant. *Nature Biotechnology* 14: 494–498. https://doi.org/10.1038/nbt0496-494.

52 Johnson, R., Narvaez, J., An, G., and Ryan, C. (1989). Expression of proteinase inhibitors I and II in transgenic tobacco plants: effects on natural defense against *Manduca sexta* larvae. *Proceedings of the National Academy of Sciences of the United States of America* 86: 9871–9875. https://doi.org/10.1073/pnas.86.24.9871.

53 Xu, D., Xue, Q., McElroy, D. et al. (1996). Constitutive expression of a cowpea trypsin inhibitor gene, CpTi, in transgenic rice plants confers resistance to two major rice insect pests. *Molecular Breeding* 2: 167–173. https://doi.org/10.1007/BF00441431.

54 Leo, F.D. (2002). PLANT-PIs: a database for plant protease inhibitors and their genes. *Nucleic Acids Research* 30: 347–348. https://doi.org/10.1093/nar/30.1.347.

55 Rawlings, N.D., Barrett, A.J., Thomas, P.D. et al. (2018). The MEROPS database of proteolytic enzymes, their substrates and inhibitors in 2017 and a comparison with peptidases in the PANTHER database. *Nucleic Acids Research* 46: D624–D632. doi: 10.1093/nar/gkx1134.

56 Solomon, M., Belenghi, B., Delledonne, M. et al. (1999). The involvement of cysteine proteases and protease inhibitor genes in the regulation of programmed cell death in plants. *The Plant Cell* 11 (3): 431–443.

57 Pernas, M., Sánchez-Monge, R., Gómez, L., and Salcedo, G. (1998). A chestnut seed cystatin differentially effective against cysteine proteinases from closely related pests. *Plant Molecular Biology* 38: 1235–1242. https://doi.org/10.1023/A:1006154829118.

58 Xiaofeng, L., Yuxian, X., and Yan, P. (1998). Roles of plant proteinase inhibitors in the resistance of plant against insects and pathogens. *Sheng wuhuaxueyu Sheng wuwu li jin Zhan* 25 (4): 328–333.

59 Gatehouse, J. (2011). Prospects for using proteinase inhibitors to protect transgenic plants against attack by herbivorous insects. *Current Protein & Peptide Science* 12: 409–416. https://doi.org/10.2174/138920311796391142.

60 A. de P.G. Gomes, Dias, S.C., Bloch Jr, C., Melo, F.R., Furtado Jr, J.R., Monnerat, R.G. et al. (2005). Toxicity to cotton boll weevil Anthonomus grandis of a trypsin inhibitor from chickpea seeds. *Comparative Biochemistry and Physiology Part B: Biochemistry and Molecular Biology*, 140, 313–319.

61 Burgess, E.P.J., Main, C.A., Stevens, P.S. et al. (1994). Effects of protease inhibitor concentration and combinations on the survival, growth and gut enzyme activities of the black field cricket, *Teleogryllus commodus*. *Journal of Insect Physiology* 40: 803–811. https://doi.org/10.1016/0022-1910(94)90010-8.

62 Outchkourov, N.S., De Kogel, W.J., Schuurman-de Bruin, A. et al. (2004). Specific cysteine protease inhibitors act as deterrents of western flower thrips, *Frankliniella occidentalis* (Pergande), in transgenic potato: transgenic plants deterrent to insects. *Plant Biotechnology Journal* 2: 439–448. https://doi.org/10.1111/j.1467-7652.2004.00088.x.

63 Schneider, V.K., Soares-Costa, A., Chakravarthi, M. et al. (2017). Transgenic sugarcane overexpressing CaneCPI-1 negatively affects the growth and development of the sugarcane weevil *Sphenophorus levis*. *Plant Cell Reports* 36: 193–201. https://doi.org/10.1007/s00299-016-2071-2.

64 Tamhane, V.A., Giri, A.P., Sainani, M.N., and Gupta, V.S. (2007). Diverse forms of Pin-II family proteinase inhibitors from *Capsicum annuum* adversely affect the growth and development of *Helicoverpa armigera*. *Gene* 403: 29–38. https://doi.org/10.1016/j.gene.2007.07.024.

65 Jongsma, M.A. and Bolter, C. (1997). The adaptation of insects to plant protease inhibitors. *Journal of Insect Physiology* 43: 885–895. https://doi.org/10.1016/S0022-1910(97)00040-1.

66 Ahn, J.-E., Salzman, R.A., Braunagel, S.C. et al. (2004). Functional roles of specific bruchid protease isoforms in adaptation to a soybean protease inhibitor: functional roles of CmCPs in insect adaptation to scN. *Insect Molecular Biology* 13: 649–657. https://doi.org/10.1111/j.0962-1075.2004.00523.x.

67 Liu, Y., Salzman, R.A., Pankiw, T., and Zhu-Salzman, K. (2004). Transcriptional regulation in southern corn rootworm larvae challenged by soyacystatin N. *Insect Biochemistry and Molecular Biology* 34: 1069–1077. https://doi.org/10.1016/j.ibmb.2004.06.010.

68 Graham, J., Gordon, S.C., and McNICOL, R.J. (1997). The effect of the CpTi gene in strawberry against attack by vine weevil (*Otiorhynchus sulcatus* F. Coleoptera: Curculionidae). *Annals of Applied Biology* 131: 133–139. https://doi.org/10.1111/j.1744-7348.1997.tb05401.x.

69 De Leo, F., Ceci, L.R., Jouanin, L., and Gallerani, R. (2001). Analysis of mustard trypsin inhibitor-2 gene expression in response to developmental or environmental induction. *Planta* 212: 710–717. https://doi.org/10.1007/s004250000474.

70 Vila, L., Quilis, J., Meynard, D. et al. (2005). Expression of the maize proteinase inhibitor (mpi) gene in rice plants enhances resistance against the striped stem borer (*Chilo suppressalis*): effects on larval growth and insect gut proteinases: insect-resistant transgenic rice. *Plant Biotechnology Journal* 3: 187–202. https://doi.org/10.1111/j.1467-7652.2004.00117.x.

71 Maheswaran, G., Pridmore, L., Franz, P., and Anderson, M.A. (2007). A proteinase inhibitor from *Nicotiana alata* inhibits the normal development of light-brown apple moth, *Epiphyas postvittana* in transgenic apple plants. *Plant Cell Reports* 26: 773–782. https://doi.org/10.1007/s00299-006-0281-8.

72 Simões, R.A., Silva-Filho, M.C., Moura, D.S., and Delalibera, I. (2008). Effects of soybean proteinase inhibitors on development of the soil mite *Scheloribates praeincisus* (Acari: Oribatida). *Experimental & Applied Acarology* 44: 239–248. https://doi.org/10.1007/s10493-008-9139-9.

73 Zavala, J.A., Giri, A.P., Jongsma, M.A., and Baldwin, I.T. (2008). Digestive duet: midgut digestive proteinases of *Manduca sexta* ingesting *Nicotiana attenuata* with manipulated trypsin proteinase inhibitor expression. *PLoS One* 3: e2008. https://doi.org/10.1371/journal.pone.0002008.

74 Luo, M., Wang, Z., Li, H. et al. (2009). Overexpression of a weed (*Solanum americanum*) proteinase inhibitor in transgenic tobacco results in increased glandular trichome density and enhanced resistance to *Helicoverpa armigera* and *Spodoptera litura*. *International Journal of Molecular Sciences* 10: 1896–1910. https://doi.org/10.3390/ijms10041896.

75 Khadeeva, N.V., Kochieva, E.Z., Tcherednitchenko, M.Y. et al. (2009). Use of buckwheat seed protease inhibitor gene for improvement of tobacco and potato plant resistance to biotic stress. *Biochemistry (Moscow)* 74: 260–267. https://doi.org/10.1134/S0006297909030031.

76 Alvarez-Alfageme, F., Maharramov, J., Carrillo, L. et al. (2011). Potential use of a serpin from Arabidopsis for pest control. *PLoS One* 6: e20278. https://doi.org/10.1371/journal.pone.0020278.

77 Gressent, F., Da Silva, P., Eyraud, V. et al. (2011). Pea Albumin 1 subunit b (PA1b), a promising bioinsecticide of plant origin. *Toxins* 3: 1502–1517. https://doi.org/10.3390/toxins3121502.

78 Carrillo, L., Martinez, M., Álvarez-Alfageme, F. et al. (2011). A barley cysteine-proteinase inhibitor reduces the performance of two aphid species in artificial diets and transgenic Arabidopsis plants. *Transgenic Research* 20: 305–319. https://doi.org/10.1007/s11248-010-9417-2.

79 Molina, D., Patiño, L., Quintero, M. et al. (2014). Effects of the aspartic protease inhibitor from *Lupinus bogotensis* seeds on the growth and development of *Hypothenemus hampei*: an inhibitor showing high homology with storage proteins. *Phytochemistry* 98: 69–77. https://doi.org/10.1016/j.phytochem.2013.11.004.

80 Cingel, A., Savić, J., Vinterhalter, B. et al. (2015). Growth and development of Colorado potato beetle larvae, *Leptinotarsa decemlineata*, on potato plants expressing the oryzacystatin II proteinase inhibitor. *Transgenic Research* 24: 729–740. https://doi.org/10.1007/s11248-015-9873-9.

81 Tanpure, R.S., Barbole, R.S., Dawkar, V.V. et al. (2017). Improved tolerance against *Helicoverpa armigera* in transgenic tomato over-expressing multi-domain proteinase inhibitor gene from *Capsicum annuum*. *Physiology and Molecular Biology of Plants* 23: 597–604. https://doi.org/10.1007/s12298-017-0456-5.

82 Arnaiz, A., Talavera-Mateo, L., Gonzalez-Melendi, P. et al. (2018). Arabidopsis kunitz trypsin inhibitors in defense against spider mites. *Frontiers in Plant Science* 9: 986. https://doi.org/10.3389/fpls.2018.00986.

83 Shamsi, T.N., Parveen, R., Ahmad, A. et al. (2018). Inhibition of gut proteases and development of dengue vector, *Aedes aegypti* by *Allium sativum* protease inhibitor. *Acta Ecologica Sinica* 38: 325–328. https://doi.org/10.1016/j.chnaes.2018.01.002.

84 Losvik, A., Beste, L., Stephens, J., and Jonsson, L. (2018). Overexpression of the aphid-induced serine protease inhibitor CI2c gene in barley affects the generalist green peach aphid, not the specialist bird cherry-oat aphid. *PLoS One* 13: e0193816. https://doi.org/10.1371/journal.pone.0193816.

85 Christova, P.K., Christov, N.K., Mladenov, P.V., and Imai, R. (2018). The wheat multidomain cystatin TaMDC1 displays antifungal, antibacterial, and insecticidal activities in planta. *Plant Cell Reports* 37: 923–932. https://doi.org/10.1007/s00299-018-2279-4.

86 Blanco-Labra, A., Chagolla-Lopez, A., Marínez-Gallardo, N., and Valdes-Rodriguez, S. (1995). Further characterization of the 12 kDa protease/alpha amylase inhibitor present in maize seeds. *Journal of Food Biochemistry* 19: 27–41. https://doi.org/10.1111/j.1745-4514.1995.tb00518.x.

87 Rane, A.S., Joshi, R.S., and Giri, A.P. (2020). Molecular determinant for specificity: differential interaction of α-amylases with their proteinaceous inhibitors. *Biochimica et Biophysica Acta (BBA) - General Subjects* 1864: 129703. https://doi.org/10.1016/j.bbagen.2020.129703.

88 Fatima Grossi De Sa, M. and Chrispeels, M.J. (1997). Molecular cloning of bruchid (*Zabrotes subfasciatus*) α-amylase cDNA and interactions of the expressed enzyme with bean amylase inhibitors. *Insect Biochemistry and Molecular Biology* 27: 271–281. https://doi.org/10.1016/S0965-1748(96)00093-8.

89 Qian, M., Haser, R., and Payan, F. (1993). Structure and molecular model refinement of pig pancreatic alpha-amylase at 2·1 Å resolution. *Journal of Molecular Biology* 231: 785–799. https://doi.org/10.1006/jmbi.1993.1326.

90 Strobl, S., Maskos, K., Betz, M. et al. (1998). Crystal structure of yellow meal worm α-amylase at 1.64 Å resolution. *Journal of Molecular Biology* 278: 617–628. https://doi.org/10.1006/jmbi.1998.1667.

91 Abe, K., Kondo, H., Watanabe, H. et al. (1991). Oryzacystatins as the first well-defined cystatins of plant origin and their target proteinases in rice seeds. *Biomedica Biochimica Acta* 50: 637–641.

92 Feng, G.H., Richardson, M., Chen, M.S. et al. (1996). α-Amylase inhibitors from wheat: amino acid sequences and patterns of inhibition of insect and human α-amylases. *Insect Biochemistry and Molecular Biology* 26: 419–426. https://doi.org/10.1016/0965-1748(95)00087-9.

93 Franco, O.L., Rigden, D.J., Melo, R., F. et al. (2000). Activity of wheat α-amylase inhibitors towards bruchid α-amylases and structural explanation of observed specificities. *European Journal of Biochemistry* 267: 2166–2173. https://doi.org/10.1046/j.1432-1327.2000.01199.x.

94 Yamagata, H., Kunimatsu, K., Kamasaka, H. et al. (1998). Rice bifunctional α-amylase/subtilisin inhibitor: characterization, localization, and changes in developing and germinating seeds. *Bioscience, Biotechnology, and Biochemistry* 62: 978–985. https://doi.org/10.1271/bbb.62.978.

95 Ishimoto, M., Sato, T., Chrispeels, M.J., and Kitamura, K. (1996). Bruchid resistance of transgenic azuki bean expressing seed α-amylase inhibitor of common bean. *Entomologia Experimentalis et Applicata* 79: 309–315. https://doi.org/10.1111/j.1570-7458.1996.tb00838.x.

96 Marshall, J.J. and Lauda, C.M. (1975). Purification and properties of phaseolamin, an inhibitor of alpha-amylase, from the kidney bean, Phaseolus vulgaris. *Journal of Biological Chemistry* 250 (20): 8030–8037.

97 Chrispeels, M.J., Fatima Grossi de Sa, M., and Higgins, T.J.V. (1998). Genetic engineering with α-amylase inhibitors makes seeds resistant to bruchids. *Seed Science Research* 8: 257–264. https://doi.org/10.1017/S0960258500004153.

98 Valencia, A., Bustillo, A.E., Ossa, G.E., and Chrispeels, M.J. (2000). α-Amylases of the coffee berry borer (*Hypothenemus hampei*) and their inhibition by two plant amylase inhibitors. *Insect Biochemistry and Molecular Biology* 30: 207–213. https://doi.org/10.1016/S0965-1748(99)00115-0.

99 Ishimoto, M. and Chrispeels, M.J. (1996). Protective mechanism of the Mexican bean weevil against high levels of [alpha]-amylase inhibitor in the common bean. *Plant Physiology* 111: 393–401. https://doi.org/10.1104/pp.111.2.393.

100 Ishimoto, M. and Kitamura, K. (1989). Growth inhibitory effects of an α-amylase inhibitor from the kidney bean, *Phaseolus vulgaris* (L.) on three species of bruchids (Coleoptera: Bruchidae). *Applied Entomology and Zoology* 24: 281–286. https://doi.org/10.1303/aez.24.281.

101 Chrispeels, M.J. and Raikhel, N.V. (1991). Lectins, lectin genes, and their role in plant defense. *Plant Cell* 3: 1–9. https://doi.org/10.1105/tpc.3.1.1.

102 Dias, R.O., Via, A., Brandão, M.M. et al. (2015). Digestive peptidase evolution in holometabolous insects led to a divergent group of enzymes in Lepidoptera. *Insect Biochemistry and Molecular Biology* 58: 1–11. https://doi.org/10.1016/j.ibmb.2014.12.009.

103 Bhide, A.J., Channale, S.M., Yadav, Y. et al. (2017). Genomic and functional characterization of coleopteran insect-specific α-amylase inhibitor gene from Amaranthus species. *Plant Molecular Biology* 94: 319–332. https://doi.org/10.1007/s11103-017-0609-5.

104 Farias, L.R., Costa, F.T., Souza, L.A. et al. (2007). Isolation of a novel *Carica papaya* alpha-amylase inhibitor with deleterious activity toward *Callosobruchus maculatus*. *Pesticide Biochemistry and Physiology* 87: 255–260. https://doi.org/10.1016/j.pestbp.2006.08.004.

105 Morton, R.L., Schroeder, H.E., Bateman, K.S. et al. (2000). Bean alpha-amylase inhibitor 1 in transgenic peas (*Pisum sativum*) provides complete protection from pea weevil (*Bruchus pisorum*) under field conditions. *Proceedings of the National Academy of Sciences of the United States of America* 97: 3820–3825. https://doi.org/10.1073/pnas.070054597.

106 Schroeder, H.E., Gollasch, S., Moore, A. et al. (1995). Bean [alpha]-amylase inhibitor confers resistance to the pea weevil (*Bruchus pisorum*) in transgenic peas (*Pisum sativum* L.). *Plant Physiology* 107: 1233–1239. https://doi.org/10.1104/pp.107.4.1233.

107 Shaikh, F.K., Gadge, P.P., Shinde, A.A. et al. (2013). Novel isoforms of proteinaceous α-amylase inhibitor (α-AI) from seed extract of *Albizia lebbeck*. *Acta Physiologiae Plantarum* 35: 901–909. https://doi.org/10.1007/s11738-012-1133-5.

108 Shainkin, R. and Birk, Y. (1970). α-amylase inhibitors from wheat isolation and characterization. *Biochimica et Biophysica Acta (BBA) - Protein Structure* 221: 502–513. https://doi.org/10.1016/0005-2795(70)90221-7.

109 Sanchez-Monge, R., Gomez, L., Garcia-Olmedo, F., and Salcedo, G. (1989). New dimeric inhibitor of heterologous alpha-amylases encoded by a duplicated gene in the short arm of chromosome 3B of wheat (*Triticum aestivum* L.). *European Journal of Biochemistry* 183: 37–40. https://doi.org/10.1111/j.1432-1033.1989.tb14893.x.

110 Bloch, C. and Richardson, M. (1991). A new family of small (5 kDa) protein inhibitors of insect α-amylases from seeds or sorghum (*Sorghum bicolor* (L) Moench) have sequence homologies with wheat γ-purothionins. *FEBS Letters* 279: 101–104. https://doi.org/10.1016/0014-5793(91)80261-Z.

111 Giri, A.P. and Kachole, M.S. (1998). Amylase inhibitors of pigeonpea (*Cajanus cajan*) seeds. *Phytochemistry* 47: 197–202. https://doi.org/10.1016/S0031-9422(97)00570-0.

112 Kadziola, A., Søgaard, M., Svensson, B., and Haser, R. (1998). Molecular structure of a barley alpha-amylase-inhibitor complex: implications for starch binding and catalysis. *Journal of Molecular Biology* 278: 205–217. https://doi.org/10.1006/jmbi.1998.1683.

113 Iulek, J., Franco, O.L., Silva, M. et al. (2000). Purification, biochemical characterisation and partial primary structure of a new α-amylase inhibitor from Secale cereale (rye). *The International Journal Of Biochemistry & Cell Biology* 32 (11-12): 1195–1204.

114 Schimoler-O'Rourke, R., Richardson, M., and Selitrennikoff, C.P. (2001). Zeamatin inhibits trypsin and α-amylase activities. *Applied and Environmental Microbiology* 67: 2365–2366. https://doi.org/10.1128/AEM.67.5.2365-2366.2001.

115 Franco, O.L., Rigden, D.J., Melo, F.R., and Grossi-de-Sá, M.F. (2002). Plant α-amylase inhibitors and their interaction with insect α-amylases: structure, function and potential for crop protection. *European Journal of Biochemistry* 269: 397–412. https://doi.org/10.1046/j.0014-2956.2001.02656.x.

116 Sen, S. and Dutta, S.K. (2012). Cloning, expression and characterization of biotic stress inducible Ragi bifunctional inhibitor (RBI) gene from *Eleusine coracana* Gaertn. *Journal of Plant Biochemistry and Biotechnology* 21: 66–76. https://doi.org/10.1007/s13562-011-0082-1.

117 Zhu, F., Zhou, Y.-K., Ji, Z.-L., and Chen, X.-R. (2018). The plant ribosome-inactivating proteins play important roles in defense against pathogens and insect pest attacks. *Frontiers in Plant Science* 9: 146. https://doi.org/10.3389/fpls.2018.00146.

118 Akkouh, O., Ng, T.B., Cheung, R.C.F. et al. (2015). Biological activities of ribosome-inactivating proteins and their possible applications as antimicrobial, anticancer, and anti-pest agents and in neuroscience research. *Applied Microbiology and Biotechnology* 99: 9847–9863. https://doi.org/10.1007/s00253-015-6941-2.

119 Choudhary, N.L., Yadav, O.P., and Lodha, M.L. (2008). Ribonuclease, deoxyribonuclease, and antiviral activity of *Escherichia coli*-expressed *Bougainvillea xbuttiana* antiviral protein 1. *Biochemistry (Moscow)* 73: 273–277. https://doi.org/10.1134/S000629790803005X.

120 Stevens, W.A., Spurdon, C., Onyon, L.J., and Stirpe, F. (1981). Effect of inhibitors of protein synthesis from plants on tobacco mosaic virus infection. *Experientia* 37: 257–259. https://doi.org/10.1007/BF01991642.

121 Wang, P. and Turner, N.E. (2000). Virus resistance mediated by ribosome inactivating proteins. In: *Advances in Virus Research*, 325–355. Elsevier https://doi.org/10.1016/S0065-3527(00)55007-6.

122 Girbes, T., Ferreras, J., Arias, F., and Stirpe, F. (2004). Description, distribution, activity and phylogenetic relationship of ribosome-inactivating proteins in plants, fungi and bacteria. *Mini Reviews in Medicinal Chemistry* 4: 461–476. https://doi.org/10.2174/1389557043403891.

123 Hartley, M.R. and Lord, J.M. (2004). Cytotoxic ribosome-inactivating lectins from plants. *Biochimica et Biophysica Acta (BBA) - Proteins and Proteomics* 1701: 1–14. https://doi.org/10.1016/j.bbapap.2004.06.004.

124 Van Damme, E.J.M., Hao, Q., Chen, Y. et al. (2001). Ribosome-inactivating proteins: a family of plant proteins that do more than inactivate ribosomes. *Critical Reviews in Plant Sciences* 20: 395–465. https://doi.org/10.1080/07352689.2001.10131826.

125 de Virgilio, M., Lombardi, A., Caliandro, R., and Fabbrini, M.S. (2010). Ribosome-inactivating proteins: from plant defense to tumor attack. *Toxins* 2: 2699–2737. https://doi.org/10.3390/toxins2112699.

126 Lord, J.M., Roberts, L.M., and Robertus, J.D. (1994). Ricin: structure, mode of action, and some current applications. *The FASEB Journal* 8: 201–208. https://doi.org/10.1096/fasebj.8.2.8119491.

127 Domashevskiy, A. and Goss, D. (2015). Pokeweed antiviral protein, a ribosome inactivating protein: activity, inhibition and prospects. *Toxins* 7: 274–298. https://doi.org/10.3390/toxins7020274.

128 Shang, C., Rougé, P., and Van Damme, E. (2016). Ribosome inactivating proteins from rosaceae. *Molecules* 21: 1105. https://doi.org/10.3390/molecules21081105.

129 Shu, S., Xie, G., Guo, X., and Wang, M. (2009). Purification and characterization of a novel ribosome-inactivating protein from seeds of *Trichosanthes kirilowii* Maxim. *Protein Expression and Purification* 67: 120–125. https://doi.org/10.1016/j.pep.2009.03.004.

130 Brown, J.E., Ussery, M.A., Leppla, S.H., and Rothman, S.W. (1980). Inhibition of protein synthesis by Shiga toxin: activation of the toxin and inhibition of peptide elongation. *FEBS Letters* 117: 84–88. https://doi.org/10.1016/0014-5793(80)80918-5.

131 Russo, L.M., Melton-Celsa, A.R., Smith, M.J., and O'Brien, A.D. (2014). Comparisons of native Shiga toxins (Stxs) type 1 and 2 with chimeric toxins indicate that the source of the binding subunit dictates degree of toxicity. *PLoS One* 9: e93463. https://doi.org/10.1371/journal.pone.0093463.

132 Dowd, P.F., Holmes, R.A., Pinkerton, T.S. et al. (2006). Relative activity of a tobacco hybrid expressing high levels of a tobacco anionic peroxidase and maize ribosome-inactivating protein against *Helicoverpa zea* and *Lasioderma serricorne*. *Journal of Agricultural and Food Chemistry* 54: 2629–2634. https://doi.org/10.1021/jf058180p.

133 Dowd, P.F., Mehta, A.D., and Boston, R.S. (1998). Relative toxicity of the maize endosperm ribosome-inactivating protein to insects. *Journal of Agricultural and Food Chemistry* 46: 3775–3779. https://doi.org/10.1021/jf980334w.

134 Dowd, P.F., Zuo, W.-N., Gillikin, J.W. et al. (2003). Enhanced resistance to *Helicoverpa zea* in tobacco expressing an activated form of maize ribosome-inactivating protein. *Journal of Agricultural and Food Chemistry* 51: 3568–3574. https://doi.org/10.1021/jf0211433.

135 Wei, G., Liu, R.-S., Wang, Q., and Liu, W.-Y. (2004). Toxicity of two type II ribosome-inactivating proteins (cinnamomin and ricin) to domestic silkworm larvae. *Archives of Insect Biochemistry and Physiology* 57: 160–165. https://doi.org/10.1002/arch.20024.

136 Gatehouse, A.M.R., Barbieri, L., Stirpe, F., and Croy, R.R.D. (1990). Effects of ribosome inactivating proteins on insect development – differences between Lepidoptera and Coleoptera. *Entomologia Experimentalis et Applicata* 54: 43–51. https://doi.org/10.1111/j.1570-7458.1990.tb01310.x.

137 Stirpe, F. (2013). Ribosome-inactivating proteins: from toxins to useful proteins. *Toxicon* 67: 12–16. https://doi.org/10.1016/j.toxicon.2013.02.005.

138 Shahidi-Noghabi, S., Van Damme, E.J.M., and Smagghe, G. (2008). Carbohydrate-binding activity of the type-2 ribosome-inactivating protein SNA-I from elderberry (*Sambucus nigra*) is a determining factor for its insecticidal activity. *Phytochemistry* 69: 2972–2978. https://doi.org/10.1016/j.phytochem.2008.09.012.

139 Bertholdo-Vargas, L.R., Martins, J.N., Bordin, D. et al. (2009). Type 1 ribosome-inactivating proteins-entomotoxic, oxidative and genotoxic action on *Anticarsia gemmatalis* (Hübner) and *Spodoptera frugiperda* (J.E. Smith) (Lepidoptera: Noctuidae). *Journal of Insect Physiology* 55: 51–58. https://doi.org/10.1016/j.jinsphys.2008.10.004.

140 Hamshou, M., Shang, C., Smagghe, G., and Van Damme, E.J.M. (2016). Ribosome-inactivating proteins from apple have strong aphicidal activity in artificial diet and in planta. *Crop Protection* 87: 19–24. https://doi.org/10.1016/j.cropro.2016.04.013.

141 Hamshou, M., Shang, C., De Zaeytijd, J. et al. (2017). Expression of ribosome-inactivating proteins from apple in tobacco plants results in enhanced resistance to *Spodoptera exigua*. *Journal of Asia-Pacific Entomology* 20: 1–5. https://doi.org/10.1016/j.aspen.2016.09.009.

142 Das, M.K., Sharma, R.S., and Mishra, V. (2012). Induction of apoptosis by ribosome inactivating proteins: importance of *N*-glycosidase activity. *Applied Biochemistry and Biotechnology* 166: 1552–1561. https://doi.org/10.1007/s12010-012-9550-x.

143 Narayanan, S., Surendranath, K., Bora, N. et al. (2005). Ribosome inactivating proteins and apoptosis. *FEBS Letters* 579: 1324–1331. https://doi.org/10.1016/j.febslet.2005.01.038.

144 Shahidi-Noghabi, S., Van Damme, E.J., Mahdian, K., and Smagghe, G. (2010). Entomotoxic action of *Sambucus nigra* agglutinin i in *Acyrthosiphon pisum* aphids and *Spodoptera exigua* caterpillars through caspase-3-like-dependent apoptosis. *Archives of Insect Biochemistry and Physiology* 75: 207–220. https://doi.org/10.1002/arch.20387.

145 Acosta-Gallegos, J.A., Quintero, C., Vargas, J. et al. (1998). A new variant of arcelin in wild common bean, *Phaseolus vulgaris* L., from southern Mexico. *Genetic Resources and Crop Evolution* 45: 235–242. https://doi.org/10.1023/A:1008636132108.

146 Lioi, L. and Bollini, R. (1989). Identification of a new arcelin variant in wild bean seeds. *Annual Report of the Bean Improvement Cooperative* 28-29: 32.

147 Osborn, T.C., Blake, T., Gepts, P., and Bliss, F.A. (1986). Bean arcelin: 2. Genetic variation, inheritance and linkage relationships of a novel seed protein of *Phaseolus vulgaris* L. *Theoretical and Applied Genetics* 71: 847–855. https://doi.org/10.1007/BF00276428.

148 Santino, A., Valsasina, B., Lioi, L. et al. (1991). Bean (*Phaseolus vulgaris* L.) seed lectins: a novel electrophoretic variant of arcelin. *Plant Physiology* 10: 7–11.

149 Minney, B.H.P., Gatehouse, A.M.R., Dobie, P. et al. (1990). Biochemical bases of seed resistance to *Zabrotes subfasciatus* (bean weevil) in *Phaseolus vulgaris* (common bean); A mechanism for arcelin toxicity. *Journal of Insect Physiology* 36: 757–767. https://doi.org/10.1016/0022-1910(90)90049-L.

150 Osborni, T.C., Alexander, D.C., Sun, S.S.M. et al. (1988). Insecticidal activity and lectin homology of arcelin seed protein. *Science* 240: 207–210. https://doi.org/10.1126/science.240.4849.207.

151 Goossens, A., Geremia, R., Bauw, G. et al. (1994). Isolation and characterisation of arcelin-5 proteins and cDNAs. *European Journal of Biochemistry* 225: 787–795. https://doi.org/10.1111/j.1432-1033.1994.0787b.x.

152 Hamelryck, T.W., Poortmans, F., Goossens, A. et al. (1996). Crystal structure of arcelin-5, a lectin-like defense protein from *phaseolus vulgaris*. *The Journal of Biological Chemistry* 271: 32796–32802. https://doi.org/10.1074/jbc.271.51.32796.

153 Goossens, A., Quintero, C., Dillen, W. et al. (2000). Analysis of bruchid resistance in the wild common bean accession G02771: no evidence for insecticidal activity of arcelin 5. *Journal of Experimental Botany* 51: 1229–1236. https://doi.org/10.1093/jexbot/51.348.1229.

154 Mourey, L., Pédelacq, J.-D., Birck, C. et al. (1998). Crystal structure of the arcelin-1 dimer from *Phaseolus vulgaris* at 1.9-Å resolution. *The Journal of Biological Chemistry* 273: 12914–12922. https://doi.org/10.1074/jbc.273.21.12914.

155 Cardona, C., Kornegay, J., Posso, C.E. et al. (1990). Comparative value of four arcelin variants in the development of dry bean lines resistant to the Mexican bean weevil. *Entomologia Experimentalis et Applicata* 56: 197–206. https://doi.org/10.1111/j.1570-7458.1990.tb01397.x.

156 Paes, N.S., Gerhardt, I.R., Coutinho, M.V. et al. (2000). The effect of arcelin-1 on the structure of the midgut of bruchid larvae and immunolocalization of the arcelin protein. *Journal of Insect Physiology* 46: 393–402. https://doi.org/10.1016/S0022-1910(99)00122-5.

157 Mirkov, T.E., Wahlstrom, J.M., Hagiwara, K. et al. (1994). Evolutionary relationships among proteins in the phytohemagglutinin-arcelin-α-amylase inhibitor family of the common bean and its relatives. *Plant Molecular Biology* 26: 1103–1113. https://doi.org/10.1007/BF00040692.

158 Fabre, C., Causse, H., Mourey, L. et al. (1998). Characterization and sugar-binding properties of arcelin-1, an insecticidal lectin-like protein isolated from kidney bean (*Phaseolus vulgaris* L. cv. RAZ-2) seeds. *Biochemical Journal* 329: 551–560. https://doi.org/10.1042/bj3290551.

159 Gerhardt, I.R., Paes, N.S., Bloch, C. et al. (2000). Molecular characterization of a new arcelin-5 gene. *Biochimica et Biophysica Acta (BBA) - Gene Structure and Expression* 1490: 87–98. https://doi.org/10.1016/S0167-4781(99)00219-5.

160 Malaikozhundan, B., Suresh, P., Seshadri, S., and Janarthanan, S. (2003). Toxicity assessment of wild bean seed protein-Arcelin on Asian armyworm, Spodoptera litura (Fabricius).

161 Zambre, M., Goossens, A., Cardona, C. et al. (2005). A reproducible genetic transformation system for cultivated *Phaseolus acutifolius* (tepary bean) and its use to assess the role of arcelins in resistance to the Mexican bean weevil. *Theoretical and Applied Genetics* 110: 914–924. https://doi.org/10.1007/s00122-004-1910-7.

162 Janarthanan, S., Suresh, P., Radke, G. et al. (2008). Arcelins from an Indian wild pulse, *Lablab purpureus*, and insecticidal activity in storage pests. *Journal of Agricultural and Food Chemistry* 56: 1676–1682. https://doi.org/10.1021/jf071591g.

163 Blair, M.W., Prieto, S., Díaz, L.M. et al. (2010). Linkage disequilibrium at the APA insecticidal seed protein locus of common bean (*Phaseolus vulgaris* L.). *BMC Plant Biology* 10: 79. https://doi.org/10.1186/1471-2229-10-79.

164 Janarthanan, S., Sakthivelkumar, S., Veeramani, V. et al. (2012). A new variant of antimetabolic protein, arcelin from an Indian bean, *Lablab purpureus* (Linn.) and its effect on the stored product pest, *Callosobruchus maculatus*. *Food Chemistry* 135: 2839–2844. https://doi.org/10.1016/j.foodchem.2012.06.129.

165 Karuppiah, H., Kirubakaran, N., and Sundaram, J. (2018). Genetic resources for arcelin, a stored product insect antimetabolic protein from various accessions of pulses of Leguminosae. *Genetic Resources and Crop Evolution* 65: 79–90. https://doi.org/10.1007/s10722-017-0510-8.

166 Grossi-de-Sá, M.F., Pelegrini, P.B., Vasconcelos, I.M. et al. (2015). Entomotoxic plant proteins: potential molecules to develop genetically modified plants resistant to insect-pests. In: *Plant Toxins* (eds. P. Gopalakrishnakone, C.R. Carlini and R. Ligabue-Braun), 1–34. Dordrecht: Springer Netherlands https://doi.org/10.1007/978-94-007-6728-7_13-1.

167 Lacerda, A.F., Vasconcelos, Ã.A.R., Pelegrini, P.B., and Grossi de Sa, M.F. (2014). Antifungal defensins and their role in plant defense. *Frontiers in Microbiology* 5 https://doi.org/10.3389/fmicb.2014.00116.

168 Chen, J.-J., Chen, G.-H., Hsu, H.-C. et al. (2004). Cloning and functional expression of a mungbean defensin VrD1 in pichia pastoris. *Journal of Agricultural and Food Chemistry* 52: 2256–2261. https://doi.org/10.1021/jf030662i.

169 Pelegrini, P.B., Lay, F.T., Murad, A.M. et al. (2008). Novel insights on the mechanism of action of α-amylase inhibitors from the plant defensin family. *Proteins* 73: 719–729. https://doi.org/10.1002/prot.22086.

170 dos Santos, I.S., Carvalho, A.d.O., de Souza-Filho, G.A. et al. (2010). Purification of a defensin isolated from *Vigna unguiculata* seeds, its functional expression in *Escherichia coli*, and assessment of its insect α-amylase inhibitory activity. *Protein Expression and Purification* 71: 8–15. https://doi.org/10.1016/j.pep.2009.11.008.

171 Carlini, C.R. and Guimarães, J.A. (1981). Isolation and characterization of a toxic protein from *Canavalia ensiformis* (jack bean) seeds, distinct from concanavalin A. *Toxicon* 19: 667–675. https://doi.org/10.1016/0041-0101(81)90104-5.

172 Holland, M.A., Griffin, J.D., Elise Meyer-Bothling, L., and Polacco, J.C. (1987). Developmental genetics of the soybean urease isozymes. *Developmental Genetics* 8: 375–387. https://doi.org/10.1002/dvg.1020080508.

173 Witte, C.-P., Tiller, S.A., Taylor, M.A., and Davies, H.V. (2002). Leaf urea metabolism in potato. Urease activity profile and patterns of recovery and distribution of 15 N after foliar urea application in wild-type and urease-antisense transgenics. *Plant Physiology* 128: 1129–1136. https://doi.org/10.1104/pp.010506.

174 Pires-Alves, M., Grossi-de-Sá, M.F., Barcellos, G.B.S. et al. (2003). Characterization and expression of a novel member (JBURE-II) of the urease gene family from jackbean [*Canavalia ensiformis* (L.) DC]. *Plant and Cell Physiology* 44: 139–145. https://doi.org/10.1093/pcp/pcg018.

175 Jennings, C.V., Rosengren, K.J., Daly, N.L. et al. (2005). Isolation, solution structure, and insecticidal activity of kalata B2, a circular protein with a twist: do Möbius strips exist in nature? *Biochemistry* 44: 851–860. https://doi.org/10.1021/bi047837h.

176 Menegassi, A., Wassermann, G.E., Olivera-Severo, D. et al. (2008). Urease from cotton (*Gossypium hirsutum*) seeds: isolation, physicochemical characterization, and antifungal properties of the protein. *Journal of Agricultural and Food Chemistry* 56: 4399–4405. https://doi.org/10.1021/jf0735275.

177 Krishna, B.L., Singh, A.N., Patra, S., and Dubey, V.K. (2011). Purification, characterization and immobilization of urease from *Momordica charantia* seeds. *Process Biochemistry* 46: 1486–1491. https://doi.org/10.1016/j.procbio.2011.03.022.

178 Stanisçuaski, F. and Carlini, C.R. (2012). Plant ureases and related peptides: understanding their entomotoxic properties. *Toxins* 4: 55–67. https://doi.org/10.3390/toxins4020055.

179 Balasubramanian, A., Durairajpandian, V., Elumalai, S. et al. (2013). Structural and functional studies on urease from pigeon pea (*Cajanus cajan*). *International Journal of Biological Macromolecules* 58: 301–309. https://doi.org/10.1016/j.ijbiomac.2013.04.055.

180 Oguis, G.K., Gilding, E.K., Huang, Y.H. et al. (2020). Insecticidal diversity of butterfly pea (Clitoriaternatea) accessions. *Industrial Crops and Products* 147: 112214.

181 Tang, Y., He, H., Qu, X. et al. (2020). RNA interference-mediated knockdown of the transcription factor Krüppel homologue 1 suppresses vitellogenesis in *Chilo suppressalis*. *Insect Molecular Biology* 29: 183–192. https://doi.org/10.1111/imb.12617.

182 Sá, C.A., Vieira, L.R., Pereira Almeida Filho, L.C. et al. (2020). Risk assessment of the antifungal and insecticidal peptide Jaburetox and its parental protein the Jack bean (*Canavalia ensiformis*) urease. *Food and Chemical Toxicology* 136: 110977. https://doi.org/10.1016/j.fct.2019.110977.

183 Bruix, M., Jimenez, M.A., Santoro, J. et al. (1993). Solution structure of .gamma.1-H and .gamma.1-P thionins from barley and wheat endosperm determined by proton NMR: a structural motif common to toxic arthropod proteins. *Biochemistry* 32: 715–724. https://doi.org/10.1021/bi00053a041.

184 De Coninck, B.M.A., Sels, J., Venmans, E. et al. (2010). *Arabidopsis thaliana* plant defensin AtPDF1.1 is involved in the plant response to biotic stress. *New Phytologist* 187: 1075–1088. https://doi.org/10.1111/j.1469-8137.2010.03326.x.

185 Pelegrini, P.B., Quirino, B.F., and Franco, O.L. (2007). Plant cyclotides: an unusual class of defense compounds. *Peptides* 28: 1475–1481. https://doi.org/10.1016/j.peptides.2007.04.025.

186 Jennings, C., West, J., Waine, C. et al. (2001). Biosynthesis and insecticidal properties of plant cyclotides: the cyclic knotted proteins from *Oldenlandia affinis*. *Proceedings of the National Academy of Sciences of the United States of America* 98: 10614–10619. https://doi.org/10.1073/pnas.191366898.

187 Conlan, B.F., Colgrave, M.L., Gillon, A.D. et al. (2012). Insights into processing and cyclization events associated with biosynthesis of the cyclic Peptide kalata B1. *The Journal of Biological Chemistry* 287: 28037–28046. https://doi.org/10.1074/jbc.M112.347823.

188 Barbeta, B.L., Marshall, A.T., Gillon, A.D. et al. (2008). Plant cyclotides disrupt epithelial cells in the midgut of lepidopteran larvae. *Proceedings of the National Academy of Sciences of the United States of America* 105: 1221–1225. https://doi.org/10.1073/pnas.0710338104.

189 Kelemu, S., Cardona, C., and Segura, G. (2004). Antimicrobial and insecticidal protein isolated from seeds of *Clitoria ternatea*, a tropical forage legume. *Plant Physiology and Biochemistry* 42: 867–873. https://doi.org/10.1016/j.plaphy.2004.10.013.

190 Carlini, C.R. and Udedibie, A.B. (1997). Comparative effects of processing methods on hemagglutinating and antitryptic activities of *Canavalia ensiformis* and *Canavalia braziliensis* seeds. *Journal of Agricultural and Food Chemistry* 45 (11): 4372–4377.

191 Ferreira DaSilva, C.T., Gombarovits, M.E.C., Masuda, H. et al. (2000). Proteolytic activation of canatoxin, a plant toxic protein, by insect cathepsin-like enzymes. *Archives of Insect Biochemistry and Physiology* 44 (4): 162–171.

192 Tomazetto, G., Mulinari, F., Stanisçuaski, F. et al. (2007). Expression kinetics and plasmid stability of recombinant *E. coli* encoding urease-derived peptide with bioinsecticide activity. *Enzyme and Microbial Technology* 41: 821–827. https://doi.org/10.1016/j.enzmictec.2007.07.006.

193 Defferrari, M.S., Demartini, D.R., Marcelino, T.B. et al. (2011). Insecticidal effect of *Canavalia ensiformis* major urease on nymphs of the milkweed bug *Oncopeltus fasciatus* and characterization of digestive peptidases. *Insect Biochemistry and Molecular Biology* 41: 388–399. https://doi.org/10.1016/j.ibmb.2011.02.008.

194 Follmer, C., Real-Guerra, R., Wasserman, G.E. et al. (2004). Jackbean, soybean and *Bacillus pasteurii* ureases: biological effects unrelated to ureolytic activity. *European Journal of Biochemistry* 271: 1357–1363. https://doi.org/10.1111/j.1432-1033.2004.04046.x.

195 Mulinari, F., Becker-Ritt, A.B., Demartini, D.R. et al. (2011). Characterization of JBURE-IIb isoform of Canavalia ensiformis (L.) DC urease. *Biochimica et Biophysica Acta (BBA)-Proteins and Proteomics* 1814 (12): 1758–1768.

196 Mulinari, F., Stanisçuaski, F., Bertholdo-Vargas, L.R. et al. (2007). Jaburetox-2Ec: an insecticidal peptide derived from an isoform of urease from the plant *Canavalia ensiformis*. *Peptides* 28: 2042–2050. https://doi.org/10.1016/j.peptides.2007.08.009.

197 Cohen, E. (1993). Chitin synthesis and degradation as targets for pesticide action. *Archives of Insect Biochemistry and Physiology* 22: 245–261. https://doi.org/10.1002/arch.940220118.

198 Collinge, D.B., Kragh, K.M., Mikkelsen, J.D. et al. (1993). Plant chitinases. *The Plant Journal* 3: 31–40. https://doi.org/10.1046/j.1365-313X.1993.t01-1-00999.x.

199 Nagpure, A., Choudhary, B., and Gupta, R.K. (2014). Chitinases: in agriculture and human healthcare. *Critical Reviews in Biotechnology* 34: 215–232. https://doi.org/10.3109/07388551.2013.790874.

200 Kitajima, S., Kamei, K., Taketani, S. et al. (2010). Two chitinase-like proteins abundantly accumulated in latex of mulberry show insecticidal activity. *BMC Biochemistry* 11: 6. https://doi.org/10.1186/1471-2091-11-6.

201 Harrison, R.L. and Bonning, B.C. (2010). Proteases as insecticidal agents. *Toxins* 2: 935–953. https://doi.org/10.3390/toxins2050935.

202 Pechan, T., Jiang, B., Steckler, D. et al. (1999). Characterization of three distinct cDNA clones encoding cysteine proteinases from maize (Zea mays L.) callus. *Plant Molecular Biology* 40 (1): 111–119.

203 Lopez, L., Camas, A., Shivaji, R. et al. (2007). Mir1-CP, a novel defense cysteine protease accumulates in maize vascular tissues in response to herbivory. *Planta* 226 (2): 517–527.

204 Jiang, B.-H., Siregar, U., Willeford, K.O. et al. (1995). Association of a 33-kD cysteine proteinase found in corn callus with the inhibition of fall armyworm larval growth. *Plant Physiology* 108: 1631–1640.

205 Pechan, T., Ye, L., Chang, Y. et al. (2000). A unique 33-kD cysteine proteinase accumulates in response to larval feeding in maize genotypes resistant to fall armyworm and other Lepidoptera. *The Plant Cell* 12: 1031–1040.

206 Mohan, S., Ma, P.W., Williams, W.P., and Luthe, D.S. (2008). A naturally occurring plant cysteine protease possesses remarkable toxicity against insect pests and synergizes Bacillus thuringiensis toxin. *PloS One* 3 (3): e1786.

207 Konno, K., Hirayama, C., Nakamura, M. et al. (2004). Papain protects papaya trees from herbivorous insects: role of cysteine proteases in latex. *The Plant Journal* 37: 370–378. https://doi.org/10.1046/j.1365-313X.2003.01968.x.

208 Upadhyay, S.K. and Singh, P.K. (2012). Receptors of garlic (Allium sativum) lectins and their role in insecticidal action. *Protein J*, 31 (6): 439–446.

209 Upadhyay, S.K. and Chandrashekar, K. (2012). Interaction of salivary and midgut proteins of Helicoverpa armigera with soybean trypsin inhibitor. *Protein J*, 31, 259–264.

3

Bioprospecting of Natural Compounds for Industrial and Medical Applications

Current Scenario and Bottleneck

Sameer Dixit[1], Akanchha Shukla[1], Vinayak Singh[1], and Santosh Kumar Upadhyay[2]

[1]*Department of Biology, University of Western Ontario, London, Ontario, Canada*
[2]*Department of Botany, Panjab University, Chandigarh, UT, India*

3.1 Introduction

All substances produced by biosynthetic pathways inside living organisms that are not essential for energy metabolism but required for their ecological fitness are termed as natural products, for example, secondary metabolites. Evolutionary pressure on biosynthetic pathways resulted in the development of such chemically diverse and biologically potent molecules [1]. These natural products can be attained from enormous biodiversity that is natural sources or organisms from diverse ecosystems and ecological complexes. Thus, the process of exploration of biodiversity for bioactive compounds with agricultural, industrial, and medical applications is known as bioprospecting [2, 3]. Since natural resources are limited, protecting genes, species, and habitats in ecosystems became necessary to prevent long-term depletion of biodiversity. A Convention on Biological Diversity agreement was made obligatory on 29 December 1993 for preservation and sustainable use of biodiversity, as well as ethical sharing of benefits with independent states and local communities [4]. Thus, a series of ethical and legal issues like patents, intellectual property rights, cultural and individual rights to privacy, cross-cultural understanding, and other aspects must be dealt before starting bioprospecting ventures [5, 6]. A successful bioprospecting program involves scientific and economic activities enabling sustainable development and economic growth by the revenue generated from royalties contributing to biodiversity conservation and safeguarding traditional medicine knowledge [2].

Due to changes in globalization and evolving environmental problems, there is increased economic relevance of biological diversity. It is difficult to estimate the actual overall biodiversity value in financial terms for all ecosystems [7]. However, it was recently estimated to be about US$2.9 trillion for the whole world [8]. Bioprospecting has contributed

[#]Authors have contributed equally.

Bioprospecting of Plant Biodiversity for Industrial Molecules, First Edition.
Edited by Santosh Kumar Upadhyay and Sudhir P. Singh.

enormously in drug discovery to treat variety of infectious diseases, cancer, autoimmune diseases, etc. [9]. In agricultural field, bioprospecting leads to the development of various pesticidal (*Bacillus thuringiensis* toxin and annonins from *Annona squamosal* plant) and herbicidal compounds and bio-fertilizers (*rhizobium*) enhancing the plant growth, productivity, and ability to withstand adverse environmental conditions [10, 11]. Apart from pharmaceutical and agricultural applications, bioprospecting products are also used in bioremediation (laccase enzyme from *Phanerochaete chrysosporium*), cosmetics (zeaxanthin from *Xanthobacter autotrophicus*, collagenases from *Clostridium histolyticum*, and keratinases from *Microsporum*), and biosensors (laccase containing electrodes) [12–14].

In this chapter, we will discuss about bioprospecting, rational bioprospecting processes, their drawbacks, and new approaches to advance utility of natural products in agricultural, medical, and industrial fields.

3.2 Why Bioprospecting Is Important

The sole purpose of bioprospecting is to explore bio-diversity to find natural product/organisms that can benefit humans. After the *Convention on Biological Diversity* 1992–1993, the useful natural product/organisms not only benefited industries involved in identification or production but also the indigenous community or host country. Thus, local agency has the control and sovereignty on identified biological resources and also expect adequate economic compensation for resource. This will enhance the cooperation between industries and local agencies and also increases the employment opportunity resulting in the improvement of socioeconomic status. It is typically considered that the impact of bioprospecting benefits will be much better if it is based on the knowledge and information from local people [15].

3.3 Major Sites for Bioprospecting

Bioprospecting can be performed anywhere with rich biological diversity such as forest, conservation areas, hotspots, ocean, etc. [16]. Extreme environments such as polar regions and hot water springs can also serve as potential site for bioprospecting [16]. Statistically, it is considered that 1 in 30,000 to 40,000 is the chance to find useful natural product [17].

Terrestrial region is most commonly used for bioprospecting programs. A classic example is the discovery of Artemisinin by TuYouyou, a nobel laureate [18]. Globally, scientists test ~240,000 compounds without any breakthrough success. TuYouyou inspired by "The handbook of prescriptions for emergency treatments," a Chinese medical handbook written by Ge Hong around 340 CE during Jindynasty. She screened ~2000 biological samples and identified that ethyl ether extract of Qinghao (Chinese common name *Artemisia* plant) significantly inhibited malaria parasites; this led to the successful discovery of Artemisinin [18]. Many national and international terrestrial bioprospecting programs have started in recent years, such as INBio-Merck (between INBio-National Biodiversity Institute of Costa Rica and Merck & Co. Ltd), Peruvian (between ICBG, USA; Bristol-Myers Squibb, Monsanto and Glaxo-Wellcome), and The TBGRI-Kani contract (between TBGRI and Arya Vaidya Pharmacy along with the help of Kani community) are some of them [19–21].

After land, oceans are the second most explored site for bioprospecting programs. About 70% of the earth is covered from ocean with one of the richest biological diversity on planet. In last four decades, ~15,000 marine bioactive natural products are identified [22]. Omega-3 fatty acid from fish oil, Cytarabine and Vidarabine from Sponge *Tethyacrypta*, Cephalosporin from Marine fungi are some of the few examples of bioactive compounds derived from marine biodiversity [23].

Polar region and Hot Spring Lake are the other important sites that are exploited for bioprospecting programs [16]. These regions have harsh climate during most of the year, temperatures reach minimum of −50 °C (Artic) to maximum 100 °C (Hot Spring). These extreme environments lead to the local and unique adaptations in biological organisms which are living in these regions [24]. Sometimes these adaptations are due to the accumulation of new compounds that can have bioactive capabilities. Due to the global warming and climate change, the ice in polar region is melting, thus proving more area to explore or may exploit. The recent example is MabCent project, a team of researchers spent a year in extreme environment and collect ~3000 pounds of biological sample (consisting microalgae, invertebrates, and many others) from the 1000 different sites around Norway's Svalbard archipelago that are being screened for their bioactive properties [24]. Identification of New Biofuel-producing bacteria from Iceland's famed hot springs, New lipase for the production of biodiesel from Taptapani Hot Spring (Odisha), and phytase-producing thermophilic fungi from West Anatolia are some of the recent examples of hot springs bioprospecting expedition [25–27].

3.4 Pipeline of Bioprospecting

Bioprospecting process is generally divided in to four major phases which starts from the collection of sample to commercialization of the product [28, 29]. Phase I consists of the identification of Bioprospecting site followed by onsite collection and preservation of biological sample. Initial processing such as washing/cleaning/freezing of biological sample also lies in this phase. Phase II starts with the chemical processing of sample followed by isolation and identification of bioactive compound/organism. If required, this phase also consists of the characterization and mass production of specific bioactive material. Phase III begins with the detailed characterization of specific compound/organism and in-depth screening or validation of biological property. Phase IV is the last stage in which bioactive compound/organism goes for product development and ready for commercialization [28, 29].

3.5 Biopiracy: An Unethical Bioprospecting

Biopiracy is considered as the exploitation or monopolization of biological material or indigenous knowledge without informing/compensating the local community or country from where the biological material or knowledge is procured [30]. Biopiracy imposes many negative effects on local biodiversity like overexploitation of endemic biological material, reduction in biodiversity or local niche, and illegal privatization of biological material, which results in adversely affecting the cultural, regional, and traditional identity or knowledge of indigenous people [31]. Biopiracy is illegal by all means but due to its commercial and financial profit, many private companies prefer it [31]. There are many international, national, and provisional law or rights are available to counter biopiracy such as Intellectual property rights, Plant breeders' rights, Plant Variety Protection Act, Convention on biological diversity act, and many more [21, 32, 33]. The well-known case of biopiracy in the context of India is Neem tree (India) against W. R. Grace and Company. Neem (*Azadirachta indica*) tree is found throughout in India and its neighboring country and traditionally known for its antifungal properties [34]. W. R. Grace and Company and USDA obtained a European patent (EP0436257B1) on *Method for controlling fungi on plants by the aid of a hydrophobic extracted neem oil* in 1994 without compensating to the local community or country [35]. Many groups from India and Europe mainly Vandana Shiva, EU Green Party, and the International Federation of Organic Agriculture Movements opposed this patent which is finally revoked in 2000 [35]. The similar case is happened in India–US basmati rice dispute. USA-based corporation RiceTec acquired a patent (US5663484C1) of Basmati Rice hybrid lines bas 867 Rt1117 and Rt112 in 1997 [36]. In 2002, Government of India challenged this patent and found that 15 of the 20 claims are not valid [37]. Underdeveloped countries like African countries are the most adversely affected countries. There is a proper need to strengthen the policy, law, and institutional framework in these countries.

3.6 Bioprospecting Derived Products in Agriculture Industry

To provide a balanced diet and nutrition-rich food products for global population is a major challenge in this era. Unfortunately, nutrition in human diet is not rich because of non-plant-based food product. So, deficiency causes malnutrition in human beings. There is a strong need of protein-rich food for human beings to combat through malnutrition. So, in this way bioprospecting is an important field for enhancing the agricultural products. Bioprospecting in agriculture industry is utilized in different ways. Some of these are neglected and underutilized wild plants for nutritional enhancement, endophytes to enhance the production of crops, and nutrient enrichment through transgenic approach.

There are many crops having nutritional-rich value, but they are treated as wild, neglected, and underutilized, although there are thousands of wild and neglected plant species that are reported to have nutritional value. Therefore, presently, food production depends only on limited plant species. Different underutilized crops have been studied. There are many genera of legumes species for important grains and agroforestry species [38]. In these, genera of legumes, chick pea (*Cicer arietinum*), cowpea (*Vigna unguiculata*), and pea (*Pisum sativum*) are utilized as major food source in case in human beings. Apart from these legumes, some are used as orphan crop like *Psophocarpus tetragonolobus* (L.) *Parkia roxburghii*, and Canavalia sp. *Psophocarpus tetragonolobus* is also called a wonder legume as it has high protein content in the seeds and therefore considered as a versatile legume. It is also known as "Soybean of tropics" due to its high protein content in each part of this legume. But there is also the presence of anti-nutrient as condensed tannin, which binds with protein of seed and edible part of plant. That creates absorption problem at the time of digestion. So to identify the best germplasm, Singh and others identified less condensed tannin (0.265 mg/g of dry weight) winged bean from 100 different global germplasm line of *Psophocarpus tetragonolobus* on the basis of physiochemical, biochemical, and genetic basis [39]. Along with this legume *Parkia* sp. and *Canavalia* sp. bioprospection is under process to develop the best variety for sustainable development [40, 41].

Endophytes secrete bioactive compound in host plant in which they survive. Those compounds are necessary for plant growth and protection. They have numerous functions to increase agricultural productivity. They sustain the agriculture industry in a proper way in terms of nutrition richness. There are many bacteria, fungi, and actinomycetes involved in the synthesis of bioactive compounds. You and others identified the *Penicillium* sp. in *Suaeda japonica* plant, which synthesize Giberellic acid and help in seed germination [42]. *Aspergillus* sp., *Cladosporium* sp., *Penicillium* sp., and many more fungi live with *panax ginseng* plant and synthesize triterpenoid- and saponin-type secondary metabolites, which provide protection and root growth to plant [43]. Seed germination in *Phaseolus vulgaris* is enhanced by bioactive compounds (proteolytic enzymes, phosphate solubilization factors, active volatile and non-volatile metabolites) synthesized by endophytes *Trichoderma atroviridae*, *T. polysporum*, and *T. harzianum* [44]. Along with these there is a long list of bacteria that are involved in plant growth and nutrition uptake. Plant bioprospecting has also been performed for the isolation of numerous insect toxic proteins for their use in agricultural industries [129] [128] [127] [126] [130].

Transgenic approach is also applied for bioprospection of value-added products. In this approach, first one is insect pest management in crop plant to avoid loss of crop production by pest. Second, to increase the level of particular micronutrients in particular plant part that is edible. For instance, protein disorder micronutrients deficiency is another major nutritional disorder in women and young children [45]. Iron is a type of micronutrient whose deficiency causes anemia in a large population in the world. To eliminate iron deficiency, there is a need of iron-rich food. Genetic engineering approaches have been taken care to Fe-fortified plants to increase iron content in rice and maize [46].

ZmZIP5 and *OsIRT1* genes are involved in iron uptake and translocation in maize and rice, respectively [47, 48]. There are many other genes, which are overexpressed/ knockdown for bio-fortified minerals in plant.

3.7 Bioprospecting Derived Products for Bioremediation

In the twenty-first century, the pollution of water and soil with toxic elements due to industry revolution is one of the major concerns. So, the removal of these pollutants from the environment is a major hurdle for sustainable development. Apart from conventional method to remove the pollutants, bioremediation has evolved as eco-friendly, cost-effective, and sustainable approach. It is processed through microorganism and plants. Bioprospection of these microorganisms and plants with better activity have been performed by several researchers. It means use of new beneficial functions from microorganisms, such as new enzymes for chemical and biochemical reactions of interest, processes to increase bioremediation.

Several microorganisms are well known for bioremediation of heavy metal pollutants such as Cd, Hg, Pb, Zn, and U. But genetically engineered bacteria draw more attention these days. Genetically modified bacteria have performed more efficient and target-based bioremediation. Lee et al reported that, *Pseudomonas putida* 06909, an antifungal rhizosphere bacterium was engineered for the purpose of producing metal binding peptides. This peptide has very high affinity with Cd [49]. An Hg (II) resistance gene (*merA*) has been expressed in *Deinococcus radiodurans* strains to remove radioactive pollutant [50]. *Mesorhizobium huakuii* B_3 bacteria was genetically modified with Phytochelatin synthase (PCS) gene expression, which eradicate Cd^{2+} from rice field [51]. Similarly the *B. subtilis* BR151 (pTOO24) has been modified with the expression of luminescent Cd sensor gene which remove Cd from polluted soils [52].

These results imply that a modified microbial load in soil can be a promising strategy for heavy metal cleanup. This could be a crucial factor for sustaining the growth of the engineered strain in the presence of the native bacterial population.

Endophytes are the most diverse group of microorganisms found in every plant on earth. They may be mainly fungi or bacteria. Entophytes were reported to be mainly isolated from plants ranging from herbs to large trees, marine grasses, and lichens. *Pseudomonas* bacteria live in *Lolium multiflorum*, a type of Italian ryegrass which releases bioactive compound alkanes [53]. Alkanes are used for diesel degradation at the site of pollution. *Burkholderia cepacia* is present in *Zea maize* plant. They release phenol and toluene, which is used in Petroleum tolerance and degradation [54]. The Co-contamination of nickel and trichloroethylene tolerance and degradation were performed by bioactive compound (Trichloroethylene) released from *Pseudomonas putida* in poplar plant [55]. Similarly, the pyrenes bioactive compounds released by Enterobacter, which lives within *Alium macrostemon* and function as Pyrenes tolerance and degradation [56].

Apart from microorganisms and bioactive molecules, some plants also perform bioremediation against toxic elements. Along with these, the transgenic plants are also designed to degrade the particular pollutant from environment.

Sarah Jamil and others revealed that *Jatropha curcas* plant helps in remediation of fly ash (FA), which is produced in thermal power plants [57]. They showed that the Fe and Mn accumulate in roots and Cu, Al, and Cr were translocated more to the shoot, when the plants were treated with fly ash of thermal decomposition.

Environmental pollution with pesticides, pharmaceuticals, and petroleum compounds is not decontaminated through conventional methods. So, transgenic strategy has been enlightened to develop the transgenic plant with Insertion of CYP450 in plants to increase the xenobiotics metabolism [58]. Over expression of CYP450 isoenzymes (CYP1, CYP1, CYP3 isolated from human and mammals) in plants has been done mainly in *Solanum tuberosum*, Oryza sativa and *Nicotiana tabaccum*. The main objective of these transgenic developments is to produce herbicide resistant plant and plants capable of enhanced metabolization of foreign metabolites. The *CYP1A1* gene from human expressed in Oryza sativa, the transgenic plant enhanced the metabolism of chlorotoluron, norflurazon in soil [59, 60]. Similarly onr gene isolated from *Enterobacter cloaceae* and expressed in *Nicotiana tabaccum*. The result showed enhancement of denitration of glycerol trinitrate (GTN) and TNT [61]. But when CYP1A1 gene with CYP2B6 gene from human was expressed in *Solanum tuberosum*, it showed different result than Oryza sativa transgenic with CYP1A1 gene. Transgenic *Solanum tuberosum* has performed resistance to sulfonylurea and other herbicides [62].

3.8 Bioprospecting for Nanoparticles Development

Nanoparticles revolutionized science and technology in recent years. They have already proven their vast application in medical, electrical, wastewater processing, construction, military and many more industries [63, 64]. Many plants and micro-organism have potential to synthesis metallic nanoparticles (Table 3.1), [64, 89, 90]. Biological organism mainly utilized two mechanisms for the synthesis of nanoparticles: (i) bioreduction and (ii) biosorption [89, 90]. In bioreduction, metal ions are biochemically reduced into inert and stable nano form. This bioreduction is mostly couple with the oxidation of specific enzymes

Table 3.1 List of nanoparticles producing bio-organism.

Nanoparticle	Bio-organism	Name	Method	Localization	References
Au	Bacteria	*Thermomonospora* sp.	Reduction	Extracellular	[65]
	Bacteria	*Rhodococcus* sp.	Reduction	Intracellular	[66]
	Bacteria	*Rhodopseudomonas capsulata*	Reduction	Extracellular	[67]
	Bacteria	*Pseudomonas aeruginosa*	Reduction	Extracellular	[68]
	Bacteria	*Delftia acidovorans*	Reduction	Extracellular	[69]
	Fungi	*Fusarium oxysporum*	Reduction	Intracellular	[70]
	Fungi	*Verticillium* sp.	Reduction	Intracellular	[71]
	Plant extract	*Cymbopogon flexuosus*	Reduction	Extracellular	[72]
	Live plant	*Medicago sativa*	—	Intracellular	[73]
Ag	Bacteria	*Bacillus licheniformis*	Reduction	Intracellular	[74]
	Bacteria	*Bacillus* sp.	Reduction	Extracellular	[75]
	Bacteria	*Klebsiella pneumonia*	Reduction	Extracellular	[76]
	Bacteria	*Escherichia coli*	Reduction	Extracellular	[76]
	Bacteria	*Enterobacter cloacae*	Reduction	Extracellular	[76]
	Bacteria	*Lactobacillus* sp.	Both	Extracellular	[77]
	Bacteria	*Enterococcus faecium*	Both	Extracellular	[77]
	Bacteria	*Lactococcus garvieae*	Both	Extracellular	[77]
	Fungi	*Pediococcus pentosaceus*	Both	Extracellular	[77]
	Fungi	*Fusarium oxysporum*	Reduction	Extracellular	[70]
	Fungi	*Aspergillus fumigatus*	Reduction	Extracellular	[78]
	Fungi	*Aspergillus flavus*	Reduction	Extracellular	[79]
	Fungi	*Coriolus versicolor*	Reduction	Both	[80]
	Plant latex	*Jatropha curcas*	Reduction	Extracellular	[81]
	Leaf extract	*Acalyphaindica*	Reduction	Extracellular	[82]
	Seed exudate	*Medicago sativa*	Reduction	Extracellular	[83]
	Leaf extract	*Magnolia kobus*	Reduction	Extracellular	[84]

(*Continued*)

Table 3.1 (Continued)

Nanoparticle	Bio-organism	Name	Method	Localization	References
Pt	Bacteria	*Escherichia coli*	Reduction	Extracellular	[85]
	Fungi	*Fusarium oxysporum*	Reduction	Extracellular	[70]
	Fungi	*Neurospora crassa*	Reduction	Both	[86]
Pd	Bacteria	*Escherichia coli*	Reduction	Extracellular	[85]
	Bacteria	*Desulfovibrio desulfuricans*	Reduction	Extracellular	[87]
U, Cu, Pb, Al, Cd	Bacteria	*Bacillus sphaericus JG-A12*	Both	Extracellular	[88]

Au, gold; Ag, silver; Pt, platinum; Pd, palladium; U, uranium; Cu, copper; Pb, lead; Al, aluminium; Cd, cadmium; both (method), reduction and absorption; both (localization), intra and extracellular.

and may also require biological energy. During biosorption, metal ions are absorbed into the bioorganism itself, such as bacterial, fungal, or plant cell. These ion–cell complexes are stable complexes and able to perform biological reaction. These bio-nanoparticles can be used in biosensors for the detection of heavy metal in soil, polyphenolic in wine, phenols and lignins in wastewater, and many more [89, 90].

3.9 Bioprospecting Derived Products in Pharmaceutical Industry

Current environmental and global hazards and the emergence of infectious diseases demand new initiatives in the development of therapeutics and their commercialization [91]. Bioprospecting can be a promising approach for drug discovery that utilizes the biodiversity. This multidisciplinary approach exploits ecology, pharmacology, and therapeutics to explore new chemical compounds from terrestrial, marine, and other ecosystems. The pharmaceutical industry depends the most on bioprospecting. Food and Drug Administration (FDA) approved approximately 30% of the new drugs that originated from natural sources from 2008 to 2012 [92]. Bioprospecting for pharmaceuticals involves identification of a disease state, development of an extrapolative biological assay that can either generally or specifically test the efficacy of compounds followed by identification, collection, and testing of biodiversity samples. New compounds can be identified through two major strategies: first, the random collections that are relatively large in number, and second, focused collections that follow phylogenetic or cultural clues, where the likelihood of success is maximum with a minimum number of samples [93].

Bioprospecting has given numerous drugs that are used in the treatment of variety of diseases and ailments (Table 3.2). Traditionally, compounds isolated from terrestrial sources, mainly plants, have contributed to the pharmaceutical industry [94]. The bioactive entities obtained from plants may be purified natural products or semi-synthetic/synthetic derivatives that show high efficacy and decreased side-effects [93]. Bioprospecting from terrestrial sources have generated multiple drugs approved by the FDA to treat various forms of cancer [95–97]. For example, taxol (isolated from the Pacific yew tree

Table 3.2 List of drugs obtained from bio-organism.

Name	Source	Function
Drugs derived from plants		
Taxol	*Taxus brevifolia*	Anticancer
Vincristine Vinblastine	*Catharanthus roseus*	Anticancer
Harringtonine	*Cephalotaxus harringtonia*	Anticancer
Homoharringtonine	*Cephalotaxus harringtonia*	Anticancer
Topotecan	*Camptotheca acuminate*	Anticancer
Irinotecan	*Camptotheca acuminate*	Anticancer
Artemether	*Artemisia annua* L	Antimalarial
Quinine	*Cinchona ledgeriana*	Antimalarial, antipyretic
Crofelemer	*Croton lechleri*	Treatment of diarrhea associated with antiretroviral HIV/AIDS therapy
Prostratin	*Homalanthus nutans*	Antiviral
Morphine Codeine	*Papaver somniferum*	Analgesic, antitussive
Atropine Hyoscyamine	*Atropa belladonna*	Anticholinergic
Digoxin Digitalin Digitoxin	*Digitalis purpurea*	Cardiotonic
Convallatoxin	*Convallaria majalis*	Cardiotonic
Acetylsalicylic acid	*Salix alba*	Analgesic
Sinecatechins	*Camellia sinensis*	Treatment of genital warts caused by human papillomavirus (HPV)
Colchicine	*Colchicum autumnale*	Anti-inflammatory
Rivastigmine	*Physostigma venenosum*	Cholinesterase inhibitors to treat Alzheimer's disease
Galantamine	*Lycoris radiata*	Cholinesterase inhibitors to treat Alzheimer's disease
Agrimophol	*Agrimonia supatoria*	Anthelmintic
Danthron	*Cassia species*	Laxative
Scopolamine	*Datura species*	Sedative
Tetrahydrocannabinol	*Cannabis sativa*	Antiemetic, decreases occular tension
Thymol	*Thymus vulgaris*	Topical antifungal
Theophylline	*Theobroma cacao*	Diuretic, bronchodilator
Drugs derived from microbes		
Bleomycin	*Streptomyces verticillus*	Anticancer
Daunorubicin Doxorubicin	*Streptomyces* species	Anticancer

(Continued)

Table 3.2 (Continued)

Name	Source	Function
Griseofulvin	*Penicillium griseofulvum*	Antifungal
Amphotericin B	*Streptomyces nodosus*	Antifungal and antileishmanial
Ivermectin	*Streptomyces avermitilis*	Antihelminthic
Ciclosporin	*Tolypocladium inflatum*	Immunosuppressant
Compactin	*Penicillium compactum*	Cholesterol-lowering agent
Penicillin	*Penicillium notatum*	Antibacterial
Fusidic Acid	*Fusidium coccineum*	Antibacterial
Cyclosporin A	*Tolypocladium inflatum*	Immunosuppressant
Lentinan	*Letinula edodes*	Anticancer, cholesterol-lowering agent, anti-infectious agent
Ergotamine	*Claviceps purpurea*	Migraine headaches
Drugs derived from marine sources		
Cytarabine	*Tethya crypta*	Anticancer
Trabectedin	*Ecteinascidia turbinate*	Anticancer
Eribulin mesylate	*Halichondria okadai*	Anticancer
Vidarabine	*Tethya crypta*	Antiviral
Ziconotide	*Conus magus*	Analgesic

Taxusbrevifolia), vincristine, and vinblastine (from Madagascar periwinkle *Catharanthusroseus*) as well as their semi-synthetic and synthetic derivatives. A variety of gymnosperm species have been studied for taxol-like activity. Bioactive compounds harringtonine and homoharringtonine were from the gymnosperm, *Cephalotaxus harringtonia* [98]. Smith-Kline Beecham and Pharmacia & Upjohn developed topotecan and irinotecan, respectively, that are semi-synthetic derivatives of the natural compound Camptothecin isolated from *Camptotheca acuminate* for the treatment of ovarian cancers and rectal cancer [93]. New Earth Biomed (NEBM), a not-for-profit cancer research organization, is currently evaluating alternative research programs to analyze complex mixtures of plant-derived entities and other natural substances used in Indian Ayurveda and traditional Chinese medicine for new anticancer treatments [99].

Besides anticancer drugs, terrestrial bioprospecting has yielded antimalarial drug Artemether, a semi-synthetic derivative of artemisinin [100, 101] isolated from the Chinese medicinal plant *Artemisia annua* L. Crofelemer (Fulyzaq™) developed and approved by FDA for the treatment of diarrhea associated with antiretroviral HIV/AIDs therapy. Crofelemer is a purified compound from the latex of *Croton lechleri* (Euphorbiaceae) [99]. Another promising drug Prostratin is isolated from a *Homalanthus nutans* tree. Prostratin is able to clear laboratory animals of virus [99]. Other examples of drugs from plants include morphine and codeine from *Papaver somniferum*, atropine and hyoscyamine from *Atropa belladonna*, digoxin from *Digitalis* spp., acetylsalicylic acid (ASA) (painkiller derived from willow bark, Salix alba), Sinecatechins obtained green tea leaves from

Camellia sinensis (active ingredient in an ointment to treat genital wart caused by human papilloma virus), colchicine (anti-inflammatory drug) to treat gout flares (obtained from *Colchicum autumnale*), the cholinesterase inhibitors rivastigmine (from *Physostigma venenosum*) and galantamine (from *Lycoris radiata*) used to treat Alzheimer's disease [93, 102–104].

Besides plants, drugs obtained from terrestrial microbial sources include bleomycin (obtained from the soil bacterium *Streptomyces verticillus*), Daunorubicin, and its 14-hydroxylated form doxorubicin (from *Streptomyces* spp.) that are used as anticancer drugs [98]. Numerous antibacterial drugs were also discovered by bioprospecting like β-lactam antibiotics, rifamycins, tetracyclines, polymyxins, aminoglycosides, and phosphonic acid antibiotics [105]. In addition to antifungal drug griseofulvin (from the soil fungus *Penicillium griseofulvum*) [106], amphotericin B (from the soil bacterium *Streptomyces nodosus*) is the antifungal and antileishmanial drug [107], ivermectin (from *Streptomyces avermitilis* soil bacterium) used as the antihelminthic drug [108], and ciclosporin (from the *Tolypocladium inflatum* soil fungus) immunosuppressant drugs used to treat rheumatoid arthritis and psoriasis [109] were also obtained from microbial sources from terrestrial microbial bioprospecting.

Marine organisms usually produce more varieties and amounts of metabolites than are presently identified in other sources. Seaweeds, salt marsh plants, and marine worms survive in the extreme temperatures, wide-ranging pressures, low energy, and absence of sunlight favoring evolution of highly developed defense system [28]. Thus, these organisms possess unique genetic pools that may have the potential of treating several diseases or ailments [110, 111]. Several FDA-approved drugs have been discovered from marine bioprospecting like anticancer drugs Cytarabine (Ara-C) trabectedin and eribulin mesylate from Sponge *Tethya crypta*, Marine tunicate *Ecteinascidia turbinate*, and Sponge *Halichondria okadai*, respectively, antiviral drug vidarabine (Ara-A) from Sponge *Tethyacrypta*, analgesic ziconotide from Cone snail *Conus magus*, etc. [112, 113]. Omega-3 fatty acids play an important role in the human diet and in human physiology preventing cardiovascular, atopic, and inflammatory diseases [114–117]. Eicosapentaenoic acid (EPA) and docosahexaenoic acid (DHA) omega-3 fatty acids are commonly found in marine fish oils [28]. Recently, the arctic having extreme environmental conditions favoring significant and exceptional adaptations in organisms that inhabit the region have emerged as promising bioprospecting site that may hold future new medicines.

3.10 Conclusion and Future Prospects

Bioprospecting is exploration of biodiversity and ethnic knowledge for the production of commercially valuable products for medicine, agriculture, and other industries [118]. Initially bioprospecting focused on plant-based entities, but now other forms of biodiversity like insects, algae, micro-organisms have been explored with significant success [119]. Experimentation of new approaches like metabolomics or genetic manipulation for the quantification of metabolites, understating of metabolic profiles of compounds associated with diverse metabolic pathways, metabolomic fingerprinting to categorize the biological sample, and relationship of metabolite biosynthesis with a particular genomic sequence have made significant developments in this field [120, 121]. Moreover, bioprospecting programs are also including nanotechnology for natural products drug delivery systems to augment their bioavailability, therapeutic effects, and reducing the multiple administration [122, 123]. Bioprospecting has benefitted the society by providing various lifesaving drugs and agricultural products; there is growing concerns of biodiversity overexploitation and biopiracy [124]. Despite various sanctioned agreements by biodiversity convention industries resort to malpractice and unethical means [124]. This is due to lack of effective monitoring and enforcement of such guidelines [91]. It is worth noting that the proper terms and conditions of bioprospecting agreements must follow transparency and clarity without ambiguity [124]. Sharing of information and accessibility to natural resources between companies/scientific organizations and bioprospecting site nations can be helpful to reduce biopiracy [91]. Besides equal profit sharing with indigenous communities, multi-national companies can also benefit local people at bioprospecting sites by providing jobs, trainings, and expertise [124, 125]. Thus, the principal aim of bioprospecting programs should be sustainable growth of the communities, balanced ecosystems, biodiversity conservation, and social benefits of the developed products [16].

Acknowledgments

We wish to apologize to all colleagues whose work, because of lack of space, could not be cited. SD is also thankful to MITACS for postdoc funding. We thank all the lab members of the Grbic laboratory situated in western university, Canada, for helpful discussions.

References

1 Moghe, G.D., Leong, B.J., Hurney, S.M. et al. (2017). Evolutionary routes to biochemical innovation revealed by integrative analysis of a plant-defense related specialized metabolic pathway. *eLife* 6: e28468.
2 Costello, C. and Ward, M. (2006). Search, bioprospecting and biodiversity conservation. *Journal of Environmental Economics and Management* 52 (3): 615–626.
3 Harvey, A.L. and Gericke, N. (2011). Bioprospecting: creating a value for biodiversity. In: *Research in Biodiversity – Models and Applications* (ed. I. Pavlinov), 323–338. IntechOpen.
4 STOCK, TAKING (1992). The Convention on Biological Diversity. http://62.160.8.20/bsp/Organisations/documents/62_18.pdf.
5 Balick, M.J., Elisabetsky, E., and Laird, S.A. (eds.) (1996). *Medicinal Resources of the Tropical Forest: Biodiversity and Its Importance to Human Health*. Columbia University Press.
6 Greaves, T. (1994). *Intellectual Property Rights for Indigenous Peoples: A Sourcebook*. Oklahoma City, OK: Society for Applied Anthropology.
7 Bhattacharya, S. (2014). Bioprospecting, biopiracy and food security in India: the emerging sides of neoliberalism. *International Letters of Social and Humanistic Sciences* 23: 49–56.

8 Ministry of Environment and Forests, Govt. of India (2002). *Biotechnology and Bioprospecting for Sustainable Development*. Cancun, Mexico: India's Presentation for the Ministerial Meeting of Megabiodiversity Countries.
9 Grabley, S. and Thiericke, R. (eds.) (1998). *Drug Discovery from Nature*. Springer Science & Business Media https://www.springer.com/gp/book/9783540669470.
10 Kanchiswamy, C.N., Malnoy, M., and Maffei, M.E. (2015). Bioprospecting bacterial and fungal volatiles for sustainable agriculture. *Trends in Plant Science* 20 (4): 206–211.
11 Sahoo, S., Sarangi, S., and Kerry, R.G. (2017). Bioprospecting of endophytes for agricultural and environmental sustainability. In: *Microbial Biotechnology*, 429–458. Singapore: Springer.
12 Gupta, P.L., Rajput, M., Oza, T. et al. (2019). Eminence of microbial products in cosmetic industry. *Natural Products and Bioprospecting* 9 (4): 267–278.
13 Pascoal, F., Magalhães, C., and Costa, R. (2020). The link between the ecology of the prokaryotic rare biosphere and its biotechnological potential. *Frontiers in Microbiology* 11: 231.
14 Upadhyay, P., Shrivastava, R., and Agrawal, P.K. (2016). Bioprospecting and biotechnological applications of fungal laccase. *3 Biotech* 6 (1): 15.
15 Dhillion, S.S., Svarstad, H., Amundsen, C., and Chr, H. (2002). Bioprospecting: effects on environment and development. *Ambio*: 491–493. https://bioone.org/journals/ambio-a-journal-of-the-human-environment/volume-31/issue-6/0044-7447-31.6.491/Bioprospecting-Effects-on-Environment-and-Development/10.1579/0044-7447-31.6.491.short.
16 Juan, B. (2017). Bioprospecting and drug development, parameters for a rational search and validation of biodiversity. *Journal of Microbial and Biochemical Technology* 9: e128.
17 Dias, D.A., Urban, S., and Roessner, U. (2012). A historical overview of natural products in drug discovery. *Metabolites* 2 (2): 303–336.
18 Youyou, T. (2015). Artemisinin – a gift from traditional Chinese medicine to the world. *Nobel lecture [serial online]*.
19 Bijoy, C.R. (2007). Access and benefit sharing from the indigenous peoples' perspective: the Tbgri-Kani model. *Law, Environment and Development Journal* 3: 1.
20 Coughlin, M.D. Jr. (1993). Using the Merck-INBio agreement to clarify the convention on biological diversity. *Columbia Journal of Transnational Law* 31: 337.
21 Guérin-McManus, M., Famolare, L., Bowles, I.A. et al. (1998). *Bioprospecting in Practice: A Case Study of the Suriname ICBG Project and Benefits Sharing under the Convention on Biological Diversity*. Case Studies on Benefit Sharing Arrangements. https://52.74.118.238/financial/bensharing/suriname-icbg.pdf
22 Mehbub, M.F., Lei, J., Franco, C., and Zhang, W. (2014). Marine sponge derived natural products between 2001 and 2010: trends and opportunities for discovery of bioactives. *Marine Drugs* 12 (8): 4539–4577.
23 Hunt, B. and Vincent, A.C. (2006). Scale and sustainability of marine bioprospecting for pharmaceuticals. *Ambio: A Journal of the Human Environment* 35 (2): 57–64.
24 Page, M. and Willahan, E. (2018). *Arctic Antibiotics – The Hunt for New Medicines in the Arctic*. Polar Research and Policy Initiative. https://polarconnection.org/antibiotics-arctic-bioprospecting.
25 Koskinen, P.E., Lay, C.H., Beck, S.R. et al. (2008). Bioprospecting thermophilic microorganisms from Icelandic hot springs for hydrogen and ethanol production. *Energy & Fuels* 22 (1): 134–140.

26 Özdemir, S.Ç. and Uzel, A. (2020). Bioprospecting of hot springs and compost in West Anatolia regarding phytase producing thermophilic fungi. *Sydowia* 72: 1.

27 Sahoo, R.K., Kumar, M., Sukla, L.B., and Subudhi, E. (2017). Bioprospecting hot spring metagenome: lipase for the production of biodiesel. *Environmental Science and Pollution Research* 24 (4): 3802–3809.

28 Bhatia, P. and Chugh, A. (2015). Role of marine bioprospecting contracts in developing access and benefit sharing mechanism for marine traditional knowledge holders in the pharmaceutical industry. *Global Ecology and Conservation* 3: 176–187.

29 Leary, D.K. (2007). *International Law and the Genetic Resources of the Deep Sea*, vol. 56. Martinus Nijhoff Publishers.

30 Park, C. and Allaby, M. (2007). A dictionary of environment and conservation. *Agenda* 21: 12.

31 Mgbeoji, I. (2014). *Global Biopiracy: Patents, Plants, and Indigenous Knowledge*. University of British Columbia Press.

32 Chen, J. (2005). The parable of the seeds: interpreting the plant variety protection act in furtherance of innovation policy. *The Notre Dame Law Review* 81: 105.

33 Peterson, K. (2001). Benefit sharing for all: bioprospecting NGOs, intellectual property rights, new governmentalities. *PoLAR* 24: 78.

34 Dubey, S. and Kashyap, P. (2014). Azadirachtaindica: a plant with versatile potential. *RGUHS Journal of Pharmaceutical Sciences* 4 (2): 39–46.

35 Locke, J.C. and Gordon, L.H. (1991). Method for controlling fungi on plants by the aid of a hydrophobic extracted neem oil. EU Patent published, 7.

36 Sarreal, E.S., Mann, J.A., Stroike, J.E. Andrews, R.D., and RiceTecInc. (1997). Basmati rice lines and grains. US Patent 5,663,484.

37 Mukherjee, U. (2008). A study of the basmati case (India-US basmati rice dispute): the geographical indication perspective. *Available at SSRN 1143209*.

38 Graham, P.H. and Vance, C.P. (2003). Legumes: importance and constraints to greater use. *Plant Physiology* 131 (3): 872–877.

39 Singh, V., Goel, R., Pande, V. et al. (2017). De novo sequencing and comparative analysis of leaf transcriptomes of diverse condensed tannin-containing lines of underutilized *Psophocarpus tetragonolobus* (L.) DC. *Scientific Reports* 7: 44733.

40 Peter, A.E., Aruna, D., Rao, P.S. et al. (2016). Canavaliavirosaroxb.: a review. *International Journal of Pharmaceutical Sciences and Research* 7 (10): 3917.

41 Sahoo, U.K. (2013). Parkiaroxburghii: an underutilized but multipurpose tree species for reclamation of jhum land. *Current Science* 104 (12): 1598.

42 You, Y.H., Yoon, H., Kang, S.M. et al. (2012). Fungal diversity and plant growth promotion of endophytic fungi from six halophytes in Suncheon Bay. *Journal of Microbiology and Biotechnology* 22 (11): 1549–1556.

43 Wu, H., Yang, H.Y., You, X.L., and Li, Y.H. (2013). Diversity of endophytic fungi from roots of *Panax ginseng* and their saponin yield capacities. *SpringerPlus* 2 (1): 107.

44 Pierre, E., Louise, N.W., Marie, T.K.R. et al. (2016). Integrated assessment of phytostimulation and biocontrol potential of endophytic *Trichoderma* spp. against common bean (*Phaseolus vulgaris* L.) root rot fungi complex in centre region, Cameroon. *International Journal of Pure and Applied Bioscience* 4 (4): 50–68.

45 Miller, J.L. (2013). Iron deficiency anemia: a common and curable disease. *Cold Spring Harbor Perspectives in Medicine* 3 (7): a011866.

46 Mulualem, T. (2015). Application of bio-fortification through plant breeding to improve the value of staple crops. *Biomedicine and Biotechnology* 3 (1): 11–19.

47 Lee, S. and An, G. (2009). Over-expression of OsIRT1 leads to increased iron and zinc accumulations in rice. *Plant, Cell & Environment* 32 (4): 408–416.

48 Li, S., Liu, X., Zhou, X. et al. (2019). Improving zinc and iron accumulation in maize grains using the zinc and iron transporter ZmZIP5. *Plant and Cell Physiology* 60 (9): 2077–2085.

49 Lee, S.W., Glickmann, E., and Cooksey, D.A. (2001). Chromosomal locus for cadmium resistance in *Pseudomonas putida* consisting of a cadmium-transporting ATPase and a MerR family response regulator. *Applied and Environmental Microbiology* 67 (4): 1437–1444.

50 Brim, H., McFarlan, S.C., Fredrickson, J.K. et al. (2000). Engineering *Deinococcus radiodurans* for metal remediation in radioactive mixed waste environments. *Nature Biotechnology* 18 (1): 85–90.

51 Sriprang, R., Hayashi, M., Ono, H. et al. (2003). Enhanced accumulation of Cd2+ by a *Mesorhizobium* sp. transformed with a gene from *Arabidopsis thaliana* coding for phytochelatin synthase. *Applied and Environmental Microbiology* 69 (3): 1791–1796.

52 Ivask, A., Dubourguier, H.C., Põllumaa, L., and Kahru, A. (2011). Bioavailability of Cd in 110 polluted topsoils to recombinant bioluminescent sensor bacteria: effect of soil particulate matter. *Journal of Soils and Sediments* 11 (2): 231–237.

53 Andria, V., Reichenauer, T.G., and Sessitsch, A. (2009). Expression of alkane monooxygenase (alkB) genes by plant-associated bacteria in the rhizosphere and endosphere of Italian ryegrass (*Lolium multiflorum* L.) grown in diesel contaminated soil. *Environmental Pollution* 157 (12): 3347–3350.

54 Wang, Y., Li, H., Zhao, W. et al. (2010). Induction of toluene degradation and growth promotion in corn and wheat by horizontal gene transfer within endophytic bacteria. *Soil Biology and Biochemistry* 42 (7): 1051–1057.

55 Weyens, N., Beckers, B., Schellingen, K. et al. (2015). The potential of the Ni-resistant TCE-degrading *Pseudomonas putida* W619-TCE to reduce phytotoxicity and improve phytoremediation efficiency of poplar cuttings on a Ni-TCE co-contamination. *International Journal of Phytoremediation* 17 (1): 40–48.

56 Sheng, X., Chen, X., and He, L. (2008). Characteristics of an endophytic pyrene-degrading bacterium of *Enterobacter* sp. 12J1 from *Allium macrostemon* Bunge. *International Biodeterioration and Biodegradation* 62 (2): 88–95.

57 Jamil, S., Abhilash, P.C., Singh, N., and Sharma, P.N. (2009). Jatropha curcas: a potential crop for phytoremediation of coal fly ash. *Journal of Hazardous Materials* 172 (1): 269–275.

58 Abhilash, P.C., Jamil, S., and Singh, N. (2009). Transgenic plants for enhanced biodegradation and phytoremediation of organic xenobiotics. *Biotechnology Advances* 27 (4): 474–488.

59 Kawahigashi, H., Hirose, S., Ohkawa, H., and Ohkawa, Y. (2007). Herbicide resistance of transgenic rice plants expressing human CYP1A1. *Biotechnology Advances* 25 (1): 75–84.

60 Kawahigashi, H., Hirose, S., Ohkawa, H., and Ohkawa, Y. (2008). Transgenic rice plants expressing human P450 genes involved in xenobiotic metabolism for phytoremediation. *Journal of Molecular Microbiology and Biotechnology* 15 (2–3): 212–219.

61 French, C.E., Rosser, S.J., Davies, G.J. et al. (1999). Biodegradation of explosives by transgenic plants expressing pentaerythritol tetranitrate reductase. *Nature Biotechnology* 17 (5): 491–494.

62 Inui, H., Kodama, T., Ohkawa, Y., and Ohkawa, H. (2000). Herbicide metabolism and cross-tolerance in transgenic potato plants co-expressing human CYP1A1, CYP2B6, and CYP2C19. *Pesticide Biochemistry and Physiology* 66 (2): 116–129.

63 Elahi, N., Kamali, M., and Baghersad, M.H. (2018). Recent biomedical applications of gold nanoparticles: a review. *Talanta* 184: 537–556.

64 Schröfel, A., Kratošová, G., Šafařík, I. et al. (2014). Applications of biosynthesized metallic nanoparticles – a review. *Acta Biomaterialia* 10 (10): 4023–4042.

65 Kasthuri, J., Kathiravan, K., and Rajendiran, N. (2009). Phyllanthin-assisted biosynthesis of silver and gold nanoparticles: a novel biological approach. *Journal of Nanoparticle Research* 11 (5): 1075–1085.

66 Park, Y., Hong, Y.N., Weyers, A. et al. (2011). Polysaccharides and phytochemicals: a natural reservoir for the green synthesis of gold and silver nanoparticles. *IET Nanobiotechnology* 5 (3): 69–78.

67 He, S., Guo, Z., Zhang, Y. et al. (2007). Biosynthesis of gold nanoparticles using the bacteria *Rhodopseudomonas capsulata*. *Materials Letters* 61 (18): 3984–3987.

68 Narayanan, K.B. and Sakthivel, N. (2010). Biological synthesis of metal nanoparticles by microbes. *Advances in Colloid and Interface Science* 156 (1-2): 1–13.

69 Johnston, C.W., Wyatt, M.A., Li, X. et al. (2013). Gold biomineralization by a metallophore from a gold-associated microbe. *Nature Chemical Biology* 9 (4): 241–243.

70 Ahmad, A., Mukherjee, P., Senapati, S. et al. (2003). Extracellular biosynthesis of silver nanoparticles using the fungus *Fusarium oxysporum*. *Colloids and Surfaces B: Biointerfaces* 28 (4): 313–318.

71 Ramanathan, R., Field, M.R., O'Mullane, A.P. et al. (2013). Aqueous phase synthesis of copper nanoparticles: a link between heavy metal resistance and nanoparticle synthesis ability in bacterial systems. *Nanoscale* 5 (6): 2300–2306.

72 Iravani, S. and Zolfaghari, B. (2013). Green synthesis of silver nanoparticles using *Pinus eldarica* bark extract. *BioMed Research International*: 1–5. https://downloads.hindawi.com/journals/bmri/2013/639725.pdf.

73 Mallikarjuna, K., Narasimha, G., Dillip, G.R. et al. (2011). Green synthesis of silver nanoparticles using *Ocimum* leaf extract and their characterization. *Digest Journal of Nanomaterials and Biostructures* 6 (1): 181–186.

74 Schlüter, M., Hentzel, T., Suarez, C. et al. (2014). Synthesis of novel palladium (0) nanocatalysts by microorganisms from heavy-metal-influenced high-alpine sites for dehalogenation of polychlorinated dioxins. *Chemosphere* 117: 462–470.

75 Lloyd, J.R., Yong, P., and Macaskie, L.E. (1998). Enzymatic recovery of elemental palladium by using sulfate-reducing bacteria. *Applied and Environmental Microbiology* 64 (11): 4607–4609.

76 Kalimuthu, K., Babu, R.S., Venkataraman, D. et al. (2008). Biosynthesis of silver nanocrystals by *Bacillus licheniformis*. *Colloids and Surfaces B: Biointerfaces* 65 (1): 150–153.

77 Shahverdi, A.R., Minaeian, S., Shahverdi, H.R. et al. (2007). Rapid synthesis of silver nanoparticles using culture supernatants of Enterobacteria: a novel biological approach. *Process Biochemistry* 42 (5): 919–923.

78 Bhainsa, K.C. and D'souza, S.F. (2006). Extracellular biosynthesis of silver nanoparticles using the fungus *Aspergillus fumigatus*. *Colloids and Surfaces B: Biointerfaces* 47 (2): 160–164.

79 Saha, S., Pal, A., Kundu, S. et al. (2010). Photochemical green synthesis of calcium-alginate-stabilized Ag and Au nanoparticles and their catalytic application to 4-nitrophenol reduction. *Langmuir* 26 (4): 2885–2893.

80 Ahmad, A., Mukherjee, P., Mandal, D. et al. (2002). Enzyme mediated extracellular synthesis of CdS nanoparticles by the fungus, *Fusarium oxysporum*. *Journal of the American Chemical Society* 124 (41): 12108–12109.

81 Thapa, D., Palkar, V.R., Kurup, M.B., and Malik, S.K. (2004). Properties of magnetite nanoparticles synthesized through a novel chemical route. *Materials Letters* 58 (21): 2692–2694.

82 Krishnaraj, C., Jagan, E.G., Rajasekar, S. et al. (2010). Synthesis of silver nanoparticles using *Acalypha indica* leaf extracts and its antibacterial activity against water borne pathogens. *Colloids and Surfaces B: Biointerfaces* 76 (1): 50–56.

83 Spadaro, D. and Gullino, M.L. (2005). Improving the efficacy of biocontrol agents against soilborne pathogens. *Crop Protection* 24 (7): 601–613.

84 Song, J.Y. and Kim, B.S. (2009). Rapid biological synthesis of silver nanoparticles using plant leaf extracts. *Bioprocess and Biosystems Engineering* 32 (1): 79.

85 Deplanche, K., Caldelari, I., Mikheenko, I.P. et al. (2010). Involvement of hydrogenases in the formation of highly catalytic Pd (0) nanoparticles by bioreduction of Pd (II) using *Escherichia coli* mutant strains. *Microbiology* 156 (9): 2630–2640.

86 Sanghi, R. and Verma, P. (2009). Biomimetic synthesis and characterisation of protein capped silver nanoparticles. *Bioresource Technology* 100 (1): 501–504.

87 Cai, J., Kimura, S., Wada, M., and Kuga, S. (2009). Nanoporous cellulose as metal nanoparticles support. *Biomacromolecules* 10 (1): 87–94.

88 Das, V.L., Thomas, R., Varghese, R.T. et al. (2014). Extracellular synthesis of silver nanoparticles by the *Bacillus* strain CS 11 isolated from industrialized area. *3 Biotech* 4 (2): 121–126.

89 Pantidos, N. and Horsfall, L.E. (2014). Biological synthesis of metallic nanoparticles by bacteria, fungi and plants. *Journal of Nanomedicine & Nanotechnology* 5 (5): 1.

90 Thakkar, K.N., Mhatre, S.S., and Parikh, R.Y. (2010). Biological synthesis of metallic nanoparticles. *Nanomedicine: Nanotechnology, Biology and Medicine* 6 (2): 257–262.

91 Zakrzewski, P.A. (2002). Bioprospecting or biopiracy? The pharmaceutical industry's use of indigenous medicinal plants as a source of potential drug candidates. *University of Toronto Medical Journal* 79 (3): 252–254.

92 Bueno, J. and Ritoré, S. (2019). Bioprospecting model for a new Colombia drug discovery initiative in the pharmaceutical industry. In: *Analysis of Science, Technology, and Innovation in Emerging Economies* (eds. C. Inés and P. Martínez), 37–63. Cham: Palgrave Macmillan.

93 McClatchey, W. and Stevens, J. (2001). An overview of recent developents in bioprospecting and pharmaceutical development. In: *Development of Plant-Based Medicines: Conservation, Efficacy and Safety* (eds. A. Cotte and Poveda), 17–45. Dordrecht: Springer.

94 Sandsborg, W.R. (1999). Natural products in drug discovery and development. In: *Bioassay Methods in Natural Product Research and Drug Development* (eds. S. Patricia and M. Fletscher), 143–149. Dordrecht: Springer.

95 Cordell, G.A., Farnsworth, N.R., Beecher, C.W.W. et al. (1993). Novel strategies for the discovery of plant-derived anti-cancer agents. In: *Human Medicinal Agents from Plants*, ACS Symposium Series, vol. 534 (eds. A.D. Kinghorn and M.F. Balandrin), 191–204. Washington, DC: American Chemical Society.

96 Cragg, G.M., Boyd, M.R., Cardellina, J.H. II et al. (1993). Role of plants in the National Cancer Institute drug discovery and development program. In: *Human Medicinal Agents from Plants*, ACS Symposium Series, vol. 534 (eds. A.D. Kinghorn and M.F. Balandrin), 80–95. Washington, DC: American Chemical Society.

97 Pezzuto, J.M. (1997). Plant-derived anticancer agents. *Biochemical Pharmacology* 53 (2): 121–133.

98 Demain, A.L. and Vaishnav, P. (2011). Natural products for cancer chemotherapy. *Microbial Biotechnology* 4 (6): 687–699.

99 Cox, P.A. and King, S. (2013). Bioprospecting.

100 Avery, M.A., McLean, G., Edwards, G., and Ager, A. (2000). Structure-activity relationships of peroxide-based artemisinin antimalarials. In: *Biologically Active Natural Products: Pharmaceuticals* (eds. S.J. Cutler and H.G. Cutler), 121–132. Boca Raton: CRC Press.

101 Haynes, R.K. and Vonwiller, S.C. (1997). From qinghao, marvelous herb of antiquity, to the antimalarial trioxaneqinghaosu and some remarkable new chemistry. *Accounts of Chemical Research* 30 (2): 73–79.

102 Buenz, E.J., Verpoorte, R., and Bauer, B.A. (2018). The ethnopharmacologic contribution to bioprospecting natural products. *Annual Review of Pharmacology and Toxicology* 58: 509–530.

103 Kumar, A., Singh, A., and Aggarwal, A. (2017). Therapeutic potentials of herbal drugs for Alzheimer's disease – an overview. *Journal: Indian Journal Of Experimental Biology* 55: 63–73. http://nopr.niscair.res.in/bitstream/123456789/40230/1/IJEB%2055%282%29%2063-73.pdf.

104 Russo, P., Frustaci, A., Del Bufalo, A. et al. (2013). Multitarget drugs of plants origin acting on Alzheimer's disease. *Current Medicinal Chemistry* 20 (13): 1686–1693.

105 Cushnie, T.T., Cushnie, B., Echeverría, J. et al. (2020). Bioprospecting for antibacterial drugs: a multidisciplinary perspective on natural product source material, bioassay selection and avoidable pitfalls. *Pharmaceutical Research* 37 (7): 1–24.

106 Beekman, A.M. and Barrow, R.A. (2014). Fungal metabolites as pharmaceuticals. *Australian Journal of Chemistry* 67 (6): 827–843.

107 de Lima Procópio, R.E., da Silva, I.R., Martins, M.K. et al. (2012). Antibiotics produced by *Streptomyces*. *The Brazilian Journal of Infectious Diseases* 16 (5): 466–471.

108 Saraiva, R.G. and Dimopoulos, G. (2020). Bacterial natural products in the fight against mosquito-transmitted tropical diseases. *Natural Product Reports* 37 (3): 338–354.

109 Borel, J.F., Kis, Z.L., and Beveridge, T. (1995). The history of the discovery and development of cyclosporine (Sandimmune®). In: *The Search for Anti-Inflammatory Drugs*, 27–63. Boston: Birkhäuser.

110 Demunshi, Y. and Chugh, A. (2009). Patenting trends in marine bioprospecting based pharmaceutical sector. *Journal of Intellectual Property Rights* 14: 122–130.

111 Lazcano-Pérez, F., Roman-Gonzalez, S.A., Sánchez-Puig, N., and Arreguín-Espinosa, R. (2012). Bioactive peptides from marine organisms: a short overview. *Protein and Peptide Letters* 19 (7): 700–707.

112 Martins, A., Vieira, H., Gaspar, H., and Santos, S. (2014). Marketed marine natural products in the pharmaceutical and cosmeceutical industries: tips for success. *Marine Drugs* 12 (2): 1066–1101.

113 Mayer, A.M., Glaser, K.B., Cuevas, C. et al. (2010). The odyssey of marine pharmaceuticals: a current pipeline perspective. *Trends in Pharmacological Sciences* 31 (6): 255–265.

114 Abdelhamid, A.S., Brown, T.J., Brainard, J.S. et al. (2020). Omega-3 fatty acids for the primary and secondary prevention of cardiovascular disease. *Cochrane Database of Systematic Reviews* 3: 1–538.

115 Lohner, S. and Decsi, T. (2013). Role of long-chain polyunsaturated fatty acids in the prevention and treatment of atopic diseases, Chapter 11. In: *Polyunsaturated Fatty Acids: Sources, Antioxidant Properties and Health Benefits* (ed. A. Catalá), 1–24. NOVA Publishers.

116 Miles, E.A. and Calder, P.C. (2012). Influence of marine n-3 polyunsaturated fatty acids on immune function and a systematic review of their effects on clinical outcomes in rheumatoid arthritis. *British Journal of Nutrition* 107 (S2): S171–S184.

117 Scorletti, E. and Byrne, C.D. (2013). Omega-3 fatty acids, hepatic lipid metabolism, and nonalcoholic fatty liver disease. *Annual Review of Nutrition* 33: 231–248.

118 Paul, S. (2015). Trips and biotechnology Vs CBD and biodiversity: is it bio-prospecting or bio-piracy of developing countries' traditional knowledge? *International Journal of Physical and Social Sciences* 5 (9): 603–631.

119 Kumar, P. and Tarui, N. (2004). Identifying the contribution of indigenous knowledge in bioprospecting for effective conservation strategy. In *Bridging Scales and Epistemelgioes Conference*, Alexandria, Ezypt (March), pp. 17–20.

120 Forseth, R.R. and Schroeder, F.C. (2011). NMR-spectroscopic analysis of mixtures: from structure to function. *Current Opinion in Chemical Biology* 15 (1): 38–47.

121 Ulrich-Merzenich, G., Zeitler, H., Jobst, D. et al. (2007). Application of the "-Omic-" technologies in phytomedicine. *Phytomedicine* 14 (1): 70–82.

122 Ansari, S.H., Islam, F., and Sameem, M. (2012). Influence of nanotechnology on herbal drugs: a review. *Journal of Advanced Pharmaceutical Technology & Research* 3 (3): 142.

123 Kumari, A., Kumar, V., and Yadav, S.K. (2012). Nanotechnology: a tool to enhance therapeutic values of natural plant products. *Trends in Medical Research* 7 (2): 34–42.

124 Sandhu, H.S. (2011). Bioprospecting: Pros and cons. Punjab Agricultural University [Updated on 5 May 2012].

125 Smith, R.B. and Kumar, P. (2002). Royalties and benefit sharing contracts in bioprospecting. Institute of Economic Growth.

126 Upadhyay, S.K., Mishra, M., Singh, H., Ranjan, A., Chandrashekar, K., Verma, P.C., Singh, P.K., and Tuli, R. (2010). Interaction of Allium sativum leaf agglutinin (ASAL) with midgut BBMV proteins and its stability in Helicoverpa armigera. *Proteomics* 10: 4431–4440.

127 Upadhyay, S.K., Saurabh, S., Singh, R., Rai, P., Dubey, N.K., Chandrashekar, K., Negi, K.S., Tuli, R., and Singh, P.K. (2011). Purification and characterization of a Lectin with high hemagglutination property isolated from Allium altaicum. *Protein J*, 30: 374–383.

128 Upadhyay, S.K. and Chandrashekar K. (2012). Interaction of salivary and midgut proteins of Helicoverpa armigera with soybean trypsin inhibitor. *Protein J*, 31: 259–264.

129 Shukla, A.K., Upadhyay, S.K., [...], Singh, P.K. (2016). Expression of an insecticidal fern protein in cotton protects against whitefly. *Nature Biotechnology*, 34: 1046–1051.

130 Upadhyay, S.K., Singh, S., Chandrashekar, K., Singh, P.K., and Tuli, R. (2012). Compatibility of garlic (Allium sativum L.) leaf agglutinin and Cry1Ac δ-endotoxin for gene pyramiding. *Applied Microbiology and Biotechnology*, 93: 2365–2375.

4

Role of Plants in Phytoremediation of Industrial Waste

Pankaj Srivastava and Nishita Giri

ICAR-Indian Institute of Soil and Water Conservation (ICAR-IISWC), Dehradun, Uttarakhand, India

4.1 Introduction

In order to increase our agriculture production, farmers, industries and other communities are using huge amount of environmental hazardous materials that are discharged into the soil, air and water. Moreover, the emergence of new industries such as electrical, pharmaceuticals and materials science industry leads to pollute our biosphere with new emerging pollutants such as cosmetics, disinfectant, plasticisers and phthalates, wood preservatives, paint additives, analgesics, nanoparticles, etc. [1–11]. The trace metals are important part of our ecosystem, but the accumulation of harmful metals in plants and human beings may be dangerous and have serious health problems. The toxic metals are difficult to remove from the biosphere as they cannot be biologically or chemically degraded in the environment.

In phytoremediation system, there is a close relation between plants, microbes and soil system for a long duration. This concept makes use of plants' ability to remove, activate and collect the materials from the atmosphere in different parts of the plants [12]. Phytoremediation is an integrated term that covers plant-related strategies and approaches for remediation of contaminated biosphere [13]. Phytoremediation includes (i) phytoextraction – the accumulation of heavy metals in plant biomass with high concentration, (ii) rhizofilteration – the removal of metals from aqueous waste streams through adsorption in plant parts, (iii) phytovolatalization – the process of volatilization includes through air and plants, (iv) phytodetoxification – where plants change the chemical species into less toxic form, (v) phytostabilization – in this process plants immobilize the contaminants physically and chemically at the site and reduce the chances of movement in surrounding areas [14]. The phytoremediation technology is still in research and advancement phase and there are some methodological barriers, which need to be still addressed. Most heavy metal accumulating plants have a small biomass and are slow growing. To make phytoremediation a feasible technique, there is an urgent need to

Bioprospecting of Plant Biodiversity for Industrial Molecules, First Edition.
Edited by Santosh Kumar Upadhyay and Sudhir P. Singh.

Table 4.1 Naturally growing plant species capable of capturing fly ash dust [17,18].

English name	Botanical name	Family	Habit
Opposite leaved fig	*Ficus hispida*	Moraceae	Tree
Gigantic swallow wort	*Calotropis gigantean*	Asclepediaceae	Shrub
Sebastian plum	*Cordia dichotoma*	Boraginaceae	Tree
Banyan tree	*Ficus benghalensis*	Moraceae	Tree
Jackfruit	*Artocarpus heterophyllus*	Moraceae	Tree
Wild sage	*Lantana camara*	Verbenaceae	Shrub
Country Mallow	*Abutilon indicum*	Malvaceae	Herb
Sweet basil	*Ocimum basilicum*	Labiatae	Herb
Jujube tree	*Ziziphus mauritiana*	Rhamnaceae	Tree
African Tulip	*Spathodea campanulata*	Bignoneaceae	Tree
Coral Jasmine	*Nyctanthus arbor-tristis*	Oleaceae	Shrub
Fig	*Ficus glomerata*	Moraceae	Tree
Vasan vel	*Cocculus hirsutus*	Manispermaceae	Shrub
French mulberry of ghats	*Calicarpa tomentosa*	Verbenaceae	Shrub
Indian Tulip tree	*Thepesia populnea*	Malvaceae	Tree
Bastard cedar	*Guazuma ulmifolia*	Sterculiaceae	Tree
Monkey face tree	*Mallotus philippinensis*	Euphorbiaceae	Tree
Teak	*Tectona grandis*	Verbenaceae	Tree
Gunpowder tree	*Trema orientalis*	Ulmaceae	Tree

either find fast growing hyperaccumulators or transgenic plants with better genes in hyperaccumulators for high metal accumulation [15].

Several plant genes encoding metal transporters have been already identified and characterized [16]. The selection of plant species is a vital task in the success of phytoremediation technology (Table 4.1). The species selected should be able to grow in high concentration of trace metals. Some of the plants belonging to Brassicaceae like Alyssum species, Thlaspi species and *Brassica juncea*, Violaceae such as *Viola calaminaria*, Leguminosae such as *Astragalus racemosus* are known to take up high concentrations of heavy metals and radionuclides. To date, there are approximately 400 known metal hyperaccumulators in the world that can be used for phytoremediation [19].

Some new concept of symbiosis between nitrogen fixing plant species and rhizobia was also developed. This system uses both advantages of plants and microbes, particularly engineered genes can be transformed to plants through infection with recombinant microbes [13]. Our atmosphere recently polluted with various pesticides, pharmaceuticals, petroleum compounds, PAHs, PCBs, etc. is a worldwide issue, and the progress of inventive phytoremediation for the decontamination of sites is therefore of paramount importance. The different mechanisms can be used for this purpose [20–28]; however, the technology has effective mechanism in decontamination of the affected areas [29–31].

Several authors demonstrated that phytoremediation potential in their studies and proved that plants have excellent mechanism in phytoremediation process [32–55], and the technique requires more wide application.

However, the process of detoxification in plants is somewhat slow, but the metals or pesticides could be released into the atmosphere [56]. Some biofuel plants like *Ricinus communis* [57], *Jatropha curcas* [58], *Miscanthus giganteus* [59], etc. have great potential to reduce the contamination and sustainable ecosystem services.

4.2 Different Toxic Materials from Industries

4.2.1 Fly Ash from Thermal Power Plants

The thermal power plants are the major source of fly ash that creates environmental problem in nature. The coal used in Indian power plants has more ash (34–35%) [60]. Fly ash particles contain the small sized particles ranging from 0.01 to 100 μm [61, 62]. The thermal power plants generate more than 100 million tons of fly ash each year, and close to 90% of this is dumped in landfills or settling ponds. Distribution of fly ash in the surrounding areas may occur either during crude transportation or from the disposal site like ash ponds or landfills where the dry fly ash become airborne along the direction of the wind [10]. The distribution of particulate matter (PM) in the surrounding area of thermal power plants is enormous when the disposal of fly ash is going near dumping sites. The finer particles deposits on the surface of the materials and plants. A number of health studies have proven adverse effect on humans due to the presence of particulate matter ($PM < 10\,\mu m$). Some smaller sized particles ($<2.5\,\mu m$) have serious effects on the respiratory system [63–67].

Fly ash has been recognized as a great environmental issue due to large consumption of coal in thermal power plants which need more than $1000\,km^2$ area in India. Globally, it would require more than $3235\,km^2$ of area for fly ash disposal and remediation [68]. Therefore, phytoremediation of fly ash disposal is the need of the hour to reduce environmental contamination and sustainable development of the community and to reduce health hazards [18, 58, 69–71]. Several plants from the families *Brassicaceae, Chenopodiaceae, Fabiaceae, Leguminoceae* and *Poaceae* are growing in fly ash with reduction in contamination. *Acacia* sp. and *Leucaena leucocephala* were established to have a high tolerance and survival in waterless, barren and metal-contaminated areas [72].

4.2.2 Heavy Metals and Pesticides in Environment

4.2.2.1 Cadmium

Cadmium (Cd) is dangerous to our environment and released into the environment though different industries waste like electroplating, plastic manufacturers, paint industry, alloy and battery preparations. One of the important toxic metals in storm water is cadmium (Cd), which may be unsafe even in very low concentration and harm human kidney and have severe health effects [73]. The appliances used in houses, automobiles and heavy trucks, agricultural, industrial and aeroplane tools and fasteners of all kinds (like nuts, bolts, screws and nails) are mostly found in Cd coating, which is available for photography, rubber curing, fungicides and for luminescent dials [74].

4.2.2.2 Arsenic

The problems of toxicity and occurrence of Arsenic (As) have drawn much attention in Indian scenario. The issue of release of As from fly ash has increased from several thermal power plants and produces environmental pollution. The heavy metal As exists in the −3, 0, +3 and +5 oxidation states [75]. Due to the existence of different oxidation speciations, the arsenic may be found in many different chemical states and forms. Therefore, the health effects of As may vary widely depending on the chemical form of the As. The burning of coal and smelting of metals are major sources of As in the air. The most probable As species are elemental (As) and oxide forms (As_2O_3) in the oxidizing flue gas in coal combustion process. However, $As_2O_3$is much more volatile than the elemental As, and researchers have concluded that As could only be present in an oxide form in the flue gas [76, 77]. In coal gasification, As_4 with traces of arsine (AsH_3) is the most probable species [78, 79]. As concentrations in the combustion residues vary widely depending on the coal quality, pH of ashes and the ignition conditions, ranging from 2 to 240 ppm in the fly ash and from 0.02 to 168 ppm in bottom ash [80]. Arsenic (As) represents an important environmental problem due to the toxic effects of this metalloid, and its accumulation through the food chain that poses long-term risks to human being [81].

4.2.2.3 Chromium

Chromium is present in different states, ranging from Cr^{+2} to Cr^{+6}, but in soils the most stable and popular are trivalent Cr (III) and hexavalent Cr (VI) forms [82], which display quite different chemical properties and affect organisms in different ways. In fact, in contrast to other metals, the hazard of chromium is dependent on its oxidation state. Hexavalent chromium is most dangerous to human health and carcinogenic in nature, while trivalent chromium is essential (in low concentrations) for human and animal nutrition, relatively water insoluble and less toxic than Cr (VI) [83]. In many countries, chromate, which is the most prevalent form of Cr (VI), is present in solid/liquid waste due to anthropogenic activities [84].

4.2.2.4 Pesticide in Environment

Chemical pesticides have contributed greatly to the increase of yields in agriculture by controlling pests and diseases and also towards checking the insect borne diseases (malaria, dengue, encephalitis, filariasis, etc.) in the human health sector [85, 86]. The world's increasing population has facing the risk of food security [87, 88]. One of the best options to increase the food production is the use of pest management practices because most of the food degraded due to pests. Due to variations in climate conditions, the tropical countries are facing more reduction in food grain production [89, 90].

However, the sporadic use has been leading to significant consequences not only to public health but also to food quality resulting in an impact load on the environment and hence the development of pest resistance [91]. Through over use and misuse there is considerable waste, adding to the cost and contributing to the adverse environmental and health consequences. Inappropriate application of pesticides affects the whole ecosystem by entering the residues in food chain and polluting the soil, air, ground and surface water [91–93]. Some important insecticides like aldrin, Orgnochlorine, DDT hexachlorocyclohexane (HCH) and dieldrin, are among the frequently used pesticides in the developing countries of Asia because of their low cost and versatility against various pests ([88, 94, 95]).

Nevertheless, because of their potential for bioaccumulation and biological effects, these compounds were banned in developed nations two and half decades ago [96–98]. Their resistance to degradation has resulted in contamination universally and found in many environmental compartments. Such residues may be composed of many substances, which include any specified derivatives such as degradation products, metabolites and congeners that are considered to be of toxicological significance. According to the Food and Agriculture Organization (FAO) inventory [99], more than 500,000 tons of unused and obsolete pesticides are threatening the environment and public health in many countries. Public concern over pesticide residue has been increasing during the last decade. Recovering from the euphoria of green revolution, India is also now battling from residual effects of extensively used chemical fertilizers and pesticides such as HCH, DDT, endosulfan, phorate, etc. [86, 87, 100].

4.2.3 Phytoremediation Technology in Present Scenario

Several phytoremediation strategies have been initiated by the different industries and organizations working in the field of phytoremediation. The concept of using green plants for phytoremediation like green belts came initially from three nations such as Kenya, Britain and USA [101–103].

The green belt has been defined as "a strip of trees of such species and such geometry that when planted around a source would considerably attenuate the pollution in air by intercepting and accumulating the toxic elements in sustainable manner" [103]. Sometimes green belts are simply called as trees; in mainstays of green belts are trees with some shrubs and ground vegetation. The purpose of green belt may vary for from industry to industry. It is important to capture several factors, viz. selection of suitable plant species depending on the agro-climatic environment of the province; nature of the pollutant to be ameliorated; tolerance of plant species against the pollutant and general landscape and topography of the area [103, 104]. Plants have taken the toxic elements form atmosphere by three methods such as through leaves where absorption and deposition of elements takes place and particulate matter on downwind side due to slow movement of air [105–110]. The green belt plantation can help in removing the air pollutants and alter the urban environment by lowering the temperature, and evapotranspiration can help in the improvement of healthy life for humans [111–113]. Vegetation in the landfills seems to be the most viable in *in situ* remediation technique that will lead to stabilization against wind and water erosion; decontamination of toxic heavy metals by plants and development of aesthetically pleasing landscape that will provide shelter and habitat for wildlife [72]. Phytoremediation of heavy metal-contaminated soil is a developing technology and has attracted much attention because it is an environment friendly and relatively cheap technique. One of the process of phytoremediation the phytoextraction in which green plants uptake the metals and pesticides from the soil and accumulate in their parts.

The fly ash grown species are recommended by various researchers are like *Calotropis procera*, *Cassia tora*, *Chenopodium album*, *Sida cardifolia*, *Blumea lacera* [114]; *Prosopis juliflora* [115]; *Acacia auriculiformis*, *Leucaena leucocephala* [72]; *Cicer arietinum* [116]; *Sesbania cannabina* [117]; *Cassia siamea* [118]; *Agropyron elongatum*, *Festuca arundinacea*, *Melilotus officinalis* [119]; *Liquidambar styraciflua*, *Platanus occidentalis* [120];

Mesembryanthemum nodiflorum, *Enchylaena tomentosa*, *Halosarcia halocnemoides*, *Halosarcia pergranulata* [121]; *Melilotus alba* [122]. Phytoremediation technique offers a useful way to check leaching of metals from fly ash dykes and growing of multipurpose tree species (Table 4.1) on these soils.

Another phenomenon, the "phytostabilization," involves the use of plants to arrest the contaminant in the rhizosphere and thus reduce its bioavailability in the environment [123, 124]. In several cases, it has been observed that despite the presence of high concentration of metals in soil, they have low mobility into the plant system [125]. For such cases, chelate-induced phytoextraction was developed, with the objective of desorbing heavy metals from soil matrix into soil solution to facilitate the transport of metals into xylem and increase the translocation of metals from the roots to shoots of some fast growing, high biomass producing plants [126, 127].

Although the latest advent of phytoremediation has provided state-of-art molecular biology to genetically modify suitable plants for enhanced tolerance, uptake and remediation of pesticides, lot of innovations and additional researches are required to establish real-time onsite phytoremediation technologies (Figure 4.1). Taking all these factors into consideration, *Jatropha curcas* tested its suitability for phytoremediation of fly ash. *J. curcas* is a large shrub or small tree belonging to the family Euphorbiaceae. It is regarded as a potential biofuel crop for future due to its low moisture demands, pure hardiness and stress

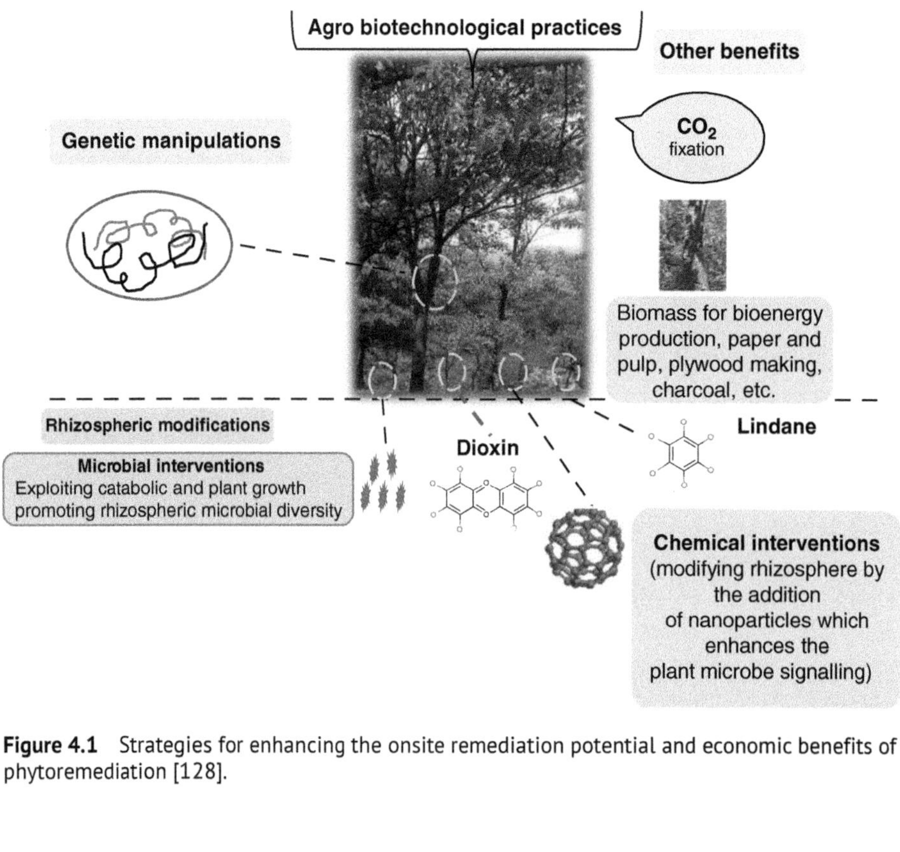

Figure 4.1 Strategies for enhancing the onsite remediation potential and economic benefits of phytoremediation [128].

handling ability [129, 130]. It grows fast with little maintenance and can reach a height of 3–8 m [129–131].

It has been identified in India as the most suitable oil bearing plant and has been recommended for plantation on wasteland as it requires minimal inputs for its establishment [130, 131]. Fly ash landfills are vast wastelands in India. Our aim is to exploit this plant for phytoremediation. First, for phytostabilization of fly ash, to reduce its blowing in nearby crop fields; and second, for phytoextraction of heavy metals through the use of chelator. This would restrict the leaching in the water bodies.

Sinha et al. [131] suggested some species for phytoremediation like *Elodea canadensis*, *Ceratophyllum demersum*, *Potamogeton* spp., *Myriophyllum* spp., *Spartina alterniflora*, *Pinus sylvestris*, *Poa alpine*, *Bouteloua gracilis* [133, 134].

Some best suited plants for phytoremediation are from the family related to Gramineae and Cyperaceae and some members such as genera Brassica, Alyssum and Thalapsi of families Brassicaceae, and Salicaceae (willow and poplar trees). Several grass species like vetiver, rye, Bermuda and tall fescues have important role in phytoremediation of metal and oil pollution [135].

Phytoremediation by vetiver (grass) is a low-cost technology as compared with conventional (engineering) methods for site remediation. It is also virtually maintenance free, the grasses regrow very quickly and its efficiency improves with age [136]. So many green plants are used in phytoremediation task. Important among them are other grasses like the Bermuda grass (*Cynodon dactylon*), Bahia grass (*Paspalum notatum*), Rhodes grass (*Chloris guyana*), the tall wheat grass (*Thynopyron elongatum*), common reed grass (*Phragmites australis*), the munj grass (*Sachharum munja*) and *Imperata cylindrica*. Other plants are the marine couch (*Sporobolus virginicus*), cumbungi (*Typha domingensis*) and *Sarcocrina* spp.

Although various remediation methods such as containment, solidification and stabilization etc. have been proposed for the decontamination of onsite As-contaminated soils, all of these methods require appropriate controls and long-term monitoring to ensure the behaviour of As through soil column [137]. In some studies, it has been reported that hyperaccumulators growing in As-contaminated soils have the ability of accumulating high concentration of arsenic [138–142]. However, the field utilization potential of most of these plants are very less due to their small biomass and slow growth rate ([143]). Rhizoremediation, involving both plants and the rhizospheric microbes, is an efficient bioremediation process for contaminant degradation, and different metal tolerance mechanisms have also been discovered in various microbes [144, 145]: exclusion, active removal, biosorption, precipitation or bioaccumulation, both in external and intracellular spaces. Some bacterial strains are also known to play an important role in the biochemical cycle of As, through its conversion to species with different solubility, mobility, bioavailability and toxicity [146, 147].

Variety of technologies (including chemical physical and biological treatment methods) are available for storm water treatment. However, these methods present different efficiencies for different metals, and they can be very expensive for the treatment of low level metal-contaminated water. In contrast, the phytofiltration has been proposed as a promising, environmentally friendly technology for removing the heavy metal concentration of the contaminated water. In phytofiltration, high metal-accumulating plants function as biofilters, which can be remarkably effective in sequestering metals from polluted waters [9]. Many aquatic plants, usually those found in polluted water bodies, have been suggested for waste water treatment, i.e. they have the ability to accumulate unusually high concentration of heavy metals, without impact on their growth and development. However, most species identified so far are not suitable for onsite phytofiltration due to their small root and shoot biomass and slow growth rate. In contrast, plants with good growth usually show low metal accumulation as well as low tolerance to heavy metals [74].

The advantage of this free-floating culture system is its easy harvesting. Further, more studies are needed to evaluate the onsite application of this free-floating phytofiltration technique [1]. There is a report that Brassica (mustard) species or varieties of *Brassica juncea* (Indian mustard) have an enhanced ability to accumulate metals from hydroponics solution into their above ground (harvestable) parts. These plants reduce the toxicity of the metals through accumulation in shoot biomass [148].

The remediation of Cr(VI)-contaminated soils, today, is essentially based on physical and other approaches, which consist of mine or pumping of infected material and added reducing agents that lead to the precipitation and sedimentation of reduced chromium [Cr(III)], less toxic than Cr(VI) and greatly insoluble. Cr(VI) remediation strategies using traditional technologies have been dealt in depth by Higgins et al. [149].

The ability of several microbial groups (bacteria, fungi, microalgae) to reduce Cr(VI) to Cr(III) has been considered of much interest in order to clean up soil/water polluted with chromate. In fact, there is no doubt that the development of an effective biological system to alleviate the environmental problems associated with hexavalent chromium is highly desirable. The ability of a bacterial strain to reduce hexavalent chromium, although the mechanism of Cr(VI)-reduction may differ from strain to strain, is an attractive feature in order to plan a biological strategy for effective chromate detoxification, but high concentrations of chromate in the environment often can repress the microbial activity and growth [150, 151]. The ability of direct Cr(VI) reduction has been found in many bacterial genera including *Pseudomonas*, *Micrococcus*, *Bacillus*, *Achromobacter*, *Microbacterium Arthrobacter*, *Corynebacterium* [151–154].

4.2.4 Conclusion

Phytoremediation is an emerging technology because conventional methods to clean up the environment are cost intensive and eco-unfriendly. In addition, microbes can modify chemical reactions in toxic metals and enhance their uptake in root zone areas to be taken up by metal hyperaccumulating plants. This technology is termed as phytoremediation and has received a lot of attention in recent years due to its cost-effectiveness, solar driven and high efficiency.

Some molecular techniques have the potential to increase the phytoremediation process with the genetic manipulations from low biomass accumulating metal hyperaccumulator plants to high biomass yielding non-accumulating plants. Phytoremediation by vetiver is a low cost technology as compared with conventional methods for phytoremediation. However, phytoremediation, especially phytoextraction, is getting popularity as a low cost and inventive method of remediation. Phytoextraction is a solar-driven technology so that it can be successfully deployed for the cleaning up of the As-contaminated soils.

References

1 Abhilash, P.C. and Singh, N. (2009). Pesticide use and application: an Indian scenario. *J. Hazard. Mater.* 165: 1–12.
2 De Oliveira, J.L., Campos, E.V.R., Bakshi, M. et al. (2014). Application of nanotechnology for the encapsulation of botanical insecticides for sustainable agriculture: prospects and promises. *Biotechnol. Adv.* 32: 1550–1561.
3 Dubey, R.K., Tripathi, V., Singh, N., and Abhilash, P.C. (2014). Phytoextraction and dissipation of lindane by *Spinacia oleracea* L. *Ecotoxicol. Environ. Saf.* 109: 22–26.
4 Tripathi, V., Dubey, R.K., Edrisi, S.A. et al. (2014). Towards the ecological profiling of a pesticide contaminated soil site for remediation and management. *Ecol. Eng.* 71: 318–325.
5 Srivastava, M., Abhilash, P.C., and Singh, N. (2011). Remediation of Lindane using engineered nanoparticles. *J. Biomed. Nanotechnol.* 7: 173–175.
6 Bakshi, M., Singh, H.B., and Abhilash, P.C. (2014). The unseen impact of nanoparticles: more or less? *Curr. Sci.* 106 (3): 350–352.
7 Vijgen, J., Abhilash, P.C., Li, Y.F. et al. (2011). HCH isomers as new Stockholm convention POPs-A global perspectives on the management of Lindane and its waste isomers. *Environ. Sci. Pollut. Res.* 18: 152–162.
8 Gavrilescu, M., Demnerová, K., Aamand, J. et al. (2015). Emerging pollutants in the environment: present and future challenges in biomonitoring, ecological risks and bioremediation. *New Biotechnol.* 32: 147–156.
9 Dushenkov, V., Kumar, P.B.A.N., Motto, H., and Raskin, I. (1995). Rhizofiltration: the use of plants to remove heavy metals from aqueous streams. *Environ. Sci. Technol.* 29: 1239–1245.
10 Jamil, S., Abhilash, P.C., Singh, A. et al. (2009a). Flyash trapping and metal accumulating capacity of plants: implication for green belt around thermal power plants. *Landsc. Urban Plan.* 92: 136–147.
11 Jamil, S., Abhilash, P.C., Singh, N., and Sharma, P.N. (2009b). Jatropha curcas: a potential crop for phytoremediation of coal fly ash. *J. Hazard. Mater.* 172: 269–275.
12 Salt, D.E. and Rauser, W.E. (1995). MgATP-dependent transport of phytochelatins across the tonoplast of oat roots. *Plant Physiol.* 107: 1293–1301.
13 Sriprang, R. and Murooka, Y. (2007). Accumulation and detoxification of metals by plants and microbes. In: *Environmental Bioremediation Technologies* (eds. S.N. Singh and R.D. Tripathi), 77–100. Berlin, Heidelberg: Springer-Verlag.
14 Eapen, S. and D'Souza, S.F. (2005). Prospects of genetic engineering of plants for phytoremediation of toxic metals. *Biotechnol. Adv.* 23: 97–114.
15 Karenlampi, S., Schat, H., Vangronsveld, J. et al. (2000). Genetic engineering in the improvement of plants for phytoremediation of metal polluted soils. *Environ. Pollut.* 107: 225–231.
16 Eide, D., Broderius, M., Fett, J., and Guerinot, M.L. (1996). A novel iron-regulated metal transporter from plants identified by functional expression in yeast. *Proc. Natl Acad. Sci. USA* 93: 5624–5628.
17 Chaphekar, S.B. and Madav, R.P. (1999). Thermal power plants and environmental management. *J. Indian Associ. Environ. Manag.* 26: 48–53.

18 Pandey, V.C., Abhilash, P.C., and Singh, N. (2009). The Indian perspective of utilizing fly ash in phytoremediation, phytomanagement and biomass production. *J. Environ. Manag.* 90: 2943–2958.

19 Reeves, R.D. and Baker, A.J.H. (2000). Metal accumulating plants. In: *Phytoremediation of Toxic Metals: Using Plants to Clean Up the Environment* (eds. I. Raskin and E.D. Ensley), 193–229. New York: Wiley.

20 Yang, H., Cho, Y., Eum, H., and Kim, E. (2007). Destruction of chlorinated organic solvents in a two stage molten salt oxidation reactor system. *Chem. Eng. Sci.* 68: 5137–5143.

21 Matsunaga, A. and Yashuhara, A. (2003). Complete dechlorination of 1-chloronaphthalene by electrochemical reduction with naphthalene radical anion as mediator. *Environ. Sci. Technol.* 37: 3435–3441.

22 Felsot, A.S. and Dzantor, E.K. (1995). Effect of alachlor concentration and an organic amendment on the soil dehydrogenase activity and pesticide degradation rate. *Environ. Toxicol. Chem.* 14: 23–28.

23 Hatakeda, K., Ikushima, Y., Sato, O. et al. (1999). Supercritical water oxidation of polychlorinated biphenyls using hydrogen peroxide. *Chem. Eng. Sci.* 54: 3079–3084.

24 Johnson, A.C., Worall, F., White, C. et al. (1997). The potential of incorporated organic matter to reduce pesticide leaching. *Toxicol. Environ. Chem.* 58: 47–61.

25 Kummling, K.E., Gray, D.J., Power, J.P., and Woodland, S.E. (2001). Gas phase chemical reduction of hexachlorocyclohexane and other chlorinated compounds: waste treatment experience and applications. *6th International HCH and Pesticides Forum*, Poznan, Poland (20–22 March 2001).

26 Lee, C.C. and Huffman, G.L. (1989). Innovative thermal destruction technologies. *Environ. Prog.* 8: 190–199.

27 Perrin-Ganier, C., Schiavon, F., Morel, J.L., and Schiavon, M. (2001). Effect of sludge-amendment or nutrition addition on the biodegradation of the herbicide isoproturon in soil. *Chemosphere* 44: 887–892.

28 Wirtz, M., Klucik, J., and Rivera, M.J. (2000). Ferredoxin-mediated electrocatalytic dechlorination of haloalkanes by cytochrome P450 cam. *J. Am. Chem. Soc.* 122: 1047–1056.

29 Macek, T., Mackova, M., and Kas, J. (2000). Exploitation of plants for the removal of organics in environmental remediation. *Biotechnol. Adv.* 18: 23–34.

30 Meagher, R.B. (2000). Phytoremediation of toxic elemental and organic pollutants. *Curr. Opin. Plant Biol.* 3: 153–162.

31 Salt, D.E., Smith, R.D., and Raskin, I. (1998). Phytoremediation. *Annu. Rev. Plant Physiol. Plant Mol. Biol.* 49: 643–668.

32 Aprill, W. and Sims, R.C. (1990). Evaluation of prairie grasses for stimulating polycyclic aromatic hydrocarbon treatment in soil. *Chemosphere* 20: 253–265.

33 Aslund, M.L.W., Zeeb, B.A., Rutter, A., and Reimer, K.J. (2007). In situ phytoextraction of polychlorinated biphenyl-(PCB) contaminated soil. *Sci. Total Environ.* 374: 1–12.

34 Gao, X., Wang, W., and Liu, X. (2008). Low-temperature dechlorination of hexachlorobenzene on solid supports and the pathway hypothesis. *Chemosphere* 71: 1093–1099.

35 Ma, X., Andrew, A.R., Burken, J.G., and Albers, S. (2004). Phytoremediation of MTBE with hybrid poplar trees. *Int. J. Phytoremed.* 4: 157–167.

36 Mattina, M.I., Eitzer, B.D., Iannucci-Berger, W. et al. (2004). Plant uptake and translocation of highly weathered, soil-bound technical chlordane residues: data from field and rhizotron studies. *Environ. Toxicol. Chem.* 23: 2756–2762.

37 Mills, T., Arnold, B., Sivakumaran, S. et al. (2006). Phytoremediation and long-term site management of soil contaminated with pentachlorophenol (PCP) and heavy metals. *J. Environ. Manag.* 79 (3): 232–241.

38 Parrish, Z.D., White, J.C., Isleyen, M. et al. (2006). Accumulation of weathered polycyclic aromatic hydrocarbons (PAHs) by plant and earthworm species. *Chemosphere* 64: 609–618.

39 Suresh, B., Sherkhane, P.D., Kale, S. et al. (2005). Uptake and degradation of DDT by hairy root cultures *Cichorium intybus* and *Brassica juncea*. *Chemosphere* 61: 1288–1292.

40 White, J.C., Mattina, M.J.I., Lee, W.Y. et al. (2003). Role of organic acids in enhancing the desorption and uptake of weathered p,p'-DDE by Cucurbita pepo. *Environ. Pollut.* 124: 71–80.

41 Gunther, H., Perner, B., and Gramss, G. (1998). Activities of phenol oxidizing enzymes of ectomycorrhizal fungi in axenic culture and in symbiosis with Scots pine (*Pinus sylvestris* L.). *J. Basic Microb.* 38: 197–206.

42 Binet, P., Portal, J.M., and Leyval, C. (2000). Dissipation of 3-6-ring polycyclic aromatic hydrocarbons in the rhizosphere of rye grass. *Soil Biol. Biochem.* 32: 2011–2017.

43 Fismes, J., Perrin-Ganier, C., Empereur-Bissonnet, P., and Morel, J.L. (2002). Soil-to-root transfer and translocation of polycyclic aromatic hydrocarbons by vegetables grown on industrial contaminated soils. *J. Environ. Qual.* 31: 1649–1656.

44 Li, H., Sheng, G., Sheng, W., and Xu, O. (2002). Uptake of trifluralin and lindane from water by ryegrass. *Chemosphere* 48: 335–341.

45 Liste, H.H. and Alexander, M. (2000). Plant promoted pyrene degradation in soil. *Chemosphere* 40: 7–10.

46 Maila, M.P. and Cloete, T.E. (2002). Germination of *Lepidium sativum* as a method to evaluate polycyclic aromatic hydrocarbons (PAHs) removal from contaminated soil. *Int. Biodeterior. Biodegrad.* 50: 107–113.

47 Mattina, M.I., Iannucci-Berger, W., and Dykas, L. (2000). Chlordane uptake and its translocation in food crops. *J. Agric. Food Chem.* 48: 1909–1915.

48 Singh, O.V. and Jain, R.K. (2003). Phytoremediation of toxic aromatic pollutants from soil. *Appl. Microbiol. Biotechnol.* 63: 128–135.

49 Sunderberg, S.E., Ellington, J.J., Evans, J.J. et al. (2003). Accumulation of perchlorate in tobacco plants: development of a plant kineticmodel. *J. Environ. Monit.* 5: 505–512.

50 Sung, K., Munster, C.L., Rhykerd, R. et al. (2003). The use of vegetation to remediate soil freshly contaminated by recalcitrant contaminants. *Water Res.* 37: 2408–2418.

51 Trapp, S., Yu, X., and Mosbaek, H. (2003). Persistence of methyl tertiary butyl ether (MTBE) against metabolism by Danish vegetation. *Environ. Sci. Pollut. Res.* 10: 357–360.

52 White, J.C. (2000). Phytoremediation of weathered p,p'-DDE residues in soil. *Int. J. Phytoremediat.* 2: 133–144.

53 White, J.C. (2001). Plant-facilitated mobilization of weathered 2, 2-bis (p-chlorophenyl)-1, 1-dichloroethylene (p,p'-DDE) from an agricultural soil. *Environ. Toxicol. Chem.* 20: 2047–2052.

54 White, J.C. (2002). Differential bioavailability of field weathered p,p'-DDE to plants of the *Cucurbita peppo* and *Cucumis* genera. *Chemosphere* 49: 143–152.

55 Yoon, J.M., Oh, B.T., Just, C.L., and Schnoor, J.L. (2002). Uptake and leaching of octahydro-1, 3, 5, 7-tetranitro-1, 3, 5, 7-tetrazocine by hybrid poplar trees. *Environ. Sci. Technol.* 36: 4649–4655.

56 Aken, B.V. (2008). Transgenic plants for phytoremediation: helping nature to clean up environmental pollution. *Trends Biotechnol.* 26: 225–227.

57 Pandey, V.C. (2013). Suitability of *Ricinus communis* L. cultivation for phytoremediation of fly ash disposal sites. *Ecol. Eng.* 57: 336–341.

58 Pandey, V.C., Singh, K., Singh, J.S. et al. (2012). *Jatropha curcas*: a potential biofuel plant for sustainable environmental development. *Renew. Sust. Energ. Rev.* 16: 2870–2883.

59 Nsanganwimana, F., Pourrut, B., Mench, M., and Francis, D.F. (2014). Suitability of *Miscanthus* species for managing inorganic and organic contaminated land and restoring ecosystem services: a review. *J. Environ. Manag.* 143: 123–134.

60 Mathur, R., Chand, S., and Tezuka, T. (2003). Optimal use of coal for the power generation in India. *Energy Policy (India)* 31 (3): 319–331.

61 Adriano, D.C., Page, A.L., Elseewi, A.A. et al. (1980). Utilization and disposal of fly ash and other coal residues in terrestrial ecosystem: a review. *J. Environ. Qual.* 9: 333–344.

62 Bhanarkar, A.D., Gavane, A.G., Tajne, D.S. et al. (2008). Composition and size distribution of particules emissions from a coal-fired power plant in India. *Fuel* 87 (10–11): 2095–2101.

63 Knutsen, S., Shavlik, D., Chen, L.H. et al. (2004). Results from the AHSMOG Study, The association between ambient particulate air pollution levels and risk of cardiopulmonary and all-cause mortality during 22 years follow-up of a non-smoking cohort. *Epidemiology* 15 (4): S45.

64 Lopez, M.T., Zuk, M., Garibay, V. et al. (2005). Health impact from power plant emissions in Mexico. *Atmos. Environ.* 39 (7): 1199–1209.

65 Pekkanen, J., Timonen, J., and Ruuskanen, J.P. (1997). Effects of ultra fine and fine particles in urban air on peak expiratory flow among children with asthmatic symptoms. *J. Environ. Res.* 74: 24–33.

66 Pope, C.A. III, Burnett, R.T., Thun, M.J. et al. (2002). Lung cancer, cardiopulmonary mortality, and long-term exposure to fine particulate air pollution. *J. Am. Med. Assoc.* 287: 1132–1141.

67 Vallack, H.W. and Chadwick, M.J. (1992). Monitoring air borne dust in a high-density coal-fired power station region in North Yorkshire. *Environ. Pollut.* 80: 177–183.

68 Maiti, S.K. and Nandhini, S. (2006). Bioavailability of metals in fly ash and their bioaccumulation in naturally occurring vegetation: a Pilot Scale study. *Environ. Monit. Assess.* 116: 263–273.

69 Abreu, C.A., Coscione, A.R., Pires, A.M., and Paz-Ferreiro, J. (2012). Phytoremediation of a oil contaminated by heavy metals and boron using castor oil plants and organic atter amendments. *J. Geochem. Explor.* 123: 3–7.

70 Pandey, V.C. (2012a). Invasive species based efficient green technology for phytoremediation of fly ash deposits. *J. Geochem. Explor.* 123: 13–18.

71 Pandey, V.C. (2012b). Phytoremediation of heavy metals from fly ash pond by *Azolla caroliniana*. *Ecotoxicol. Environ. Saf.* 82: 8–12.

72 Cheung, K.C., Wong, J.P.K., Zhang, Z.Q. et al. (2000). Revegetation of lagoon ash using the legume species *Acacia auriculiformis* and *Leucaena leucocephala*. *Environ. Pollut.* 109: 75–82.

73 Hapke, H.J. (1991). Effects of metals on domestic animals. In: *Metals and Their Compounds in the Environment: Occurrence, Analysis, and Biological Relevance* (ed. E. Merian), 531–546. Weinheim, Germany: VCH Verlagsgesellschaft mbH.

74 Kirkham, M.B. (2006). Cadmium in plants on polluted soils: effects of soil factors, hyperaccumulation, and amendments. *Geoderma* 137: 19–32.

75 Smedley, P.L. and Kinniburgh, D.G. (2002). A review of the source, behaviour and distribution of arsenic in natural waters. *Appl. Geochem.* 17: 517–568.

76 Dismukes, E.B. (1994). Trace element control in electrostatic precipitators and fabric filter. *Fuel Process. Technol.* 39: 403–416.

77 Winter, R.M., Mallepalli, R.R., Hellem, K.P., and Szydlo, S.W. (1994). Determination of As, Cd, Cr and Pb species formed in a combustion environment. *Combust. Sci. Technol.* 101: 45–58.

78 Clark, L.B. and Sloss, L.L. (1992). *Trace Elements-Emissions from Coal Combustion and Gasification*, 21–72. London: IEA Coal Research, IEACR.

79 Helble, J.J., Mojtahedi, W., Lyyranen, J. et al. (1996). Trace element partitioning during coal gasification. *Fuel* 75: 932–939.

80 Eary, L.E., Rai, D., Mattigod, S.V., and Anisworth, C.C. (1990). Geochemical factors controlling the mobilization of inorganic constituents from fossil fuel combustion residues I: review of the major elements. *J. Environ. Qual.* 19: 202–214.

81 Williams, P.N., Price, A.H., Raab, A. et al. (2005). Variation in arsenic speciation and concentration in paddy rice related to dietaryexposure. *Environ. Sci. Technol.* 39: 5531–5540.

82 Fendorf, S.E. (1995). Surface reactions of chromium in soils and waters. *Geoderma* 67: 55–71.

83 Francisco, R., Alpoim, M.C., and Morais, P.V. (2002). Diversity of chromium-resistant and – reducing bacteria in a chromium-contaminated activated sludge. *J. Appl. Microbiol.* 92: 837–843.

84 Kamaludeen, S.P., Megharaj, M., Juhasz, A.L. et al. (2003). Chromium microorganism interactions in soils: remediation implications. *Rev. Environ. Contam. Toxicol.* 178: 93–164.

85 Bhatnagar, V.K. (2001). Pesticides pollution: trends and perspectives. *ICMR Bull.* 31: 87–88.

86 Rekha, Naik, S.N., and Prasad, R. (2006). Pesticide residue in organic and conventional food-risk analysis. *J. Chem. Health Saf.* 13: 12–19.

87 Agoramoorthy, G. (2008). Can India meet the increasing food A demand by 2020? *Futures* 40: 503–506.

88 FAO (2005). *Proceedings of the Asia Regional Workshop, Regional Office for Asia and the Pacific*, Bangkok.

89 Kannan, K., Tanabe, S., Ramesh, A. et al. (1992). Persistent orgnochlorine residues in food stuffs from India and their implications on human dietary exposure. *J. Agric. Food Chem.* 40: 518–524.

90 Lakshmi, A. (1993). Pesticides in India: risk assessment to aquatic ecosystems. *Sci. Total Environ.* 134: 243–253.

91 Agnihotri, N.P. (1999). *Pesticide Safety and Monitoring, All India Coordinated Research Project on Pesticides Residues*. New Delhi, India: Indian Council of Agricultural Research.

92 ICAR (1967). *Report of the Special Committee on Harmful Effects of Pesticides*, 78. New Delhi: ICAR.

93 UN/DESA (2002). Changing unsustainable patterns of consumption and production (Chapter III). In: *Johannesburg Plan on Implementation of the World summit on Sustainable Development*. Johannesburg.
94 Gupta, P.K. (2004). Pesticide exposure-Indian scene. *Toxicology* 198: 83–90.
95 Lallas, P. (2001). The Stockholm convention on persistent organic pollutants. *Am. J. Int. Law.* 95: 692–708.
96 Rotterdam Convention Secretariat (2005). Guideline on the Development of National Laws to Implement the Rotterdam Convention, Rome/Geneva, 2004 (revised 2005).
97 Thullner, F. (1997). Impact of pesticide resistance and network for global pesticide management based on a regional structure. *World Anim. Health Rev.* 89: 41.
98 World Bank (2002). *Toxics and Poverty: The Impact of Toxic Substances on the Poor in Developing Countries*. Washington, DC: FAO/UNEP/OECD/SIB, Baseline Study on the Problem of Obsolete Pesticides Stocks, Rome, 2001.
99 FAO/UNEP/OECD/SIB (2001). Baseline Study on the Problem of Obsolete Pesticides Stocks, Rome.
100 Government of India (2008). *Eleventh Five-Year Plan:(2008–2012)*. New Delhi: Planning Commission of India http://planningcommission.nic.in/plans/planrel/fiveyr/welcome.html.
101 Andy, C. (1991). *Dictionary of Environment and Development*. London: Earth Scan Publications Ltd.
102 Gareth, J., Alan, R., Jean, F., and Graham, H. (1992). *Environmental Science*, 564. Harper Perennial (A Division of Harper Collins Publishers).
103 Ruth, A.E. and William, R.E. (1994). *The Encyclopedia of the Environment. The Rene Dubor Centre for Human Environments*. USA: Houghton Miffin Company.
104 Khan, F.I. and Abbasi, S.A. (2000). Attenuation of gaseous pollutants by greenbelts. *Environ. Monit. Assess.* 64: 457–475.
105 Pal, A., Kulshreshtha, K., Ahmad, K.J., and Yunus, M. (2000). Changes in leaf surface structures of two avenue tree species caused by auto-exhaust pollution. *J. Environ. Biol.* 21 (1): 15–21.
106 Rawat, J.S. and Banerjee, S.P. (1996). Urban forestry for improvement of environment. *J. Energy Environ. Monit.* 12 (2): 109–116.
107 Spitsyna, N.T. and Skripal̂shchikova, L.N. (1991). Phytomass and dust accumulation of birch forests near open-pit mines. *Sov. J. Ecol.* 22: 354–359.
108 Tomasevic, M., Vukmirovic, Z., Rajsic, S. et al. (2005). Characterization of trace metal particles deposited on some deciduous tree leaves in an urban area. *Chemosphere* 61: 753–760.
109 Turner, K., Lefler, L., and Freedman, B. (2005). Plant communities of selected urbanized areas of Halifax, Nova Scotia, Canada. *Landsc. Urban Plan.* 71: 191–206.
110 Varshney, C.K. and Mitra, I. (1993). Importance of hedges in improving urban air quality. *Landsc. Urban Plan.* 25 (1-2): 85–93.
111 Tzoulas, K., Korpela, K., Venn, S. et al. (2007). Promoting ecosystem and human health in urban areas using Green Infrastructure: a literature review. *Landsc. Urban Plan.* 81: 167–178.
112 Whitford, V., Ennos, A.R., and Handley, J.F. (2001). City form and natural processes: indicators for the ecological performance of urban areas and their application to Merseyside, UK. *Landsc. Urban Plan.* 20 (2): 91–103.

113 Escobedo, F.J. and Nowak, D.J. (2009). Spatial heterogeneity and air pollution removal by an urban forest. *Landsc. Urban Plan* 90: 102–110.
114 Gupta, A.K. and Sinha, S. (2008). Decontamination and/or revegetation of fly ash dykes through naturally growing plants. *J. Hazard. Mater.* 153: 1078–1087.
115 Rai, U.N., Pandey, K., Sinha, S. et al. (2004). Revegetating fly-ash landfills with *Prosopis juliflora* L.: impact of different amendments & *Rhizobium* inoculation. *Environ. Int.* 30: 293–300.
116 Gupta, D.K., Rai, U.N., Sinha, S. et al. (2004). Role of *Rhizobium* (CA-1) inoculation in increasing growth and metal accumulation in *Cicer arietinum* L. growing under fly-ash stress condition. *Bull. Environ. Contam. Toxicol.* 73: 424–431.
117 Sinha, S. and Gupta, A.K. (2005). Translocation of metals from fly-ash amended soil in the plant of *Sesbania cannabina* L. Ritz: effect on antioxidants. *Chemosphere* 61: 1204–1214.
118 Tripathi, R.D., Vajpayee, P., Singh, N. et al. (2004). Efficacy of various amendments for amelioration of fly-ash toxicity: growth performance and metal composition of *Cassia siamea* Lamk. *Chemosphere* 54: 1581–1588.
119 Mulhern, D.W., Robel, R.J., Furness, J.C., and Hensley, D.L. (1989). Vegetation of waste disposal areas of a coal fired power plant in Kansas. *J. Environ. Qual.* 18: 285–292.
120 Carlson, C.L. and Adriano, D.C. (1991). Growth and elemental content of two tree species growing on abandoned coal fly ash basins. *J. Environ. Qual.* 20: 581–587.
121 Jusaitis, M. and Pillman, A. (1997). Revegetation of waste fly-ash lagoons. I. Plant selection and surface amelioration. *Waste Manag. Res.* 15: 307–321.
122 Gutenmann, W.H., Bache, C.A., Youngs, W.D., and Lisk, D.J. (1976). Selenium in fly ash. *Science* 174: 966–967.
123 Garbisu, C., Hernández-Allica, J., Barrutia, O. et al. (2002). Phytoremediation: a technology that uses green plants to remove contaminants from polluted areas. *Rev. Environ. Health* 17 (3): 173–188.
124 Wei, S., Teixeira da Silva, J.A., and Zhou, Q. (2008). Agro-improving method of phyto extracting heavy metal contaminated soil. *J. Hazard. Mater.* 150 (3): 662–668.
125 Alkorta, I., Hernandez-Allica, J., Becerril, J.M. et al. (2004). Recent findings on the phytoremediation of soils contaminated with environmentally toxic heavy metals and metalloids such as zinc, cadmium, lead, and arsenic. *Rev. Environ. Sci. Biotechnol.* 3: 71–90.
126 Madrid, F., Liphadzi, M.S., and Kirkham, M.B. (2002). Heavy metal displacement in chelate irrigated soil during phytoremediation. *J. Hydrol.* 272: 107–119.
127 Nowack, B., Schulin, R., and Robinson, B.H. (2006). A critical assessment of chelant enhanced metal phtyoextraction. *Environ. Sci. Technol.* 40: 5225–5232.
128 Abhilash, P.C., Dubey, R.K., Tripathi, V., Srivastava, P., Verma, J.P., and Singh, H.B. (2013). Remediation and management of POPs-contaminated soils in a warming climate: challenges and perspectives, *Environmental Science and Pollution Research* 20, 5879–5885. https://doi.org/10.1007/s11356-013-1808-5
129 Kaushik, N., Kumar, K., Kumar, S. et al. (2007). Genetic variability and divergence studies in seed traits and oil content of *Jatropha* (*Jatropha curcas* L.) accessions. *Biomass Bioenergy* 31: 497–502.
130 Singh, S.K. (2007). *Global Agriculture Information Network (GAIN).* Report IN7047, India Biofuels Annual, pp. 5–12.

131 Gunaseelan, V.N. (2009). Biomass estimates, characteristics, biochemical methane potential, kinetics and energy flow from *Jatropha curcus* on dry lands. *Biomass Bioenergy* 33: 589–596.

132 Sinha, R.K., Herat, S., and Tandon, P.K. (2007). Phytoremediation: role of plants in contaminated site management. In: *Environmental Bioremediation Technologies* (eds. S.N. Singh and R.D. Tripathi), 77–100. Berlin: Springer-Verlag.

133 Rice, P.J., Anderson, T.A., Anhalt, J.C., and Coats, J.R. (1997). Phytoremediation of atrazine and metachlor contaminated water with submerged and floating aquatic plants. *Proceedings of the 12th Annual Conference on Hazardous Waste Research*, Kansas City, Missouri (19–22 May 1997). pp. 52.

134 Watanabe, M. (1997). Phytoremediation on the brink of commercialization. *Environ. Sci. Technol.* 31: 182A–186A.

135 Kim, I. (1996). Harnessing the green clean. *J. Chem. Eng.* 103: 39–41.

136 Truong, P. (1999). Vetiver Grass Technology for Mine Rehabilitation. Technical Bull. No. 1999/2; Pacific Rim Vetiver Network; Bangkok, Thailand.

137 Chang, Y.C., Nawata, A., Jung, K., and Kikuchi, S. (2011). Isolation and characterization of an arsenate-reducing bacterium and its application for arsenic extraction from contaminated soil. *J. Ind. Microbiol. Biotechnol.* 39: 37–44. https://doi.org/10.1007/s10295-011-0996-6.

138 Francesconi, K., Visoottiviseth, P., Sridokchan, W., and Goessler, W. (2002). Arsenic species in an arsenic hyperaccumulating fern, *Pityrogramma calomelanos*: a potential phytoremediator of arsenic-contaminated soils. *Sci. Total Environ.* 284: 27–35.

139 Ma, L.Q., Komar, K.M., Tu, C. et al. (2001). A fern that hyper-accumulates arsenic. *Nature* 409: 579.

140 Singh, N., Raj, A., Khare, P.B. et al. (2010). Arsenic accumulation patternin 12 Indian ferns and assessing the potential of *Adiantum capillus*-veneris, incomparison to *Pteris vittata*, as arsenic hyperaccumulator. *Bioresour. Technol.* 101: 8960–8968.

141 Srivastava, M., Ma, L.Q., Singh, N., and Singh, S. (2005). Antioxidant responses of hyper-accumulator and sensitive fern species to arsenic. *J. Exp. Bot.* 56: 1335–1342.

142 Xie, Q.E., Yan, X.L., Liao, X.Y., and Li, X. (2009). The arsenic hyperaccumulator fern *Pteris vittata* L. *Environ. Sci. Technol.* 43: 8488–8495.

143 Garbisu, C. and Alkorta, I. (2001). Phytoextraction: a cost-effective plant-based technology for the removal of metals from the environment. *Bioresour. Technol.* 77: 229–236.

144 Haferburg, G. and Kothe, E. (2007). Microbes and metals: interactions in the environment. *J. Basic Microbiol.* 47: 453–467.

145 Harrison, J.J., Ceri, H., and Turner, R.J. (2007). Multimetal resistance and tolerance in microbial biofilms. *Nat. Rev. Microbiol.* 5: 928–938.

146 Jiang, C.Y., Sheng, X.F., Qian, M., and Wang, Q.Y. (2008). Isolation and characterization of aheavy metal-resistant *Burkholderia* sp. from heavy metal-contaminated paddy field soil and its potential in promoting plant growth and heavy metal accumulation in metal-polluted soil. *Chemosphere* 72: 157–164.

147 Joshi, D.N., Flora, S.J.S., and Kalia, K. (2009). *Bacillus* sp. strain DJ-1, potent arsenic hypertolerant bacterium isolated from the industrial effluent of India. *J. Hazard. Mater.* 166: 1500–1505.

148 Kumar, P.B.A.N., Dushenkov, V., Motto, H., and Raskin, I. (1995). Phytoextraction: the use of plants to remove heavy metals from soils. *Environ. Sci. Technol.* 29: 1232–1238.

149 Higgins, T.E., Halloran, A.R., Dobbins, M.E., and Pittignano, A.J. (1998). In situ reduction of hexavalent chromium in alkaline soils enriched with chromite ore processing residue. *J. Air Waste Manage. Assoc.* 48: 1100–1110.

150 Lovley, D.R. and Coates, J.D. (1997). Bioremediation of metal contamination. *Curr. Opin. Biotechnol.* 8: 285–289.

151 Megharaj, M., Avudainayagam, S., and Naidu, R. (2003). Toxicity of hexavalent chromium and its reduction by bacteria isolated from soil contaminated with tannery waste. *Curr. Microbiol.* 47: 51–54.

152 McLean, J. and Beveridge, T.J. (2001). Chromate reduction by a pseudomonad isolated from a site contaminated with chromated copper arsenate. *Appl. Environ. Microbiol.* 67: 1076–1084.

153 McLean, J.S., Beveridge, T.J., and Phipps, D. (2000). Isolation and characterization of chromium reducing bacterium from a chromated copper arsenate-contaminated site. *Environ. Microbiol.* 2: 611–619.

154 Pattanapipitpaisal, P., Brown, N.L., and Macaskie, L.E. (2001). Chromate reduction and 16S rRNA identification of bacteria isolated from a Cr(VI)-contaminated site. *Appl. Microbiol. Biotechnol.* 57: 257–261.

5

Ecological Restoration and Plant Biodiversity

Shalini Tiwari[1] *and Puneet Singh Chauhan*[1,2]

[1] *CSIR-National Botanical Research Institute (CSIR-NBRI), Lucknow, India*
[2] *Academy of Scientific and Innovative Research (AcSIR), Ghaziabad, India*

5.1 Introduction

An organized and systematized exploration of socially and economically valuable products obtained from biological resources, viz. microorganisms, plants, animals, etc., is called bioprospecting. It involves the search, collection and derivation of genetic material from biodiversity samples that can be further developed for commercialization of pharmaceutical, agricultural, industrial products and hence, overall for societal benefits [1, 2]. It is carried out by a diverse range of industries including botanical medicines, pharmaceuticals, crop protection, agricultural seeds, horticulture, cosmetics, environmental monitoring and manufacturing [3]. Biodiversity is variable between living organisms from all sources including terrestrial and aquatic ecosystems. This includes intraspecific and interspecific diversity and thus can be divided into three parts: (i) Genetic biodiversity: difference in the types and number of genes and chromosomes present in different species; (ii) Species biodiversity: variation in the number as well as species richness within a region and (iii) Ecosystem biodiversity: interaction among species that live together in a given area [4]. Considering terrestrial habitat, 5–30 million species contains thousands of genes. Still, approximately known two million species have limited knowledge of the global distribution [2]. Therefore, it is rational to use the application of new technologies to explore the largely unidentified and unexplained species that will provide even more benefits to mankind. Hence, it can be assumed that for bioprospecting, biodiversity is the essential resource, but it is hardly ever possible to predict the specific genes, species as well as ecosystems will turn out to be valuable in the future for bioprospecting.

Bioprospecting thus has generated numerous valuable products from different various ecosystems, viz. tropical forests, temperate forests, temperate grasslands, arid and semiarid regions, freshwater ecosystems, along with cold and warm oceans. Among these ecosystems, tropical forests are a major source for countless products due to the species richness in its environment. Unfortunately, for many diverse purposes, unregulated and rigorous collection of samples by humans for hundreds of years, the United Nations has established

Bioprospecting of Plant Biodiversity for Industrial Molecules, First Edition.
Edited by Santosh Kumar Upadhyay and Sudhir P. Singh.

a framework to preserve the world's biodiversity while promoting the sustainable use of biodiversity. Consequently, the conservation and restoration of biodiversity present in all ecosystems would provide various prospects for future bioprospecting. Therefore, this chapter includes the area of bioprospecting, followed by its conservation and ecological restoration.

5.2 Major Areas of Bioprospecting

Bioprospecting comprises systematic search for genes, natural compounds, as well as organisms of wild resources to develop product by bio-physicochemical and genetic methods without disrupting nature and natural resources. Thus, bioprospecting includes three major areas [3]: (i) Chemical prospecting, (ii) Gene prospecting, and (iii) Bionic prospecting (Figure 5.1).

5.2.1 Chemical/Biochemical Prospecting

Biochemical prospecting includes investigation of novel and valuable chemicals from organisms [5]. Recently, high-performance chemical screening and automated bioassay platforms for identification, isolation and characterization of novel bioactive compounds from wild bio-resources (members from lower to higher plant kingdom, insects and invertebrates from animal kingdom) have shed the light in the area of natural product, drug and pharmaceuticals research [6]. Chemical prospecting from wild-type plant species is widely applicable in agro-chemistry (bio-pesticides), pharmaceuticals, enzymes, proteins, food additives, cosmetics and other industrially useful chemical products.

5.2.2 Gene/Genetic Prospecting

Molecular systematics aims to identify and characterize genetic variants within a species or a population at the DNA level by widely using DNA fingerprinting techniques [3]. Such types of approaches are gaining momentum nowadays and are useful for gene- and

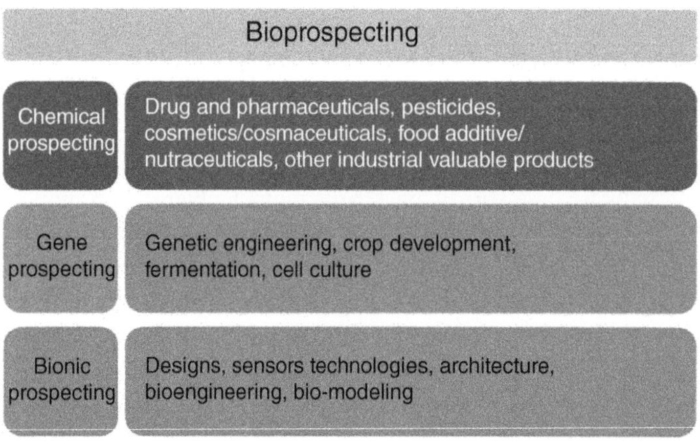

Figure 5.1 Major areas of bioprospecting.

biodiversity-rich poor countries. But due to the new patent laws, development of genetically modified organisms, hybrids and transgenic plants may rise serious implications on the jurisdiction of biodiversity-rich poor nations. Therefore, it is essential for such countries to develop appropriate research programmes to identify, characterize and evaluate their own genetic bio-resources, specifically of endemic species and perform molecular characterization to prevent biopiracy or genepiracy issues [7]. Modern molecular approaches such as DNA recombinant techniques and transgenic technologies facilitate transfer of the desirable agronomic traits or chemicals from plant to bacteria. The resulting transgenic bacteria are potential candidate for chemical factories by synthesizing specific products such as proteins, enzymes and other related biomolecules at large scale.

5.2.3 Bionic Prospecting

Bionics also known as biologically inspired engineering is a new area of bioprospecting in which new patterns, models, designs and techniques are developed and built on natural biodiversity. Bionic prospecting also includes sensor technologies, bio-engineering, bio-modelling and architecture.

5.3 Bioprospecting: Creating a Value for Biodiversity

The nature's library has potential to abundantly provide food, nutrients, medicines and various valuable industrial products. Apart from this information there are still various unknown potential in genes, species, population as well as ecosystems that embodies incredibly never-ending biological horizons. Years ago researchers focused only on the domesticated plant species of biodiversity, and the information about most of the worthy bioresources was challenging and unavailable. But, with the help of bioprospecting, the majority of the bioresources have been explored and their genetic, chemical as well as economic potential has been evaluated using modern technologies. These recent approaches thus have capability to identify and isolate potent and essential bioactive compounds at massive scale.

Biochemical techniques are one of the approaches that help to isolate and identify bioactive chemicals from plant extracts. It includes initial screening on the basis of biological activity of compounds using in vitro enzyme methodology and further isolation and characterization of chemicals via modern chromatographic techniques such as high-performance liquid chromatography (HPLC), high-performance thin-layer chromatography (HPTLC), gas liquid chromatography (GLC), medium pressure liquid chromatography (MPLC) and liquid chromatography mass spectrometry (LCMS).

Apart from biochemical techniques, biotechnological approaches are also powerful tools that not only help to comprehend the structure and function of living cells, tissues or organisms but also help in manipulating or editing the genome of the organism to produce modified or novel products. Hence, it called the most promising area of bioprospecting named gene prospecting. It provides facts about the retaining and transmitting of genetic information of living beings, their response against environmental, physical and chemical changes, etc. The improvements made in older biotechnology tools include DNA fingerprinting; it has now become possible to study the species genetic diversity in an easy manner. Hence, biotechnology applications provide various benefits to humankinds in the form of foods, medicines, drugs and other healthcare products as well as economic goods.

5.4 Conservation and Ecological Restoration for Sustainable Utilization of Resources

Uncontrolled exploitation of biodiversity for food, fodder, fibre, fuels, medicines, oils, perfumes, dyes, gums, resins, pesticides, insecticides, phytochemicals, proteins and genes causes biodiversity loss. Hence, special attention is required to protect economic, ecological and existence value of biodiversity. Conservation of biodiversity needs greater importance as humans will confront serious threats in near future due to the problems that arise via environmental degradation, including loss of biological diversity. Approximately 60,000 plant species out of the total 2,87,655 known species are on the verge of extinction due to countless reasons, in which overexploitation is one of the major issues [8]. Being a priceless and renewable resource, biodiversity exhibit potential values to humankind and interestingly have self-sustaining ecological system for maintaining the integrity and potency of biosphere. Thus, this devastating condition of biodiversity contrives numerous national and international initiatives, strategies, action plans, policies and legal frameworks aiming to prevent the continuing damage and loss of plant genetic resources [7]. The Global Strategy for Plant Conservation (GSPC) is among those strategies introduced by conference of parties (COP) of CBD at its sixth meeting held in Hague, 2002. Its aim is to conserve plant diversity by in situ and ex situ or by combination of both methods. In recent years, the GSPC committee members updated their strategies of plant conservation for the year 2011–2020 (https://www.cbd.int/gspc/). There are several other plant conservation programmes organized by government officials at the national level to halt the loss of plant resources. In India, the National Biodiversity Strategy and Action Plan identifies ex situ plant diversity conservation as a priority area of action in India and suggests strengthening and enhancing the role of botanical gardens, home gardens and other ex situ conservation networks.

The National Biodiversity Strategy and Action Plan (NBSAP) defines 'ex situ plant diversity conservation' as a priority action to ensure the social, environmental and economic development of the country (https://sustainabledevelopment.un.org/). It emphasizes and enhances the role of botanical gardens, home gardens and other ex situ conservation setups within individual countries as effective ways for biodiversity conservation. Recently in 2019, NBSAP prepared the first Global Sustainable Development Report (GSDR) by the independent group of researchers and scientists appointed by the United Nations Secretary-General and laid the foundation for 2030 Agenda and the Sustainable Development Goals (SDGs). In India, a special assistance program is being implemented at huge level by the Ministry of Environment & Forests (MOEF) to provide fund to botanical gardens to develop and maintain their infrastructure and other resources to support ex situ plant diversity conservation programmes as India supports 8.1% of the world's biodiversity with its 2.4% of world landmass (http://moef.gov.in/wp-content/uploads/2019/04/EFC.pdf). India has a strong network of research and development institutions funded by the ministries and departments of the central as well as the state government. Research programmes carried out by both national and international institutes to conserve plant resources include maintenance of catalogue of rare endangered threatened (RET) species, their assessment and distribution, in situ and ex situ conservation of RET species via micro-propagation, genetic diversity assessment, gene bank, bioprospecting and sustainable utilization [6].

Ecological restoration varies from conservation as its effort is to restore the ecosystem in its original appearance. The conservation focuses on sustainable use and preservation of the original form, while the restoration, on the other hand, involves restoring the ecosystem to its original form that once existed. Although, the diversity of the restored species is usually reduced in comparison to the former habitat [3]. Hence, restoration is strictly necessary across the globe, as a result of national and local government policies demanding the repair and maintenance of destroyed habitats such as abandoned industrial and mining sites, polluted and contaminated farm land as well as surface water by rigorous human activities. Ecological restoration is possible by proper harvesting or offsite plantation and further transfers to the restored site of plants (species) of neighbouring or comparable ecosystems [6, 9, 10]. It requires an in-depth knowledge of species, their ecological behaviour and their interaction with each other and within the ecosystem. Sometime, species seems extinct at a particular location, but in actual they survive in dormant or hidden forms. These species could be easily restored via removing the inhibiting factors [11]. For example, via preventing excessive burning or grazing, the woody plants of degraded land of tropical savanna restored by their own due to the presence of hidden rootstocks in the soil. Another species that reappeared after an absence of almost 70 years at wetland site of Britain is fen violet *Viola persicifolia* [12]. It became possible due to long dormancy period of its seeds. Alternatively, agroforestry is also an example of ecological restoration and nowadays it became an important practice in this field as the presence of trees in land confers several advantages. Numerous agroforestry approaches have been performed till date to restore and increase the productivity of farm land. For example, trees stabilize the soil, and used in terracing, strip-cropping and contour cultivation to prevent soil erosion and increase water availability. Trees are also able to fix nitrogen and thus increase soil fertility [13]. These suitable soil conditions help species in their restoration and conservation of biodiversity. Food and non-food products obtained from trees also helps farmers by providing nutritional security, enhancement in income, better livelihoods and hence combating poverty [14]. About concerning the land degradation and biodiversity loss, an international committee takes a global initiative in 2011 to restore 150 mha area of deforested land by 2020. The New York Declaration on Forests in 2014 promises and raised up the restoration target up to 350 mha area by 2030 [15].

5.5 Biodiversity Development Agreements

Besides several advantages of bioprospecting, there are some demerits of it. Bioprospecting needs to evaluate the frequent occurrence of biopiracy and most importantly the enhancement in biodepletion that might led to mass extinction [16]. Biodiversity development agreements (BDA) intend to distribute the benefits of biodiversity to those who bear the conservation costs through bioprospecting efforts. BDAs are agreements between biodiversity holders (usually of developing country) and biodiversity users (usually of private firm) to share bioresources and the benefits from the development of new products [7]. In general, the prospecting efforts for biodiversity have two objectives: (i) to discover and use certain genetic resources of unknown species that could have potential benefit for agriculture and industry and (ii) to offer local residents a return on biodiversity conservation. These agreements are proposed for the help of countries as well as local people that have right to share greater benefits of resource conservation. Unfortunately, compared with the different opportunity aspects, necessary attention is not paid to the potential of BDAs. For example, the consequences of food prices, demographic demand, hunger and land-use policies by the government have a significant impact on the returns to competing forest land use. From this perspective, excessive demand and exploitation for agricultural land leads to loss of species richness and biodiversity.

5.6 Conclusions

Bioprospecting will accomplish several goals such as (i) generate revenues for local communities, lands and environmental projects; (ii) develop scientific and technological approaches to maintain biodiversity; (iii) raise awareness for the commercial and non-commercial value of biodiversity and (iv) sustainable resource management. Overall, the biodiversity is a valuable natural resource that must be protected and used sustainably for the good of both present and future human generations. But excessive exploitation of biodiversity for benefits of humankind causes biodiversity loss at mass level. At present, a pandemic disease COVID-19 is negatively impacting human societies of most of the countries in the world, however reduced human activities due to lockdown in most of the countries is benefiting biodiversity [17]. This situation also points towards the need of conservation and restoration of biodiversity in present. Hence, the conservation and ecological restoration is an optimistic, prosperous vision in which human activities promote the sustainability of plant life such as maintenance of plant genetic diversity, endurance of plant species and their habitat as well as related ecosystems which in turn strengthen our livelihoods and well-being. Thus, the remarkable capability of humans will definitely help in the field of biodiversity by deciphering the complexities of nature, developing novel technologies and controlled utilization for the sustainable use of the limited bioresources.

References

1 Blaustein, R. (2006). Genetic resources and the Convention on Biological Diversity. *BioScience* 56 (7): 560–563.

2 Sawarkar, A., Sharma, R.K., and Gautam, V. (2019). Bioprospecting: creating value for biodiversity. *The Pharma Innovation Journal* 8 (4): 256–265.

3 Pushpangadan, P., George, V., Ijinu, T.P., and Chithra, M.A. (2018). Biodiversity, bioprospecting, traditional knowledge. sustainable development and value added products: a review. *Journal of Traditional Medicine & Clinical Naturopathy* 7 (1): 1–7.

4 Frisvold, G. and Day-Rubenstein, K. (2008). Bioprospecting and biodiversity conservation: what happens when discoveries are made. *Arizona Law Review* 50: 545.
5 Schuster, J.C. (1999). Chemical prospecting: an evolutionary-biogeographical approach-mesoamerican cloud forests as an example. *International Conference on Biodiversity and Bioresources. Consérvation and Utilization* Phuket, Thailand (November 1999).
6 Beattie, A.J., Hay, M., Magnusson, B. et al. (2011). Ecology and bioprospecting. *Austral Ecology* 36 (3): 341–356.
7 Day-Rubenstein, K. and Frisvold, G.B. (2001). Genetic prospecting and biodiversity development agreements. *Land Use Policy* 18 (3): 205–219.
8 Sangeetha, J., Thangadurai, D., Prathima, P. et al. (2019). *In situ* and *ex situ* conservation of biodiversity in India: strategies and approaches for environmental sustainability. In: *Biodiversity and Conservation* (eds. J. Sangeetha, D. Thangadurai, H.C. Goh and S. Islam), 289–340. Apple Academic Press.
9 Harker, D., Libby, G., Harker, K. et al. (2001). *Landscape Restoration Handbook*. Boca Raton: Lewis Publishers.
10 Whisenant, S.G. (2001). *Repairing Damaged Wildlands*. Cambridge: Cambridge University Press.
11 Hamilton, A. (2013). *Plant Conservation: An Ecosystem Approach*. Routledge.
12 Marren, P. (1999). *Britain's Rare Flowers*. London, UK: Academic Press.
13 Hillbrand, A., Borelli, S., Conigliaro, M., and Olivier, E. (2017). Agroforestry for landscape restoration: exploring the potential of agroforestry to enhance the sustainability and resilience of degraded landscapes. Food and Agriculture Organization of the United Nations (FAO), www.fao.org/publications.
14 Acharya, A.K. and Kafle, N. (2009). Land degradation issues in Nepal and its management through agroforestry. *Journal of Agriculture and Environment* 10: 133–143.
15 Climate Summit (2014). *Forests: Action Statements and Action Plans*. New York, USA: United Nations.
16 Harvey, A.L. and Gericke, N. (2011). Bioprospecting: creating a value for biodiversity. In: *Research in Biodiversity-Models and Applications* (ed. I.Y. Pavlinov), 323–338. IntechOpen.
17 Corlett, R.T., Primack, R.B., Devictor, V. et al. (2020). Impacts of the coronavirus pandemic on biodiversity conservation. *Biological Conservation* 246: 108571.

6

Endophyte Enzymes and Their Applications in Industries

Rufin Marie Kouipou Toghueo and Fabrice Fekam Boyom

Antimicrobial and Biocontrol Agents Unit (AmBcAU), Laboratory for Phytobiochemistry and Medicinal Plants Studies, Department of Biochemistry, Faculty of Science, University of Yaoundé I, Yaoundé, Cameroon

6.1 Introduction

Produced by living cells to bring about specific biochemical reactions, enzymes are biocatalysts, providing the lower-energy pathway between reactants and products. These biomolecules are the primary instruments for the expression of gene action, required for both syntheses and breakdown reactions. Enzymes are highly specific in their action on substrates and often act as a consortium to catalyse, by concerted action, the sequence of metabolic reactions performed by the living cells [1–3].

The use of enzymes by humans is thousands of years old. The earliest known record of enzymes is found in Homer's Greek epic poems dating from about 800 BCE, mentioning their use in the production of cheese. In Japan, documents from more than a 1000 years ago denote the use of naturally occurring enzymes in the production of fermented products such as sake (Japanese rice wine). Other ancient documents mention the use of enzymes for brewing, baking and alcohol production [3, 4]. Nowadays, the enzyme industry has a great market because of their increasing demand in multiple industries, including food manufacturing, animal nutrition, cosmetics, as well as for research and development and mostly as therapeutic agents against many diseases, to name but a few [5, 6]. These biomolecules are found in animals, plants and particularly microbes which are the source of most commercial enzymes [7].

Indeed, microorganisms are the greatest source of enzymes, bacteria and fungi being the best suited for industrial-scale production of most of the enzymes used in various industries [3, 5]. Their favourable features include the ability to be cultured in large quantities in simple and inexpensive substrates, to be easily handled and grown in huge tanks without light, to have a very high growth rate and shorter generation times, and to be genetically modified more easily to increase enzyme production yield [8, 9]. Moreover, microbial enzymes are preferred in multiple industries because they are biochemically diverse and have a broad range of activities inherent to their resilience towards variation

Bioprospecting of Plant Biodiversity for Industrial Molecules, First Edition.
Edited by Santosh Kumar Upadhyay and Sudhir P. Singh.

of environmental parameters such as pH, temperature and salinity. In addition, compared to some plant and animal enzymes, they are more active and stable and remain active under a wide range of varied physicochemical conditions [9–11]. However, the absolute and relative amounts of the various individual enzymes produced vary markedly between genus and even between strains of the same species. Therefore, selecting the appropriate microbial strains with the capacity for producing the highest amounts of particular or desired enzymes is of great importance. Of note, industries are nowadays looking for new microbial strains with abilities to produce specific enzymes to fulfil the current enzyme requirements. The search for novel active enzymes is currently concentrated on microorganisms living in diverse environmental conditions, or inhabiting unique and exceptional biotopes. Endophytic microorganisms able to survive under quite extreme and inhospitable conditions have recently seen a huge jump in scientific attention and interest [12]. Living asymptomatically inside plants for at least part of their life cycle, endophytes have emerged as a valuable source of novel metabolites and industrially important enzymes [13]. This group of microorganism has been meticulously investigated and used as a base for sustainable research and development of new enzymes [14]. Within this scope, multiple enzymes, including amylase, asparaginase, cellulase, lipase, laccase, protease, xylanase, etc., have already been reported from endophytes [15, 16]. This chapter presents recent updates on the potential of endophytes as enzyme sources, with particular focus on investigations reporting the purification and characterisation of enzymes produced by endophytic microbes.

6.2 The Rationale for Bioprospecting Endophytes for Novel Industrial Enzymes

The rationale behind investigating endophytes for novel industrial enzymes is guided by the recognition that endophytic microorganisms are capable of producing enzymes for their inherent natural needs. Of note, endophytes produce enzymes to (i) facilitate their penetration in their host plant tissues, (ii) evade the host defence mechanism and (iii) protect and promote the growth of the host plant (Figure 6.1).

Indeed, for the successful infection of their host plant, endophytes produce a series of hydrolytic and oxidative enzymes that are necessary for the degradation of lignocellulosic components from the cells of the host plant to facilitate the penetration and

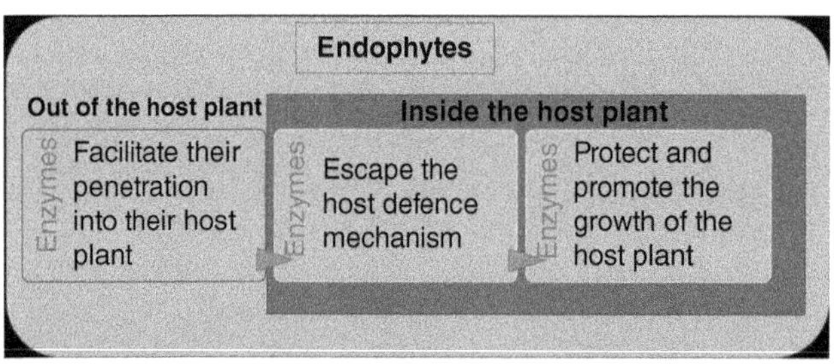

Figure 6.1 Attributes of endophytes making them suitable for enzymes bioprospecting.

colonization. Endophytes have been reported to possess a hydrolytic system responsible for polysaccharides degradation consisting mainly of xylanases and cellulases and a unique oxidative ligninolytic system comprising mainly laccases, ligninases and peroxidases which degrade lignin and open the phenyl rings [14]. On the other hand, it is widely recognised that to enter the host plant, endophytes have to pass through the first line of defence of the plant's immune system. To this end, endophytes produce enzymes such as amylases, lipases and proteases, as part of their mechanism to overcome the defence set in place by the host against any microbial invasion [17]. Fungal endophytes for instance have been reported to modulate the plant's immune system by secreting chitin deacetylases, which deacetylate chitosan oligomers and, hence, prevent from being recognised by chitin-specific receptors [18, 19]. Moreover, to counteract the oxidative burst of reactive oxygen species (ROS) generated by the plant as a defence mechanism, endophytes also produce multiple enzymes such as superoxide dismutases, catalases, peroxidases, alkyl hydroperoxide reductases and glutathione S-transferases [20]. In case the host plant produces antimicrobial secondary metabolites as a defence mechanism against microbial colonisation, endophytes might secrete the matching detoxification enzymes, such as cellulases, lactases, xylanases and proteases, to decompose these secondary metabolites before penetrating through the defence systems of the resided host plants [21].

Besides helping endophytes to escape the plant defence system, the production of enzymes is also needed to help initiate the host symbiosis process [22].Once inside the host plant, the symbiosis very well established between endophytes and their host might condition the survival and health of plants [23, 24]. Some endophytes play an imperative role in protecting or preparing the plant against abiotic and biotic stresses and also help enhance the growth and yields [25, 26]. Chen et al. [27] reported that some endophytic fungi could promote the growth and fitness of the host plants by activating the expression of certain enzymes and genes. Corroborating this statement, Sherameti et al. [28] showed that the endophyte *Piriformospora indica* increased the growth of tobacco roots by stimulating the expression of nitrate reductase and the starch-degrading enzyme (glucan-water dikinase). These extracellular enzymes produced by endophytes target various macromolecules such as carbohydrates, lignin, organic phosphate, proteins and sugar-based polymers to breakdown into transportable products throughout the cells and to continue heterotopic metabolism [29]. Overall, the obvious ability of endophytes to produce a wide variety of enzymes, including lipases, phytases, amylases, cellulases, proteases, xylanases, etc., that are capable of degrading complex macromolecules make them an attractive source for bioprospecting for industrially important enzymes.

6.3 Endophytes as a Source of Industrial Enzymes

Endophytic microbes inhabiting plants from various environmental settings have been reported to produce various kinds of extracellular enzymes (Table 6.1). In the present section, we are discussing enzymes characterised from diverse endophytes and their current industrial application. This is an attempt to catalogue endophytic enzyme producers that have the potential to be further investigated for practical application.

Table 6.1 Selected enzymes, endophyte-producing strains and the optimal conditions for production of enzymes.

Enzymes	Endophytes	Host plant	Fermentations condition	Yields	References
Amylases	*G. pulicaris*	NM	Medium composition: 2 g $NaNO_3$, 0.4 g yeast extract, 0.5 g KH_2PO_4, 0.5 g KCl, 0.5 g $MgSO_4 \cdot 7H_2O$, 0.01 g $FeSO_4 \cdot 7H_2O$ and 60 g raw potato; T: 30 °C; Ip: 5 days; shake: 120 rpm	260 U/mg	[30]
	Cylindrocephalum sp.	*A. calcarata*	Medium composition: g/L (3 g $NaNO_3$; 0.5 g $MgSO_4 \cdot 7H_2O$; 5 g KCl; 1 g KH_2PO_4; 0.01 g $FeSO_4 \cdot 7H_2O$; 0.1 g $CaCl_2$) with 1.5% maltose and 0.3% sodium nitrate; pH: 7.0; T: 30 °C; Ip: 7 days	0.035–0.25 U/mL	[31]
	P. minima	*E. longifolia*	Medium composition: 4% soluble starch-asparagine; pH: 5.0; T: 25 °C; Ip:5 days; shake: 180 rpm	212 U/mL	[32]
	T. cecidicola URM7826	*T. catimbauensis*	Medium composition: Czapex Dox's brothsupplemented with L-asparagine; T: 30 °C; IS: 10^8 spores/mL; pH: 6; Ip: 4 days; shake 120 rpm	2.30 U/g	[33]
	Penicillium sp. 4URM7827			1.28 U/g	
	F. proliferatum	*A. altissima*	Medium composition: modified Czapex dox's broth (glucose (2.0 g/L), L-asparagine (10.0 g/L), KH_2PO_4 (1.52 g/L), KCl (0.52 g/L), $MgSO_4 \cdot 7H_2O$ (0.52 g/L), $CuNO_3 \cdot 3H_2O$ (0.001 g/L), $ZnSO_4 \cdot 7H_2O$ (0.001 g/L), $FeSO_4 \cdot 7H_2O$ (0.001 g/L)); T: 36 °C; Ip: 5 days; shake 120 rpm	0.492 U/mL	[34]
Cellulases	*P. glabrum*	*Espeletia* spp.	Medium composition: w/v (1% NH_4NO_2, 1% K_2HPO_4, 0.5% $MgSO_4$, 0.1% NaCl, 0.13% $CaCl_2$, 0.01% $MnSO_4$, 0.01% boric acid, 0.001% $CuSO_4$, 0.2% $FeSO_4$, 0.001% $ZnSO_4$, 10 g/L palm oil empty fruit bunch, 5 g/L glucose and 6 g/L peptone); T: 25 °C; shake: 150 rpm and Ip: 20 days	0.04–1.13 U/mL	[35]
	Strain Tahrir-25	*L. corticata*	Medium composition: sugarcane bagasse-corncob (2 : 1); pH: 5.5; T: 30 °C; Mc: 60%; and IS:10^6	450 U/gds	[36]
Laccases	*M. verrucaria*	*Cajanus cajan L.*	Medium composition: g/L (200 potato; 3 KH_2PO_4; 1.5 $MgSO_4 \cdot 7H_2O$; 0.01 VB_1; and 0.25 $CuSO_4 \cdot 5H_2O$) with 10 g/L peptone and 2% glucose; Ip: 5 days, T: 30 °C; and pH: 6.22	16.52 U/mL	[37]

Table 6.1 (Continued)

Enzymes	Endophytes	Host plant	Fermentations condition	Yields	References
	N. luteum	Eucalyptus trees	Medium composition: modified Czapek Dox broth (KH_2PO_4, 1 g/L; $C_4H_{12}N_2O_6$, 2 g/L; yeast extract, 1 g/L; $MgSO_4 \cdot 7H_2O$, 0.5 g/L; KCl, 0.5 g/L; and 1 mL of a mineral solution with trace elements: $Na_2B_4O_7$, 0.092 mg/L; $MnSO_4 \cdot H_2O$, 0.1 mg/L; $(NH_4)_6Mo_7O_2 \cdot 4H_2O$, 0.01 mg/L; $FeSO_4 \cdot 7H_2O$, 0.5 mg/L; $ZnSO_4 \cdot 7H_2O$; 0.1 mg/L; and $CuSO_4 \cdot 5H_2O$, 0.01 mg/L) with 1% (v/v) ethanol and 0.30 mM $CuSO_4$; T: 22 °C; shake: 120 rpm; Ip: 20 days	0.28 U/mL	[38]
Lipases	*C. kikuchii*	*Tithonia diversifolia*	Medium composition: Vogel's minimum medium with 2% soybean oil; T: 30 °C; shake: 120 rpm; Ip: 6 days	14.87 U/mL	[39]
	F. oxysporum PTM7	*C. oblongifolius*	Medium composition: (w/v) (1% peptone, 0.15% KH_2PO_4, 0.5% $NaNO_3$, 0.05% NaCl, 0.05% $MgSO_4$, 0.025% $CaCl_2$, 0.0001% $FeSO_4$, 0.0001% $ZnSO_4$, 0.0001% $CuSO_4$ and 1% (v/v) olive oil); T: 30 °C; pH:7; Ip: 6 days	4.3 U/mL	[40]
	C. guillermondi	*R. communis*	Medium composition: 15 g/L $(NH_4)_2SO_4$, 0.15 g/L manganese sulfate, 10 g/L brewer's yeast, 80 g/L Tween, 15 g/L glycerol and 30 g/L soybean oil, dissolved in 25 mmol/L phosphate buffer; pH: 6.5; T: 30 °C; Ip: 3 days and shake: 180 rpm	18 U/mL	[41, 42]
Proteases	*Penicillium* sp. Morsy1	*D. hemprichi*	Medium composition : rice straw supplemented with g/L (0.3 $(NH_4)_2SO_4$; 0.5 NaCl; 0.5 KH_2PO_4; 0.4 $CaCl_2$); T: 26 °C, IS: 10^5 spores/mL; pH: 6; Ip: 5 days and Mc: 80%	3560 U/g	[43]
	Acremonium *sp.*	*Saraca asoca*	Medium composition: (g/L) (skimmed milk 20, yeast extract 0.7, 1.9 KH_2PO_4, KH_2PO_4 0.85, 0.0015 $MgSO_4 \cdot 7H_2O$) with fructose and $(NH_4)_2SO_4$; T: 37 °C; pH: 7; Ip: 3 days	3.34 U/mL	[44]
Xylanases	*A. terreus*	*M. peregrina*	Medium composition: 1% wheat bran as a carbon source; T: 30 °C; pH: 6; shake: 100 rpm and Ip: 48 h	54 U/mg	[45]

IS, inoculum size (spores/mL); Mc, moisture content (v/w); Ip, incubation time; T, temperature; rpm, rotation per minute; NM, not mentioned.

6.3.1 Amylases

Amylases including α-amylase (1,4 α-glucan glucanohydrolase, EC 3.2.1.1), β-amylase (1,4 α-glucan maltohydrolase, EC 3.2.1.2), and glucoamylase (1,4 α-glucan glucohydrolase, EC 3.2.1.3) represent one of the most important groups of enzymes for biotechnological application [46, 47]. In a variety of industries (e.g. food, chemical, detergent and textile), microbial amylolytic enzymes are used to convert starch into different sugar solutions through the hydrolysis of α-1,4-glycosidic linkages. Indeed, in today's processing industries, microbial amylases have completely replaced the chemical hydrolysis of starch, because of their multiple advantages, such as higher yield and specificity, the better control over amylolysis with the complete elimination of the neutralisation steps and undesirable reactions, as well as the reduction of the probabilities of contamination because of the thermostability of these enzymes. In addition, the products generated are also very stable and the overall process is cost-effective [7]. Among the amylases produced by diverse microorganisms, the α-amylases from *Bacillus* sp. have been the most extensively studied given especially their extreme thermostability [48], while *Aspergillus* sp. and *Rhizopus* sp. are often used as sources of industrial production of glucoamylases [46].

There is a growing interest in exploring endophytes for the production of amylolytic enzymes that are active at low temperatures and alkaline conditions. In this respect, Marlida et al. [49] screened endophytic isolates of *Gibberella pulicaris*, *Acremonium* sp., *Synnematous* sp. and *Nodilusporium* sp., for their ability to produce raw starch degrading enzymes in submerged fermentation. Their results showed that enzymes produced by *Acremonium* sp. had a broad activity towards both small and large granule sized raw starches, whereas the enzyme from the other isolates showed high activity only toward starches of smaller granule size. Moreover, the thin layer chromatography analysis of the end products revealed that enzymes from *G. pulicaris*, *Acremonium* sp. and *Nodilusporium* sp. hydrolysed raw sago starch to produce solely glucose, while the amylase from *Synnematous* sp. produced glucose and maltose. In subsequent investigations, they studied the effect of different nitrogen and carbon sources on the growth of *Acremonium* sp. and the yield of amylase enzyme. Their data showed that using peptone and sodium nitrate as nitrogen sources and raw sago starch as carbon source each at the optimum concentration of 0.5, 3 and 20 g/L, respectively, the fungal growth and enzyme activity levels significantly improved. Moreover, at the optimum pH of 5.0 and incubation temperature of 30 °C, the enzyme production significantly increased by 19- to 22-fold compared to the activity obtained in the original basal medium [50]. Further investigation of *G. pulicaris* led Marlida et al. [30] to show the suitability of raw potato for amylase production and the highest amylase production (260 U/mg protein) to be achieved when the concentration of raw potato starch was increased to 60 g/L. After several endophytic fungi were screened for amylolytic activity, *Fusicoccum* sp. BCC4124 exhibited great potential when grown on multi-enzyme induction-enriched media and agro-industrial substrates. It was also found that both α-amylase and α-glucosidase activities were highly induced in the presence of maltose and starch as carbon sources. An α-amylase with strong hydrolytic activity ($kcat/Km$:6.47×10^3 min/(mL/mg)) on soluble starch was purified and exhibited characteristics such as high activity at 70 °C and resistance to inhibition by glucose of up to 1 mol/L, making it a very good candidate for potential biotechnological applications [51]. An endophytic *Cylindrocephalum* sp. strain isolated from the medicinal plant *Alpinia calcarata* (Haw.) Roscoe was also reported to exhibit maximal amylase production when cultured in submerged fermentation at 30 °C and pH 7.0. Various carbon and nitrogen sources were tested to identify maltose (1.5%) and sodium nitrate (0.3%) as the best combination of carbon and nitrogen sources for the production of enzymes [31]. An α-amylase produced by the endophytic *Preussia minima* from the Australian native plant, *Eremophilia longifolia*, was purified to homogeneity through a fractional acetone precipitation and Sephadex G-200 gel filtration, followed by DEAE-Sepharose ion-exchange chromatography. This enzyme with a molecular weight of 70 kDa exhibited maximum activity at optimum temperature and pH of 25 °C and 9, respectively, and was activated and stabilised by Mn^{2+} and Ca^{2+}. The investigation of the effect of different carbon and nitrogen sources on the enzyme production revealed that with starch as the carbon source and L-asparagine as the nitrogen source the highest production of 138 U/mg was achieved. Moreover, bioreactor studies showed that the enzymatic activity was comparable to that obtained in shaker cultures, which indicates likelihood of scale-up fermentation for the mass production of this enzyme [32].

6.3.2 Asparaginase

L-Asparaginase also known as L-asparagine amidohydrolase (EC 3.5.1.1) is an enzyme catalysing the deamidation of L-asparagine to L-aspartic acid and ammonia. Both the substrate and product of this enzyme are necessary to living organisms because of their important role in various metabolic processes. L-Asparagine is essential for a cell's existence. For instance, L-asparagine is the most abundant metabolite for the storage and transport of nitrogen in plants [52]. This amino acid is required for the fast growth of cells and can be synthesised by the human body when needed without any external supplementation required [7]. Some cancer cells, especially lymphocytes, do not have the ability to synthesise L-asparagine because their lack of the enzyme aspartate ligase. Therefore, their survival entirely depends on the external uptake of asparagine. The depletion of L-asparagine from the surrounding environments of cancer cells by the L-asparaginase supplied exogenously has proven its efficacy as a therapeutic strategy for cancer treatment [7, 52]. Over the past decades, L-asparaginase became a powerful anti-tumour medicine and researchers are looking for microbial strains able to produce more potent and safer L-asparaginase at the industrial level [53]. Therefore, endophytic microbes have been investigated as a potential source of this important therapeutic enzyme.

Fifty endophytic microbes (14 bacteria, 22 actinomycetes and 14 fungi) isolated from rhizomes of five medicinal plants of the Zingiberaceae family, including *Alpinia galanga*, *Curcuma amada*, *Curcuma longa*, *Hedychium coronarium*, and *Zingiber officinale*, were screened for L-asparaginase activity. Of these microorganisms, 31 (62%) exhibited L-asparaginase activity in the range of 54.17–155.93 U/mL in an unoptimised medium. Unfortunately, further experiments with an isolate *Talaromyces pinophilus* from *Curcuma amada* showed a decrease in enzymatic activity from 108.95 to 80 U/mL due to strain degeneration [54] possibly caused by the inadaptation of the endophyte isolate to the laboratory medium. Twenty endophytic isolates from *Tillandsia catimbauensis* were screened for L-asparaginase activity to identify 10 (50%) potential enzyme (0.50–2.30 U/g) producers.

Two isolates of *Talaromyces cecidicola* URM7826 (2.30 U/g) and *Penicillium* sp. 4URM7827 (1.28 U/g) were the most potent [33]. Similarly, 20 endophytic fungi isolated from leaves of the Brazilian medicinal plant *Myracrodruon urundeuva* were screened for L-asparaginase production out of which 17 exhibited L-asparaginase activity ranging from 0.57 to 2.41 U/g. The fungus *Diaporthe* sp. URM 7793 (2.41 U/g) was the best L-asparaginase producer, followed by *Diaporthe* sp. URM 7779 (2.0 U/g), *Talaromyces* sp. URM 7785 (1.91 U/g) and *Diaporthe* sp. URM 7792 (1.47 U/g) [55].

In another screening of bacterial and fungal endophytes isolated from *Ocimum tenuiflorum* (Tulasi) from NIT Warangal, Telangana, India, four strains produced a high amount of L-asparaginase. The enzyme produced by the most potent strain, *Bacillus stratosphericus*, was further purified. The docking studies showed that in comparison to L-Glutamine having 67.7 kcal/mol, the free energy binding efficiency of receptor towards L-asparagine was good with lesser energy −71.6 kcal/mol, revealing the high affinity of this enzyme to L-asparagine [56]. In another *in silico* study to investigate L-asparaginase-producing endophytic bacteria, an approach consisting of protein-protein basic local alignment search tool with *Escherichia coli* and *Erwinia chrysanthemi* asparaginase enzyme sequences against 262 endophytic bacteria was adopted. The results showed that some of the asparaginase enzymes produced by endophytic bacteria possess more suitable immunological indices when compared with asparaginase enzymes of *E. coli* and *E. chrysanthemi*. For instance, *Herbaspirillum rubrisubalbicans* were predicted to produce a non-allergen and non-antigen asparaginase enzyme, and the number of antigenic determinants was also predicted to be lower in asparaginase enzymes produced by *Bacillus amyloliquefaciens*, *H. rubrisubalbicans* and *Herbaspirillum seropedicae*. Moreover, the number of high-scored B-cell epitopes was lower in enzyme sequences related to the aforementioned bacteria and *Paenibacillus polymyxa*. The number of discontinuous epitopes and the number of T-cell epitopes were lower in *B. amyloliquefaciens*-produced enzymes. Therefore, the therapeutic use of these enzymes is possible because they are predicted to be very potent with lesser or no side effect [57].

Endophytic fungi isolated from four anticancer medicinal plants, including *Murraya koenigii, Oldenlandia diffusa, Pereskia bleo* and *Cymbopogon citratus*, were investigated for their ability to produce L-asparaginase. Four isolates showed potency among which *Fusarium oxysporum* and *Penicillium simplicissimum* were the most potent with mean L-asparaginase activity of 0.013 and 0.019 μM/mL/min, respectively [58]. In another screening of 33 endophytic fungi from plants collected in moist and evergreen forests of Western Ghats, India, 31 isolates were L-asparaginase positive, with 19 isolates able to produce glutaminase-free L-asparaginase. Further investigation led to the identification of 11 promising fungi belonging to *Acremonium, Alternaria, Aspergillus, Botrytis, Cylindrocladium, Cladosporium, Corynespora, Fusarium, Sordaria, Lasiodiplodia* and *Pestalotiopsis* genera. The most potent isolate was *Alternaria* sp., exhibiting a glutaminase-free L-asparaginase specific activity of 1.65 U/mg [59]. Similarly, the L-asparaginase screening of endophytic fungi from the cactus plant *Cereus jamacaru* showed that species belonging to *Aspergillus, Fusarium* and *Penicillium* exhibited L-asparaginase activity with *Aspergillussydowii* (29.02 U/mL) as the most potent [60]. In a recent study conducted to isolate L-asparaginase-producing endophytic fungi from medicinal plants of the Asteraceae family, a total of 84 isolates were screened using a qualitative plate assay on modified Czapex dox's agar medium and L-asparaginase further quantified by the nesslerisation method. Of the 84 isolates screened, 38 were able to produce L-asparaginase and their activities ranged between 0.019 and 0.492 U/mL with *Fusarium proliferatum* being the most potent followed by *Plenodomus tracheiphilus* isolated from *Anthemis altissima* (0.481 U/mL). Overall, this study indicated that L-asparaginase-producing endophytes belong to *Plectosphaerella, Fusarium, Stemphylium, Septoria, Alternaria, Didymella, Phoma, Chaetosphaeronema, Sarocladium, Nemania, Epicoccum, Ulocladium* and *Cladosporium* genera. This study further demonstrates the diversity of fungi that can be investigated as alternative source for the production of this anticancer enzyme (L-asparaginase) [34].

6.3.3 Cellulases

In nature, cellulose is found in combination with lignin and hemicellulose resulting in a complex and recalcitrant solid called lignocellulose. Cellulases namely endoglucanase (1,4-D-glucan-4-glucanohydrolase, EC 3.2.1.4), exo-cellobiohydrolase (1,4-D-glucan glucohydrolase, EC 3.2.1.74) and β-glucosidase (D-glucoside glucohydrolase, EC 3.2.1.21) play a predominant role in the hydrolysis and conversion of cellulosic substrates into monomeric products. In general, while the endoglucanase hydrolases the glycosidic bonds of cellulosic substrates, exo-cellobiohydrolase is attacking the crystalline ends to yield cellobiose [61] which under the action of β-glucosidase is converted into glucose molecules [62]. These enzyme-based conversions have the advantage of using low-cost substrates mainly consisting of cellulose and hemicelluloses (75–80%).In fact, cellulosic biomass, deriving from non-food sources, such as trees and grasses, is being explored as raw material for the production of biofuels, especially bioethanol for the transportation sector, with cellulases being the main group of enzymes used [63]. In addition, cellulases can also be used for the extraction of olive oil, treatment of wines and the clarification of fruit juices and to improve the quality of bakery products in food industries. In the textile industry, they are used for the stoning of denim and the polishing of cellulosic fibre. Cellulases are also used for the recycling of waste paper in the paper and pulp industry and to improve the digestibility of animal feed [64]. Therefore, investigations have been carried out to identify new microbial species capable of producing cellulases with high yield and specificity.

Two endophytic fungi isolates, *Colletotrichum* sp. and *Alternaria* sp., were evaluated for their potential as biodiesel feedstock. The results showed that in solid-state fermentation, the two endophytes grew successfully on a combined rice straw and wheat bran as substrate leading to the production of cellulase (1.21–2.51 U/g dry substrate (gds)) and a substantial amount of lipid (60.32–84.30 mg/gds) [65]. Earlier, Peng and Chen [66] screened 149 endophytic fungi isolated from 7 plants for their ability to contain intercellular lipid bodies and 26 showed great potency. In solid-state fermentation, these isolates produce cellulase enzymes and microbial oil with the yields of 0.31–0.69 filter paper unit and 19–42 mg/g initial dry substrate, respectively. This study also demonstrated the ability of endophytes to produce simultaneously a substantial amount of lipid and cellulase enzymes. In another study, 100 endophytes isolated from *Espeletia* spp. were evaluated for their cellulolytic potential on the saccharification of the palm oil empty fruit bunch using filter paper (FPA) followed by the assessment of carboxymethyl cellulase (CMCase), exoglucanase and β-glucosidase activities. The results showed that four isolates were potently active among which *Penicillium glabrum* had the highest cellulolytic activity (CMCase (1.87 U/mL), FPA (0.035 U/mL), β-glucosidase (0.06 U/mL) and exoglucanases (0.87 U/mL)). The partially purified enzymes exhibited maximal CMCase, exoglucanase and β-glucosidase activities of 44.52, 48.36 and 0.45 U/mL, respectively [35]. Another study reported the ability of β-1,4-endoglucanase from a citrus endophytic *Bacillus pimulus* to hydrolyse cellulose under *in vitro* conditions. This enzyme was reported to have a specific activity of 8.25 U/mg protein, an optimum pH range of 5–8 and remarkable heat stability (50–60 °C), retaining more than 85% activity even after a 24-hour incubation at pH 6–8.6. Ions Ca^{2+}, Na^{+}, Mn^{2+} and Mg^{2+} slightly stimulated, while Zn^{2+}, Cu^{2+}, K^{+}, Li^{+} and Co^{2+} inhibited its activity [67]. Another endophytic Bacillus isolate, *Bacillus amyloliquefaciens*, was also reported to produce β-1,3-glucanase[68].

The cellobiohydrolase encoding gene from a fungal endophyte *Fusiccoccum* sp. BCC4124 identified as a potent cellulase producer was expressed in *Pichia pastoris* KM71. Further investigations showed that the recombinant enzyme obtained could exhibit the ability to degrade Avicel, filter paper and 4-methyl umbelliferyl β-D-cellobioside but was unable to hydrolyse carboxymethylcellulose. This enzyme exhibited optimal activity at temperature and pH of 40 °C and 5.0, respectively, and had activity parameters (*K*m and *V*max) values of 0.57 mM and 3.086 nmoL/min/mg protein, respectively. The enzyme was stable at pH ranging from 3 to 11 and retained 50% of its activity at 70–90 °C for 30 minutes [69]. In a similar investigation by Harnpicharnchai et al. [70], a thermotolerant β-glucosidase enzyme was reported from an endophytic fungus, *Periconia* sp. BCC2871, and its full-length gene was cloned in *Pichia pastoris* KM71 strain. The recombinant enzyme obtained showed optimal activity at pH 5 and 6 and a temperature of 70 °C. More interestingly, the enzyme maintained 60% activity after a 90-minute incubation at 70 °C, remaining 100% active after 120 minutes incubation at pH≥8. Both investigations [69, 70] showcase the enzymatic ability of recombinant enzymes and their stability in a wide range of pH and temperatures and encourage further investigations for various biotechnological applications. In another study, several endophytes from Egyptian marine sponge *Latrunculia corticata* were screened for cellulase activity, and *Trichoderma* sp. Merv6, *Penicillium* sp. Merv2 and *Aspergillus* sp. Merv70 exhibited high cellulase activity while using different lignocellulosic substrates in solid-state fermentation. By applying mutagenesis and intergeneric protoplast fusion, the authors obtained a recombinant strain named Tahrir-25 that exhibited higher cellulases (Exo-β-1,4-glucanase, endo-β-1,4-glucanase and β-1,4-glucosidase) activity. Moreover, by optimising parameters such as initial moisture content at 60% (v/w), inoculum size at 10^6 spores/mL, the average substrate particle size at 1.0 mm, carbon source (a mixture of sugarcane bagasse and corncob (2 : 1) supplemented with carboxymethyl cellulose and corn steep solids), fermentation time (7 days), medium pH (5.5) and temperature (30 °C), El-Bondkly and El-Gendy [36] afforded an activity of 450, 191 and 225 U/g dry substrate (gds) of CMCase, filter-paperase (FPase) and β-glucosidase, respectively. The authors suggested that these marine endophytes could be exploited for the cost-effective production of cellulases for second-generation bioethanol processes.

Fifty endophytic fungi from Cameroonian medicinal plants were also screened for their cellulase activity using a qualitative assay. A promising fungus named *Penicillium* sp. 51 exhibited FPase and CMCase activities of 0.36 and 0.44 U/mg. respectively, with the optimal pH and temperature for the activity being, respectively, 5–6 and 40–50 °C [71]. From the screening of 20 endophytic fungi isolated from leaves of the ornamental plant *Pachystachys lutea* for cellulase activity, five endophytes including *Diaporthe anacardii* PL01, *Diaporthe* sp. PL03, fungal endophyte PL35, *Xylaria berteroi* PL36 and *Diaporthe* sp. PL67 were active with cellulase production ranging between 0.87 and 1.60 μmoL/min [72]. In another investigation, four endophytes *Hypoxylon* sp. CI4A, *Hypoxylon* sp. EC38, *Hypoxylon* sp. CO27 and *Daldinia eschscholzii* EC12 were evaluated for their consolidated bioprocessing (CBP) potential. The analysis of their genomes indicated that these endophytes have a rich reservoir of biomass-deconstructing carbohydrate-active enzymes (CAZys), which includes enzymes active on both polysaccharides and lignin, as well as terpene synthases (TPSs), enzymes that may produce fuel-like molecules. In addition, the analysis of their cellulase activity revealed that these endophytes actively produce endoglucanases, exoglucanases and β-glucosidases [73]. The authors concluded that looking at the richness of CAZymes as well as TPSs identified in these four endophytic fungi, they are great candidates to pursue development into platform CBP organisms. More recently, *Lysinibacillus xylanilyticus*, a Gram-positive cellulose-decomposing endophytic bacterium was isolated from medicinal plant *Chiliadenus montanus*. Further analysis showed that this isolate was capable of degrading 58% of cellulosic filter paper (100 g/L) after 15 days of incubation. In another experiment, the authors showed that maximum cellulase activity (0.18 U/min) was detected after 12 days of incubation, while the maximum release of soluble sugars (11.85 mg/mL) was detected after 15 days of incubation [74].

6.3.4 Chitinases

Chitinases are enzymes belonging to the glycosyl hydrolase family. They are used to reduce chitin into monomer and oligomer units via the hydrolysis of the 1 → 4 β-glycoside bonds of *N*-acetyl-D-glucosamine. They are classified into two categories, endochitinases and exochitinases. Endochitinases (EC 3.2.1.14) are known to randomly cleave chitin at internal sites to produce chitotriose, chitotetraose and diacetylchitobiose. Exochitinases on the other hand are further classified into two subcategories including (i) chitobiosidases (EC 3.2.1.29) and (ii) β-1,4-*N*-acetyl glucosaminidases (EC 3.2.1.30). The former release the diacetylchitobiose from the non-reducing end of chitin, while the latter hydrolases the products of endochitinases and chitobiosidases, to generate units of *N*-acetyl-D-glucosamine [7]. This group of enzymes has a huge pharmaceutical and agricultural potential since they are used in morphogenesis of insects and yeasts, to potentiate the activity of antifungal drugs, and as fungicide and insecticide agents [75]. They are also used for the production of proteins from shellfish waste, chitooligosaccharides and glucosamines, which have wide applications in the pharmaceutical industry [76].

In a search for chitinases, Purushotham et al. [77] reported the presence of four chitinases named *Sp*ChiA, *Sp*ChiB, *Sp*ChiC and *Sp*ChiD from a genome analysis of the endophyte *Serratia proteamaculans* 568. The heterologous expression and characterisation of *Sp*ChiA, *Sp*ChiB and *Sp*ChiC revealed their optimal activity at pH 6.0–7.0 and 40 °C. Additional investigation showed that the three *S. proteamaculans* chitinases displayed the highest activity binding to β-chitin and showed a broad range of substrate specificities. In fact, these enzymes released dimers as major end products from oligomeric and polymeric substrates, and tetramers from colloidal chitin substrates. However, the longest incubation time was required for the hydrolysis of the trimer. Overall, enzymes *Sp*ChiA and *Sp*ChiB were processive chitinases, while *Sp*ChiC was a non-processive chitinase. This study is encouraging and suggests that further investigation of endophytes could lead to the identification of several and more potent chitinase enzymes.

6.3.5 Laccases

Laccases (EC 1.10.3.2) are phenol oxidases which oxidise phenolic and aromatic compounds, including some amines, ethers and esters. This multicopper blue oxidase couples the electron transfer, causing the oxidation of water molecules. Laccases have been investigated for use in various biotechnological processes because of characteristics such as the wide range of substrate specificity, the use of oxygen as a final electron acceptor, and no requirement for a cofactor or peroxide for their catalytic activity [6, 7]. In fact, laccases are currently used in various industries such as food and beverage stabilisation, cosmetics, pulp and paper, textile for polymer synthesis, petrochemical, pharmaceutical as catalysts for the manufacture of anticancer drugs, and so forth [6]. They are also used for the degradation of pesticides and herbicides from agricultural fields [78], the treatment of wastewater that contains dyes from textile industries [78], the bioremediation of soil contaminated with oil hydrocarbons [79] in order to protect the environment from hazardous compounds, and for bio pulping and biobleaching processes [80, 81]. This wide spectrum of applications has raised the interest of researchers to investigate new sources of laccase enzymes.

In this scope, the endophytic fungus *Myrothecium verrucaria* was isolated from pigeon pea. Using response surface methodology to optimise several fermentation parameters, laccase activity reached 16.52 U/mL after five days, at 30 °C and pH 6.22. This enzymatic activity was relatively stable within pH (4.5–6.5) and in the temperature range of 35–55 °C. Additionally, the laccase showed effective decolourisation capability towards several synthetic dyes such as Congo red, Methyl orange, Methyl red and Crystal violet in the presence of the redox mediator ABTS (2,2′-azino-bis(3-ethylbenzothiazoline-6-sulfonic acid)), with more than 70% of dyes decolorising after 24 hours of incubation [37]. In another investigation, the *in vitro* culture of adventitious roots of *Ipomoea hederifolia* and its endophytic fungus *Cladosporium cladosporioides* successfully decolourised Navy Blue HE2R at a concentration of 20 ppm up to 83.3 and 65%, respectively, within 96 hours. Whereas the combination of *C. cladosporioides* and adventitious roots decolourised the dye more efficiently and gave 97% removal within 36 hours. Further analysis showed the significant induction of enzymes tyrosinase, laccase and riboflavin reductase in *C. cladosporioides*. The combination was tested in a rhizoreactor leading to 97% decolourisation after 36 hours [82]. These findings suggest that laccase-producing endophytes could provide a promising alternative for future industrial and environmental applications, including eco-friendly remediation purposes.

New laccase-producing endophytic fungi isolated from eucalyptus trees in Spain were assessed for their ability to enhance the saccharification of *Eucalyptus globulus* wood. Overall, endophytic *Ulocladium* sp. and *Hormonema* sp. produced greater enhancements in saccharification of *E. globulus* than *Trametes* sp. I-62 with an increase in sugar yields of 8.5, 8.0 and 6.0 fold, respectively. This study demonstrated the potential of these new laccase-producing fungi to enhance saccharification [83]. In another investigation, 127 endophytic fungi isolated from Eucalyptus trees were screened for laccases activity. From all the studied fungal strains, only five, namely *Neofusicoccum luteum*, *Neofusicoccum australe*, *Pringsheimiasmilacis*, *Hormonema* sp. and *Ulocladium* sp., were able to exhibit laccase activity, reaching maximum of 0.01–0.04 U/mL by the fifth day in all cases. At room temperature and pH 4, the laccase activity was 0.25, 0.15, 0.15 and 0.05 U/mL for *N. luteum*, *N. australe*, *Hormonema* sp. and *P. smilacis*, respectively. The maximum laccase activity was obtained at pH 2 for *Hormonema* sp., *N. australe* and *N. luteum*, while the enzymatic activity decreased significantly at pH values higher than 4 for *N. australe* and *Hormonema* sp. The investigation of the effect of temperature showed that the higher the temperature, the greater laccase activity was for the four strains. At 80 °C, *Hormonema* sp. reached maximum activity, while *P. smilacis* showed an abrupt decrease in laccase activity [38].

6.3.6 Lipases

Lipases (EC 3.1.1.3) are hydrolytic enzymes also called triacylglycerol ester hydrolases, catalysing the hydrolysis of triglycerides to glycerol and free fatty acids [84]. Besides the hydrolysis reaction, these enzymes are also able to catalyse esterification, interesterification, transesterification, acidolysis, alcoholysis and aminolysis reactions under proper conditions [85, 86]. They possess broad substrate specificity with optimal activity over a wide range of temperatures. This makes lipases a most versatile biocatalysts. In fact, interests for lipases are predicted to be the fastest-growing given their wide range of applications in various industrial sectors [7]. Microbial lipases that are mainly obtained from certain species of bacteria and filamentous fungi are the most sought after because of their outstanding characteristics including high stability in organic solvents, non-requirement of cofactors and their ability to act over an ample variety of substrates [84, 87, 88]. Their Industrial applications are various including detergent processing, flavour improvement by the synthesis of esters of short-chain fatty acids and alcohols, resolution of racemic mixtures and amino acid derivatives, and also the construction of biosensors that are used as diagnostic tools for the detection of various diseases [87]. Therefore, the identification of novel microorganisms capable of producing highly potent lipases with desirable physicochemical properties suitable for industrial applications is of interest. Endophytes have been investigated for such purpose and interesting findings were reported.

Within this framework, Torres et al. [89] screened endophytic fungi isolated from several Mediterranean plants and identified lipase from *Rhizopus oryzae* fungus with the ability to catalyse the esterification of fatty acids in isooctane. Further investigation of various factors such as water content, temperature and pH on the enzyme activity reveals that the catalytic activity inversely correlates with the water content. Moreover, the enzyme was active over a wide pH range (3–8) with maximal activity at pH 4 and 7. The enzyme was also thermostable, with maximal activity at 60 °C. In another investigation performed by Costa-Silva et al. [39] to evaluate the effect of various conditions on the production and stabilisation of a lipase enzyme produced by the endophytic fungus *Cercospora kikuchii*, it was noted that the maximum enzyme production (14.87 U/mL) was obtained after six days in a medium supplemented with 2% soybean oil. As well, after drying lipases with 10% (w/v) of each different adjuvants, including lactose, β-cyclodextrin, maltodextrin, mannitol, gum arabic and trehalose, the residual enzyme activity ranged from 63 to 100%. However, the absence of adjuvant led to the loss of the enzyme activity. After eight months of storage, the enzymatic activity was maintained to at least 50% at 5 °C and 40% at 25 °C with the lipase dried with 10% of β-cyclodextrin retaining 72% of its activity at 5 °C. Finally, the lipases were separated in a butyl-Sepharose column into four pools, among which one pool was partially purified (33.1%; 269.5 U/mg protein). This fraction was also spray-dried in 12% maltodextrin De10 that maintained its activity (27 U/mL) at 100%.

Sixty-five endophytic fungi were isolated from *Croton oblongifolius* Roxb. (Plao yai) and also screened for extracellular lipase activity. Ten out of the investigated fungi exhibited enzyme activity, with *Fusarium oxysporum* PTM7 isolated from the plant leaf being the most potent. Further investigation indicated that carbon, organic and inorganic nitrogen sources such as 1% (v/v) olive oil, 1% (w/v) peptone and 0.5% (w/v) sodium nitrate, respectively, could elicit increased lipase activity of *F. oxysporum*. The purified lipase had a molecular mass of 37.4 kDa and showed optimal activity at pH 8 and 30 °C, with reasonable stability at 40 °C over a wide range of pH (8.0–12). Further analysis showed that divalent cations such as Ca^{2+}, Mg^{2+} and Mn^{2+} could stimulate the enzyme, while Cu^{2+}, Fe^{2+}, Hg^{2+} and Zn^{2+} inhibited its activity. Further determination of activity parameters showed that the enzyme has a *Km* and *V*max values of 2.78 mM and 9.09 μmoL/min/mg protein, respectively, for the substrate *p*-nitrophenyl palmitate. However, the enzyme presented low transesterification activity [40].

The endophytic yeast *Candida guillermondi* was isolated from castor leaves (*Ricinus communis* L.) and used for the production of lipase under submerged fermentation in a medium containing soybean oil as the main nutrient. The obtained enzyme was partially purified and freeze-dried before immobilisation on agarose and silica gel supports [41]. In further investigations, the partially purified enzyme was characterised and showed to present high activity (26.8 U/mL) in the presence of 5 mmol/L NaCl at 30 °C and pH 6.5. The scale-up production of this enzyme in submerged fermentation in a 14-L bioreactor led to an activity of 18 U/mL. The enzyme was applied in the oleic acid esterification in different solvents (hexane, cyclohexane and cyclohexanone) and different acid: alcohol molar ratios. As results, higher ester conversion rate (81%) was achieved using hexane and methanol with a molar ratio of 1 : 9 [42]. These studies suggest the potential for the development of endophytic yeast to produce biocatalysts via the submerged fermentation using agro-industrial residues as culture media.

In a recent study, *Lasiodiplodia theobromae* VBE1, the coconut kernel-associated fungus, was grown on coconut cake with added coconut oil as lipase inducer under solid-state fermentation conditions. Two extracellular lipases were purified including lipase A (68 kDa) and lipase B (32 kDa). Both lipases were shown to exhibit optimal activity at pH 8.0 and 35 °C, and when activated by Ca^{2+}, they exhibited the highest specificity towards coconut oil and *p*-nitrophenyl palmitate, and were stable in isooctane and hexane. Ethanol supported higher lipase activity, while *n*-butanol inactivated both lipases. Additional studies showed that crude lipase immobilised by entrapment within 4% (w/v) calcium alginate beads was more stable than the crude-free lipase preparation within the pH and temperature range of 2.5–10.0 and 20–80 °C, respectively. The immobilised lipase preparation was used to catalyse the transesterification/methanolysis of coconut oil to biodiesel (fatty acyl methyl esters (FAMEs)). The principal FAMEs were laurate (46.1%), myristate (22.3%), palmitate (9.9%) and oleate (7.2%), with trace amounts of caprylate, caprate and stearate. The FAME profile was comparatively similar to NaOH-mediated transesterified biodiesel from coconut oil but distinctly different from petroleum-derived diesel [90]. The authors concluded that *Lasiodiplodia theobromae* VBE1-lipases have the potential for biodiesel production using coconut oil as a substrate. Previously, 212 endophytic fungi isolated from leaves and fruits of Amazonian plants were screened for lipase activity. The results showed that about 87% of the isolates could hydrolyse a tributyrin substrate, among which 30% were able to growth in lipase amended media. The evaluation of the esterification and transesterification ability of these isolates in organic solvents led to the selection of nine isolates that were further submitted to lipase activity in (*R*, *S*)-2-octanol resolution reaction. One isolate named endophyte UEA_115 was the most ingenuous biocatalyst, demonstrating a conversion ability of more than 90% in esterification reactions and enhanced aptitude for the resolution of (*R*, *S*)-2-octanol (ees 29%; eep 99%; c 22%; $E > 200$). This investigation revealed the potent nature of endophytic fungi as lipase suppliers in lipid biocatalysis [91].

6.3.7 Proteases

Proteases also called proteinases, or proteolytic enzymes, are a group of enzymes belonging to the hydrolases family, hydrolysing peptide bonds of proteins. They are subdivided into two major groups thatare, exo- and endopeptidases. Proteases are ubiquitous enzymes present in all living organisms and generate small peptides and amino acids that are essential for cell signalling, differentiation, cell growth and metabolism [7]. In addition to their physiological importance, proteases are also widely employed for commercial, industrial and health purposes [92, 93]. In fact, proteases are widely used in bioremediation processes and recycling of waste [94, 95], wastewater treatment [96], catalysis of peptide synthesis in organic solvents [97, 98], fortification of fruit juices and preparation of protein-enriched diets [99], leather processing [100], manufacturing of laundry detergents [101], meat tenderisation and cheese preparation [10], production of methionine-enriched protein [102], and treatment of diseases such as inflammation and cancer [103]. Industrially, microbial proteases have very high commercial value compared to animal and plant proteases. Therefore, there is a growing interest in investigating microorganisms as a source for novel proteases for industrial needs. Few studies have reported the ability of endophytes to produce proteolytic enzymes.

Among those, the screening of endophytic fungi from *Dendronephthya hemprichi* plant led to the identification of fungal strain *Penicillium* sp. Morsy1, exhibiting the highest keratinase activity (1600 U/g) when grown on different agricultural and poultry wastes using solid-state fermentation technology. The maximum production of keratinase was observed when the fungus was grown for five days on rice straw as the carbon source, at 26 °C, pH 6 and moisture content of 80%. Two types of keratinases named Ahm1 and Ahm2 with a respective molecular weight of 19 and 40 kDa were purified and characterised by precipitation with ammonium sulphate, DEAE-Sepharose and gel exclusion chromatography in a sequential manner. The optimal condition for the hydrolysis of azokeratin was pH range 7.0–8.0, at 50 °C, stable in the pH range of 6.0–8.0 for Ahm1 and pH range 10.0–11.0, at 60–65 °C, stable in the pH range of 6.0–11.0 for Ahm2. Both enzymes were inhibited by chelating agents such as EGTA and ethylenediaminetetraacetic acid (EDTA) and activated by Ca^{2+}, Mg^{2+} and Mn^{2+} [43]. The endophytic fungus *Fusarium* sp. CPCC 480097 isolated from *chrysanthemum* stems produces a very potent fibrinolytic enzyme, characterised by a molecular weight of 28 kDa, an isoelectric point of 8.1 and maximal activity at 45 °C and pH 8.5 [104]. In another investigation, a novel fibrinolytic enzyme from the endophytic bacterium *Pa. polymyxa* EJS-3 was purified with ammonium sulphate precipitation, hydrophobic chromatography, ion-exchange and gel filtration chromatography. This new enzyme has a molecular weight of 63.3 kDa with the optimum temperature and pH value of 37 °C and 7.5, respectively. Additional investigations revealed that the enzyme rapidly hydrolysed the α-chain of fibrinogen, followed by the β-chains, and also hydrolysed the γ-chains, but more slowly. The enzyme was also activated by metal ions such as Zn^{2+}, Mg^{2+} and Fe^{2+} and was inhibited by Ca^{2+} and Cu^{2+}. Further studies showed that the enzyme exhibited a higher specificity for the synthetic substrate *N*-succinyl-Ala-Ala-Pro-Phe-pNA for chymotrypsin, indicating that the enzyme is a chymotrypsin-like serine protease and displayed significant anticoagulant activity *in vitro* [105]. These proteases may have potential applications in thrombolytic therapy and in thrombosis prevention. In fact, although *Bacillus* from traditional fermented foods has been so far the most important source of fibrinolytic enzymes discovered from microorganisms [106], outcomes from the above two investigations demonstrate the potential of endophytic fungi and bacteria as fibrinolytic enzymes-producing organisms.

An endophytic *Acremonium* sp. isolated from leaves of *Saraca asoca* exhibited protease activity when grown on skim milk medium. For the highest protease production, fructose and ammonium sulphate were the preferred carbon and nitrogen sources. The evaluation of the enzymatic activity of the partially purified enzyme showed that the optimum enzyme activity of 3.34 U/mL was recorded at pH 7.0 with the protein content of 20 μg/mL, and the specific activity being 0.167 U/μg [44]. Zaferanloo et al. [107] investigated the protease-producing potential of three fungal endophytes *Alternaria alternata*, *Phoma herbarum* and an unidentified fungus, isolated from the Australian native plant, *E. longifolia*. From the results, *A. alternata* showed the greatest protease activity in a wide range of pH (3–9), while the broadest activity between 9 and 50 °C was observed at pH 7. Overall, the optimum conditions were 37 °C and pH 7 with a maximum specific activity value of 69.86 U/mg. Mayerhofer et al. [108] also reported the enzymatic ability of roots endophytes *Umbelopsis isabellina* and *Meliniomyces variabilis* from Birch tree (*Betula* spp.) and *Phialocephala fortinii* from *Pseudotsuga menziesii* plant. *U. isabellina* was shown to efficiently utilised protein, while *M. variabilis* and *Ph. fortinii* exhibited intermediate levels of protein utilisation, demonstrating their ability to produce proteolytic enzymes. In a more recent investigation, *Bacillus halotolerans* isolated from Tunisian potatoes was identified as a protease-producing strain. The proteolytic enzyme was purified by ammonium sulphate precipitation, sephacryl S-200 gel filtration and SP-Sepharose cation-exchange chromatography and demonstrated an optimal activity at pH 9 and temperature of 50 °C. The proteolytic activity was enhanced by Ca^{2+} and Mn^{2+} ions and exhibited high stability in the presence of commercial detergents including liquid (Auchan, Ariel, Nadhir and Det) and powder (Ariel, Dixan and Omo) formulations. This enzyme showed broad substrate specificity towards both synthetic and natural substrates, with a high capacity to hydrolyse legume seed proteins. The electrophoretic analysis reveals a molecular weight of around 250 kDa with two major subunits, showing important homologies with serine proteases belonging to the subtilisin-like serine proteases. The *K*m and *V*max values were 10 mg/mL and 50,000 U/mg, respectively. The *K*cat and *K*cat/*K*m were also 5263.15×10^3/min and 526.31×10^3/min/(mg/mL), respectively [109]. The authors concluded that this new enzyme displayed relevant properties suitable for various industrial and biotechnological applications.

In an investigation designed to identify microorganisms with ability to efficiently degrade polyester polyurethane in both solid and liquid suspensions, Russell et al. [110] screened several endophytic fungi and identified two *Pestalotiopsis microspora* isolates capable of growing on polyester polyurethane as the sole carbon source under both aerobic and anaerobic conditions. From their investigation, the authors concluded that since the main enzyme responsible for the degradation of polyester polyurethane is a serine protease, these endophytes can be used to produce these enzymes at the industrial level for bioremediation purposes.

6.3.8 Xylanases

Xylanases (EC 3.2.1.8), also named as endo-β-1,4-xylan-xylanohydrolase, belong to the glycoside hydrolases family, catalysing the hydrolysis of xylan into xylooligosaccharides, xylobiose and D-xylose. In general, three major enzymes, including endoxylanases, exoxylanases and β-xylosidases, act synergistically to breakdown the xylan backbone into hemicellulose. In fact, endoxylanases (EC 3.2.1.8) cleaves the β-1,4 bonds of xylan backbone, while exoxylanases (EC 3.2.1.37) hydrolyse the β-1,4 bonds of xylan from the non-reducing ends to release xylooligosaccharides and β-xylosidases release xylose from the cleavage of xylobiose and xylooligosaccharides [111]. In nature, xylanases are produced by microorganisms to cleave xylans, a major constituent of hemicellulose. Xylanases have received a great deal of attention because of their biotechnological potential in various industrial sectors. Xylanases are used to improve the quality of bread. In fact, some evidence showed that the use of xylanases decreases the water absorption and thus reduces the amount of added water needed in baking leading to a more stable dough [3]. Xylanases are applied in bleaching of kraft pulps, which liberate lignin fragments by hydrolysing residual xylan and improve the pulp fibrillation, and restoration of bonding, which increases the freeness in recycled fibres and improves the bleaching of wood pulp. Overall, this considerably reduces the need for chlorine-based bleaching chemicals [112]. Xylanases are also used to treat hemicelluloses waste and clarify fruit and vegetable juices, for the pre-treatment of forage crops to improve the digestibility of ruminant feed, for the production of ethanol and xylitol, and for the degumming of bast fibres such as flax, hemp, jute and ramie [113]. One of the industrial producing organisms is *Trichoderma* sp. from which xylanase has already been purified, crystallised and improved by design-oriented mutagenesis [114]. To supply the industry with novel and more potent xylanase-producing strains, microorganisms from diverse ecological niches including endophytes have been investigated.

In this respect, 54 endophytic fungi isolated from leaves of *C. oblongifolius* were screened for xylanase activity. Thirty isolates were positive at the primary and secondary screenings. The most potent isolate, named PTRa9, was further investigated for purification and characterisation of the enzyme produced. The optimal conditions for xylanase production were 96 hours of incubation in a medium composed of 2% (w/v) rice bran and 0.1% (w/v) ammonium sulfate ($(NH_4)_2SO_4$), as sources of carbon and nitrogen, respectively. The enzyme had a molecular weight of 54.8 kDa and specific activity of 161.1 U/mg protein. It was active at a broad range of temperature (−20 to 45 °C) and pH (3–11) with the optimum being 45 °C and 5.0, respectively. The enzyme exhibited a *K*m of 0.421 μg/mL and *V*max of 0.826 U/mg protein and was sensitive to Cu^{2+}, Hg^{2+} and EDTA [115]. In another investigation, a thermotolerant xylanase produced by *Aspergillus terreus* isolated from *Memora peregrina* was purified and characterised. When grown on wheat bran as a carbon source at 30 °C for 48 hours, the production yield was higher. The xylanase with a molecular weight of 23 kDa exhibited optimal enzyme activity at 55 C and pH 4.5. The enzyme was tolerant to a temperature of 45 and 50 °C, with a half-life of 55 and 36 minutes, respectively, and was activated by ions K^+ and Mn^{2+}. Enzyme kinetic studies revealed the *K*m and *V*max values of 22 and 625 μg/mL, respectively [45].

6.3.9 Other Enzymes Produced by Endophytes

6.3.9.1 AHL-Lactonase

Biofilm formation is one of the strategies used by pathogenic bacteria and fungi to resist antibiotic treatment. Several approaches have been investigated to fight against pathogens biofilms, including the use of enzymes. In their investigation of the antibiofilm potential of endophytes, Rajesh and Rai [116] reported that the ability of the endophytic bacterium *Enterobacter aerogenes* VT66 to quench the *N*-acyl homoserine lactone (AHL) molecules produced by *Pseudomonas aeruginosa* PAO1 was due to the production of an AHL-lactonase enzyme. The purified AHL-lactonase with a molecular weight of about 30 kDa was able to inhibit *in vitro* more than 70% of the biofilm formed by *P. aeruginosa* PAO1. These results are encouraging and suggest that the AHL-lactonase enzyme produced by *E. aerogenes* VT66 could be further developed as an antibiofilm agent for biomedical applications.

6.3.9.2 Agarase

Agar is a polysaccharide extracted from the cell walls of some macro-algae. Most of the reported agarases come from the marine environment. There is a need to investigate new sources of this enzyme. In this line, Song et al. [117] reported the identification of seven agar-degrading endophytic bacteria belonging to three genera, *viz. Paenibacillus, Pseudomonas* and *Klebsiella*. Further studies showed that the seven endophytes could grow on several polysaccharides such as araban, carrageenan, chitin, starch and xylan and could also produce agarase in the presence of different polysaccharides other than agar. Further analysis of the extracellular agarase produced revealed a molecular weight of 75 kDa, which greatly differentiated from previously reported agarases.

6.3.9.3 Chromate Reductase

The phytoremediation industry is in continuous expansion. Endophytic bacteria from *Albizzia lebbeck*, a chromium hyperaccumulator plant, were investigated for their chromium detoxification ability. Out of four different groups of endophytic bacteria investigated, comprising *Pseudomonas, Rhizobium, Bacillus* and *Salinicoccus*, three *Bacillus* spp. exhibited not only remarkable chromium accumulation ability but also high chromium reductase activity. The most potent isolate *Bacillus* sp. DGV 019 was selected to purify the enzyme chromate reductase to homogeneity. The enzyme had a molecular weight of 34.2 kDa and was stable at temperatures and pH ranging from 20 to 60 °C and 4.0 to 9.0, respectively. Further analysis showed that the enzyme activity was enhanced with the electron donors NADH, followed by NADPH, and not affected by glutathione and ascorbic acid. In addition, ion Cu^{2+} enhanced its activity, while Zn^{2+} and EDTA inhibited it. This chromate reductase enzyme showed versatile adaptability that can be used for chromium remediation[118].

6.3.9.4 β-Mannanase

The bacterial endophyte *Paenibacillus* sp. CH-3 isolated from *Robinia pseudoacacia* was identified as a highly potent β-mannanase producer. The β-mannanase gene (manB) from this bacterium was cloned, expressed, purified and characterised to reveal that the gene encoding 327 amino acid residues, which belongs to family 5 of glycosyl hydrolase. The enzyme with a molecular weight of 50.4 kDa had an optimal activity level at pH 7.0 and 45 °C and was stable in the pH range of 4–9. Further study revealed that the optimal temperature, isopropyl β-D-thiogalactoside concentration and induction time were 28 °C, 0.05 mM and 12 hours, respectively, with the highest activity reaching 1054.17 U/mL. In addition, the enzyme has *Km* and catalytic efficiency (*k*cat/*Km*) values, respectively, of 7.30 and 176.31 mL/(mg/s) for locust bean gum, 8.69 and 116.69 mL/(mg/s) for guar gum, and 27.17 mg/mL and 115.49 mL/(mg/s) for konjac glucomannan. The hydrolysis of glucomannans by this enzyme produced mannobiose, mannotetrose and a higher manno-oligosaccharide [119]. The authors suggested that this enzyme could be useful as an additive in the animal feed industry.

6.4 Overview of the Methods Used to Investigate Endophytes as Sources of Enzymes

The investigation of endophytes for the production of novel products including but not limited to bioactive secondary metabolites and enzymes starts with the collection of plant material based on relevant criteria, followed by the isolation and purification of endophytic fungi or bacteria on agar plates. To achieve microbial enzyme isolation, fungal or bacterial isolates are subsequently cultured on agar medium supplemented with a specific discriminatory substrate as a function of the enzyme of interest. For instance, starch is used as substrate for amylases, carboxymethylcellulose is used for cellulases, casein or gelatin for proteases, etc. Potent microbial isolates identified based on their ability to hydrolyse the specific substrate and form a measurable inhibition halo are thereafter selected for submerged or solid-state fermentation[15]. At this stage, several agro-industrial residues, such as raw potato, cassava bagasse, sugarcane bagasse, sugar beet pulp/husk, orange bagasse, oil cakes, apple pomace, grape juice, grape seed, coffee husk, wheat bran, coir pith, etc., are often used as adequate substrates for fermentation. The successful fermentation is directly followed by extraction, concentration and purification of enzyme. Methods reported to be adequately used for enzyme purification in several investigations include precipitation, centrifugation, ion-exchange chromatography, affinity chromatography, hydrophobic chromatography, electrophoresis, and ultrafiltration and gel filtration [3]. The proof of concept of the purified enzymes activity is subsequently achieved using a broad range of substrates and diverse experimental conditions, including varied pH, temperature, divalent cations, different adjuvants (e.g. lactose, β-cyclodextrin, maltodextrin, mannitol, gum arabic and trehalose), etc., to identify the optimal conditions for enzyme activity. Parameters for enzymatic activity such as Km, Vmax and catalytic efficiency ($k$cat/Km) constant are thereafter determined.

6.5 Strategies Applied to Improve the Production of Enzymes by Endophytes

It is obvious from the data presented in this chapter that endophytic microorganisms have the ability to produce a wide range of industrially relevant enzymes. Despite their outstanding potential as candidates for enzyme production, it is noted that the amount of enzyme produced is generally low, not allowing comprehensive enzyme profiling. Moreover, some endophytes often lose the vigour of enzyme production in axenic cultures under laboratory conditions as exemplified by the investigation of Krishnapura and Belur [54]. This could be due to the absence of native environmental conditions found in the host plants where the production of metabolites is mediated by multiple signals from the surrounding environment [120]. To overcome these limitations, several methodologies have been developed to improve the production of enzymes by endophytes.

One of such approaches that are currently used with some extent of success is the variation of culture conditions, including but not limited to composition of culture media (carbon and nitrogen sources), pH and temperature of incubation, aeration rate, agitation or not [121]. For instance, Marlida et al. [30, 50] used this strategy to improve the production yield of amylase by *Acremonium* sp. by 19- to 22-fold and also reported an optimal amylase production with raw potato starch as carbon source. Similarly, Sunitha et al. [31] optimised the production of amylase by *Cylindrocephalum* sp. by varying temperature of incubation, pH of growth medium and carbon and nitrogen sources. Although time-consuming, this approach is easily applicable and has the potential to improve the yield of enzymes production by endophytes when the growth conditions are successfully identified.

In addition to varying growth conditions, the supplementation of the culture medium with inducers such as metals ions, small organic chemicals or organic solvents [122] has also been proposed as a vital strategy for improving the yield of endophytic enzyme production. In fact, Fillat et al. [38] have demonstrated in their study that the production yield of laccases by endophytes *N. luteum*, *N. australe* and *Hormonema* sp. significantly improved with the addition of 1% ethanol (v/v) and 0.30 mM $CuSO_4$ in the growth medium. Several other authors have reported similar trends with microorganism growth media supplementation ethanol [123–125]. However, the addition of inducers has not shown to elicit the same yield trend in every endophytic strain tested. In fact, the addition of ethanol in the growth medium of *P. smilacis* did not improve the production of laccase but rather reduced the yield when the incubation time was prolonged [38]. This observation reveals that the influence of inducers on enzyme production is highly dependent on the microorganism under study. This further strongly denotes the absolute necessity to carefully plan screening to identify a proper inducer for each microorganism that is being investigated [122].

Finally, besides culture-dependent approaches used to improve the production of enzymes by endophytes, genetic manipulation of producing strains has also been applied with great success. This approach has been used to improve the production of cellulases by the endophytic fungi *Fusiccoccum* sp. [69] and *Periconia* sp. BCC2871 [70]. Moreover, El-Bondkly [126] used a shuffling of the genome to improve xylanase production by endophyte *Aspergillus* sp. NRCF5. Genetic variability was induced on this isolate using different combinations and doses of mutagens. Ultraviolet irradiation (five minutes) and *N*-methyl *N*-nitro-*N*-nitrosoguanidine (NTG, 100 mg/mL) for 30 (UNA) and 60 (UNB) minutes along with NTG (100 mg/mL) and ethidium bromide (250 mg/mL) for 30 (NEA) and 60 (NEB) minutes were used in combination as mutagens which led to the increase in xylanase activity of mutants strains. Four rounds of genome shuffling led to the rise of seven high xylanase producing fungal strains. The recombinant xylanase obtained from the most potent strain showed 6.13 higher folds of xylanase activity when compared with starting strain NCRF5 with 427.5 U/mL xylanase [126]. Mutation technology has been used in industry to generate hyper-producer strains including *Trichoderma reesei* RUT C-30, one of the best cellulase producers for decades [3]. Moreover, recombinant enzymes from various microorganisms are used today for the production of a wide range of products at various industrial scale [7]. Overall, different actions could be performed depending on available facilities to efficiently improve the production of enzymes by endophytic microbes at the laboratory or industrial levels. These methods can be fermentation media optimisation using multiple micronutrients or genetic modification of strains or the combination of both methods.

6.6 Conclusion

Microorganisms represent a reservoir of important enzymes with multiple applications in various industries. Their multiple characteristics including the easy multiplication under controlled conditions, short generation times, easy genetic manipulation and high production yield make microorganisms the first choice for industrial production of enzymes. Discovering novel of such sources of enzymes is of high interest for various industrial sectors. In this chapter, an attempt was made to share an update regarding the investigation of endophytic microbes as a potent source of industrial enzymes. An emphasis was made on the producing strains, the characteristics of enzymes reported so far (Table 6.2) and the possible industrial application. This insightful look into literature revealed that a huge number of endophytic microbes are still to be examined for their enzymatic ability. In fact, only limited number of bacteria and actinomycetes endophytes have already been reported as enzymes producers. This leaves the room for a wide area of investigation and highlighting the need to expand such investigations to increase the possibility for new discoveries. It also came out from this survey that the strategies such as variation of growth conditions, supplementation of growth media with inducers and genetic manipulation have been successfully applied to improve the production of enzymes by endophytes. Indeed, these strategies are already very well established as necessary tools to activate silent genes clusters in endophytic microbes and induce the production of structurally diverse secondary metabolites exhibiting various biological activity. Therefore, investigations in this direction are urgently needed to fully harness the potential of endophytic microbes to produce useful

Table 6.2 Selected endophyte enzymes and their characteristics.

Enzymes	Endophytes	MW (kDa)	Enzyme activity	Optimal pH	Optimal T (°C)	*Km*	*Vmax*	Other characteristics	References
Amylases	*Fusicoccum* sp. BCC4124	52.58	838.9 U/mg	7	70	5.2 mg/mL	640.8 U/mg protein	Resistance to inhibition by glucose of up to 1 mol/L and activated by Ca^{2+}	[51]
	P. minima	70	138 U/mg	9	25	ND	ND	Activated by Mn^{2+} and Ca^{2+}	[32]
Cellulases	*Bacillus pumilus*	71.3	8.25 U/mg protein	5–8	60	ND	ND	Retain 85% of activity after a 24 h at pH range of 6–8.6	[67]
	Fusiccoccum sp. BCC4124	50	4.37 U/mg protein	5.0	40	0.57 mM	3.086 nmol/min/mg	Stable at pH (3–11) and retained 50% of its activity at 70–90 °C for 30 min	[69]
	Penicillium sp. 51	55	0.44 U/mg	5–6	40–50	ND	ND	None	[71]
Laccases	*M. verrucaria*	ND	16.52 U/mL	6.22	30	ND	ND	Stable at pH (4.5–6.5) and 35–55 °C	[37]
Lipases	*F. oxysporum* PTM7	37.4	156.3 U/mg protein	8	30	2.78 mM	9.09 μmoL/min/mg	Activated by Ca^{2+}, Mg^{2+}, and Mn^{2+}	[40]
	L. theobromae VBE1	68	1981 U/mg protein	8.0	35	ND	ND	Activated by Ca^{2+}, Mg^{2+} and Mn^{2+}	[90]
		32	1440 U/mg protein	8.0	35	ND	ND		
Proteases	*Penicillium* sp. Morsy1	19	8.06 U/mg	7–8	50	ND	ND	Activated by Mn^{2+}, Ca^{2+} and Mg^{2+} and inhibited by EGTA and EDTA	[43]
		40	39.593 U/mg	10–11	60–65	ND	ND		
	Fusarium sp. CPCC 480097	28	76,111 U/mg	8.5	45	ND	ND	Isoelectric point of 8.1; activated by Ca^{2+} and inhibited by Co^{2+}, Cu^{2+}, and Zn^{2+}	[104]
	Acremonium sp.	ND	3.4 U/mL	7.0	37	ND	ND	Specific activity: 0.167 U/μg	[44]

	B. halotolerans	250	773.4 U/mg	9	50	10 mg/mL	50,000 U/mg	Enhanced by Ca^{2+}, Mn^{2+}; 10% ethanol and hexane	[109]
Xylanases	Strain PTRa9	54.8	161.1 U/mg	5.0	45	0.421 μg/mL	0.826 U/mg	Sensitive to Cu^{2+}, Hg^{2+}, and EDTA	[115]
	A. terreus	23	6.92 U/mg	4.5	55	22 mg/mL	625 U/mg/min	Activated by K^{+} and Mn^{2+}	[45]
Chromate reductase	*Bacillus* sp.	34.2	ND	4–9	20–60	ND	ND	Enhanced by NADH, NADPH and Cu^{2+}	[118]
β-Mannanase	*Paenibacillus* sp. CH-3	50.4	1054.17 U/mL	7.0	45	7.30–27.17 mg/mL	6.09–18.83 μmol/min	Activated by Ca^{2+}, Ba^{2+}, Mg^{2+}, Mn^{2+} and Co^{2+} and inhibited by Cu^{2+}, Zn^{2+} and EDTA	[119]

ND, not determined; T, temperature; MW, molecular weight.

enzymes. Therefore, available technologies such as metagenomics, genomics, proteomics, efficient expression systems and recombinant DNA technology should be used as tools to improve endophyte enzymes already identified and find new enzyme-producing strains. Moreover, attempts should be made to further the investigation on potent endophyte strains identified so far to make them suitable for the industrial production of enzymes. This can only be achieved by the combination of research efforts in a strong and open partnership between academics and industries.

Acknowledgements

The authors are grateful to Íñigo Zabalgogeazcoa, Senior scientist at the Instituto de Recursos Naturales y Agrobiología de Salamanca (IRNASA), Consejo Superior de Investigaciones Científicas (CSIC), Salamanca, Spain, for his critical review of the manuscript.

References

1 Underkofler, L.A., Barton, R.R., and Rennert, S.S. (1957). Production of microbial enzymes and their applications. Microbiological Process Report. *Appl. Microbiol.* 6: 212–221.
2 Nigam, P. (2013). Microbial enzymes with special characteristics for biotechnological applications. *Biomolecules* 3 (3): 597–611.
3 Patel, A.K., Singhania, R.R., and Pandey, A. (2017). Production, purification, and application of microbial enzymes. In: *Biotechnology of Microbial Enzymes* (ed. G. Brahmachari), 13–41. Academic Press http://dx.doi.org/10.1016/B978-0-12-803725-6.00002-9.
4 Cornish-Bowden, A. (2011). History of enzyme chemistry. *eLS* https://doi.org/10.1002/9780470015902.a0003466.
5 Yamaguchi, S. (2017). The quest for industrial enzymes from microorganisms. *Biosci. Biotechnol. Biochem.* 81: 54–58. https://doi.org/10.1080/09168451.2016.1248371.
6 Peralta, R.M., da Silva, P.B., Côrrea, G.C.R. et al. (2017). Enzymes from basidiomycetes – peculiar and efficient tools for biotechnology. In: *Biotechnology of Microbial Enzymes*, 119–149. Academic Press http://dx.doi.org/10.1016/B978-0-12-803725-6.00005-4.
7 Singh, R.S., Singh, T., and Pandey, A. (2019). Microbial enzymes – an overview. In: *Biomass, Biofuels, Biochemicals: Advances in Enzyme Technology*, 1–40. Elsevier https://doi.org/10.1016/B978-0-444-64114-4.00001-7.
8 Demain, A.L. and Adrio, J.L. (2008). Contributions of microorganisms to industrial biology. *Mol. Biotechnol.* 38: 41–55.
9 Anbu, P., Gopinath, S.C., Chaulagain, B.P. et al. (2015). Microbial enzymes and their applications in industries and medicine 2014. *Biomed. Res. Int.* 2015: 816419.
10 Adrio, J.L. and Demain, A.L. (2014). Microbial enzymes: tools for biotechnological processes. *Biomolecules* 4: 117–139.
11 Singh, R., Kumar, M., Mittal, A., and Mehta, K.P. (2016). Microbial enzymes: industrial progress in 21st century. *3 Biotech* 6: 174. https://doi.org/10.1007/s13205-016-0485-8.
12 Strobel, G. (2018). The emergence of endophytic microbes and their biological promise. *J. Fungi.* 4: 57. https://doi.org/10.3390/jof4020057.

13 Toghueo, R.M.K. and Boyom, F.F. (2020). Endophytic *Penicillium* species and their agricultural, biotechnological and pharmaceutical applications. *3 Biotech* 10: 107.
14 Corrêa, R.C.G., Rhoden, S.A., Mota, T.R. et al. (2014). Endophytic fungi: expanding the arsenal of industrial enzyme producers. *J. Ind. Microbiol. Biotechnol.* 41: 1467–1478.
15 Sunitha, V.H., Devi, N.D., and Srinivas, C. (2013). Extracellular enzymatic activity of endophytic fungal strains isolated from medicinal plants. *World J. Agric. Res.* 9: 1–9.
16 Toghueo, R.M.K., Zabalgogeazcoa, I., Vázquez de Aldana, B.R., and Boyom, F.F. (2017). Enzymatic activity of endophytic fungi from the medicinal plants *Terminalia catappa*, *Terminalia mantaly* and *Cananga odorata*. *S. Afr. J. Bot.* 109: 146–153.
17 Compant, S., Clement, C., and Sessitsch, A. (2010). Plant growth-promoting bacteria in the rhizo- and endosphere of plants: their role, colonization, mechanisms involved and prospects for utilization. *Soil Biol. Biochem.* 42: 669–678.
18 Cord-Landwehr, S., Melcher, R.L.J., Kolkenbrock, S., and Moerschbacher, B.M. (2016). A chitin deacetylase from the endophytic fungus *Pestalotiopsis sp.* efficiently inactivates the elicitor activity of chitin oligomers in rice cells. *Sci. Rep.* 6: 38018. https://doi.org/10.1038/srep38018.
19 Khare, E., Mishra, J., and Arora, N.K. (2018). Multifaceted interactions between endophytes and plant: developments and prospects. *Front. Microbiol.* 9: 2732.
20 Zeidler, D., Zahringer, U., Gerber, I. et al. (2004). Innate immunity in *Arabidopsis thaliana*: lipopolysaccharides activate nitric oxide synthase (NOS) and induce defense genes. *Proc. Natl. Acad. Sci. U. S. A.* 101: 15811–15816.
21 Jia, M., Chen, L., Xin, H.-L. et al. (2016). A friendly relationship between endophytic fungi and medicinal plants: a systematic review. *Front. Microbiol.* 7: 906. https://doi.org/10.3389/fmicb.2016.00906.
22 Hallmann, J., Qualt-Hallmann, A., Mahaffee, W.F., and Kloepper, J.W. (1997). Bacterial endophytes in agricultural crops. *Can. J. Microbiol.* 43: 895–914.
23 Hardoim, P.R., van Overbeek, L.S., Berg, G. et al. (2015). The hidden world within plants: ecological and evolutionary considerations for defining functioning of microbial endophytes. *Microbiol. Mol. Biol. Rev.* 79: 293–320. https://doi.org/10.1128/MMBR.00050-14.
24 Potshangbam, M., Devi, S.I., Sahoo, D., and Strobel, G.A. (2017). Functional characterization of endophytic fungal community associated with *Oryza sativa* L. and *Zea mays L. Front. Microbiol.* 8: 325. https://doi.org/10.3389/fmicb.2017.00325.
25 Tanaka, A., Tapper, B.A., Popay, A. et al. (2005). A symbiosis expressed non-ribosomal peptide synthetase from a mutualistic fungal endophyte of perennial ryegrass confers protection to the symbiotum from insect herbivory. *Mol. Microbiol.* 57: 1036–1050. https://doi.org/10.1111/j.1365-2958.2005.04747.x.
26 Lugtenberg, B.J.J., Caradus, J.R., and Johnson, L.J. (2016). Fungal endophytes for sustainable crop production. *FEMS Microbiol. Ecol.* 92: fiw194. https://doi.org/10.1093/femsec/fiw194.
27 Chen, X., Guo, S., and Wang, C. (2005). Effects of four endophytic fungi on the growth and polysaccharide content of *Anoectochilus roxburghii* (Wall.) Lindl. *Chin. Pharm. J.* 40: 13–16.
28 Sherameti, I., Shahollari, B., Venus, Y. et al. (2005). The endophytic fungus *Piriformospora indica* stimulates the expression of nitrate reductase and the starch-degrading enzyme glucanwater dikinase in tobacco and *Arabidopsis* roots through a homeodomain

transcription factor that binds to a conserved motif in their promoters. *J. Biol. Chem.* 280: 26241–26247. https://doi.org/10.1074/jbc.M500447200.

29 Wingender, J., Neu, T.R., and Flemming, H.C. (2012). *Microbial Extracellular Polymeric Substances: Characterization, Structure and Function*. Springer Science & Business Media.

30 Marlida, Y., Saari, N., Radu, S., and Bakar, F.A. (2000c). Production of an amylase-degrading raw starch by *Gibberella pulicaris*. *Biotechnol. Lett.* 22: 95–97.

31 Sunitha, V.H., Ramesha, A., Savitha, J., and Srinivas, C. (2012). Amylase production by endophytic fungi *Cylindrocephalum* sp. isolated from medicinal plant *Alpinia calcarata* (Haw.) Roscoe. *Braz. J. Microbiol.* 43: 1213–1221.

32 Zaferanloo, B., Bhattacharjee, S., Ghorbani, M.M. et al. (2014). Amylase production by *Preussia minima*, a fungus of endophytic origin: optimization of fermentation conditions and analysis of fungal secretome by LC-MS. *BMC Microbiol.* 14: 55.

33 Silva, L.F., Freire, K.T.L.S., Araújo-Magalhães, G.R. et al. (2018). *Penicillium* and *Talaromyces* endophytes from *Tillandsia catimbauensis*, a bromeliad endemic in the Brazilian tropical dry forest, and their potential for L-asparaginase production. *World J. Microbiol. Biotechnol.* 34 (11): 162. https://doi.org/10.1007/s11274-018-2547-z.

34 Hatamzadeh, S., Rahnama, K., Nasrollahnejad, S. et al. (2020). Isolation and identification of L-asparaginase-producing endophytic fungi from the Asteraceae family plant species of Iran. *PeerJ* 8: e8309. https://doi.org/10.7717/peerj.8309.

35 Cabezas, L., Calderon, C., Medina, L.M. et al. (2012). Characterization of cellulases of fungal endophytes isolated from *Espeletia* spp. *J. Microbiol.* 50 (6): 1009–1013. https://doi.org/10.1007/s12275-012-2130-5.

36 El-Bondkly, A.M. and El-Gendy, M.M. (2012). Cellulase production from agricultural residues by recombinant fusant strain of a fungal endophyte of the marine sponge *Latrunculia corticata* for production of ethanol. *Antonie Van Leeuwenhoek* 101 (2): 331–346.

37 Sun, J., Guo, N., Niu, L.L. et al. (2017). Production of laccase by a new *Myrothecium verrucaria* MD-R-16 isolated from pigeon pea [*Cajanus cajan* (L.) Millsp.] and its application on dye decolorization. *Molecules* 22 (4): E673.

38 Fillat, Ú., Martín-Sampedro, R., MacayaSanz, D. et al. Screening of eucalyptus wood endophytes for laccase activity. *Process Biochem.* 51: 589–598. http://dx.doi.org/10.1016/j.procbio.2016.02.006.

39 Costa-Silva, T.A., Nogueira, M.A., Souza, C.R.F. et al. (2011). Lipase production by endophytic fungus *Cercospora kikuchii*: stability of enzymatic activity after spray drying in the presence of carbohydrates. *Dry. Technol.* 29 (9): 1112–1119.

40 Panuthai, T., Sihanonth, P., Peapukiew, J. et al. (2012). An extracellular lipase from the endophytic fungi *Fusarium oxysporum* isolated from the Thai medicinal plant, *Croton oblongifolius* Roxb. *Afr. J. Microbiol. Res.* 6: 2622–2638.

41 Oliveira, A.C.D., Farion Watanabe, F.M., Vargas, J.V.C. et al. (2012). Production of methyl oleate with a lipase from an endophytic yeast isolated from castor leaves. *Biocatal. Agric. Biotechnol.* 1: 295–300.

42 Oliveira, A.C., Fernandes, M.L., and Mariano, A.B. (2015). Production and characterization of an extracellular lipase from *Candida guilliermondii*. *Braz. J. Microbiol.* 45 (4): 1503–1511.

43 El-Gendy, M.M.A. (2010). Keratinase production by endophytic *Penicillium* spp. Morsy1 under solid-state fermentation using rice straw. *Appl. Biochem. Biotechnol.* 162: 780–794.

44 Jain, P., Aggarwal, V., Sharma, A., and Pundir, R.K. (2012). Isolation, production and partial purification of protease from an endophytic *Acremonium* sp. *J. Agric. Technol.* 8: 1979–1989.

45 Sorgatto, M., Guimarães, N.C.A., Zanoelo, F.F. et al. (2012). Purification and characterization of an extracellular xylanase produced by the endophytic fungus, *Aspergillus terreus*, grown in submerged fermentation. *Afr. J. Biotechnol.* 11: 8076–8084.

46 Pandey, A., Nigam, P., Soccol, C.R. et al. (2000). Advances in microbial amylases. *Biotechnol. Appl. Biochem.* 31: 135–152.

47 Sundarram, A. and Murthy, T.P.K. (2014). α-Amylase production and applications: a review. *J. Appl. Environ. Microbiol.* 2: 166–175.

48 Roy, J. and Mukherjee, A.K. (2013). Applications of a high maltose alkalophilic *Bacillus licheniformes* strain AS08e in food and laundry detergent industries. *Biochem. Eng. J.* 77: 220–230.

49 Marlida, Y., Saari, N., Hassan, Z., and Radu, S. (2000a). Raw starch degrading enzyme from newly isolated strains of endophytic fungi. *World J. Microbiol. Biotechnol.* 16: 573–578.

50 Marlida, Y., Saari, N., Hassan, Z., and Radu, S. (2000b). Improvement in raw sago starch degrading enzyme production from *Acremonium* sp. endophytic fungus using carbon and nitrogen sources. *Enzyme Microb. Technol.* 27: 511–515.

51 Champreda, V., Kanokratana, P., Sripang, R. et al. (2007). Purification, biochemical characterization, and gene cloning of a new extracellular thermotolerant and glucose tolerant maltooligosaccharide-forming amylase from an endophytic ascomycete *Fusicoccum* sp. BCC 4124. *Biosci. Biotechnol. Biochem.* 71: 2010–2020.

52 Paul, T., Mondal, A., and Bandyopadhyay, T.K. (2019). Isolation, purification, characterisation and application of L-ASNase: a review. *Recent Pat. Biotechnol.* 13 (1): 33–44.

53 Basha, N.S., Rekha, R., Komala, M., and Ruby, S. (2009). Production of extracellular anti-leukaemic enzyme L-asparaginase from marine actinomycetes by solid state and submerged fermentation: purification and characterisation. *Trop. J. Pharm. Res.* 8: 353–360.

54 Krishnapura, P.R. and Belur, P.D. (2016). Isolation and screening of endophytes from the rhizomes of some Zingiberaceae plants for L-asparaginase production. *Prep. Biochem. Biotechnol.* 46 (3): 281–287. https://doi.org/10.1080/10826068.2015.1031385.

55 Lopes de Pádua, A.P.S., de Sousa Freire, K.T.L., de Oliveira, T.G.L. et al. (2018). Fungal endophyte diversity in the leaves of the medicinal plant *Myracrodruon urundeuva* in a Brazilian dry tropical forest and their capacity to produce L-asparaginase. *Acta Bot. Bras.* 33 (1): 39–49. https://doi.org/10.1590/0102-33062018abb0108.

56 Pola, M., Rajulapati, S.B., Potla Durthi, C. et al. (2018). *In silico* modelling and molecular dynamics simulation studies on L-Asparaginase isolated from bacterial endophyte of *Ocimum tenuiflorum*. *Enzym. Microb. Technol.* 117: 32–40.

57 Zadeh Hosseingholi, E., Neghabi, N., Molavi, G. et al. (2020). *In silico* identification and characterization of antineoplastic asparaginase enzyme from endophytic bacteria. *IUBMB Life* 72 (5): 991–1000. https://doi.org/10.1002/iub.2237.

58 Chow, Y. and Ting, A.S.Y. (2015). Endophytic L-asparaginase-producing fungi from plants associated with anticancer properties. *J. Adv. Res.* 6: 869–876.

59 Nagarajan, A., Thirunavukkarasu, N., Suryanarayanan, T.S., and Gummadi, S.N. (2014). Screening and isolation of novel glutaminase free L-asparaginase from fungal endophytes. *Res. J. Microbiol.* 9: 163.

60 Santos, M.G.S., Bezerra, J.D.P., Svedese, V.M. et al. (2015). Screening of endophytic fungi from cactus of the Brazilian tropical dry forest according to their L-asparaginase activity. *Sydowia* 67: 147–156.

61 Saranraj, P., Stella, D., and Reetha, D. (2012). Microbial cellulases and its applications: a review. *Int. J. Biochem. Biotech. Sci.* 1: 1–12.

62 Wang, M., Liu, K., Dai, L. et al. (2013). The structural and biochemical basis for cellulose biodegradation. *J. Chem. Technol. Biotechnol.* 88: 491–500.

63 Yoon, L.W., Ang, T.N., Ngoh, G.C., and Chua, A.S.M. (2014). Fungal solid-state fermentation and various methods of enhancement in cellulase production. *Biomass Bioenergy* 67: 319–338.

64 Zhang, X.Z. and Zhang, Y.H.P. (2013). Cellulases: characteristics, sources, production, and applications. In: *Bioprocessing Technologies in Biorefinery for Sustainable Production of Fuels, Chemicals, and Polymers* (eds. S.-T. Yang, H.A. El-Enshasy and N. Thongchul), 131–146. Wiley.

65 Dey, P., Banerjee, J., and Maiti, M.K. (2011). Comparative lipid profiling of two endophytic fungal isolates – *Colletotrichum* sp. and *Alternaria* sp. having potential utilities as biodiesel feedstock. *Bioresour. Technol.* 102 (10): 5815–5823.

66 Peng, X.-W. and Chen, H.-Z. (2007). Microbial oil accumulation and cellulase secretion of the endophytic fungi from oleaginous plants. *Ann. Microbiol.* 57: 239.

67 Lima, A.O., Quecine, M.C., Fungaro, M.H. et al. (2005). Molecular characterization of a beta-1,4-endoglucanase from an endophytic *Bacillus pumilus* strain. *Appl. Microbiol. Biotechnol.* 68 (1): 57–65.

68 Abraham, A., Narayanan, S.P., Philip, S. et al. (2013). In silico characterization of a novel β-1,3-glucanase gene from *Bacillus amyloliquefaciens* – a bacterial endophyte of *Hevea brasiliensis* antagonistic to *Phytophthora meadii*. *J. Mol. Model.* 19 (3): 999–1007. https://doi.org/10.1007/s00894-012-1645-3.

69 Kanokratana, P., Chantasingh, D., Champreda, V. et al. (2008). Identification and expression of cellobiohydrolase (CBHI) gene from an endophytic fungus, *Fusicoccum* sp. (BCC4124) in *Pichia pastoris*. *Protein Expr. Purif.* 58: 148–153.

70 Harnpicharnchai, P., Champreda, V., Sornlake, W., and Eurwilaichitr, L. (2009). A thermotolerant β-glucosidase isolated from an endophytic fungi, *Periconia* sp., with a possible use for biomass conversion to sugars. *Protein Expr. Purif.* 67: 61–69.

71 Toghueo, R.M.K., Ejiya, I.E., Sahal, D. et al. (2017). Production of cellulolytic enzymes by endophytic fungi isolated from Cameroonian medicinal plants. *Int. J. Curr. Microbiol. App. Sci.* 6 (2): 1264–1271.

72 da Silva Ribeiro, A., Polonio, J.C., Costa, A.T. et al. (2018). Bioprospection of culturable endophytic fungi associated with the Ornamental plant *Pachystachys lutea*. *Curr. Microbiol.* 75: 588–596. https://doi.org/10.1007/s00284-017-1421-9.

73 Wu, W., Davis, R.W., Tran-Gyamfi, M.B. et al. (2017). Characterization of four endophytic fungi as potential consolidated bioprocessing hosts for conversion of lignocellulose into advanced biofuels. *Appl. Microbiol. Biotechnol.* 101 (6): 2603–2618.

74 Yousef, N., Mawad, A., and Abeed, A. (2019). Enhancement the cellulase activity induced by endophytic bacteria using calcium nanoparticles. *Curr. Microbiol.* 76 (3): 346–354.

75 Dahiya, N., Tewari, R., and Hoondal, G.S. (2006). Biotechnological aspects of chitinolytic enzymes: a review. *Appl. Microbiol. Biotechnol.* 71: 773–782.

76 Karthik, N., Binod, P., and Pandey, A. (2017). Chitinases. In: *Current Developments in Biotechnology and Bioengineering: Production, Isolation and Purification of Industrial Products*, vol. VII (eds. A. Pandey, S. Negi and C.R. Soccol), 335–368. Amsterdam: Elsevier.

77 Purushotham, P., Sarma, P.V., and Podile, A.R. (2012). Multiple chitinases of an endophytic *Serratia proteamaculans* 568 generate chitin oligomers. *Bioresour. Technol.* 112: 261–269.

78 Guimaraes, L.R.C., Woicieehowski, A.L., Karp, S.G. et al. (2017). Laccase. In: *Current Developments in Biotechnology and Bioengineering: Production, Isolation and Purification of Industrial Products*, vol. VII (eds. A. Pandey, S. Negi and C.R. Soccol), 199–216. Amsterdam: Elsevier.

79 D'Annibale, A., Quaratino, D., Federici, F., and Fenice, M. (2006). Effect of agitation and aeration on the reduction of pollutant load of olive mill wastewater by the white-rot fungus *Panus tigrinus*. *Biochem. Eng. J.* 29: 243–249.

80 Martin-Sampedro, R., Eugenio, M.E., and Villar, J.C. (2012). Effect of steam explosion and enzymatic pre-treatments on pulping and bleaching of *Hesperaloe funifera*. *Bioresour. Technol.* 111: 460–467.

81 Sharma, A., Thakur, V.V., Shrivastava, A. et al. (2014). Xylanase and laccase based enzymatic kraft pulp bleaching reduces adsorbable organic halogen (AOX) in bleach effluents: a pilot scale study. *Bioresour. Technol.* 169: 96–102.

82 Patil, S.M., Chandanshive, V.V., Rane, N.R. et al. (2016). Bioreactor with *Ipomoea hederifolia* adventitious roots and its endophyte *Cladosporium cladosporioides* for textile dye degradation. *Environ. Res.* 146: 340–349.

83 Martín-Sampedro, R., Fillat, Ú., Ibarra, D., and Eugenio, M.E. (2015). Use of new endophytic fungi as pretreatment to enhance enzymatic saccharification of *Eucalyptus globulus*. *Bioresour. Technol.* 196: 383–390. https://doi.org/10.1016/j.biortech.2015.07.088.

84 Jaeger, K. and Reetz, M.T. (1998). Microbial lipases form versatile tools for biotechnology. *Trends Biotechnol.* 16: 396–403.

85 Diaz, J.C.M., Rodríguez, S., Roussos, J. et al. (2006). Lipases from the thermotolerant fungus *Rhizopus homothallicus* is more thermostable when produced using solid state fermentation than liquid fermentation procedures. *Enzym. Microb. Technol.* 39: 1042–1050.

86 Aragão, V.C., Anschau, A., Porciuncula, B.D.A. et al. (2009). Síntese enzimática do butirato de isoamila empregando lipases microbianas comerciais. *Quim Nova* 32: 2268–2272.

87 Hasan, F., Shah, A.A., and Hameed, A. (2006). Industrial applications of microbial lipases. *Enzym. Microb. Technol.* 39: 235–251.

88 Contesini, F.J., Lopes, D.B., Macedo, G.A. et al. (2010). *Aspergillus* sp. lipases: potencial biocatalyst for industrial use. *J. Mol. Catal. B Enzym.* 67: 163–171.

89 Torres, M., Dolcet, M.M., Sala, N., and Canela, R. (2003). Endophytic fungi associated with Mediterranean plants as a source of mycelium-bound lipases. *J. Agric. Food Chem.* 51 (11): 3328–3333.

90 Venkatesagowda, B., Ponugupaty, E., Barbosa-Dekker, A.M., and Dekker, R.F.H. (2018). The purification and characterization of Lipases from *Lasiodiplodia theobromae*, and their immobilization and use for biodiesel production from coconut oil. *Appl. Biochem. Biotechnol.* 185 (3): 619–640. https://doi.org/10.1007/s12010-017-2670-6.

91 Zanotto, S.P., Romano, I.P., Lisboa, L.U.S. et al. (2009). Potential application in biocatalysis of mycelium-bound lipases from Amazonian fungi. *J. Braz. Chem. Soc.* 20: 1046–1059.

92 Mahajan, R.T. and Badgujar, S.B. (2010). Biological aspects of proteolytic enzymes: a review. *J. Pharm. Res.* 3: 2048–2068.

93 Li, Q., Marek, P., and Iverson, B.L. (2013). Commercial proteases: present and future. *FEBS Lett.* 587: 1155–1163.

94 Barrett, A.J., Rawlings, N.D., and Woessner, J.F. (2003). *The Handbook of Proteolytic Enzymes*, 2e. London: Academic Press.

95 Verma, A., Singh, H., Anwar, S. et al. (2016). Microbial keratinases: industrial enzymes with waste management potential. *Crit. Rev. Biotechnol.* 13: 1–16.

96 Jisha, V.N., Smitha, R.B., Pradeep, S. et al. (2013). Versatility of microbial proteases. *Adv. Enzyme Res.* 1: 39–51.

97 Motyan, J.A., Toth, F., and Tozser, J. (2013). Research applications of proteolytic enzymes in molecular biology. *Biomolecules* 3: 923–942.

98 Anbu, P. (2016). Enhanced production and organic solvent stability of a protease from *Brevibacillus laterosporus* strain PAP04. *Braz. J. Med. Biol. Res.* 49 (4): e5178. doi: 10.1590/1414-431X20165178.

99 Amore, A. and Faraco, V. (2016). Enzymes for food and beverage industries: current situation, challenge and perspectives. In: *Advances in Food Biotechnology* (ed. V. Ravishankar Rai), 165–177. Wiley.

100 Jridi, M., Lassoued, I., Nasri, R. et al. (2014). Characterization and potential use of cuttlefish skin gelatin hydrolysates prepared by different microbial proteases. *Biomed. Res. Int.* 2014: 461728.

101 Bialkowska, A., Gromek, E., Florczak, T. et al. (2016). Extremophilic proteases: developments of their special function, potential resources and biotechnological applications. In: *Biotechnology of Extremophiles* (ed. P.H. Rampelotto), 399–444. New York: Springer International.

102 Wang, H., Yang, L., Ping, Y. et al. (2016). Engineering of a *Bacillus amyloliquefaciens* strain with high neutral protease producing capacity and optimization of its fermentation conditions. *PLoS One* 11 (1): e0146373. https://doi.org/10.1371/journal.pone.0146373.

103 Petra, K. and Monika, C. (2016). Spontaneous regression of tumour and the role of microbial infection-possibilities for cancer treatment. *Anti-Cancer Drugs* 27: 269–277.

104 Wu, B., Wu, L., Chen, D. et al. (2009). Purification and characterization of a novel fibrinolytic protease from *Fusarium* sp. CPCC 480097. *J. Ind. Microbiol. Biotechnol.* 36: 451–459.

105 Lu, F., Lu, Z., Bie, X. et al. (2010). Purification and characterization of a novel anticoagulant and fibrinolytic enzyme produced by endophytic bacterium *Paenibacillus polymyxa* EJS-3. *Thromb. Res.* 126 (5): e349–e355. https://doi.org/10.1016/j.thromres.2010.08.003.

106 Peng, Y., Yang, X., and Zhang, Y. (2005). Microbial fibrinolytic enzymes: an overview of source, production, properties, and thrombolytic activity *in vivo*;. *Appl. Microbiol. Biotechnol.* 69: 126–132.

107 Zaferanloo, B., Quang, T.D., Daumoo, S. et al. (2014). Optimization of protease production by endophytic fungus, *Alternaria alternata*, isolated from an Australian native plant. *World J. Microbiol. Biotechnol.* 30 (6): 1755–1762.

108 Mayerhofer, M.S., Fraser, E., and Kernaghan, G. (2015). Acid protease production in fungal root endophytes. *Mycologia* 107 (1): 1–11. https://doi.org/10.3852/14-106.

109 Dorra, G., Ines, K., Imen, B.S. et al. (2018). Purification and characterization of a novel high molecular weight alkaline protease produced by an endophytic *Bacillus halotolerans* strain CT2. *Int. J. Biol. Macromol.* 111: 342–351.

110 Russell, J.R., Huang, J., Anand, P. et al. (2011). Biodegradation of polyester polyurethane by endophytic fungi. *Appl. Environ. Microbiol.* 77: 6076–6084.
111 Sukumaran, R.K. (2009). Bioethanol from lignocellulosic biomass, Part II. Production of cellulases and hemicellulases. In: *Handbook of Plant-Based Biofuels* (ed. A. Pandey), 141–157. Boca Raton, FL, USA: CRC Press, Taylor & Francis Group.
112 Thomas, L., Arumugam, M., and Pandey, A. (2013). Production, purification characterization and overexpression of xylanases from actinomycetes. *Indian J. Exp. Biol.* 51 (6): 875–884.
113 Kumar, D., Kumar, S.S., Kumar, J. et al. (2017). Xylanases and their industrial applications: a review. *Biochem. Cell. Arch.* 17: 353–360.
114 Fenel, F., Leisola, M., Janis, J., and Turunen, O. (2004). A *de novo* designed *N*-terminal disulphide bridge stabilizes the *Trichoderma reesei* endo-1,4-beta-xylanase II. *J. Biotechnol.* 108 (2): 137–143.
115 Wipusaree, N., Sihanonth, P., Piapukiew, J. et al. (2011). Purification and characterization of a xylanase from the endophytic fungus *Alternaria alternata* isolated from the Thai medicinal plant, *Croton oblongifolius* Roxb. *Afr. J. Microbiol. Res.* 5: 5697–5712.
116 Rajesh, P.S. and Rai, V.R. (2015). Purification and antibiofilm activity of AHL-lactonase from endophytic *Enterobacter aerogenes* VT66. *Int. J. Biol. Macromol.* 81: 1046–1052.
117 Song, T., Zhang, W., Wei, C. et al. (2015). Isolation and characterization of agar-degrading endophytic bacteria from plants. *Curr. Microbiol.* 70 (2): 275–281.
118 Manikandan, M., Gopal, J., Kumaran, R.S. et al. (2016). Purification and characterization of a highly active chromate reductase from endophytic *Bacillus* sp. DGV19 of *Albizzia lebbeck* (L.) Benth. actively involved in phytoremediation of tannery effluent-contaminated sites. *Prep. Biochem. Biotechnol.* 46 (2): 192–199.
119 Zhang, J.X., Chen, Z.T., Meng, X.L. et al. (2017). Gene cloning, expression, and characterization of a novel β-mannanase from the endophyte *Paenibacillus* sp. CH-3. *Biotechnol. Appl. Biochem.* 64 (4): 471–481. https://doi.org/10.1002/bab.1510.
120 Khan, A.L., Al-Harrasi, A., Al-Rawahi, A. et al. (2016). Endophytic fungi from Frankincense tree improves host growth and produces extracellular enzymes and indole acetic acid. *PLoS One* 11: e0158207.
121 Bode, H.B., Bethe, B., Höfs, R., and Zeek, A. (2002). Big effects from small changes: possible ways to explore nature's chemical diversity. *Chembiochem* 3: 619–627.
122 Toghueo, R.M.K., Sahal, D., and Boyom, F.F. (2020). Recent advances in inducing endophytic fungal specialized metabolites using small molecule elicitors including epigenetic modifiers. *Phytochemistry* 174: 112338.
123 Lee, I.-Y., Jung, K.-H., Lee, C.-H., and Park, Y.-H. (1999). Enhanced production of laccase in *Trametes vesicolor* by the addition of ethanol. *Biotechnol. Lett.* 21: 965–968.
124 Lomascolo, A., Record, E., Herpoël-Gimbert, I. et al. (2003). Overproduction of laccase by a monokaryotic strain of *Pycnoporuscinnabarinus* using ethanol as inducer. *J. Appl. Microbiol.* 94: 618–624.
125 Manavalan, T., Manavalan, A., Thangavelu, K.P., and Heese, K. (2013). Characterization of optimized production, purification and application of laccase from *Ganoderma lucidum*. *Biochem Eng.* 70: 106–114.
126 El-Bondkly, A.M.A. (2012). Molecular identification using ITS sequences and genome shuffling to improve 2-deoxyglucose tolerance and xylanase activity of marine-derived fungus, *Aspergillus* sp. NRCF5. *Appl. Biochem. Biotechnol.* 167: 2160–2173.

7

Resource Recovery from the Abundant Agri-biomass

Shilpi Bansal[1], Jyoti Singh Jadaun[2], and Sudhir P. Singh[3]

[1]*Vegetable Science Division, ICAR-Indian Agricultural Research Institute, Pusa, New Delhi, India*
[2]*Department of Botany, Dayanand Girls Postgraduate College, Kanpur, Uttar Pradesh, India*
[3]*Center of Innovative and Applied Bioprocessing (CIAB), Mohali, India*

7.1 Introduction

Unpredictable increase in human population leads to a rapid depletion of natural resources and consequently load of waste products is also expanded on the earth. Globally, a massive increase in waste quantity is becoming a threat to maintaining a worthy living environment for the future generation. Before the onset of the difficult adverse situation, we have to think over the management of different waste sources by using a sustainable development approach for maintaining the balance between economy, society and environment. Bio-economy may be a suitable alternative for waste or biomass valorization, which will support in securing the need of the growing population. Bio-economy is defined as the process in which renewable biological resources are utilized for generation of food, feed, bioenergy and other value-added products [1]. Renewable biological resources include crops, forestry, algae, microorganism and also waste products generated by them [2]. Use of biomass for production of bioenergy and other valuable products will reduce the negative impact on the environment as it will help to minimize the problem of landfill disposal and emission of greenhouse gases due to combustion of crop residues [3]. Simultaneously, it will boost up the economy, and it will be a fundamental step towards achieving sustainable development goals. Agri-biomass includes all the waste materials produced through agricultural practices, for example, cereal straw, rice husk, corn stover, sugarcane bagasse, crop residues and animal manure, etc. Each year approximately 140 Gt of agricultural residues are produced [4]. Crop straw constituted more than half of total crop residues, and burning of crop straw is becoming a significant challenge for environmentalist [5, 6]. Different technologies have been applied to convert these renewable sources into profitable products for society, but the advent of synthetic biology built a more suitable platform for

[#] Equal authorship

Bioprospecting of Plant Biodiversity for Industrial Molecules, First Edition.
Edited by Santosh Kumar Upadhyay and Sudhir P. Singh.

this purpose [7, 8]. There are several successful case studies showing the synthesis of valuable products from low-cost substrates [9–11]. The expanding market of bio-based products is now focusing on End-of-Use (EoU)/End-of-Life (EoL) alternatives which will support resources for establishing the circular economy [12]. Agri-biomass is mainly composed of cellulose, hemicellulose and lignin, and in the near future, it will be an appropriate alternative for the production of bio-energy and industrially valuable chemicals [13]. Lactic acid, a valuable chemical, is produced through saccharification and fermentation process of various agriculture feedstocks such as alfalfa fibre and soya fibre corncob (soft) and wheat straw by applying *Lactobacillus delbrueckii* and *Lactobacillus plantarum* [14]. Lignocellulosic part of different agri-residues such as corn cob, rice straw, rice husk, wheat straw, bagasse, sorghum stalk, jatropha pruning cotton stalk, mustard stalk, corn stover have good potential for generation of bioenergy [15, 16]. Use of biomass for bioenergy production serves a dual purpose as it can combat energy crises in a sustainable manner with environment protection [17]. Biofuels have the capability to displace fossil fuels like gasoline, coal and natural gas, etc. [18]. Lignin shows resistance to chemical and biochemical degradation; therefore, pre-treatment either through physical, chemical or biological methods is required (Figure 7.1). Thereafter, cellulose and hemicellulose are hydrolyzed into constituent sugars which undergo fermentation reaction to produce biofuel [19]. This is the peak time to think over the problem of agri-biomass utilization in an advantageous manner; globally researchers are trying to develop sustainable technologies for consuming the agri-biomass in a creative manner with minimization of its negative impact on environment and society. Although technologies are progressing vigorously in this direction, but still we have to walk a long distance for implementation of these technologies globally.

In this chapter, we have described different ways to utilize the agricultural residues for the recovery of valuable resources. Globally, utilization of agri-biomass is performed to

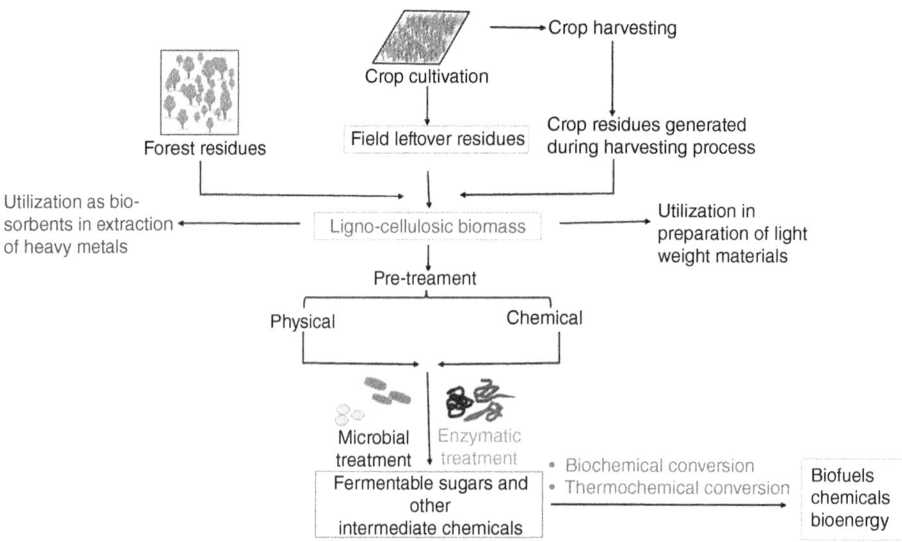

Figure 7.1 Schematic illustration of different routes for utilization of agri-biomass to produce valuable products.

generate energy. Recent technologies are gaining attention to develop methodologies for the extraction of valuable chemicals from crop residues. We have also discussed the novel utility of agri-biomass in manufacturing of light-weighted materials that are useful in construction work and other industrial products.

7.2 Potential of Agri-biomass to Produce Different Products

7.2.1 Conversion of Agri-biomass into Valuable Chemicals

Agri-biomass serves as a reservoir of chemicals/metabolites that possess large scale applicability. The so-called waste has immense potential to be used as a source of sugars, fibres, vitamins, minerals, pharmaceuticals, nutraceuticals, etc. [20]. Preference of people towards natural products over synthetic products triggered the development of new techniques and new resources (wastes) for the production of bioactive compounds [21]. One such compound includes dietary fibres that are complex and heterogeneous polysaccharides which impart various health benefits. Discarded peel, pulp, pomace of fruits such as apple, banana, guava, etc. are known to be a rich source of dietary fibres [22]. By utilization of a novel technique, dietary fibres are extracted from olive mill wastewater, a by-product from olive industry [23]. Bioactive molecules are another important class of chemicals that hold importance as therapeutics and are produced as secondary metabolites *via* different pathways. Regardless of the mode of extraction (chemical, physical, biological), agri-biomass provides a wide range of bioactive compounds. Several reports ensure the presence of various carotenoids, flavonoids, polyphenols, solanesol, saponarin, glucosinolates, capsaicin, isovitexin, etc. from the horticulture crop-based residues (broccoli, pepper, cucumber, tomato) which possess anti-oxidant and antimicrobial properties [24–26]. Thousand tons of starch waste is produced in the form potato peel, rice bran and husk, wheat bran, maize bran and barley bran. Utilizing different extraction protocols and solvents, research community successfully isolated a pool of compounds which included vitamin E, tocopherols, β-glucan, carotenoids (lutein, zeaxanthin, β-carotene), phenolic acids (p-couramic acid, ferulic acid, protocatechuic acid, vanillic acid, gallic acid), steroidal alkaloids (solanidine, demissidine) from these starch waste [27, 28]. Phenolics and flavonoids isolated from waste have been known to increase the shelf life of foods. It prevents them from browning and getting spoiled by the action of microbes. As a natural additive, these molecules increase the antioxidant capacity and prevent freshly cut fruits and vegetables against microbial attack [29]. Conte et al. showed that the application of lemon extracts on mozzarella cheeses increased its longevity without affecting its natural properties [30].

These by-products/waste have the potential to replace synthetic additives used in food items. They can be used as colouring, flavouring, anti-browning agents. Additives produced from the by-products offer the advantage of being non-toxic, easily available and cost effective. Anthocyanin belongs to an important class of food colourants that can be extracted from the wastes of red cabbage, red reddish, purple sweet potato, aronia, cherry, blackberry and red-fleshed potato [31]. Another potential candidate for food colour is lycopene. It has been established that pomace and lyco-oleoresin extract from tomato is rich in lycopene that may impart yellow to red colour to food items including baked goods, spreads, dairy products, beverages, etc. and is approved by European food safety authority [32].

Enzymes are the proteins which act upon the substrate to catalyse the biological reaction to yield products. Enzyme accounts for widespread application at home and industrial level, which involves brewing, baking, fruit juice and beverage processing, meat industry, in oil and fats and dairy industry. Despite its excessive utility, enzymes are highly expensive, and raw materials required for its production is one of the critical factor determining the overall production cost. Hence, agricultural waste from rice, wheat, barley, potato, etc. which are rich sources of carbon is being immensely used for enzyme production in order to lower the production cost. Alpha amylases are continuously being used in paper, pulp, pharmaceutical, bakery industry and has been successfully produced from *Bacillus amyloliquefaciens* using wheat bran, rice bran and potato peel as substrates with maximum enzyme production from wheat bran with specific activity 1.2 U/μg [33]. Similarly, experiments conducted to optimize alpha-amylase production by *Bacillus subtilis* using banana peel yielded maximum activity (9.06 IU/mL/min) at 35 °C, pH 7 [34]. Molasses and sugar beet pulp were employed by a scientist for beta-amylase production by a novel strain *Paenibacillus chitinolyticus* CKS1 [35]. Increase in dependency on bioethanol demands cheaper methods for its production, which implies in the identification of novel methods for cellulase production, an enzyme involved in bioethanol synthesis. Different waste products like wheat straw, paddy straw, corn straw, sorghum husk, water hyacinth and sugarcane bagasse were evaluated for their ability to produce cellulases from a bacteria *Amycolatopsis* sp. GDS, among which maximum yield was obtained with wheat straw. Further analysis showed that the enzyme is thermostable up to 70 °C and expression increases on the addition of NaCl and ionic liquid (10%) [36]. *Trichoderma reesei* produced significant levels of cellulase when the pH of the medium was adjusted in the range of 5–6 by supplying the furfural residues as the carbon source. Treatment of furfural residues with H_2O_2 to remove lignin further improved the yield of the enzyme [37]. Substantial amount of xylanase was procured from different bacterial and fungal strains by utilizing different agri-waste as carbon source such as *Ncosartorya spinosa* D19, *Fomes selerodemeus* (wheat bran), *Aspergillus terrcus* (palm waste), *Coprinellus disseminates* (wheat bran), *A. carneus* M34 (soyabean shell, rice bran), *A. awamori* (grape pomace), *T. harzianum* 1073 D3 (apple pomace, melon peel and hazelnut shell), etc. [38, 39].

7.2.2 Energy Production Using Agri-biomass

Since ancient times energy is the fundamental requirement of humans to sustain. Energy sources can be classified as renewable and non-renewable. Fast rate of depletion of non-renewable sources and lack of acceptance and economic factors associated with renewable sources have led to a surge for alternate sources of energy. The alternate sources should generate energy that is low at cost, high at productivity and can be managed within the limited land area available to meet the demands of the vast population. This marks the concept of waste reuse. Agro waste has enormous potential to be exploited as a source of energy when linked with proper technological support which otherwise would be discarded as waste only causing severe environmental issues.

Both the field-based residues (corn corb, stalks, cotton sticks, mustard stalks, wheat straw and husks) and process-based residues (bagasse, rice husks and hulls, coconut shells, fibre, by-products from olive oil industry, potato peel) are extensively used for energy. Fuels generated by these sources are categorized as first- and second-generation fuels. The carbohydrate present in the by-product is of three types: lignocellulosic, starch and sugar. Different processing techniques are applied for energy generation (i) physical, (ii) thermochemical (incineration, gasification, pyrolysis) and (iii) biological (fermentation, anaerobic digestion) (Figure 7.1).

Fermentation of sugar and starch is easier; however, fermentation of lignocellulosic biomass requires an additional step of pre-treatment prior to fermentation as lignocellulosic waste is more stable. This is mediated by the process of hydrolysis either acid or enzymatic, which converts cellulose to glucose and hemicellulose to pentose and hexoses. These simple sugars are acted upon by microbes to produce biofuels [40].

It was observed that when banana pseudostem underwent alkali and microbial treatment by *A. ellipticus* and *A. fumigatus* under co-culture fermentation to release simple sugar molecules the hydrolysate (4.1 g %) so obtained produced ethanol (17.1 g/L) on fermentation by *Saccharomyces cerevisiae* NCIM 3570 after 72 hours [40]. *Clostridium beijerinckii* P260-mediated fermentation of corn stover hydrolysate produced 16.00 and 18.04 g/L of acetone–butanol–ethanol (ABE) on dilution with water (2 : 1) and wheat straw hydrolysate (1 : 1), respectively. Further treatment of hydrolysate with calcium hydroxide to remove inhibitors elevated the production level up to 26.27 g/L which was much higher than the control reaction with glucose as carbon source that yielded 21.06 g/L ABE [41]. An ample amount of study has been conducted on sugarcane bagasse that reveals that the by-product has huge potential to produce ethanol using different micro-organisms such as *E. coli*, *S. cerevisiae D5A*, *Candida tropicalis JH030*, *Pachysolen tannophilus DW06*, *Zymomonas mobilis* under varying hydrolysis and fermentation conditions [42]. Cai et al. demonstrated the importance of every part of corn stalk in ABE production [43]. NaOH pretreated individual parts were subjected to enzymatic hydrolysis followed by fermentation with *C. acetobutylicum* ABE 1301. It was shown that 1 kg of stem, leaf, cob, flower and husk yields 116.4, 126.3, 169.1, 143.7, and 107.7 g butanol, respectively.

7.2.3 Role of Agri-biomass in Heavy Metal Decontamination

Need not to mention that water is indispensable for humans. However, the significant population of the country is still fighting for it due to the fast depletion of water resources. The condition is further worsened due to the pollution of water. The major factors responsible for water pollution are heavy metals released by various industries. The industries releasing these contaminants directly or indirectly to water bodies include textile, fertilizers, pesticides, metallurgy, tanneries, aerospace, photography, mining, electroplating, batteries, finishing, energy and fuel, refining ores, etc. The most common metals that are responsible for water toxicity include arsenic, mercury, cadmium, lead, chromium, nickel, zinc, cobalt and copper. The heavy metals accumulate in an individual's body system and impart various detrimental effects. These are non-biodegradable, have the ability of bio-magnification, interferes with various regulatory systems of the body such as the cell cycle, circulatory system, nervous system, etc. Thus poses critical health issues that many times prove to be fatal. Chemical precipitation, ion-exchange, reverse osmosis, ultra-filtration, reverse dialysis are the conventional techniques opted for heavy metal removal. However, they have their own constraints linked with them that involve high cost, lower efficiency, large scale production of sludge and its disposal, etc. To overcome these problems, a novel concept of biosorption is rapidly emerging, which involves the binding of metals to biological adsorbents. The literature cites numerous examples of agricultural waste as a biosorbant comprising of rice husk and bran, wheat husk and bran, fruits peel, sugar beet pulp, cottonseed hulls, corn cob, sugarcane bagasse and many more. Chromium (IV) adsorption efficiency of naturally occurring rice husk and its activated form showed promising results which were comparable to commercially available adsorbents [44]. These biosorbants are primarily composed of cellulose, hemicellulose, lignin along with lipids, starch, sugars, ash and many more [45]. These sequester metals with the aid of functional groups (hydroxyl, carbonyl, carboxyl, amino, amido, acetamido, sulphydryl, phenolics, esters) attached to it which either replaces the hydrogen ion with metal ions or donate lone electron pair to form complexes with metal ions [45, 46]. Copper is yet another toxic metal for which various agricultural waste product has been investigated for their adsorption potential under optimized conditions; wheat shell, peanut hulls, barley straw, rice husk, fly ash are among the few candidates [47–49]. Experiments have been conducted to validate the finding, and in one such study, it was observed that the functional groups influenced the copper-binding ability of activated carbon obtained from walnut shell. It was observed that the presence of oxygen-containing groups facilitated the copper uptake (77.68% increase), which was further enhanced by oxidation reaction, whereas amino groups have a lesser role to play [50]. Similarly, nitric acid-mediated oxidation of carbon in wheat and barley straw led to increasing in carboxylic groups by 11-folds as seen by FITR spectra and Boehm titration. Comparison of oxidized and unoxidized carbon for their Cu binding capacity showed better performance than oxidized carbon [51]. Blockage of carboxyl group led to 32.8, 58.5, 65.3% decrease in Cd(II), Cu(II) and Zn(II) adsorption capacity while blockage of hydroxyl group caused 30.9, 27.5, 46.1% decline in adsorption for the three metals thereby confirming the importance of functional groups [52]. However, these functional groups do not hold the sole responsibility for the efficacy of metal sequestration. They function well in association with various other factors including physical and chemical properties of biosorbant and metal ions, availability of binding sites, ease of access to these sites, the affinity between the binding site and metal ion, pH, temperature, initial concentration of both and contact time, etc. [53].

The rate of biosorption is higher during the initial period of contact that could be attributed to the presence of a large number of vacant active sites initially and, thereafter, a saturation stage reaches where the metal needs to be diffused into the inner sections of the biomass [54, 55]. Another determining factor is the pH that alters the surface charge of biosorbant and regulates the competition between the hydrogen ion and metal ion. At lower pH due to higher concentration of hydrogen ion, most of the active sites are occupied. Increase in pH shifts the sites towards negative charge, making functional groups available for binding of metal ions. Further higher pH leads to metal precipitation, thereby restricting the biosorption [52–56]. In the case of maize stalk sponge, the adsorption of Pb increased with increasing pH with maxima at 6.0 ± 0.2 becomes stable up to pH 8.0 and then Pb precipitation started [54]. The effect of temperature is controversial as certain reports favour an increase in temperature and relate it to improve in its surface activity while other reports are reverse to previous studies and link it to structural damage of active sites. Thus, the selection of optimum temperature is preferential that is dependent on the type of functional group, metal ion and other factors [53]. Increase in initial metal ion concentration above a threshold level is inversely proportional to the removal efficiency of metal ions, and this may be attributed to the non-availability of active binding sites as it has been already occupied by metal ions and a state of saturation is achieved. Metal ion removal by the banana peel is a good example of it where metals at a concentration of 150 mg/L, the removal efficiencies of Cu(II), Pb(II), Zn(II) and Ni(II) were noted as 92.52, 79.55, 63.23 and 68.10% while at a lower concentration of 25 mg/L the corresponding values were 94.80, 86.81, 84.63 and 82.36%, respectively [57].

This issue could be resolved by the addition of more of the biosorbants. However, the increase in biosorbants elevates the removal percentage of metal ions but decrease the maximum adsorption capacities due to a decrease in the total surface area caused by the overlap of adsorption sites.

Agricultural waste biomass and metal ions each have their individual properties. It is their amalgamation under optimized conditions that leads to successful metal adsorption.

7.2.4 Manufacturing of Lightweight Materials

The construction and building materials sector is the largest user of natural resources and concrete forms the backbone of the sector. Generally, concrete consists of 12% cement, 80% aggregates and 8% water by mass. Both water and aggregates are natural resources, whereas cement is produced by a cement manufacturing plant. The cement industry is one of the most polluting industries of the world, wherein 480 kg of CO_2 is released in producing 1 m^3 of cement. One strategy is to reduce the cement dosage in concrete formation by using fly ash which is a waste generated by the burning of coal used in the power generation sector or any other industry that uses boilers for steam formation [58]. To reduce the impact of construction on natural resources, the best option is to use materials that are produced from agricultural waste. Some of the agricultural waste that can be used as an aggregate in the concrete are groundnut shell, sugarcane bagasse ash, sawdust, rice husk ash, giant reed ash, tobacco waste and cork. The major difference among these wastes is the place from which they are collected and the process by which they are converted into aggregates.

Many studies have been conducted by using these wastes in concrete. By adding 10% of sugarcane bagasse ash in concrete, its compressive strength increases, but further addition of the ash resulted in a decrease in strength [59]. Almeida et al. found that the strength of mortar was not affected by adding sugarcane bagasse ash [60]. When groundnut shell was added to the control concrete, the maximum compressive strength could be achieved with 5% mix. The authors also found that the concrete mixed with groundnut shell was not suitable for structural members exposed to water as the moisture affects the strength and weight [61]. With giant reed ash, the compressive strength of the concrete increased till 7.5% replacement of further aggregate increase in percentage led to decrease in strength, but it was still higher than the minimum requirement. The authors argue that silica present in the ash is the reason behind the increase in strength [62]. Rice husk ash and limestone were used to form cylindrical concrete test samples [63]. The authors found very low compressive strength for 100% replacement of rice husk ash, whereas the strength was maximum for limestone at 10% replacement. The decrease in compressive strength was attributed to greater porosity of rice husk ash and limestone, which was indicated by higher water requirement. Moreover, the authors found that rice husk ash was able to fill the microvoids in the cement particle in an improved manner. Nóvoa et al. found that the compressive strength of cork containing polymer mortar was higher than normal mortar [64]. Moreira et al. found the compressive strength of screed containing expanded cork. The authors observed that the strength increases as the cement content increases, and as the cork granules percentage increases the strength decreases [65]. Öztürk and Bayrakl found that using a small amount of tobacco waste increased in compressive strength of concrete [66]. The authors suggested that the lightweight concrete can be used as an insulating material for coating and separating materials in construction.

Shafana and Venkatasubramani studied the tensile strength of concrete with the addition of sugarcane bagasse ash and found optimal strength at 10% replacement of aggregate with bagasse ash [67]. Similarly, Mageswari and Vidivelli found improvement in tensile strength up to 20% replacement of aggregate with sawdust ash further addition resulted in a reduction in tensile strength [68]. The flexural strength of reed-based concrete was carried out by Ismail and Jaeel [62]. The authors found improvement in strength till 7.5% replacement of aggregate with reed further replacement resulted in a decrease in strength even below the concrete strength levels. The authors attributed the increase in strength to the presence of organic bindings in the reed in acidic or alkali environment, which decomposes to pentose and hexose. These monosaccharides are easily dissolved in water resulting in the formation of hydrophilic adsorption layers on cement grains. Similarly, the addition of cork to mortar resulted in an increase in flexural strength and higher percentage resulted in a decrease in the flexural strength [64].

Ismail and Jaeel [62] also studied the effect of alkali on concrete containing reed ash and reed fibre. The authors found that the concrete containing giant reed fibre was affected faster than the control concrete. The authors argued that the complex lignin molecules in the fibre were affected, causing it to fracture. Almeida et al. used colorimetric treatment for studying the effect of chloride penetration in bagasse ash containing mortar [60]. The authors observed higher resistance to chloride penetration with bagasse containing mortar in comparison to control concrete. The water absorption and porosity of hardened concrete is an important indicator of its durability. If these values are low, then the life and serviceability of concrete will increase many folds. Modani and Vyawahare found that the sorptivity coefficient increases with an increase in bagasse ash content in the concrete [59]. Ismail and Jaeel found a higher percentage of water absorption with reed fibres containing concrete as compared to reed ash and control concrete [62]. Moreover, the increase in the reed ash percentage in concrete the water absorption percentage also increases.

7.3 Case Studies

7.3.1 Utilization of Paddy Waste

Rice is the second most eaten staple food after wheat with annual growth around 680 million tons. It is rich in vitamins, minerals and many other nutrients. Viable rice growth is dependent on the continuous availability of nitrogen, silicon and water. It is cultivated all over the world with distinct genetics, morphology and nutrient composition. Not all the rice grown in the field is as such ready for consumption, but approximately 40% of it is total rice produced is lost in the form of by-products. During its journey from field to house, the rice undergoes a series of processes which includes harvesting, transport, reception and pre-cleaning, drying, storage, shelling, milling/polishing, selection and classification. The amount and type of by-product produced is a function of various factors such as type of rice, milling techniques, processing techniques used which results in the production of husk, straw, bran, broken rice and ash. These products previously treated as waste has now high economic and nutritional value as it embraces a plethora of useful compounds in itself. Rice straw is produced during harvesting and accounts for approximately 50% of the dry weight of rice. It is composed of low levels of lignin but a high concentration of silica oxide [69]. It has been established that rice straw is a good source of nitrogen yielding about 2.5 kg nitrogen per ton straw and can serve as a matrix for the cultivation of numerous vegetable crops like tomato, cucumber and okra [70]. Addition of paddy straw in soil positively affects the physicochemical properties of soil. It helps in retaining Fe, Cu, Mn, Zn content and prevents nitrate leaching. It promotes the growth of microbial fauna, thereby activating enzymatic machinery in soil and alleviating the levels of macro and micronutrients [71]. Rice husk or rice hull is a protective coating developed during the course of seed formation [72]. Primarily it constitutes of cellulose (40–50%), hemicellulose (25%), lignin (20–30%) and ash (17–18%). It is rich in organic compounds (94.99% by weight) than inorganic compounds with ash contributing around 94% of total silica. High silica content lowers its decomposition rate and allows recovery of rice husk ash [73, 74]. In today's era of technology, the role of silicon oxide cannot be negated as it is the basic component of electronic circuits, diodes, insulators, solar panels, semi-conductors and also has broadened its area of application to polymers and ceramics. Synthetic production of silica is expensive, thereby generating a need for alternate sources of silica production. Methods have been developed to extract silica from rice husk ash as it is rich in silica [75]. This extracted silica has been used as a support for the synthesis of a number of catalysts where metals usually transition metals are impregnated onto the silica surface via different techniques (sol–gel, ion exchange, deposition–precipitation) [76]. In a developing country like India, energy generation is always an issue leading to exhaustion of natural non-renewable energy sources. Keeping this in mind, rice husk has been used by researches to identify alternate sources of energy production which includes power, fuels, heat, syngas, etc. In order to utilize husk for power generation, the rice husk was heated at high temperature causing decomposition of material and production of combustible gases which was then burnt to produce steam/heat which leads to activation of turbines producing electricity. In many countries, most of the rice husk is consumed by rice processing companies for electricity, making it economical and environment friendly [77, 78]. Roy et al. determined the use of rice husk fuel in the process of parboiling of rice. In a study, for the production of bioethanol, rice husk in combination with orange peel wastes were used [79]. To it, water was added, and substrate (11 kg) was prepared for hydrolysis by sulfuric acid to produce sugar at 121 °C. Further *Saccharomyces cerevisiae*-mediated fermentation led to the production of 22.77 g of ethanol per liter of the substrate [80].

Rice bran constitutes an important class of by-product from rice due to its richness in carbohydrates, lipids, proteins, fibres, minerals and vitamins. About 10% of rice grain mass comes from bran which contains 15–20% of bran oil. Similar to any medicinal plant rice bran exhibits a number of therapeutic properties. Compounds like ferulic acid and tricin induce apoptosis and inhibit cancer cell proliferation [81].

7.3.2 Utilization of Mustard Waste

Mustard commonly called as rai or kali sarso belongs to the member of *Brassicaceae* family. It is mainly cultivated in northern plains of India including Rajasthan, Madhya Pradesh, Gujarat, Haryana, Uttar Pradesh, etc. India is the third largest producer of mustard in the world. The seeds of mustard are used for oil extraction and seasoning in food while the leaves are used as a vegetable. The large scale cultivation of mustard and its utilization lead to large scale production of waste from it. However, in the present time, these waste products are no more a waste product but are efficiently utilized for different purposes. Higher levels of astaxanthin, a keto-carotenoid with a beneficial role in food, cosmetics and medical industry, were produced by using mustard waste media in comparison with standard media in yeast strain *Xanthophyllomyces dendrorhous* [82]. Further mustard stalk and straw was used as a substrate by fungus *Termitomyces clypeatus* to produce various lignocellulolytic enzymes like beta –glucosidase, beta xylanase, beta xylosidase. Also, this waste, after saccharification with the enzymes of *T. clypeatus*, was used for the production of bioethanol [83]. Sahoo et al. showed the isolation of lipase, which was thermostable in different conditions from *Anoxybacillus* sp. ARS-1 using mustard cake by solid substrate fermentation [84]. Similarly, protease, an important enzyme used widely in industries, was produced from the mustard cake with a concentration of 4% under optimized conditions of pH-11, 40 °C and submerged fermentation for 70 hours [85]. Pollution caused by heavy metals is a major concern nowadays, which poses a threat to all forms of living beings. Waste biomass obtained after extraction of oil from seeds of white mustard have shown to possess biosorption property for heavy metals. It was seen that biosorbant at a concentration of 5 g/L at 5.5 pH and 55 °C was very effective in adsorption of zinc, lead and cadmium [86]. Husk and sticks from mustard are continuously supplied to the bricks industry by farmers. Also, a combination of mustard waste along with leaves and grasses and wood waste leads to the formation of briquettes which is a potential source of energy [87]. Deforestation has led to a shortage of wood supply which in turn is affecting the furniture industry. Substitution of wood chips with white mustard straw wholly or partially has provided a solution to this problem. General boards and interior boards which are used under dry conditions can be prepared using mustard waste [88].

7.3.3 Utilization of Maize Waste

Maize, also known as corn, belongs to the family of *Gramineae*. It is believed that maize first originated in central Mexico 7000 years ago in the form of wild grass. The natives transformed maize into a better source of food. It contains nearly 72% starch, 10% protein and 4% fat, supplying an energy density of 365 kCal/100 g [89]. Maize is grown throughout the world, according to the Food and Agricultural Organization of the United Nations in the year 2018, nearly 193 million hectares were harvested by the world and produced nearly 1147 million tons of maize. United States of America, China and Brazil were the top producers of maize. United States produced nearly 392 million tons, China produced 257 million tons and Brazil produced 82 million tons of maize in the year 2018 [90]. Apart from food and fodder, it is possible to process maize into a variety of food and industrial products such as sweeteners, starch, corn oil, glue, beverages and ethanol. In the last ten years, nearly 40% of the maize produced in the United States is utilized in producing fuel-grade ethanol [89].

Apart from the grain, the maize stover consisting of leaves, stalks, cobs and husk is left in the field after harvesting [91]. After harvesting of the grain, the stover contains nearly 15% cob, 8% husk, 21% leaf and 56% stalk. The stover can be used as a litter for animals, fuel, soil conditioner and fodder for ruminants. In the United States, the maize stover is also used for producing ethanol due to its higher lignocellulose content [92]. Maize stalks can be used as a fuel; it has a higher heating value of 17.2–18.5 MJ/kg. It has moderate ash content and low moisture content. As compared with other solid fuels, its sulphur content is low, and oxygen content is high [93]. Maize stalks can be used as fodder for ruminants, but its quality is low due to insufficient protein content. To increase the protein content, various methods are available such as fermentation with yeasts and fungi, alkali treatment and ensilage [94]. The stalks can also be used as soil conditioners for replenishing the organic matter in the field for the next crop. The maize stalks contain hydroxycinnamic acids, namely ferulic and *p*-coumaric acids, as they are rich in cell wall materials. These acids are the main components for cell wall shape, integrity and defence against pathogenic ingress. As bi-functional molecules with phenolic and carboxylic bonding sites, the ferulic acid provides the linkage between esters in the primary cell wall polysaccharides and the ethers linking the lignin content [95]. With alkaline treatment, the lignin is dissolved by the cleavage of ester linkages in the lignin-polysaccharides complex, thereby releasing the phenolic acids. Both the hydroxycinnamic acids have great importance in the form of antioxidants as they can preserve food due to their ability to inhibit fatty acid peroxidation, as flavoring and aroma compounds, as feedstock for high-value vanillin bio-production and as anti-cancer or a potent anti-inflammatory substance [96, 97].

Maize cob left after extracting the grain can also be used as a fuel, as its higher heating value is close to the maize stalk. It also has low protein content, but it has high fibre content, due to their wide availability they are also used as fodder for ruminants. Compared with maize stalks they have rather higher hydroxycinnamic acid content. Therefore, maize cob is considered as an important source of ferulic and *p*-coumaric acids along with wheat bran and rice husk [98]. Due to their higher abrasiveness and absorbency, they are an ideal biomass-based material for industrial applications such as absorbing oils and chemicals, cleaning industrial and environmental spills and also for polishing and blasting of materials. Foo attempted to utilize corn cobs as an absorbent of carbofuran [99]. Carbofuran is a pesticide used against a wide range of insects. However, its incessant use has resulted in its reaching the potable drinking water causing a wide variety of human health issues, including reproductive failure, teratogenicity, neurotoxicity, carcinogenicity and mutagenicity. Tsai et al. converted corn cobs into activated carbon using potassium hydroxide/potassium carbonate and carbon dioxide [100]. The authors studied the efficacy of using the activated carbon as an absorbent of organic compounds and found the adsorbent had large surface area and higher pore volume. Synthetic organic dyes used in textiles, printing, leather tanning, food processing, rubber, cosmetics, photography, etc. are a big source of water pollution. Berber-Villamar et al. assessed the potential of removing Direct Yellow 27 dye from the wastewater using corn cobs [101]. The authors found that the adsorption capability of the cob was dependent upon the particle size of corn cob, pH of the solution, contact time and initial dye concentration, and most of the dye could be easily removed from the solution. Patil et al. used corn cob for treating sewage water and found that the biochemical oxygen demand reduced from 460 to 187 mg/L, chemical oxygen demand reduced from 466 to 194 mg/L, total suspended solids reduced from 514 to 220 mg/L. Silica extracted from corn cob can also be used for wastewater treatment [102]. Shim et al. found that 84–88% copper ion and 83–87% cadmium ion can be removed within 24 hours from the contaminated water [103]. Corn cob can be used for making particle boards, which can provide acoustic insulation [104] and thermal insulation [105]. It can also be used as an agglomerate for concrete production [106].

7.3.4 Utilization of Horticulture Waste

Fruits and vegetables as compared with edible content have a higher proportion of residues and wastes. Table 7.1 shows the type of waste and the percentage of waste in various fruits and vegetables after processing. It is seen that the citrus fruits contain the maximum waste of 50%, followed by mango, which has 45% waste and onion and guava processing gives the lowest waste of 10%. Banana peels account for 30% of the fruit, and it contains high levels of flavonoids, phenolics, proanthocyanidins and other antioxidants [108, 109]. It was found that the seed fractions and peel of some fruits such as grape seed, hawthorn peel, pomegranate peel, lychee seeds are a rich source of natural antioxidants and have high antioxidant activity [110]. In some materials, it is seen that the by-products contain a major amount of bioactive compounds in comparison to the major counterparts. For example, cashew apple and Surinam cherry by-products contain a high amount of total phenolic

Table 7.1 Fruits and vegetable processing wastes [107].

Fruit/vegetable	Nature of waste	Approximate waste percentage
Citrus	Peel, rag, seed	50
Banana	Peel	35
Pineapple	Skin, core	33
Mango	Peel, stones	45
Apple	Peel, pomace, seed	25
Grape	Stem, seed, skin	20
Tomato	Skin, core, seed	20
Guava	Peel, core, seed	10
Pea	Shell	40
Potato	Peel	15
Onion	Outer leaves	10

compounds and anthocyanins as compared to their pulps [111]. Therefore, processing waste vegetables and fruits can help recover natural antioxidants that can replace synthetic antioxidants such as butylated hydroxyanisole (BHA) and butylated hydroxytoluene (BHT) as sources of health benefits [112, 113]. Some of the bioactive compound resources are shown in Table 7.2.

In the laboratory, microorganisms are grown for microbial studies with the help of a suitable culture media. Commercially available media such as MacConkey agar, Nutrient agar and Cetrimide agar are used for the growth of microorganism; however, these media are highly costly [121]. For the culture of both fungi and bacteria, nutrient agar has been formed from carrot, cabbage, tomato, gooseberry, pumpkin, etc. [122]. To reduce the cost of microbial media, cowpea, black gram and green gram have also been used as starch and protein substitutes. Agar made from the peel of Dragon fruit has also been used as a media for growing microbes [123]. Banana, grapefruit and melon peels contain carbohydrates that can be used as a substrate for growing amylase [124]. Fungi growth has also been carried out using banana peel which is an economical and effective media [125].

Fruits and vegetable peels can also be used to make biosorbents that can remove heavy metals present in the water [126]. Malachite green was removed in batch mode using sponge gourd peel which is an inexpensive natural biosorbent [127]. Rhodamine-B is a cationic water-soluble dye which was removed using banana peels [128]. A study was also carried out in which a biosorbent was developed using banana peel-activated carbon, methylated banana peel and natural banana peel for treating palm oil mill effluent [129].

Table 7.2 Vegetables and fruit wastes containing bioactive compounds.

Source	Residue	Bioactive compounds	References
Papaya, olive	Leaves	Flavonoids, phenolic compounds, saponins, proanthocyanidins, tyrosol, gallic acid, oleuropein	[114, 115]
Grape	Stalks	Flavonol glycosides, anthocyanins, catechins, procyanidins, phenolic acids	[116]
Guava, mango, apple, cashew, sapodilla, papaya, macadamia, potato	Peels or skins	Syringic acid, Gallic acid, mangiferin, gentisyl-protocatechuic, anthocyanins, quercetin, ellagic acid, caffeic acids, carotenoids, chlorogenic, flavonoids, proto-catechuic, phenolics, proanthocyanidins	[108, 117, 109]
Acerola, olive, grape, soursop, guava, papaya, apricot, passion fruit	Seeds	Hydroxybenzoic acid, protocatechuic acid, caffeic acid, ferulic acid, syringic acid, sinapic acid, quercetin, rutin, cinnamic acid, hesperidin, persipan	[111, 109]
Pistachio	Hulls	Phenolic compounds	[118]
Apple, grape, olive, apple, tomato	Pomace	Flavonol glycosides, bioactive polar lipids, anthocyanins, catechins, phenolic acids, stilbenes and lycopene, syrigin and quercetin	[109, 119, 120]

7.4 Conclusion and Future Perspectives

Recent studies revealed that agriculture residues could be a source of valuable chemicals that can be used pharmaceuticals, nutraceuticals industry. Organic matter of biomass can be diverted into various ways to produce desired compounds as after hydrolysis of cellulosic biomass, glucose is produced, which is a platform chemical for generation diverse sugars and alcohols, etc. Novel technology shows the use of agri-residues in the preparation of construction materials also. Paddy straw and corn cobs are actively utilized in the production of bioenergy in many countries, and this process significantly supports the energy supplies to society. Many agri-residues have adequate capability to extract heavy metals from industrial effluents and contaminated water. Sustainable use of agricultural residues will support a circular economy as well as it can avoid several issues related to the environment, so it will provide dual benefit to society.

As rapid depletion of non-renewable resources is a major challenge for any country so globally, scientists are looking for the development of renewable resources for fulfilling the need of the growing population. In future, we will see more investment in this sector as bio-economy is becoming a trend in the global market. To maintain the criteria of sustainable development, constructive use of biomass is the urgent requirement to support the economy of a country.

References

1 Lago, C., Herrera, I., Caldés, N., and Lechón, Y. (2019). Nexus bioenergy–bioeconomy. In: *The Role of Bioenergy in the Bioeconomy* (eds. C. Lago, N. Caldés and Y. Lechón), 3–24. Academic Press.

2 Sánchez, J., Curt, M.D., Robert, N., and Fernández, J. (2019). Biomass resources. In: *The Role of Bioenergy in the Bioeconomy* (eds. C. Lago, N. Caldés and Y. Lechón), 25–111. Academic Press.

3 Wei-Hsin, C., Lee, K.T., and Ong, H.C. (2019). Biofuel and bioenergy technology. *Energies* 12 (2): 290.

4 Centore, M., Hochman, G., and Zilberman, D. (2014). Worldwide survey of biodegradable feedstocks, waste-to-energy technologies, and adoption technologies. In: *Modelling, Dynamics, Optimization and Bioeconomics I. Springer Proceedings in Mathematics and Statistics* (eds. A. Pinto and D. Zilberman), 73. Berlin: Springer.

5 Gadde, B., Bonnet, S., Menke, C., and Garivait, S. (2009). Air pollutant emissions from rice straw open field burning in India, Thailand and the Philippines. *Environmental Pollution* 157 (5): 1554–1558.

6 Bhuvaneshwari, S., Hettiarachchi, H., and Meegoda, J.N. (2019). Crop residue burning in India: policy challenges and potential solutions. *International Journal of Environmental Research and Public Health* 16 (5): 832.

7 Chen, Z.Y., Zhang, Z.L., and Bai, X.H. (2007). Comprehensive exploitation and utilization of crop straw. *Journal of Agricultural Mechanization Research* 5: 194–196.

8 Narnoliya, L.K., Jadaun, J.S., and Singh, S.P. (2018). Management of agro-industrial wastes with the aid of synthetic biology. In: *Biosynthetic Technology and Environmental Challenges* (eds. S. Varjani, B. Parameswaran, S. Kumar and S. Khare), 11–28. Singapore: Springer.

9 Patel, S.N., Singh, V., Sharma, M. et al. (2018). Development of a thermo-stable and recyclable magnetic nanobiocatalyst for bioprocessing of fruit processing residues and D-allulose synthesis. *Bioresource Technology* 247: 633–639.

10 Jadaun, J.S., Narnoliya, L.K., Agarwal, N., and Singh, S.P. (2019). Catalytic biosynthesis of levan and short-chain fructooligosaccharides from sucrose-containing feedstocks by employing the levansucrase from *Leuconostoc mesenteroides* MTCC10508. *International Journal of Biological Macromolecules* 127: 486–495.

11 Sharma, M., Sangwan, R.S., Khatkar, B.S., and Singh, S.P. (2020). Development of a prebiotic oligosaccharide rich functional beverage from sweet *Sorghum* stalk biomass. *Waste and Biomass Valorization*: 1–12. https://doi.org/10.1007/s12649-020-01156-5.

12 Briassoulis, D., Pikasi, A., and Hiskakis, M. (2019). End-of-waste life: Inventory of alternative end-of-use recirculation routes of bio-based plastics in the European Union context. *Critical Reviews in Environmental Science and Technology* 49 (20): 1835–1892.

13 Jørgensen, H., Kristensen, J.B., and Felby, C. (2007). Enzymatic conversion of lignocellulose into fermentable sugars: challenges and opportunities. *Biofuels, Bioproducts and Biorefining* 1 (2): 119–134.

14 Sreenath, H.K., Moldes, A.B., Koegel, R.G., and Straub, R.J. (2001). Lactic acid production from agriculture residues. *Biotechnology Letters* 23 (3): 179–184.

15 Raj, T., Kapoor, M., Gaur, R. et al. (2015). Physical and chemical characterization of various Indian agriculture residues for biofuels production. *Energy and Fuels* 29 (5): 3111–3118.

16 Mohammed, N.I., Kabbashi, N., and Alade, A. (2018). Significance of agricultural residues in sustainable biofuel development. In: *Agricultural Waste and Residues* (ed. A. Aladjadjiyan), 71. IntechOpen. doi: 10.5772/intechopen.78374.

17 Sandhya, D., Jogam, P., Narnoliya, L.K. et al. (2020). Sustainable production of biofuels through synthetic biology approach. In: *Biorefinery Production Technologies for Chemicals and Energy* (eds. A. Kuila and M. Mukhopadhyay), 289. Scrivener Publishing LLC.

18 Pandiyan, K., Singh, A., Singh, S. et al. (2019). Technological interventions for utilization of crop residues and weedy biomass for second generation bio-ethanol production. *Renewable Energy* 132: 723–741.

19 Pérez, J., Munoz-Dorado, J., De la Rubia, T.D.L.R., and Martinez, J. (2002). Biodegradation and biological treatments of cellulose, hemicellulose and lignin: an overview. *International Microbiology* 5 (2): 53–63.

20 Da Silva, L.M.R., De Figueiredo, E.A.T., Ricardo, N.M.P.S. et al. (2014). Quantification of bioactive compounds in pulps and by-products of tropical fruits from Brazil. *Food Chemistry* 143: 398–404.

21 Zabot, G.L. and Cárdenas-Toro, F.P. (2017). Recovering bioactive compounds from other agricultural wastes. In: *Recovering Bioactive Compounds from Agricultural Wastes* (ed. V.T. Nguyen), 197–219. Wiley.

22 Cui, J., Lian, Y., Zhao, C. et al. (2019). Dietary fibers from fruits and vegetables and their health benefits via modulation of gut microbiota. *Comprehensive Reviews in Food Science and Food Safety* 18 (5): 1514–1532.

23 Galanakis, C.M., Tornberg, E., and Gekas, V. (2010). A study of the recovery of the dietary fibres from olive mill wastewater and the gelling ability of the soluble fibre fraction. *LWT-Food Science and Technology* 43 (7): 1009–1017.

24 Domínguez-Perles, R., Martínez-Ballesta, M.C., Carvajal, M. et al. (2010). Broccoli-derived by-products – a promising source of bioactive ingredients. *Journal of Food Science* 75 (4): C383–C392.

25 Taylor, M.A. and Fraser, P.D. (2011). Solanesol: added value from Solanaceous waste. *Phytochemistry* 72: 1323–1327.

26 Santana-Méridas, O., González-Coloma, A., and Sánchez-Vioque, R. (2012). Agricultural residues as a source of bioactive natural products. *Phytochemistry Reviews* 11 (4): 447–466.

27 Maldonado, A.F.S., Mudge, E., Gänzle, M.G., and Schieber, A. (2014). Extraction and fractionation of phenolic acids and glycoalkaloids from potato peels using acidified water/ethanol-based solvents. *Food Research International* 65: 27–34.

28 Khong, T.T. and Nguyen, V.T. (2017). Recovering bioactive compounds from starch wastes. In: *Recovering Bioactive Compounds from Agricultural Wastes* (ed. V.T. Nguyen), 173–194. Wiley.

29 Ayala-Zavala, J.F., Rosas-Domínguez, C., Vega-Vega, V., and González-Aguilar, G.A. (2010). Antioxidant enrichment and antimicrobial protection of fresh-cut fruits using their own byproducts: looking for integral exploitation. *Journal of Food Science* 75 (8): 175–181.

30 Conte, A.M., Scrocco, C.A., Sinigaglia, M., and Del Nobile, M.A. (2007). Innovative active packaging systems to prolong the shelf life of mozzarella cheese. *Journal of Dairy Science* 90 (5): 2126–2131.

31 Ayala-Zavala, J., Vega-Vega, V., Rosas-Domínguez, C. et al. (2011). Agro-industrial potential of exotic fruit byproducts as a source of food additives. *Food Research International* 44 (7): 1866–1874.

32 Kong, K.W., Khoo, H.E., Prasad, K.N. et al. (2010). Revealing the power of the natural red pigment lycopene. *Molecules* 15 (2): 959–987.

33 Mojumdar, A. and Deka, J. (2019). Recycling agro-industrial waste to produce amylase and characterizing amylase–gold nanoparticle composite. *International Journal of Recycling of Organic Waste in Agriculture* 8 (1): 263–269.

34 Kokab, S., Asghar, M., Rehman, K. et al. (2003). Bio-processing of banana peel for α-amylase production by *Bacillus subtilis*. *International Journal of Agriculture and Biology* 5 (1): 36–39.

35 Mihajlovski, K.R., Radovanović, N.R., Veljović, Đ.N. et al. (2016). Improved β-amylase production on molasses and sugar beet pulp by a novel strain *Paenibacillus chitinolyticus* CKS1. *Industrial Crops and Products* 80: 115–122.

36 Kshirsagar, S.D., Saratale, G.D., Saratale, R.G. et al. (2016). An isolated *Amycolatopsis* sp. GDS for cellulase and xylanase production using agricultural waste biomass. *Journal of Applied Microbiology* 120 (1): 112–125.

37 Zhao, C.H., Liu, X., Zhan, T., and He, J. (2018). Production of cellulase by *Trichoderma reesei* from pretreated straw and furfural residues. *RSC Advances* 8 (63): 36233–36238.

38 Martins, D.A.B., Prado, H.F.A., Leite, R.S.R. et al. (2011). Agroindustrial wastes as substrates for microbial enzymes production and source of sugar for bioethanol production. *Integrated Waste Management* 2: 319–361.

39 Ravindran, R., Hassan, S.S., Williams, G.A., and Jaiswal, A.K. (2018). A review on bioconversion of agro-industrial wastes to industrially important enzymes. *Bioengineering* 5 (4): 93.

40 Ingale, S., Joshi, S.J., and Gupte, A. (2014). Production of bioethanol using agricultural waste: banana pseudo stem. *Brazilian Journal of Microbiology* 45 (3): 885–892.

41 Qureshi, N., Saha, B.C., Dien, B. et al. (2010). Production of butanol (a biofuel) from agricultural residues: part I–use of barley straw hydrolysate. *Biomass and Bioenergy* 34 (4): 559–565.

42 Chandel, A.K., da Silva, S.S., Carvalho, W., and Singh, O.V. (2012). Sugarcane bagasse and leaves: foreseeable biomass of biofuel and bio-products. *Journal of Chemical Technology and Biotechnology* 87 (1): 11–20.

43 Cai, D., Li, P., Luo, Z. et al. (2016). Effect of dilute alkaline pretreatment on the conversion of different parts of corn stalk to fermentable sugars and its application in acetone–butanol–ethanol fermentation. *Bioresource Technology* 211: 117–124.

44 Bishnoi, N.R., Bajaj, M., Sharma, N., and Gupta, A. (2004). Adsorption of Cr (VI) on activated rice husk carbon and activated alumina. *Bioresource Technology* 91 (3): 305–307.

45 Sud, D., Mahajan, G., and Kaur, M.P. (2008). Agricultural waste material as potential adsorbent for sequestering heavy metal ions from aqueous solutions – a review. *Bioresource Technology* 99 (14): 6017–6027.

46 Demirbas, A. (2008). Heavy metal adsorption onto agro-based waste materials: a review. *Journal of Hazardous Materials* 157 (2–3): 220–229.

47 Pehlivan, E., Altun, T., and Parlayıcı, S. (2009). Utilization of barley straws as biosorbents for Cu^{2+} and Pb^{2+} ions. *Journal of Hazardous Materials* 164 (2–3): 982–986.

48 Basci, N., Kocadagistan, E., and Kocadagistan, B. (2004). Biosorption of copper (II) from aqueous solutions by wheat shell. *Desalination* 164 (2): 135–140.

49 Johnson, P.D., Watson, M.A., Brown, J., and Jefcoat, I.A. (2002). Peanut hull pellets as a single use sorbent for the capture of Cu (II) from wastewater. *Waste Management* 22 (5): 471–480.

50 Xie, R., Jin, Y., Chen, Y., and Jiang, W. (2017). The importance of surface functional groups in the adsorption of copper onto walnut shell derived activated carbon. *Water Science and Technology* 76 (11): 3022–3034.

51 Chand, R., Watari, T., Inoue, K. et al. (2009). Chemical modification of carbonized wheat and barley straw using HNO_3 and the adsorption of Cr (III). *Journal of Hazardous Materials* 167 (1–3): 319–324.

52 Nguyen, T.A.H., Ngo, H.H., Guo, W.S. et al. (2013). Applicability of agricultural waste and by-products for adsorptive removal of heavy metals from wastewater. *Bioresource Technology* 148: 574–585.

53 Park, D., Yun, Y.S., and Park, J.M. (2010). The past, present, and future trends of biosorption. *Biotechnology and Bioprocess Engineering* 15 (1): 86–102.

54 García-Rosales, G. and Colín-Cruz, A. (2010). Biosorption of lead by maize (*Zea mays*) stalk sponge. *Journal of Environmental Management* 91 (11): 2079–2086.

55 Abdel-Ghani, N.T., Hegazy, A.K., and El-Chaghaby, G.A. (2009). Typha domingensis leaf powder for decontamination of aluminium, iron, zinc and lead: biosorption kinetics and equilibrium modeling. *International Journal of Environmental Science and Technology* 6 (2): 243–248.

56 Feng, N., Guo, X., Liang, S. et al. (2011). Biosorption of heavy metals from aqueous solutions by chemically modified orange peel. *Journal of Hazardous Materials* 185 (1): 49–54.

57 Ashraf, M.A., Wajid, A., Mahmood, K. et al. (2011). Low cost biosorbent banana peel (*Musa sapientum*) for the removal of heavy metals. *Scientific Research and Essays* 6 (19): 4055–4064.

58 Prusty, J.K., Patro, S.K., and Basarkar, S.S. (2016). Concrete using agro-waste as fine aggregate for sustainable built environment–a review. *International Journal of Sustainable Built Environment* 5 (2): 312–333.
59 Modani, P.O. and Vyawahare, M.R. (2013). Utilization of bagasse ash as a partial replacement of fine aggregate in concrete. *Procedia Engineering* 51: 25–29.
60 Almeida, F.C., Sales, A., Moretti, J.P., and Mendes, P.C. (2015). Sugarcane bagasse ash sand (SBAS): Brazilian agroindustrial by-product for use in mortar. *Construction and Building Materials* 82: 31–38.
61 Sada, B.H., Amartey, Y.D., and Bakoc, S. (2013). An investigation into the use of groundnut as fine aggregate replacement. *Nigerian Journal of Technology* 32 (1): 54–60.
62 Ismail, Z.Z. and Jaeel, A.J. (2014). A novel use of undesirable wild giant reed biomass to replace aggregate in concrete. *Construction and Building Materials* 67: 68–73.
63 Sua-Iam, G. and Makul, N. (2013). Utilization of limestone powder to improve the properties of self-compacting concrete incorporating high volumes of untreated rice husk ash as fine aggregate. *Construction and Building Materials* 38: 455–464.
64 Nóvoa, P.J.R.O., Ribeiro, M.C.S., Ferreira, A.J.M., and Marques, A.T. (2004). Mechanical characterization of lightweight polymer mortar modified with cork granulates. *Composites Science and Technology* 64 (13–14): 2197–2205.
65 Moreira, A., António, J., and Tadeu, A. (2014). Lightweight screed containing cork granules: mechanical and hygrothermal characterization. *Cement and Concrete Composites* 49: 1–8.
66 Öztürk, T. and Bayrakl, M. (2005). The possibilities of using tobacco wastes in producing lightweight concrete. *Agricultural Engineering International: CIGR Journal* 7: 1–9. Manuscript BC 05 006.
67 Shafana, T. and Venkatasubramani, R. (2014). A study on the mechanical properties of concrete with partial replacement of fine aggregate with sugarcane bagasse ash. *International Journal of Advanced Structures and Geotechnical Engineering* 3 (1): 34–39.
68 Mageswari, M. and Vidivelli, B. (2009). The use of sawdust ash as fine aggregate replacement in concrete. *Journal of Environmental Research and Development* 3 (3): 720–726.
69 Kadam, K.L., Forrest, L.H., and Jacobson, W.A. (2000). Rice straw as a lignocellulosic resource: collection, processing, transportation, and environmental aspects. *Biomass and Bioenergy* 18 (5): 369–389.
70 El-Dewany, C., Awad, F., and Zaghloul, A.M. (2018). Utilization of rice straw as a low-cost natural by-product in agriculture. *International Journal of Environmental Pollution and Environmental Modelling* 1 (4): 91–102.
71 Devedee, A. and Parihar, R. (2018). Agricultural waste/stubble management (waste to wealth). *Alternative to Burning* 16: 43–45.
72 Moraes, C.A., Fernandes, I.J., Calheiro, D. et al. (2014). Review of the rice production cycle: by-products and the main applications focusing on rice husk combustion and ash recycling. *Waste Management and Research* 32 (11): 1034–1048.
73 Chaudhary, D.S., Jollands, M.C., and Cser, F. (2002). Understanding rice hull ash as fillers in polymers: a review. *Silicon Chemistry* 1 (4): 281–289.
74 Singh, D., Kumar, R., Kumar, A., and Rai, K.N. (2008). Synthesis and characterization of rice husk silica, silica-carbon composite and H_3PO_4 activated silica. *Ceramica* 54 (330): 203–212.

75 Adam, F. and Fua, H.K. (2008). Production of silica from biogenic material. Malaysian Patent MY-136715-A.

76 Kumar, A., Sengupta, B., Dasgupta, D. et al. (2016). Recovery of value added products from rice husk ash to explore an economic way for recycle and reuse of agricultural waste. *Reviews in Environmental Science and Bio/Technology* 15 (1): 47–65.

77 Kumar, S., Sangwan, P., Dhankhar, R.M.V., and Bidra, S. (2013). Utilization of rice husk and their ash: a review. *Research Journal of Chemical and Environmental Sciences* 1 (5): 126–129.

78 Shafie, S.M., Mahlia, T.M., Masjuki, H.H., and Rismanchi, B. (2012). Life cycle assessment (LCA) of electricity generation from rice husk in Malaysia. *Energy Procedia* 14: 499–504.

79 Roy, P., Shimizu, N., Shiina, T., and Kimura, T. (2006). Energy consumption and cost analysis of local parboiling processes. *Journal of Food Engineering* 76 (4): 646–655.

80 Taghizadeh-Alisaraei, A., Abbaszadeh-Mayvan, A., and Hosseini, S.H. (2019). Bio-ethanol production from a mixture of rice hull and orange peel wastes. *Biofuels*: 1–5.

81 Henderson, A.J., Ollila, C.A., Kumar, A. et al. (2012). Chemopreventive properties of dietary rice bran: current status and future prospects. *Advances in Nutrition* 3 (5): 643–653.

82 Tinoi, J., Rakariyatham, N., and Deming, R.L. (2006). Utilization of mustard waste isolates for improved production of astaxanthin by *Xanthophyllomyces dendrorhous*. *Journal of Industrial Microbiology and Biotechnology* 33 (4): 309–314.

83 Pal, S., Banik, S.P., and Khowala, S. (2013). Mustard stalk and straw: a new source for production of lignocellulolytic enzymes by the fungus *Termitomyces clypeatus* and as a substrate for saccharification. *Industrial Crops and Products* 41: 283–288.

84 Sahoo, R.K., Das, A., Gaur, M. et al. (2020). Parameter optimization for thermostable lipase production and performance evaluation as prospective detergent additive. *Preparative Biochemistry & Biotechnology* 50 (6): 578–584.

85 Badhe, P.S., Damale, M.A., and Adivarekar, R. (2016). Use of mustard oil cake for protease production by *Bacillus subtilis*. *International Journal of Current Microbiology and Applied Sciences* 5: 845–853.

86 Nemeş, L.N. and Bulgariu, L. (2016). Optimization of process parameters for heavy metals biosorption onto mustard waste biomass. *Open Chemistry* 14 (1): 175–187.

87 Upadhyay, K. and Harshwardhan, K. (2017). Effective utilization of agricultural waste–review paper. *International Journal of Engineering Research and Technology* 6 (9): 52–59.

88 Dukarska, D., Łęcka, J., and Szafoni, K. (2011). Straw of white mustard (*Sinapis Alba*) as an alternative raw material in the production of particle boards resinated with UF resin. *Acta Scientiarum Polonorum, Silvarum Colendarum Ratio et Industria Lignaria* 10: 21–28.

89 Ranum, P., Peña-Rosas, J.P., and Garcia-Casal, M.N. (2014). Global maize production, utilization, and consumption. *Annals of the New York Academy of Sciences* 1312 (1): 105–112.

90 FAOSTAT (2019). Food and agricultural organization of the United Nations. http://www.fao.org/faostat/en/#data/QC/visualize (accessed 27 September 2020)

91 Shinners, K.J. and Binversie, B.N. (2007). Fractional yield and moisture of corn stover biomass produced in the Northern US Corn Belt. *Biomass and Bioenergy* 31 (8): 576–584.

92 AgroCycle (2017). Characterisation of agricultural waste co- and by-products. http://www.agrocycle.eu/files/2017/10/D1.2_AgroCycle.pdf.

93 Vassilev, S.V., Baxter, D., Andersen, L.K., and Vassileva, C.G. (2010). An overview of the chemical composition of biomass. *Fuel* 89 (5): 913–933.

94 Darwish, G.A., Bakr, A.A., and Abdallah, M.M.F. (2012). Nutritional value upgrading of maize stalk by using *Pleurotus ostreatus* and *Saccharomyces cerevisiae* in solid state fermentation. *Annals of Agricultural Sciences* 57 (1): 47–51.

95 Sun, R.C., Sun, X.F., and Zhang, S.H. (2001). Quantitative determination of hydroxycinnamic acids in wheat, rice, rye, and barley straws, maize stems, oil palm frond fiber, and fast-growing poplar wood. *Journal of Agricultural and Food Chemistry* 49 (11): 5122–5129.

96 Torre, P., Aliakbarian, B., Rivas, B. et al. (2008). Release of ferulic acid from corn cobs by alkaline hydrolysis. *Biochemical Engineering Journal* 40 (3): 500–506.

97 Nagasaka, R., Chotimarkorn, C., Shafiqul, I.M. et al. (2007). Anti-inflammatory effects of hydroxycinnamic acid derivatives. *Biochemical and Biophysical Research Communications* 358 (2): 615–619.

98 Leffingwell, J.C. and Leffingwell, D. (2015). Flavours and fragnances: recent advances in biotechnology. *Flavours and Fragnances* 35: 32–34.

99 Foo, K.Y. (2016). Value-added utilization of maize cobs waste as an environmental friendly solution for the innovative treatment of carbofuran. *Process Safety and Environmental Protection* 100: 295–304.

100 Tsai, W.T., Chang, C.Y., Wang, S.Y. et al. (2001). Utilization of agricultural waste corn cob for the preparation of carbon adsorbent. *Journal of Environmental Science and Health, Part B: Pesticides, Food Contaminants, and Agricultural Wastes* 36 (5): 677–686.

101 Berber-Villamar, N.K., Netzahuatl-Muñoz, A.R., Morales-Barrera, L. et al. (2018). Corncob as an effective, eco-friendly, and economic biosorbent for removing the azo dye Direct Yellow 27 from aqueous solutions. *PLoS One* 13 (4): e0196428.

102 Patil, L.P., Patil, R.S., Deore, K.S. et al. (2019). Treatment of sewage waste water using an agricultural waste corn cobs. *International Journal of Advance Research, Ideas and Innovations in Technology* 5 (3): 51–56.

103 Shim, J., Lim, J.-M., Sheab, P.J., and Oh, B.-T. (2014). Simultaneous removal of phenol, Cu and Cd from water with corn cob silica-alginate beads. *Journal of Hazardous Materials* 272: 129–136.

104 Faustino, J., Pereira, L., Soares, S. et al. (2012). Impact sound insulation technique using corn cob particleboard. *Construction and Building Materials* 37: 153–159.

105 Paiva, A., Pereira, S., Sá, A. et al. (2012). A contribution to the thermal insulation performance characterization of corn cob particleboards. *Energy and Buildings* 45: 274–279.

106 Adesanya, D.A. and Raheem, A.A. (2009). A study of the workability and compressive strength characteristics of corn cob ash blended cement concrete. *Construction and Building Materials* 23 (1): 311–317.

107 Hiloidhari, M., Bhuyan, N., Gogoi, N. et al. (2020). Agroindustry wastes: biofuels and biomaterials feedstocks for sustainable rural development. In: *Refining Biomass Residues for Sustainable Energy and Bioproducts* (eds. R. Praveen Kumar, E. Gnansounou, J.K. Raman and G. Baskar), 357–388. Academic Press.

108 Dailey, A. and Vuong, Q.V. (2015). Effect of extraction solvents on recovery of bioactive compounds and antioxidant properties from macadamia (Macadamia tetraphylla) skin waste. *Cogent Food and Agriculture* 1 (1): 1115646.

109 Schieber, A., Stintzing, F.C., and Carle, R. (2001). By-products of plant food processing as a source of functional compounds – recent developments. *Trends in Food Science and Technology* 12 (11): 401–413.

110 Guo, C., Yang, J., Wei, J. et al. (2003). Antioxidant activities of peel, pulp and seed fractions of common fruits as determined by FRAP assay. *Nutrition Research* 23 (12): 1719–1726.

111 da Silva, A.C. and Jorge, N. (2014). Bioactive compounds of the lipid fractions of agro-industrial waste. *Food Research International* 66: 493–500.

112 Llorach, R., Espin, J.C., Tomas-Barberan, F.A., and Ferreres, F. (2002). Artichoke (*Cynara scolymus* L.) byproducts as a potential source of health-promoting antioxidant phenolics. *Journal of Agricultural and Food Chemistry* 50 (12): 3458–3464.

113 Moure, A., Cruz, J.M., Franco, D. et al. (2001). Natural antioxidants from residual sources. *Food Chemistry* 72 (2): 145–171.

114 Goldsmith, C.D., Vuong, Q.V., Sadeqzadeh, E. et al. (2015). Phytochemical properties and anti-proliferative activity of *Olea europaea* L. leaf extracts against pancreatic cancer cells. *Molecules* 20 (7): 12992–13004.

115 Vuong, Q.V., Hirun, S., Roach, P.D. et al. (2013). Effect of extraction conditions on total phenolic compounds and antioxidant activities of *Carica papaya* leaf aqueous extracts. *Journal of Herbal Medicine* 3 (3): 104–111.

116 García-Pérez, J.V., Cárcel, J.A., Benedito, J. et al. (2009). Extraction of antioxidant compounds from grape stalk dried at different conditions. *Defect and Diffusion Forum* 283: 604–609.

117 Garcia-Mendoza, M.P., Paula, J.T., Paviani, L.C. et al. (2015). Extracts from mango peel by-product obtained by supercritical CO_2 and pressurized solvent processes. *LWT-Food Science and Technology* 62 (1): 131–137.

118 Goli, A.H., Barzegar, M., and Sahari, M.A. (2005). Antioxidant activity and total phenolic compounds of pistachio (*Pistachia vera*) hull extracts. *Food Chemistry* 92 (3): 521–525.

119 Bai, X.L., Yue, T.L., Yuan, Y.H., and Zhang, H.W. (2010). Optimization of microwave-assisted extraction of polyphenols from apple pomace using response surface methodology and HPLC analysis. *Journal of Separation Science* 33 (23–24): 3751–3758.

120 Karantonis, H.C., Tsantila, N., Stamatakis, G. et al. (2008). Bioactive polar lipids in olive oil, pomace and waste byproducts. *Journal of Food Biochemistry* 32 (4): 443–459.

121 Jadhav, P., Sonne, M., Kadam, A. et al. (2018). Formulation of cost effective alternative bacterial culture media using fruit and vegetables waste. *International Journal of Current Research and Review* 10 (2): 6.

122 Deivanayaki, M. and Antony, I.P. (2012). Alternative vegetable nutrient source for microbial growth. *International Journal of Biosciences (IJB)* 2 (5): 47–51.

123 Wasas, A.D., Huebner, R.E., and Klugman, K.P. (1999). Use of Dorset egg medium for maintenance and transport of Neisseria meningitidis and Haemophilus influenzae type b. *Journal of Clinical Microbiology* 37 (6): 2045–2046.

124 Siddiqui, A., Salahuddin, T., and Riaz, A. (2014). Production of amylase from *Bacillus* sp. AY3 using fruit peels as substrate. *FUUAST Journal of Biology* 4 (2): 213–215.

125 Kindo, A.J., Tupaki-Sreepurna, A., and Yuvaraj, M. (2016). Banana peel culture as an indigenous medium for easy identification of late-sporulation human fungal pathogens. *Indian Journal of Medical Microbiology* 34 (4): 457.

126 Kumar, H., Bhardwaj, K., Sharma, R. et al. (2020). Fruit and vegetable peels: utilization of high value horticultural waste in novel industrial applications. *Molecules* 25 (12): 2812.

127 Ng, H.W., Lee, L.Y., Chan, W.L. et al. (2016). Luffa acutangula peel as an effective natural biosorbent for malachite green removal in aqueous media: equilibrium, kinetic and thermodynamic investigations. *Desalination and Water Treatment* 57 (16): 7302–7311.
128 Singh, S., Parveen, N., and Gupta, H. (2018). Adsorptive decontamination of rhodamine-B from water using banana peel powder: a biosorbent. *Environmental Technology and Innovation* 12: 189–195.
129 Mohammed, R.R. and Chong, M.F. (2014). Treatment and decolorization of biologically treated Palm Oil Mill Effluent (POME) using banana peel as novel biosorbent. *Journal of Environmental Management* 132: 237–249.

8

Antimicrobial Products from Plant Biodiversity

Pankaj Kumar Verma[1], *Shikha Verma*[2], *Nalini Pandey*[1], *and Debasis Chakrabarty*[2]

[1] *Department of Botany, University of Lucknow, Lucknow, Uttar Pradesh, India*
[2] *Molecular Biology and Biotechnology Division, Tissue Culture and Transformation Lab, CSIR- National Botanical Research Institute, Lucknow, India*

8.1 Introduction

Medicinal plants are natural resources and contain a number of products that perform significant roles in human health benefits [1]. In the current scenario, only a few plants (about 1–10% of the total estimated plants on earth 2.5×10^4 to 5.0×10^4) were identified for their medicinal properties and used to treat human diseases. However, the demand for medicinal plants to isolate the bioactive compounds has been increased rapidly. In past few years, researcher shifted their attention toward medicinal plants to demonstrate the promising role of plant-derived antimicrobial substances to treat many diseases caused by microbes. These plant extracts may have antimicrobial properties and may be used for various applications, including natural therapeutics, processed food preservation, alternative medication, and pharmaceuticals. Several communities of Asian countries widely used plant products to treat microbial infection as traditional medicine. For the preparation of drugs from phytochemical compounds extracted from plants, it is important to understand the structural composition and its pharmacological properties. Hence, the development of novel scientific approaches for the identification, characterization, and validation of different plant-derived antimicrobial components, extracts, and chemical or biological compounds is important in order to develop efficient and cost-effective herbal medicines. Therefore, attention must be focused on the propagation and preservation of medicinally important plant populations, which may be active sources for more effective herbal medicine. The medicinal use of plants is studied under ethnopharmacology for a long time. It involves the identification of raw materials for the preparation of advanced medicines and the examination of bioactive isolates from natural resources. Natural compounds such as secondary metabolites may constitute a vital source of chemical diversity. Various studies described the significance of natural plant isolates to cure human diseases. Due to the effectiveness of these natural compounds, around 69% new antibacterial and 21% antifungal drugs were approved between 1981 and 2006 to treat various diseases caused by microbes.

Bioprospecting of Plant Biodiversity for Industrial Molecules, First Edition.
Edited by Santosh Kumar Upadhyay and Sudhir P. Singh.

According to a recent study, the utilization of plant-derived medicines may be classified as a three-generation of the progressive development of plant-derived medicines. Since the ancient age, the crude extract of medicinal plants was used by traditional societies to manage infectious disease grouped in the first generation; progressively the active constituents were extracted from plants through scientific processing grouped as the second generation, and the third generation of phytotherapeutic agents was examined for comprehensive pharmacological/biochemical studies. Different plants of *Dracenaceae*, *Anacardiaceae*, *Pinaceae*, *Apiaceae*, *Burseraceae*, *Palmaceae*, *Cupressaceae*, *Euphorbiaceae*, and *Fabaceae* families are considered a rich source of secondary metabolites. These metabolites may have antimicrobial activity, thus may be used as potential antimicrobial agents to treat a number of virulent including drug-resistant microbes, therefore used as a superior and cost-effective substitute for antibiotics. For example, various species of *Hypericum* spp. (Common: St John's Wort, Hindi: Choli phulya) known as the world's leading herbal plant, which contained several medicinal compounds including *hyperenone A*, *hypercalin B*, *acylphloroglucinol hyperforin*, *xanthones*, and *naphthodianthrone*, etc., having an antibacterial, antiviral, antifungal, anti-inflammatory, anticancer, and antioxidant activity [2]. Similarly, *Moringa oleifera* contains flavonoids, alkaloids, gallic acid, catechol-type saponins, and tannins, which have antimicrobial and anti-inflammatory properties and thus helps in maintaining cholesterol level and blood pressure [3]. Additionally, extracts derived from medicinal plants such as *Kaempferia pandurata* and *Senna alata* possess antimicrobial properties for *Staphylococcus aureus* [4]. Similarly, *Anogeissus schimperi* and *Anacardium occidentale* against *Pseudomonas aeruginosa* and *Camellia sinensis* against *Listeria monocytogenes* [5]. Further, phenolics and alkaloids derived from *Cameroonian* plants own antimicrobial activity (MIC $< 10\,\mu g/mL$) [6]. Wangchuk and coworkers identified various classes of phytochemicals from a number of Bhutanese medicinal plants having a broad range of biological activities [7]. They suggested that these plants having flavonoids, alkaloids, and tannins that may be used to treat diseases caused by bacterial, fungal, malarial parasites, and also used to treat African sleeping sickness caused by *Trypanosoma brucei rhodesiense*. Additionally, an aromatic plant caper (*Capparis* sp.) found in Mediterranean Sea Basin has been used as a traditional medicine to eliminate many diseases, such as seeds extract of *Capparis decidua* possesses antileishmanial, antibacterial, and antifungal activity due to its quaternary ammonium and glucosinolate [8]. Phytochemical studies indicated the presence of different compounds, such as carotenoids, spermidine, tocopherol, rutin, campesterol, stigmasterol, quercetin, and kaempferol, which are responsible for antimicrobial, antifungal, antiviral, antioxidative, and anti-inflammatory properties.

Plants-derived antimicrobial compounds possess enormous therapeutic potential with lesser side effects and very less chance to develop resistance for medicine. Therefore, there is a need to investigate essential antimicrobial substances from plants and the development of novel drugs to combat microbial infections. In this chapter, an attempt has been done to summarize the current state of knowledge regarding the antimicrobial properties of essential substances derived from plants.

8.2 Use of Plant Products as Antimicrobials: Historical Perspective

Earth has a rich diversity of flora and fauna, around 350,000–500,000 plant species are recorded on Earth. From ancient times, humans learn the use of plant products to cure a disease by their experience. The oldest record for using medicinal plants was found in Indian holy books

Vedas, which stated about the use of different plant species to cure various diseases. Similar evidence also found from different countries, such as in China there was evidence of use of dried plant parts to treat diseases; some are still in use such as *Podophyllum*, *Rhei rhizoma*, *Theae folium*, *Ephedra*, *Cinnamon* bark, etc. The *Ebers Papyrus* written in 1500 BCE was considered as the most important medicinal book of ancient Egypt. It has a collection of 700 formulas and remedies from plants to treat various diseases. Additionally, two Homeric epics, the Iliad and the Odyssey (800 BCE) reported 63 plant species that were used for pharmacotherapy. In 459–370 BCE, Greek physician Hippocrates used 300 plants according to its pharmacological behavior, such as *Centaurium umbellatum Gilib* was used to treat fever, garlic was used for intestinal parasites and as a diuretic, also the deadly nightshade, henbane, opium, and mandrake as narcotics. Theophrastus (371–287 BCE), "the father of Botany," wrote two books "*De Causis Plantarium*" and "*De Historia Plantarium*," in which he classified and described more than 500 medicinal important plants. Pedanius Dioscorides "The father of pharmacognosy," wrote his work "De Materia Medica" in 77 CE, in which he described various plant-based drugs. He stated that, domestic plants such as garlic, chamomile, onion, coriander, sea onion, ivy, parsley, and marshmallow could be used as drugs, for example, chamomile used to treat wounds, burns, stings, and ulcers. Pliny the Elder described the use of about 1000 medicinal plants in his book "Historia Naturalis" in 23–79 CE and covered all the medicinally important plants from Germany and Spain. In 131–200 CE, Roman physician, Glen explained drugs identical to dioscorides, while introducing several new plants-based drugs, such as *Uvae ursi folium* used as uro-antiseptic and diuretic even also using currently. Slavic natives (seventh century CE) used several plants including *Veratrum album*, *Cucumis sativus*, *Rosmarinus officinalis*, *Allium sativum. Ocimum basilicum*, *Urtica dioica*, and *Mentha viridis* to treat several diseases. Charles the Great (742 CE–814) quoted 100 different plants in his work "Capitularies," which are also using in the modern world such as mint, poppy, sage, iris, marshmallow, onion, etc. The Arabs added various new plants such as coffee, aloe, ginger, saffron, cinnamon, pepper, etc. from India in pharmacotherapies, which continue to be used in the present world also. In the seventeenth century, many medicinal plants were introduced in European countries, such as *Cortex Chinae* from the bark of *Cinchona succirubra* Pavon. In the eighteenth century, Linnaeus described and classified the species in his work "Species Plantarium," a binominal system where each species consists of a genus name written with an initial capital letter and a species name with an initial small letter [9].

In the early nineteenth century, use and understanding of medicinal plants were more explored with the discovery, isolation, and validations of drugs. Alkaloids, glycosides, and quinine isolation from poppy and other plants initiated the start of scientific pharmacy. With improving a chemical process, other compounds were also discovered from medicinal plants such as tannins, vitamins, hormones, saponosides, etc. In the early twentieth century, a study was conducted for stabilization and cultivation of plants with medicinal properties. For the physiological, chemical, and clinical studies, various drugs producing extinct plants were restored for pharmacies such as *Opium*, *Punica granatum*, *Filix mas*, *Colchicum*, *Stramonium*, *Styrax*, *Ricinus*, and many more. Some plant drugs have a long history such as quinine used in the cure of malaria; currently, quinine analog chloroquine is a widely prescribed drugs for malaria. Similarly, higher plants also used in cancer therapies, for example, antileukemic alkaloids were isolated from *Catharanthus roseus* syn. *Vinca roseus*. There are a lot of examples of this kind; therefore, to explore the utility of many more ancient plants-based drugs, there is a need to study the old manuscripts and books to use as a potential source in pharmacotherapy.

8.3 Major Groups of Plants-Derived Antimicrobial Compound

Plant-derived substances that are used as medicines are principally secondary metabolites produced by the plants and the basic roles of these compounds are in defense mechanisms. These compounds impart flavor, pigmentation, and medicinal properties as well as possess a broad range of activities according to plant species, country origin climate, and topography. Their antimicrobial action was modified by variations in their chemical composition. Therefore, random screening of medicinal plants for active compounds is needed to evaluate their pharmacological activity. Phytochemical extracted from medicinal plants can be typically divided into several distinct categories, as described in the next sections (Table 8.1).

8.3.1 Simple Phenols and Phenolic Acids

Phenolic compounds possess a broad range of bioactive natural substances and are utilized for various medicinal purposes. These substances perform a significant role in protecting plants from diverse microbes and pathogens through several mechanisms [59]. They have potential antimicrobial and antioxidative properties [60]. Phenols have a vast group of aromatic compounds including flavonoids, flavones, and flavonols typically consisting of one carbonyl group, whereas quinines contain two carbonyl groups, tannins with polymeric phenolic substances, and coumarins contain fused benzene and pyrone groups [61]. All groups of phenolic compounds acquire antimicrobial properties [62]. Wide ranges of phenylpropane-derived compounds have higher oxidation states such as cinnamic and caffeic acids. Traditional herbs, thyme, and tarragon contain caffeic acid, which possesses antiviral, antibacterial, and antifungal properties. Catechol and pyrogallol are hydroxylated phenols, catechol has two hydroxyl groups, and pyrogallol contains three hydroxyl groups. These hydroxylated phenols have antimicrobial properties based on-site and the number of hydroxyl groups on the phenolic ring, with toxicity against microbes, increases with increased hydroxylation. The possible mechanisms of action of these oxidized compounds usually include microbial enzyme inhibition, mainly by reaction with sulfhydryl groups or via nonspecific interaction with specific proteins. While low-molecular-weight phenolics possess antimicrobial activity by diffusing across the microbial membrane, which leads to cytoplasmic acidification, causing cell death.

8.3.1.1 Flavonoids

Flavonoids are hydroxylated phenolic structures found in photosynthetic plants such as vegetables, fruits, stem, flowers, seeds, etc. The structural composition of flavonoid is 2-phenylbenzopyrane or flavone nucleus, comprising two benzene rings connected through a heterocyclic pyrene ring. On the basis of chemical composition and substituent position on rings, flavonoids are divided into 14 classes. The flavonoid compounds can form a complex with bacterial membranes as well as with both soluble and extracellular proteins result in antibacterial properties [63]. They additionally maintain inhibitory action against various viruses [64]. Several investigations demonstrated the efficiency of flavonoids such as glycyrrhizin, chrysin, and swertifrancheside against HIV. The studies found that flavonoid compounds have inhibitory action toward respiratory syncytial viruses (RSV). Another study

Table 8.1 Major classes of antimicrobial compounds from plants.

Class	Compounds	Plant species	Mechanisms of action	Active against	References
Phenolics	Resveratrol	*Vitis vinifera, Vaccinium corymbosum*	Efflux pump inhibitor, anti-biofilm	*B. cereus, E. faecalis, P. aeruginosa, M. smegmatis, C. jejuni*, MRSA	[10]
	Tannin/ tannic acid	*Sorghum* spp.		*S. aureus, S. typhimurium, A. niger, A. flavus, S. cerevisae*	[11]
	Baicalein	*Scutellaria baicalensis* Georgi	Efflux pump inhibitor anti biofilm	*M. smegmatis, C. albicans, S. epidermidis, P. aeruginosa, E. coli*, MRSA	[12]
	Biochanin A	*Trifolium pratense* L.	Efflux pump inhibitor	*Bifidobacterium* spp., *Clostridium* spp.	[13]
	Luteolin	*Cuminum cyminum, Bacopa monnieri, Achillea millefolium*	Biofilm inhibitor	*S. aureus, L. monocytogenes*	[14]
	Kaempferol	*Polygonum tinctorium* Lour	Efflux pump inhibitor, biofilm inhibitor	*S. aureus, H. pylori*	[15, 16]
	Quercetin	*A. japonica*	Efflux pump inhibitor	*S. aureus*	[17]
	Catechol	*Acacia catechu*	Antibiofilm	*E. coli, P. mirabilis, P. aeruginosa*	[18]
	Coumarin	*Dipteryx odorata, Galium odoratum*	Quorum sensing and biofilm formation inhibition	*E. aerogenes, E. coli, S. aureus, S. typhimurium*	[19, 20]
	Gallic acid	*Hamamelis virginiana, Camellia sinensis*	Cell membrane disruption, biofilm inhibition	*E. coli, L. monocytogenes, P. mirabilis, P. aeruginosa*	[21]
	Caffeic acid	*Eucalyptus globulus*	Destruction of cellular membrane	*B. cereus, E. coli, P. aeruginosa, S. aureus*	[22, 23]
	Catechin	*Acacia catechu, Camellia sinensis*	Reducing antioxidant capacity	*B. cereus, B. subtilis, S. aureus, E. faecalis, E. coli, Salmonella*	[24]
	Apigenin	*P. oleracea L.*		*P. aeruginosa, S. typhimurium, P. mirabilis, K. pneumoniae, E. aerogenes*	[25]
	Naringenin	Citrus	Genomic DNA-binding	*S. aureus*	[26]
	Curcumin	*Curcuma* longa	Efflux pump inhibitor	*P. aeruginosa*	[27]

(Continued)

Table 8.1 (Continued)

Class	Compounds	Plant species	Mechanisms of action	Active against	References
Alkaloids	Reserpine	*Rauvolfia serpentina*	Biofilm inhibitor	*P. aeruginosa, S. aureus*	[28, 29]
	Piperine	*Piper nigrum*	Efflux pump inhibitor	*B. subtilis, E. coli, K. pneumoniae,* MRSA	[30, 31]
	Berberine	*Berberis vulgaris*	Cell division inhibitor, protein and DNA synthesis inhibitor	*E. coli, M. luteus, P. aeruginosa,* MRSA	[32, 33]
	Conessine	*Holarrhena floribunda*	Efflux pump inhibitor	*P. aeruginosa*	[34]
	Tomatidine	*Solanaceae*	Inhibitor of macromolecule biosyntheses	*S. aureus, B. cereus, B. subtilis, L. monocytogenes*	[35]
	Chelerythrine	*Chelidonium majus, Toddalia asiatica*	Destruction of bacterial cell wall and cell membrane, inhibition of protein biosynthesis	*B. subtilis, E. coli, K. pneumonia, S. pyogene*	[36, 37]
	Sanguinarine	*Sanguinaria canadensis*	Antibiofilm	*K. pneumonia, P. aeruginosa, S. pyogenes, C. albicans,* MRSA	[38–40]
Coumarin	Aegelinol	*Ferulago campestris*	DNA gyrase inhibitor	*S. enterica* serovar Typhi, *E. aerogenes, E. cloacae, S. aureus*	[41]
	Agasyllin	*Ferulago campestris*	DNA gyrase inhibitor	*S. enterica serovar Typhi, E. aerogenes, E. cloacae, S. aureus*	[41]
	4′-Senecioil-oxyosthol	*Prangos hulusii*	DNA gyrase inhibitor	*B. subtilis*	[42]
	Osthole	*Cnidium monnieri, Angelica pubescens*	DNA gyrase inhibitor	*B. subtilis, S. aureus, K. pneumoniae, MSSA*	[43]
	Novobiocin	*Citrullus colocynthis*	DNA gyrase inhibitor	*A. baumannii*	[44]
	Bergamottin epoxide	Grapefruit	Efflux pump inhibitor	*MSRA*	[45]
	Galbanic acid	*Ferula* species	Efflux pump inhibitor	*S. aureus, E. coli*	[46]

Terpene	Farnesol	*Usnea longissima*	Antibiofilm	*S. aureus*	[47]
	Nerolidol	*Momordica charantia*	Antibiofilm	*S. aureus*	[48]
	Dehydro-abietic acid	*Picea abies*, *Larix decidua*	Antibiofilm	*S. aureus*, *E. coli*	[49]
	Carvone	*Lippia alba*, *Mentha spicata* L.	Antibiofilm	*S. aureus*	[50]
	Thymol	*Thymus vulgaris*	Efflux pump inhibition, cell membrane disturbance, inhibits H(+)-ATPase in the cytoplasmic membrane, antibiofilm	*C. albicans*, *C. glabrata*, *C. krusei*, *A. niger*, *A. fumigatus*, *A. flavus*, *A. ochraceus*, *A. alternata*, *B. cinerea*, *Cladosporium* spp., *P. citrinum*, *P. chrysogenum*, *F. oxysporum*, *Rhizopus oryzae*, *E. coli*, *E. aerogenes*, *S. aureus*, *P. aeruginosa*, *S. typhimurium*, *S. enteritidis*, *S. saintpaul*	[51, 52]
	Carvacrol	*Anabasis setifera*	Cell membrane disturbance, efflux pump inhibition, antibiofilm	*A. niger*, *A. fumigatus*, *A. flavus*, *A. ochraceus*, *A. alternata*, *B. cinerea*, *Cladosporium* spp., *P. citrinum*, *P. chrysogenum*, *F. oxysporum*, *Rhizopus oryzae*, *E. coli*, *E. aerogenes*, *S. aureus*, *P. aeruginosa*	[53]
	Eugenol	*Eugenia aromaticum*, *Ocimum sanctum* L.	Cell membrane disturbance, binding to quorum sensing receptors	*A. niger*, *A. fumigatus*, *A. flavus*, *A. ochraceus*, *A. alternata*, *B. cinerea*, *Cladosporium* spp., *P. citrinum*, *P. chrysogenum*, *F. oxysporum*, *Rhizopus oryzae*, *P. aeruginosa*, MRSA, MSSA	[54–56]
	Menthol	*Mentha arvensis L.*	Cell membrane disturbance	*A. niger*, *A. fumigatus*, *A. flavus*, *A. ochraceus*, *A. alternata*, *B. cinerea*, *Cladosporium* spp., *P. citrinum*, *P. chrysogenum*, *F. oxysporum*, *Rhizopus oryzae*	[54]
	Cinnam-aldehyde	*Cinnamo-mum zeylanicum*	Cell membrane disturbance, antibiofilm	*Streptococcus* spp., *S. aureus*, *S. epidermidis*, *Campylobacter* spp.	[57, 58]

demonstrated the flavonoids' inhibitory activity toward different viruses such as poliovirus type1, RSV, simplex virus type 1(HSV-1), and parainfluenza virus type 3 and reported that hesperetin, quercetin, and catechin hinder the viral activity in diverse ways. Quercetin and catechin reduce the infectivity of the virus while hesperetin reduces the intracellular replication of viruses [65]. The study proposed that small structural differences played a critical role in their activity. Some flavonoids such as phloretin possess activity to counter a massive variety of microorganisms, Galangin from *Helichrysum aureonitens* showed activity against a broad range of gram-positive bacteria, fungi, and viruses. Catechins comprise a reduced form of C3 unit in flavonoid compounds and have been widely explored because of its presence in green tea [66]. It was observed that tea consists of a mixture of catechin and epicatechin, possess antimicrobial activity. These compounds suppressed *Streptococcus* mutants, *Vibrio cholera*, *Shigella*, and other microbial activity. *In vitro* study demonstrated that catechins inhibited glucosyltransferases in *S. mutans* and inactivated cholera toxin in *V. cholera, in vivo;* activity also confirmed the same [67]. Flavonoids, chalcones displayed diverse biological activities due to multiple substitutions in its structure. It is investigated against *E. coli* and *S. aureus* and found that their antimicrobial activity is due to the energy variation between two molecular orbitals. There is a conflict for the possible mechanism of action of flavonoids. Several reports found that flavonoids lack –OH groups on their B rings and have more antimicrobial activity than those that occupy two –OH groups; this supports the findings that mechanism of action of these flavonoids is a microbial membrane. Whereas various authors found contradict results, mean more hydroxylation, higher the antimicrobial activity. Therefore, it requires further study to correlate between the degree of hydroxylation and the effectiveness of the antimicrobial activity of flavonoid compounds.

8.3.1.2 Quinones

Quinones are a group of secondary metabolites, derived from aromatic rings having two ketone substitutions and contain important antimicrobial properties. These compounds cause brown color reaction in cut or injured vegetables and fruits and also an intermediate between melanin synthesis pathways in humans. The exchange between hydroquinone and quinine takes place by oxidation and reduction reactions. In many biological systems, the significant role of quinone–hydroquinone pair is very crucial, like the role of coenzyme Q (ubiquinone) in mammalian electron transport system. Additionally, they form stable free radicals that complex irreversibly with nucleophilic amino acids of microbial proteins, which leads to inactivation and loss of protein function. This is the reason for a broad spectrum of the antimicrobial effect of quinone. It targets the microbial cell's surface-exposed adhesions, membrane-bound proteins, and cell wall peptides, causing the death of microorganisms. A report illustrated anthraquinone from *Cassia italic*, having bacteriostatic property on *Bacillus anthracis*, *P. aeruginosa*, and *Corynebacterium pseudodiphthericum*, and bactericidal property against *Pseudomonas pseudomalliei* [68]. Another anthraquinone, hypericin from *Hypericum perforatum*, has antidepressant and antimicrobial properties [69].

8.3.1.3 Tannins

Tannins are a group of polymeric phenolic compounds found in all parts of plants such as wood, bark, roots, leaves, and fruits [70]. The molecular weight of tannin ranges from 500 to 3000. Tannins are classified into two groups: hydrolyzable tannin and condensed tannin.

Hydrolyzable tannins are multiple esters with D-glucose based on gallic acid while the condensed tannins are called proanthocyanidins, derivative of flavonols. Tannins possess antimicrobial activity by inactivating microbial adhesions proteins, transporter proteins, and enzymes. A group of Gallotannin-rich plant extracts inhibited the diverse microbial activity through the inactivation of membrane-bound proteins. In plants, the cytoplasmic leaflet of endoplasmic reticulum synthesizes anthocyanidin and accumulated in vacuoles. Typically two groups of proanthocyanidins (A and B) are discovered on the basis of the type of bonds formed between the oligomer forming anthocyanidin molecules. In A-type proanthocyanidins, two bonds are present between carbons of 2b-7 and 4b-8, while in B type only one 4b-8 bond is present between oligomer-forming molecules.

The positive results of anthocyanin have been very well-known from sixteenth century, where blackberry juice was applied to cure eye and mouth infections in humans. Though, relatively few studies have demonstrated the antimicrobial action of these substances. The study described the anthocyanin profile of different berries, containing stilbenoid resveratrol, possessing activity on gram-positive bacteria. Corilagin extracted from *Terminalia chebula* showed antimicrobial activity for *Acinetobacter baumannii*, *S. aureus*, *E. coli*, and *Candida albicans* [71]. Similarly, tannic acid, punicalagin, geraniin, and castalagin demonstrated inhibitory action against *E. coli*, *S. aureus*, and *Vibrio* [72]. In addition, punicalin, gallic acid, ellagic acid, and punicalagins isolated from *P. granatum* showed antimicrobial activity against methicillin-resistant *S. aureus* (MRSA), *Mycobacterium intracellulare*, *E. coli*, *Aspergillus fumigatus*, *P. aeruginosa*, *Cryptococcus neoformans*, and *C. albicans* [73]. Tannic acid contains significant antimicrobial action and is additionally used as a safe food additive in cocoa beans and tea. It indicated inhibitory action by binding to iron to chelate and make it unavailable for several microorganisms such as *Enterobacter cloacae*, *Bacteroides fragilis*, *E. coli*, and *Clostridium perfringens* [74]. Hydrolyzable tannins also showed antiviral activity such as casuarinin extracted from *Terminalia arjuna* showed anti-HSV-2 activity through the prevention of attachment and penetration [75]. Chebulagic acid and punicalagin isolated from *T. chebula* showed action against anti-HSV-1, measles virus, cytomegalovirus, dengue virus, hepatitis C virus, and respiratory syncytial virus [76]. Silver nanoparticles intrinsically possess superior anti-microbial property, in this way; tannic acid modified with silver nanoparticles had higher anti-HSV-2 activity [77].

8.3.1.4 Coumarins

Coumarins are heterocyclic phenolic compounds naturally found in many plants and microorganisms [78]. Coumarins are found free or as heterosides in various plant families such as *Fabaceae*, *Apiaceae*, *Rosaceae*, *Asteraceae*, *Moraceae*, *Solanaceae*, *Rubiaceae*, and *Rutaceae*, as well as *Gramineae* and orchids, which contain enormous amounts of coumarins. It is mostly synthesized in leaves, but its concentration is higher in fruits followed by roots and stems. These compounds exhibit a broad range of biological activities such as anticancer, antimicrobial, antiviral, antioxidant, antidiabetic, antihypertensive, antiparasitic, and anti-inflammatory activities. All compounds exhibited activity against *E. coli*, *S. aureus*, and *Bacillus subtilis*. There are several studies that demonstrated the antimicrobial activity of both natural coumarins and synthetic derivatives of coumarins. Coumarins and pyranocoumarins isolated from *Ferulago campestris* roots naturally have antimicrobial and antioxidant activities against both Gram-negative and positive bacteria [19]. Warfarin, a well-known coumarin, was the first used compound as oral anticoagulant and rodenticide equally retaining antiviral activity [79]. Scopoletin, a well-identified coumarin and two chalcones were extracted from *Fatoua pilosa* having antitubercular activity [80]. El-Seedi discovered new aryl coumarin glucoside, asphodelin A 4′-O-β-D-glucoside as well as asphodelin A from *Asphodelus microcarpus* [81]. A study investigated clorobiocin, novobiocin, and coumermycin A1 derived from distinct species of *Streptomyces*, containing antibiotic activity [82]. This study displayed that noviosyl sugar moiety in the compound is essential for its biological activity, in addition to the coumarin portion [83]. They also demonstrated that coumarins act as strong inhibitors of DNA topoisomerase type II, called DNA gyrase. Coumarins' antimicrobial activity is due to a passive diffusion mechanism to facilitate cellular penetration, particularly for Gram-positive bacteria. Additionally, Sardari and coworkers [84] demonstrated that in coumarin nucleus, free 6-hydroxyl and 7-hydroxyl have significant role in antifungal and antibacterial activity, respectively. The study suggested that coumarins having methoxy group at C-7, in addition to –OH moiety at either C-6 or C-8 invariably possessed antibacterial activity against a broad spectrum of bacteria. While the presence of aromatic dimethoxy array generated compounds against microorganisms that involve special growth factors including *Streptococcus pneumonia*, beta-hemolytic *Streptococcus*, and *Haemophilus influenza*. Collinin, which was extracted from *Zanthoxylum schinifolium* exhibited anti-HBV (hepatitis B virus) activity [85]. The antimalarial activity was observed in daphnetin, isolated from the plants of the genus Daphne, and also found in dentatin and clausarin, extracted from *Clausena harmandiana* [86].

8.3.2 Terpenes and Essential Oils

Terpenes are compounds called isoprenoids and their chemical structure is $C_{10}H_{16}$. Terpenes are also present as diterpenes, triterpenes, tetraterpenes (C20, C30, and C40), as well as hemiterpenes (C5) and sesquiterpenes (C15). Terpenes biosynthesis occurs mainly via two pathways, namely the Mevalonic acid pathway, which synthesized sesquiterpenes, sterols, and ubiquinones; and Methyl erythritol phosphate pathway, which primarily produces diterpenes, triterpenes, and hemiterpenes. Typical examples of terpenoids are camphor (monoterpenes), artemisinin (sesquiterpenoids) methanol, and farnesol. Artemisinin and its derivatives are equally known as qinghaosu, currently used as an antimalarial drug. Nerolidol, sesquiterpene found in several plants and used as scent products and food flavoring agent, exhibited, antimicrobial, anthelmintic, and antimalarial activity [87–90]. It was found that approximately 60% of essential oil derivatives observed to date are able to inhibit fungal growth, while 30% essential oils are able to inhibit bacteria growth. The mechanism of action of terpenes involves membrane disruption by the lipophilic compounds [90]. Terpenes showed activity against bacteria, fungi, viruses, and protozoa as well as a triterpenoid, betulinic acid able to prevent HIV infection [91]. Cichewicz and Thorpe showed that capsaicin improved the growth of *Candida albicans* but also acts as bactericidal to *Helicobacter pylori* [92]. Diterpene, a framodial from Cameroonian spice, is broad-spectrum antifungal. Guimarães and coworkers screened 33 terpenes and found that only 16 had antimicrobial activity [93]. Eugenol displayed antimicrobial activity against *Salmonella enteric*, and terpineol exhibited significant bactericidal action against *S. aureus* strains. Furthermore, carveol, geraniol, and citronellol showed antimicrobial activity against *E. coli* [94]. This report also demonstrated that terpenes containing hydroxyl groups increase the antimicrobial activity, whereas hydrocarbons decrease its activity. Citrus, bergamot oil, and terpenes also showed bactericidal action against various bacterial strains [95].

8.3.3 Alkaloids

Alkaloids are naturally occurring organic compounds that predominantly contain nitrogen atoms, possessing antimicrobial activities. Morphine, isolated and identified in 1805, was the first medically used alkaloid from *Papaver somniferum* [96]. Diterpenoid alkaloids isolated from *Ranunculaceae* plants consist of antimicrobial properties [97]. Berberine, an important alkaloid, extracted from the root and stem of Berberis species widely used as traditional medicine, has antibacterial, antiviral, antifungal, and antiprotozoal activity [98]. It is an excellent DNA intercalator and target on nucleic acid, RNA polymerase, topoisomerase IV, and gyrase actively, thus showing activity against several microorganisms. Ungeremine is methanol extract isolated from *Pancratium Illyricum* L. bulbs, possesses antimicrobial activities [99]. Solamargine, a glycoalkaloid, and other alkaloids isolated from *Solanum aculeastrum* berries might be useful against cancer-related to AIDS [100]. Lupanine, 13α-hydroxylupanine, albine, angustifoline, and 13α-tigloyloxylupanine isolated from the Lupinus genus displayed excellent activity towards *P. aeruginosa* and *Klebsiella pneumonia* [101]. Many alkaloids were extracted from *Chelidonium majus* (*Papaveraceae*) exhibit antimicrobial activity, such as chelerythrine, which was mainly effective toward *P. aeruginosa* and *C. albicans*, while sanguinarine showed activity against *S. aureus* [102]. Quinoline alkaloids, such as kokusagine, dictamnine, and maculine, extracted from bark of *Teclea afzelii* displayed effective antimicrobial activity [103]. Reserpine, an indole alkaloid isolated from *Rauwolfia serpentina*, possesses potent antibacterial activity [104]. Tomatidine, a steroidal alkaloid extracted from plants of *Solanaceae* displayed efficient antibacterial activity against *S. aureus* [35]. Apart from this, chanoclavine is classified as tricyclic ergot alkaloid, identified from *Ipomoea muricata*, exerts synergistic effects upon coadministered with tetracycline against *E. coli* [105]. *H. antidysenterica* bark contains steroidal alkaloid conessine displayed possible antimicrobial activity [106].

8.4 Mechanisms of Antimicrobial Activity

The antibacterial property of a substance is primarily based on two mechanisms; first, including chemical interference in the production or function of fundamental components of bacteria; and second, preventing traditional antimicrobial resistance mechanisms. These compounds provide multiple targets on bacterial cells that include (i) cell wall biosynthesis, (ii) cell membrane destruction, (iii) protein biosynthesis, (iv) metabolic-pathway inhibition, and (v) DNA replication and repair mechanism. In response to this, bacterial cells also show various resistance mechanisms against antimicrobial agents. Therefore, it is critical to find out the mechanism responsible for evolution of antibiotic resistance, primarily involving the activation of efflux pump, modifying, and destruction of antibacterial reagents with the help of enzymes and alteration of the target structures. It is equally significant that one sort of mechanism behind the resistance to antibacterial agents may function alone or together with other mechanisms. So, in the next sections, different mode of action of plants-derived products will be discussed.

8.4.1 Plant Extracts with Efflux Pump Inhibitory Activity

Bacterial multidrug resistance by the expression of the efflux pump represents a crucial problem in clinical trials. Therefore, an effective approach should be required to identify the molecules or compounds interfering with the bacterial efflux process. It is sufficiently known that plant-derived compounds have antimicrobial activity against Gram-positive bacteria by inhibiting the efflux pump while Gram-negative bacteria have innate multidrug resistance to many antimicrobial compounds due to efflux pumps. Several studies described a number of bacterial efflux pump inhibitors from different plants. Plant alkaloids such as berberine, reserpine, isoflavones, and methoxylated flavones have putative efflux inhibitory activity. Similarly, gallotannin extracted from *Terminalia chebula* demonstrated antimicrobial activity for *E. coli* and MDR. Additionally, an essential oil extract from the Corsican plant, *Helichrysum italicum*, consists of geraniol, demonstrated activity against different Gram-negative bacteria with chloramphenicol. Also, falcarindiol, isolated from *Levisticum officinale*, displayed synergistic activity with ciprofloxacin against Gram-negative bacteria. An efflux pump inhibitor, phenylalanine arginine b naphthylamide (PAbN) enhances *Dichrostachys glomerata* extract activity against *Klebsiella pneumonia*, *E. coli*, and *Providencia stuartii*. *Acer saccharum* extract showed efflux pump inhibitory potentials against *P. aeruginosa*, *E. coli*, and *P. mirabilis* [107]. Ursolic acid isolated from *Eucalyptus tereticornis* acts as a putative efflux pump inhibitor against MDR *E. coli* (KG4) [108]. Similarly, berberine and palmatine extracted from roots and rhizomes of *Berberis vulgaris* showed efficient efflux pump inhibitory efficacies against *P. aeruginosa* [109]. Most of the studies focused on identifying the efflux pump inhibitors against Gram-positive bacteria with very few reports on Gram-negative strains because Gram-negative bacteria are difficult to target due to the presence of more powerful efflux pumps and other efficient membrane barriers [110]. A study displayed the inhibition of EmrD3 pump-mediated drug efflux of *Vibrio cholera* by allyl sulfide, a bioactive compound present in garlic extract [111]. Considerably, more investigation is required to target Gram-negative bacteria by exploring different plant-derived compounds or molecules to act as potent efflux pump inhibitors.

8.4.2 Plant Extracts with Bacterial Quorum Sensing Inhibitory Activity

Bacterial populations collaborate and communicate through diffusible signal molecules and detection of these signal molecules known as quorum sensing (QS). The quorum sensing regulates the expression of specific bacterial genes, which modify the local host environment favoring the invasion of a pathogen. Therefore, recognizing the fact that numerous bacteria utilize QS to regulate their virulence, the QS system becomes a target for antimicrobial activity. The perfect QS inhibitor (QSI) could have low molecular weight, able to decrease the expression of QS-controlled genes, and have high specificity toward the target QS molecule. Several studies described the effect of plant-derived compounds on QS and used in clinical approaches. For example, products extracted from marine red algae *Delisea pulchra*, halogenated furanones were the first identified anti-QS compounds. In addition, plant extracts from garlic, *Capsicum chinensis*, carrot, yellow pepper, propolis inhibit *P. aeruginosa* QS [112]. This study also discussed that garlic extract contains three separate inhibitors of QS and reduced pathogenicity of *Caenorhabditis elegans* and *P. aeruginosa*. Phytoalexin resveratrol (3,5,4′-trihydroxystilbene) has direct antimicrobial activity against *Neisseria meningitides* and *Neisseria gonorrhoeae* [113]. The extracts isolated from *Citrus sinensis*, *Elettaria cardamomum*, *Laurus nobilis*, *Coriandrum sativum*, and *Allium cepa* exhibited inhibition of QS, so suppressing the virulence of *P. aeruginosa* [114]. Further, polyphenol of *Rosa rugosa* tea extract inhibited the QS regulated violacein production in *Chromobacterium violaceum* [114]. Currently, it was also observed that *Apium graveolens* oleoresin has an anti-QS activity against *Chromobacterium violaceum* CV12472 and *Pseudomonas aeruginosa* PAO1 [115]. Tannins isolated from *Terminalia bellirica*, *Phyllanthus emblica*, *S. cumini*, *Mangifera indica*, *Punica granatum*, and *Terminalia chebula* exhibited broad-spectrum anti-QS activity [116].

8.4.3 Plant Extracts with Biofilm Inhibitory Activity

Microbial biofilms are cell arrangement, attached to the substrate and enclosed by self-produced extra polymeric substance (EPS) matrix. Biofilm base infection causes various diseases in human being and its eradication with typical antibiotics is extremely difficult. For instance, *Staphylococci* can adhere and form biofilm on both abiotic surfaces and eukaryotic cells and perform a significant role in virulence hence difficult to treat. Compounds isolated from *Macleaya cordata*, *Aesculus hippocastanum*, *Krameria lappacea*, and *Chelidonium majus* show clinically significant inhibitory action on Staphylococci. Proanthocyanidin isolated from *A. hippocastanum* and chelerythrine extracted from *Macleaya cordata* downregulate the different proteins involved in various pathways and inhibit biofilm formation. The sanguinarine and chelerythrine act on bacterial proteins involved in synthesis of heat shock response including surface-exposed lipids and methoxymycolic acid. They also act as bacterial cytoskeletal protein inhibitor; therefore, they function as a beneficial compound for the development of novel drugs [117]. Tannin, hamamelitannin, isolated from *Hamamelis virginiana*, appreciably reduce biofilm metabolic activity of diverse microbes. Carvacrol, a natural biocide, inhibits the biofilm formation of *S. aureus* and *Salmonella enteric*. Collectively with thymol, interaction with lipid bilayer membranes causes loss of integrity of bacterial cells by affecting its structural and functional properties. Similarly, A 1-deoxynoijirimycin isolated from *Morus alba* inhibited biofilm formation activity of *Streptococcus mutans*, which lowers bacterial extracellular polysaccharide secretion. *Propionibacterium acnes*, with the ability to form biofilm and cause acne vulgaris, is susceptible to compounds such as salidroside, icariin, and resveratrol extracted from plants. Extract isolated from *Piper betle*, 4-chromanol acts as an effective antibacterial and antibiofilm agent against oral pathogens [118].

8.5 Conclusions and Future Prospects

In the last few years, researchers explored the importance and effectiveness of plant-derived compounds as antimicrobial agents. Plants are a rich source of a broad range of secondary metabolites, such as alkaloids, flavonoids, phenols; tannins, etc., thus have great potential to be a source of novel antimicrobial agents. To date, thousands of phytochemicals have shown inhibitory activity against different types of microorganisms. Various compounds showed both antibacterial and antibiotic modifying resistance activity. While some are not effective as antibiotics alone, however, when co-administrated with antibiotics, they significantly inhibited the antibiotic resistance mechanism in bacteria. Hence, it is essential to investigate the exact molecular mechanism and synergetic effects of these compounds with antibiotics. Furthermore, it is also important to know antimicrobial plant mechanisms for developing a novel clinical approach and also to understand the safety of antimicrobial phytochemicals.

The spread of drug-resistant microbes withstands a massive challenge to the successful treatment of microbial diseases. Therefore, there is an urgent need to discover novel compounds characterized by varied chemical structures and mechanisms of action. The uses of several plant-derived compounds as antimicrobial and antifungal agents represent an unusual approach for discovering bioactive products that could become helpful as therapeutic tools. Therefore, continued researches should be carried out to improve the understanding of exact mechanisms and also explore the pharmacodynamic and pharmacokinetic properties of the molecules.

References

1 Bernardini, S., Tiezzi, A., Laghezza Masci, V., and Ovidi, E. (2018). Natural products for human health: an historical overview of the drug discovery approaches. *Natural Product Research* 32: 1926–1950.

2 Marrelli, M., Statti, G., and Conforti, F. (2020). *Hypericum* spp.: an update on the biological activities and metabolic profiles. *Mini Reviews in Medicinal Chemistry* 20: 66–87.

3 Dhakad, A.K., Ikram, M., Sharma, S. et al. (2019). Biological, nutritional, and therapeutic significance of *Moringa oleifera* Lam. *Phytotherapy Research* 33: 2870–2903.

4 Wikaningtyas, P. and Sukandar, E.Y. (2016). Evaluation of the combined antibacterial activity of *Kaempferia pandurata* rhizome and *Senna alata* leaves against methicillin-resistant *Staphylococcus aureus*. *Evaluation* 9: 112–114.

5 Kudi, A.C. (2000). *Antiviral and Antibacterial Screening of Some Nigerian Medicinal Plant Extracts*. University of Leicester.

6 de Dieu Tamokou, J., Chouna, J.R., Fischer-Fodor, E. et al. (2013). Anticancer and antimicrobial activities of some antioxidant-rich Cameroonian medicinal plants. *PLoS One* 8: e55880.

7 Wangchuk, P., Sastraruji, T., Taweechotipatr, M. et al. (2016). Anti-inflammatory, anti-bacterial and anti-acetylcholinesterase activities of two isoquinoline alkaloids–scoulerine and cheilanthifoline. *Natural Product Communications* 11: 1801–1804.

8 Nazer, M.R., Jahanbakhsh, S., Ebrahimi, K. et al. (2020). Cytotoxic and antileishmanial effects of various extracts of *Capparis spinosa* L. *Turkish Journal of Pharmaceutical Sciences* doi:10.4274/tjps.galenos.2020.87259.

9 Petrovska, B.B. (2012). Historical review of medicinal plants' usage. *Pharmacognosy Reviews* 6: 1.

10 Bostanghadiri, N., Pormohammad, A., Chirani, A.S. et al. (2017). Comprehensive review on the antimicrobial potency of the plant polyphenol resveratrol. *Biomedicine and Pharmacotherapy* 95: 1588–1595.

11 Sulieman, A.M.E., Issa, F.M., and Elkhalifa, E.A. (2007). Quantitative determination of tannin content in some sorghum cultivars and evaluation of its antimicrobial activity. *Research Journal of Microbiology* 2: 284–288.

12 Kolouchová, I., Maťátková, O., Paldrychová, M. et al. (2018). Resveratrol, pterostilbene, and baicalein: plant-derived anti-biofilm agents. *Folia Microbiologica* 63: 261–272.

13 Sklenickova, O., Flesar, J., Kokoska, L. et al. (2010). Selective growth inhibitory effect of biochanin A against intestinal tract colonizing bacteria. *Molecules* 15: 1270–1279.

14 Qian, W., Liu, M., Fu, Y. et al. (2020). Antimicrobial mechanism of luteolin against *Staphylococcus aureus* and *Listeria monocytogenes* and its antibiofilm properties. *Microbial Pathogenesis* 142: 104056.

15 Ming, D., Wang, D., Cao, F. et al. (2017). Kaempferol inhibits the primary attachment phase of biofilm formation in *Staphylococcus aureus*. *Frontiers in Microbiology* 8: 2263.

16 Kataoka, M., Hirata, K., Kunikata, T. et al. (2001). Antibacterial action of tryptanthrin and kaempferol, isolated from the indigo plant (*Polygonum tinctorium* Lour.), against *Helicobacter pylori*-infected Mongolian gerbils. *Journal of Gastroenterology* 36: 5–9.

17 Hirai, I., Okuno, M., Katsuma, R. et al. (2010). Characterisation of anti-*Staphylococcus aureus* activity of quercetin. *International Journal of Food Science and Technology* 45: 1250–1254.

18 Maisuria, V.B., Hosseinidoust, Z., and Tufenkji, N. (2015). Polyphenolic extract from maple syrup potentiates antibiotic susceptibility and reduces biofilm formation of pathogenic bacteria. *Applied and Environmental Microbiology* 81: 3782–3792.

19 Hussain, M.I., Syed, Q.A., Khattak, M.N.K. et al. (2019). Natural product coumarins: biological and pharmacological perspectives. *Biologia* 74: 1–26.

20 Nitiema, L.W., Savadogo, A., Simpore, J. et al. (2012). in vitro; antimicrobial activity of some phenolic compounds (coumarin and quercetin) against gastroenteritis bacterial strains. *International Journal of Microbiological Research* 3: 183–187.

21 Borges, A., Ferreira, C., Saavedra, M.J., and Simoes, M. (2013). Antibacterial activity and mode of action of ferulic and gallic acids against pathogenic bacteria. *Microbial Drug Resistance* 19: 256–265.

22 Meyuhas, S., Assali, M., Huleihil, M., and Huleihel, M. (2015). Antimicrobial activities of caffeic acid phenethyl ester. *Journal of Molecular Biochemistry* 4: 21–31.

23 Kępa, M., Miklasińska-Majdanik, M., Wojtyczka, R.D. et al. (2018). Antimicrobial potential of caffeic acid against *Staphylococcus aureus* clinical strains. *BioMed Research International* 2018: 7413504.

24 Kajiya, K., Hojo, H., Suzuki, M. et al. (2004). Relationship between antibacterial activity of (+)-catechin derivatives and their interaction with a model membrane. *Journal of Agricultural and Food Chemistry* 52: 1514–1519.

25 Nayaka, H.B., Londonkar, R.L., Umesh, M.K., and Tukappa, A. (2014). Antibacterial attributes of apigenin, isolated from *Portulaca oleracea* L. *International Journal of Bacteriology* 2014: 175851.

26 Wang, L.-H., Wang, M.-S., Zeng, X.-A. et al. (2017). Membrane and genomic DNA dual-targeting of citrus flavonoid naringenin against *Staphylococcus aureus*. *Integrative Biology* 9: 820–829.

27 Negi, N., Prakash, P., Gupta, M.L., and Mohapatra, T.M. (2014). Possible role of curcumin as an efflux pump inhibitor in multi drug resistant clinical isolates of *Pseudomonas aeruginosa*. *Journal of Clinical and Diagnostic Research: JCDR* 8: DC04.

28 Parai, D., Banerjee, M., Dey, P., and Mukherjee, S.K. (2020). Reserpine attenuates biofilm formation and virulence of *Staphylococcus aureus*. *Microbial Pathogenesis* 138: 103790.

29 Parai, D., Banerjee, M., Dey, P. et al. (2018). Effect of reserpine on *Pseudomonas aeruginosa* quorum sensing mediated virulence factors and biofilm formation. *Biofouling* 34: 320–334.

30 Khameneh, B., Iranshahy, M., Ghandadi, M. et al. (2015). Investigation of the antibacterial activity and efflux pump inhibitory effect of co-loaded piperine and gentamicin nanoliposomes in methicillin-resistant *Staphylococcus aureus*. *Drug Development and Industrial Pharmacy* 41: 989–994.

31 Zarai, Z., Boujelbene, E., Salem, N.B. et al. (2013). Antioxidant and antimicrobial activities of various solvent extracts, piperine and piperic acid from *Piper nigrum*. *Lwt-Food Science and Technology* 50: 634–641.

32 Wen, S.-Q., Jeyakkumar, P., Avula, S.R. et al. (2016). Discovery of novel berberine imidazoles as safe antimicrobial agents by down regulating ROS generation. *Bioorganic and Medicinal Chemistry Letters* 26: 2768–2773.

33 Jeyakkumar, P., Zhang, L., Avula, S.R., and Zhou, C.-H. (2016). Design, synthesis and biological evaluation of berberine-benzimidazole hybrids as new type of potentially DNA-targeting antimicrobial agents. *European Journal of Medicinal Chemistry* 122: 205–215.

34 Siriyong, T., Srimanote, P., Chusri, S. et al. (2017). Conessine as a novel inhibitor of multidrug efflux pump systems in *Pseudomonas aeruginosa*. *BMC Complementary and Alternative Medicine* 17: 405.

35 Guay, I., Boulanger, S., Isabelle, C. et al. (2018). Tomatidine and analog FC04–100 possess bactericidal activities against *Listeria*, *Bacillus* and *Staphylococcus* spp. *BMC Pharmacology and Toxicology* 19: 7.

36 Miao, F., Yang, X.-J., Zhou, L. et al. (2011). Structural modification of sanguinarine and chelerythrine and their antibacterial activity. *Natural Product Research* 25: 863–875.

37 He, N., Wang, P., Wang, P. et al. (2018). Antibacterial mechanism of chelerythrine isolated from root of *Toddalia asiatica* (Linn) Lam. *BMC Complementary and Alternative Medicine* 18: 261.

38 Zhong, H., Hu, D.-D., Hu, G.-H. et al. (2017). Activity of sanguinarine against *Candida albicans* biofilms. *Antimicrobial Agents and Chemotherapy* 61: 1–9.

39 Hamoud, R., Reichling, J., and Wink, M. (2015). Synergistic antibacterial activity of the combination of the alkaloid sanguinarine with EDTA and the antibiotic streptomycin against multidrug resistant bacteria. *Journal of Pharmacy and Pharmacology* 67: 264–273.

40 Obiang-Obounou, B.W., Kang, O.-H., Choi, J.-G. et al. (2011). The mechanism of action of sanguinarine against methicillin-resistant *Staphylococcus aureus*. *The Journal of Toxicological Sciences* 36: 277–283.

41 Basile, A., Sorbo, S., Spadaro, V. et al. (2009). Antimicrobial and antioxidant activities of coumarins from the roots of *Ferulago campestris* (Apiaceae). *Molecules* 14: 939–952.

42 Tan, N., Bilgin, M., Tan, E., and Miski, M. (2017). Antibacterial activities of pyrenylated coumarins from the roots of *Prangos hulusii*. *Molecules* 22: 1098.

43 Zhang, Z.-R., Leung, W.N., Cheung, H.Y., and Chan, C.W. (2015). Osthole: a review on its bioactivities, pharmacological properties, and potential as alternative medicine. *Evidence-Based Complementary and Alternative Medicine* 2015: 919616.

44 Jara, L.M., Pérez-Varela, M., Corral, J. et al. (2016). Novobiocin inhibits the antimicrobial resistance acquired through DNA damage-induced mutagenesis in *Acinetobacter baumannii*. *Antimicrobial Agents and Chemotherapy* 60: 637–639.

45 Abulrob, A.-N., Suller, M.T., Gumbleton, M. et al. (2004). Identification and biological evaluation of grapefruit oil components as potential novel efflux pump modulators in methicillin-resistant *Staphylococcus aureus* bacterial strains. *Phytochemistry* 65: 3021–3027.

46 Kasaian, J., Iranshahy, M., and Iranshahi, M. (2014). Synthesis, biosynthesis and biological activities of galbanic acid – a review. *Pharmaceutical Biology* 52: 524–531.

47 Jabra-Rizk, M., Meiller, T., James, C., and Shirtliff, M. (2006). Effect of farnesol on *Staphylococcus aureus* biofilm formation and antimicrobial susceptibility. *Antimicrobial Agents and Chemotherapy* 50: 1463–1469.

48 Braca, A., Siciliano, T., D'Arrigo, M., and Germanò, M.P. (2008). Chemical composition and antimicrobial activity of *Momordica charantia* seed essential oil. *Fitoterapia* 79: 123–125.

49 Fallarero, A., Skogman, M., Kujala, J. et al. (2013). (+)-Dehydroabietic acid, an abietane-type diterpene, inhibits *Staphylococcus aureus* biofilms in vitro;. *International Journal of Molecular Sciences* 14: 12054–12072.

50 Porfírio, E.M., Melo, H.M., Pereira, A.M.G. et al. (2017). in vitro; antibacterial and antibiofilm activity of *Lippia alba* essential oil, citral, and carvone against *Staphylococcus aureus*. *The Scientific World Journal* 2017: 4962707.

51 Sharifzadeh, A., Khosravi, A., Shokri, H., and Shirzadi, H. (2018). Potential effect of 2-isopropyl-5-methylphenol (thymol) alone and in combination with fluconazole against clinical isolates of *Candida albicans*, *C. glabrata* and *C. krusei*. *Journal de Mycologie Medicale* 28: 294–299.

52 Amaral, V.C., Santos, P.R., da Silva, A.F. et al. (2015). Effect of carvacrol and thymol on *Salmonella* spp. biofilms on polypropylene. *International Journal of Food Science and Technology* 50: 2639–2643.

53 Trevisan, D.A.C., Silva, A.F.D., Negri, M. et al. (2018). Antibacterial and antibiofilm activity of carvacrol against *Salmonella enterica* serotype Typhimurium. *Brazilian Journal of Pharmaceutical Sciences* 54: e17229.

54 Abbaszadeh, S., Sharifzadeh, A., Shokri, H. et al. (2014). Antifungal efficacy of thymol, carvacrol, eugenol and menthol as alternative agents to control the growth of food-relevant fungi. *Journal de Mycologie Medicale* 24: e51–e56.

55 Yadav, M.K., Chae, S.-W., Im, G.J. et al. (2015). Eugenol: a phyto-compound effective against methicillin-resistant and methicillin-sensitive *Staphylococcus aureus* clinical strain biofilms. *PLoS One* 10: e0119564.

56 Rathinam, P., Vijay Kumar, H., and Viswanathan, P. (2017). Eugenol exhibits anti-virulence properties by competitively binding to quorum sensing receptors. *Biofouling* 33: 624–639.

57 He, Z., Huang, Z., Jiang, W., and Zhou, W. (2019). Antimicrobial activity of cinnamaldehyde on *Streptococcus mutans* biofilms. *Frontiers in Microbiology* 10: 2241.

58 Yu, H.H., Song, Y.J., Yu, H.S. et al. (2020). Investigating the antimicrobial and antibiofilm effects of cinnamaldehyde against *Campylobacter* spp. using cell surface characteristics. *Journal of Food Science* 85: 157–164.

59 Alamgir, A. (2018). Secondary metabolites: secondary metabolic products consisting of C and H; C, H, and O; N, S, and P elements; and O/N heterocycles. In: *Therapeutic Use of Medicinal Plants and Their Extracts: Volume 2* (ed. K.D. Rainsford), 165–309. Springer.

60 Sytar, O., Hemmerich, I., Zivcak, M. et al. (2018). Comparative analysis of bioactive phenolic compounds composition from 26 medicinal plants. *Saudi Journal of Biological Sciences* 25: 631–641.

61 Boztas, M., Taslimi, P., Yavari, M.A. et al. (2019). Synthesis and biological evaluation of bromophenol derivatives with cyclopropyl moiety: ring opening of cyclopropane with monoester. *Bioorganic Chemistry* 89: 103017.

62 Cherian, C., Vennila, J.J., and Sharan, L. (2019). Marine bromophenols as an effective inhibitor of virulent proteins (peptidyl arginine deiminase, gingipain R and hemagglutinin A) in *Porphyromas gingivalis*. *Archives of Oral Biology* 100: 119–128.

63 Parihar, A., Grotewold, E., and Doseff, A.I. (2015). Flavonoid dietetics: mechanisms and emerging roles of plant nutraceuticals. In: *Pigments in Fruits and Vegetables* (ed. C. Chen), 93–126. Springer.

64 Zakaryan, H., Arabyan, E., Oo, A., and Zandi, K. (2017). Flavonoids: promising natural compounds against viral infections. *Archives of Virology* 162: 2539–2551.

65 Rojas, Á., Del Campo, J.A., Clement, S. et al. (2016). Effect of quercetin on hepatitis C virus life cycle: from viral to host targets. *Scientific Reports* 6: 1–9.

66 Fournier-Larente, J., Morin, M.-P., and Grenier, D. (2016). Green tea catechins potentiate the effect of antibiotics and modulate adherence and gene expression in *Porphyromonas gingivalis*. *Archives of Oral Biology* 65: 35–43.

67 Xu, X., Zhou, X.D., and Wu, C.D. (2011). The tea catechin epigallocatechin gallate suppresses cariogenic virulence factors of *Streptococcus mutans*. *Antimicrobial Agents and Chemotherapy* 55: 1229–1236.

68 Eyong, K.O., Folefoc, G.N., Kuete, V. et al. (2006). Newbouldiaquinone A: a naphthoquinone–anthraquinone ether coupled pigment, as a potential antimicrobial and antimalarial agent from *Newbouldia laevis*. *Phytochemistry* 67: 605–609.

69 Milosevic, T., Solujic-Sukdolak, S., and Sukdolak, S. (2007). in vitro; study of ethanolic extract of *Hypericum perforatum* L. on growth and sporulation of some bacteria and fungi. *Turkish Journal of Biology* 31: 237–241.

70 Serrano, J., Puupponen-Pimiä, R., Dauer, A. et al. (2009). Tannins: current knowledge of food sources, intake, bioavailability and biological effects. *Molecular Nutrition and Food Research* 53: S310–S329.

71 Li, N., Luo, M., Fu, Y.J. et al. (2013). Effect of corilagin on membrane permeability of *Escherichia coli, Staphylococcus aureus* and *Candida albicans*. *Phytotherapy Research* 27: 1517–1523.

72 Kim, T., Silva, J., Kim, M., and Jung, Y. (2010). Enhanced antioxidant capacity and antimicrobial activity of tannic acid by thermal processing. *Food Chemistry* 118: 740–746.

73 Reddy, K., Reddy, C., and Muralidharan, K. (2009). Potential of botanicals and biocontrol agents on growth and aflatoxin production by *Aspergillus flavus* infecting rice grains. *Food Control* 20: 173–178.

74 Chung, K.-T., Lu, Z., and Chou, M. (1998). Mechanism of inhibition of tannic acid and related compounds on the growth of intestinal bacteria. *Food and Chemical Toxicology* 36: 1053–1060.

75 Kesharwani, A., Polachira, S.K., Nair, R. et al. (2017). Anti-HSV-2 activity of *Terminalia chebula* Retz extract and its constituents, chebulagic and chebulinic acids. *BMC Complementary and Alternative Medicine* 17: 110.

76 Lin, L.-T., Chen, T.-Y., Lin, S.-C. et al. (2013). Broad-spectrum antiviral activity of chebulagic acid and punicalagin against viruses that use glycosaminoglycans for entry. *BMC Microbiology* 13: 187.

77 Orlowski, P., Tomaszewska, E., Gniadek, M. et al. (2014). Tannic acid modified silver nanoparticles show antiviral activity in herpes simplex virus type 2 infection. *PLoS One* 9: e104113.

78 Huang, W.-Y., Cai, Y.-Z., and Zhang, Y. (2009). Natural phenolic compounds from medicinal herbs and dietary plants: potential use for cancer prevention. *Nutrition and Cancer* 62: 1–20.

79 Menezes, J.C. and Diederich, M. (2019). Translational role of natural coumarins and their derivatives as anticancer agents. *Future Medicinal Chemistry* 11: 1057–1082.

80 Keri, R.S., Sasidhar, B., Nagaraja, B.M., and Santos, M.A. (2015). Recent progress in the drug development of coumarin derivatives as potent antituberculosis agents. *European Journal of Medicinal Chemistry* 100: 257–269.

81 El-Seedi, H.R. (2007). Antimicrobial arylcoumarins from *Asphodelus microcarpus*. *Journal of Natural Products* 70: 118–120.

82 Garneau, S., Dorrestein, P.C., Kelleher, N.L., and Walsh, C.T. (2005). Characterization of the formation of the pyrrole moiety during clorobiocin and coumermycin A1 biosynthesis. *Biochemistry* 44: 2770–2780.

83 Chen, H. and Walsh, C.T. (2001). Coumarin formation in novobiocin biosynthesis: β-hydroxylation of the aminoacyl enzyme tyrosyl-S-NovH by a cytochrome P450 NovI. *Chemistry and Biology* 8: 301–312.

84 Sardari, S., Mori, Y., Horita, K. et al. (1999). Synthesis and antifungal activity of coumarins and angular furanocoumarins. *Bioorganic & Medicinal Chemistry* 7: 1933–1940.

85 Chang, C.-T., Doong, S.-L., Tsai, I.-L., and Chen, I.-S. (1997). Coumarins and anti-HBV constituents from *Zanthoxylum schinifolium*. *Phytochemistry* 45: 1419–1422.

86 Asif, M. (2015). Pharmacological activities and phytochemistry of various plant containing coumarin derivatives. *Current Science Perspectives* 1: 77–90.

87 Saito, A.Y., Rodriguez, A.A.M., Vega, D.S.M. et al. (2016). Antimalarial activity of the terpene nerolidol. *International Journal of Antimicrobial Agents* 48: 641–646.

88 Arruda, D.C., D'Alexandri, F.L., Katzin, A.M., and Uliana, S.R. (2005). Antileishmanial activity of the terpene nerolidol. *Antimicrobial Agents and Chemotherapy* 49: 1679–1687.

89 Silva, M.P., Oliveira, G.L., De Carvalho, R.B. et al. (2014). Antischistosomal activity of the terpene nerolidol. *Molecules* 19: 3793–3803.

90 Inoue, Y., Shiraishi, A., Hada, T. et al. (2004). The antibacterial effects of terpene alcohols on *Staphylococcus aureus* and their mode of action. *FEMS Microbiology Letters* 237: 325–331.

91 Ríos, J.L. and Máñez, S. (2018). New pharmacological opportunities for betulinic acid. *Planta Medica* 84: 8–19.

92 Cichewicz, R.H. and Thorpe, P.A. (1996). The antimicrobial properties of chile peppers (*Capsicum* species) and their uses in Mayan medicine. *Journal of Ethnopharmacology* 52: 61–70.

93 Guimarães, A.C., Meireles, L.M., Lemos, M.F. et al. (2019). Antibacterial activity of terpenes and terpenoids present in essential oils. *Molecules* 24: 2471.

94 Kollanoor Johny, A., Darre, M., Donoghue, A. et al. (2010). Antibacterial effect of trans-cinnamaldehyde, eugenol, carvacrol, and thymol on *Salmonella enteritidis* and *Campylobacter jejuni* in chicken cecal contents in vitro;. *Journal of Applied Poultry Research* 19: 237–244.
95 Mancuso, M., Catalfamo, M., Laganà, P. et al. (2019). Screening of antimicrobial activity of citrus essential oils against pathogenic bacteria and *Candida* strains. *Flavour and Fragrance Journal* 34: 187–200.
96 Krishnamurti, C. and Rao, S.C. (2016). The isolation of morphine by Serturner. *Indian Journal of Anaesthesia* 60: 861.
97 Liang, X., Chen, L., Song, L. et al. (2017). Diterpenoid alkaloids from the root of *Aconitum sinchiangense* WT Wang with their antitumor and antibacterial activities. *Natural Product Research* 31: 2016–2023.
98 Neag, M.A., Mocan, A., Echeverría, J. et al. (2018). Berberine: botanical occurrence, traditional uses, extraction methods, and relevance in cardiovascular, metabolic, hepatic, and renal disorders. *Frontiers in Pharmacology* 9: 557.
99 Casu, L., Cottiglia, F., Leonti, M. et al. (2011). Ungeremine effectively targets mammalian as well as bacterial type I and type II topoisomerases. *Bioorganic and Medicinal Chemistry Letters* 21: 7041–7044.
100 Burger, T., Mokoka, T., Fouché, G. et al. (2018). Solamargine, a bioactive steroidal alkaloid isolated from *Solanum aculeastrum* induces non-selective cytotoxicity and P-glycoprotein inhibition. *BMC Complementary and Alternative Medicine* 18: 137.
101 Romeo, F.V., Fabroni, S., Ballistreri, G. et al. (2018). Characterization and antimicrobial activity of alkaloid extracts from seeds of different genotypes of *Lupinus* spp. *Sustainability* 10: 788.
102 Zielińska, S., Wójciak-Kosior, M., Dziągwa-Becker, M. et al. (2019). The activity of isoquinoline alkaloids and extracts from *Chelidonium majus* against pathogenic bacteria and *Candida* sp. *Toxins* 11: 406.
103 Musiol, R., Magdziarz, T., and Kurczyk, A. (2011). Quinoline scaffold as a privileged substructure in antimicrobial drugs. In: *Science Against Microbial Pathogens: Communicating Current Research and Technological Advances* (ed. A. Mendez-Vilas), 72–83. Badajoz, Spain: Formatex.
104 Deshmukh, S.R., Ashrit, D.S., and Patil, B.A. (2012). Extraction and evaluation of indole alkaloids from *Rauwolfia serpentina* for their antimicrobial and antiproliferative activities. *International Journal of Pharmacy and Pharmaceutical Sciences* 4: 329–334.
105 Dwivedi, G.R., Maurya, A., Yadav, D.K. et al. (2019). Synergy of clavine alkaloid "chanoclavine" with tetracycline against multi-drug-resistant *E. coli*. *Journal of Biomolecular Structure and Dynamics* 37: 1307–1325.
106 Siriyong, T., Chusri, S., Srimanote, P. et al. (2016). Holarrhena antidysenterica extract and its steroidal alkaloid, conessine, as resistance-modifying agents against extensively drug-resistant *Acinetobacter baumannii*. *Microbial Drug Resistance* 22: 273–282.
107 Maisuria, V.B. and Nerurkar, A.S. (2015). Interference of quorum sensing by *Delftia* sp. VM4 depends on the activity of a novel *N*-acylhomoserine lactone-acylase. *PLoS One* 10: e0138034.
108 Seukep, A.J., Kuete, V., Nahar, L. et al. (2019). Plant-derived secondary metabolites as the main source of efflux pump inhibitors and methods for identification. *Journal of Pharmaceutical Analysis* 10: 277–290.

109 Aghayan, S.S., Mogadam, H.K., Fazli, M. et al. (2017). The effects of berberine and palmatine on efflux pumps inhibition with different gene patterns in *Pseudomonas aeruginosa* isolated from burn infections. *Avicenna Journal of Medical Biotechnology* 9: 2.

110 Prasch, S. and Bucar, F. (2015). Plant derived inhibitors of bacterial efflux pumps: an update. *Phytochemistry Reviews* 14: 961–974.

111 Shriram, V., Khare, T., Bhagwat, R. et al. (2018). Inhibiting bacterial drug efflux pumps via phyto-therapeutics to combat threatening antimicrobial resistance. *Frontiers in Microbiology* 9: 2990.

112 Savoia, D. (2012). Plant-derived antimicrobial compounds: alternatives to antibiotics. *Future Microbiology* 7: 979–990.

113 Docherty, J.J., Fu, M.M., and Tsai, M. (2001). Resveratrol selectively inhibits *Neisseria gonorrhoeae* and *Neisseria meningitidis*. *Journal of Antimicrobial Chemotherapy* 47: 243–244.

114 Al-Haidari, R.A., Shaaban, M.I., Ibrahim, S.R., and Mohamed, G.A. (2016). Anti-quorum sensing activity of some medicinal plants. *African Journal of Traditional, Complementary and Alternative Medicines* 13: 67–71.

115 Nagar, N., Aswathanarayan, J.B., and Vittal, R.R. (2020). Anti-quorum sensing and biofilm inhibitory activity of *Apium graveolens* L. oleoresin. *Journal of Food Science and Technology* 57: 2414–2422.

116 Shukla, V. and Bhathena, Z. (2016). Broad spectrum anti-quorum sensing activity of tannin-rich crude extracts of indian medicinal plants. *Scientifica* 2016: 5823013.

117 Artini, M., Papa, R., Barbato, G. et al. (2012). Bacterial biofilm formation inhibitory activity revealed for plant derived natural compounds. *Bioorganic and Medicinal Chemistry* 20: 920–926.

118 Teanpaisan, R., Kawsud, P., Pahumunto, N., and Puripattanavong, J. (2017). Screening for antibacterial and antibiofilm activity in Thai medicinal plant extracts against oral microorganisms. *Journal of Traditional and Complementary Medicine* 7: 172–177.

9

Functional Plants as Natural Sources of Dietary Antioxidants

Ao Shang[1], Jia-Hui Li[2], Xiao-Yu Xu[1], Ren-You Gan[3], Min Luo[1], and Hua-Bin Li[1]

[1] *Guangdong Provincial Key Laboratory of Food, Nutrition and Health, Department of Nutrition, School of Public Health, Sun Yat-Sen University, Guangzhou, China*
[2] *School of Science, The Hong Kong University of Science and Technology, Hong Kong, China*
[3] *Research Center for Plants and Human Health, Institute of Urban Agriculture, Chinese Academy of Agricultural Sciences, Chengdu, China*

9.1 Introduction

Reactive oxygen species (ROS), or oxygen-free radicals, are produced in cell metabolism. However, excessive production of ROS can lead to oxidative stress, which can cause the destruction of normal functions of cellular lipids, proteins, and DNA [1]. Numerous studies have demonstrated that oxidative stress is associated with aging and the development of a variety of diseases in human, such as neurodegenerative diseases, cardiovascular diseases, cancers, liver diseases [2–6]. Antioxidants are substances that can prevent the generation of free radicals or capture and neutralize free radicals. Hence, using antioxidants is considered as a strategy of eliminating harmful effects of oxidative stress on the body [7]. On the other hand, antioxidants contribute to the food preservation and shelf life extension as the food additive. For instance, antioxidants can delay or reduce the oil oxidation during food processing and preservation, which can prevent food from browning or loss of nutrients due to oxidation [8, 9].

Plants are rich in polyphenols, polysaccharides, saponins, lycopene, carotenoids and glucosinolates, and other bioactive compounds, and plants are also regarded as an important source of dietary antioxidants [10–13]. According to the researches, many kinds of vegetables, fruits, medicinal plants, cereals, flowers, microalgae, and teas have antioxidant activity, which can be excellent participants of natural antioxidants [14–16].

Bioprospecting of Plant Biodiversity for Industrial Molecules, First Edition.
Edited by Santosh Kumar Upadhyay and Sudhir P. Singh.

9.2 Evaluation of the Antioxidant Activity

The antioxidant activity needs to be evaluated by accurate scientific methods. Currently, according to the mechanisms, the evaluation methods of in vitro antioxidant activity can be divided into the free radicals scavenging and inhibiting capacity, reducing capacity, metal ion-chelating capacity, and inhibiting oxidase capacity. The common methods of detecting the antioxidant activity are 2,2′-azinobis-3-ethylbenzotiazoline-6-sulfonic acid ($ABTS^{\bullet+}$) free radical scavenging/Trolox equivalent antioxidant capacity (TEAC) assay and ferric ion reducing antioxidant power (FRAP) assay.

In ABTS free radical scavenging method, $ABTS^{\bullet+}$ free radicals are oxidized to blue-green $ABTS^{\bullet+}$ under the action of oxidants, and the presence of antioxidants can inhibit the generation of $ABTS^{\bullet+}$, resulting in a decolorization. This reaction can be observed as the change of absorbance at 743 nm, because 734 nm is the wavelength having the maximum absorbance of the mixture [17].

FRAP assay is often used to measure the total antioxidant capacity, and it evaluates the ferric ion reducing capacity of the antioxidants. At a low pH, the ferric iron (Fe(III)) and tripyridyltriazine (TPTZ) form a complex, and it appears blue when it is reduced to a complex of ferrous iron (Fe(II))-TPTZ, which has a maximum absorption at 593 nm [18].

In addition, phenols are compounds widely found in plants with potential health-promoting effects, which are often found in the peels, roots, leaves, and fruits of plants. The phenolic compounds in plants are known as natural antioxidants with free radical scavengers' capacity [19]. Therefore, the total phenolic content (TPC) can reflect the antioxidant activity of plants to some extent, which is usually detected by Folin–Ciocalteu method. The principle of this method is that the reaction of Folin–Ciocalteu reagent with phenolic compounds exhibits blue color, and the depth of blue color is proportional to the number of phenolic groups. The mixture has the maximum absorption at 760 nm [20, 21].

9.3 Antioxidant Activity of Functional Plants

9.3.1 Vegetables

Vegetables are one of the essential foods in the daily diet, which can provide a variety of nutrients necessary for the human body, and are also rich in antioxidants. According to the report of Deng et al., the antioxidant activity of 56 commonly consumed vegetables is assessed by FRAP and TEAC assays, and the TPC of these vegetables is also measured [22]. The FRAP values of hydrophilic fraction, lipophilic fraction, and whole of these vegetables respectively are in the range of 0.08–21.6, 2.44–39.30, 2.69–60.9 μmol Fe(II)/g, while the total TEAC values of hydrophilic fraction, lipophilic fraction, and total are in the range of 0.05–19.01, 5.57–18.14, 6.93–33.63 μmol Trolox/g, respectively. The TPC of hydrophilic fraction, lipophilic fraction, and whole of these vegetables are measured by Folin–Ciocalteu method, and the values are in the range of 0.02–8.16, 4.83–15.11, and 4.99–23.27 mg gallic acid equivalents (GAE)/g, respectively. The study demonstrates that the antioxidant capacity of different vegetables varied greatly, among which the antioxidant effects of Chinese toon bud, loosestrife, perilla leaf, cowpea, caraway, lotus root, sweet potato leaf, soy bean (green), pepper leaf, ginseng leaf, chives, and broccoli are the highest, while marrow squash and eggplant (purple) have quite low values (Table 9.1). At the same time, there is a positive correlation between antioxidant activity and TPC, suggesting that phenolic compounds may be the main contributor to the antioxidant activity. The compounds such as chlorogenic acid, gallic acid, and galangin are identified as prevalent in these vegetables.

In addition, the contents of antioxidants in vegetables are influenced by storage and processing. For example, the phenolic compounds are greatly affected by the environmental conditions around the harvest. The proper storage conditions are helpful to increase the phenolic content, while vitamin C and phenols are more likely to lose during processing [23].

9.3.2 Fruits

Fruits contain abundant nutrients that are beneficial for human health. The antioxidant activity of 62 fruits is evaluated by FRAP and TEAC assays, and there are obvious differences in the antioxidant capacity of different fruits [24]. The FRAP values of all 62 fruits range from 0.11 to 72.11 μmol Fe(II)/g, with an average of 8.76 μmol Fe(II)/g. The TEAC values range from 0.84 to 80.68 μmol Trolox/g, with an average of 6.01 μmol Trolox/g. The TPC values range from 0.12 to 5.85 mg GAE/g. The fruits with the highest antioxidant activity and TPC are Chinese date, pomegranate, guava, sweetsop, persimmon, Chinese wampee, and plum. On the other hand, grape is also a commonly consumed fruit, and the antioxidant property of 30 different grape varieties is measured by FRAP, TEAC assays, and their TPC and total flavonoid contents (TFC) are also detected [25]. Some grape varieties including Pearl Black Grape, Summer Black Grape, Pearl Green Grape, Seedless Green Grape, and Seedless Red Grape possess the most significant antioxidant activity and highest TPC. The antioxidants in those grapes are identified as caffeic acid, catechin gallate, epicatechin, gallic acid, protocatechuic acid, and rutin.

Another study paid attention to the antioxidant activity of 56 wild fruits from South China [26]. Wild fruits have strong capacities of scavenging free radicals and reducing oxidants, and there is a certain correlation between the FRAP value and TEAC values. The FRAP, TEAC, and TPC values vary in the range of 1.28–502 μmol Fe(II)/g, 3.38–1140 μmol Trolox/g, and 0.49–54.8 mg GAE/g, respectively. The wild fruits with the strongest antioxidant activity and the highest TPC rank as *Eucalyptus Robusta*, *Euryanitida*, *Melastoma sanguineum*, *Melaleuca leucadendron*, *Lagerstroemia indica*, *Caryota mitis*, *Lagerstroemia speciosa*, and *Gordonia axillaris*. As a result, wild fruits appear to have higher antioxidant capacity than common fruits, suggesting that they have the potential to be a source of natural antioxidants.

Moreover, some fruits wastes such as peels and seeds also show strong antioxidant activity, suggesting that the fruits without edible parts are also valuable [27]. The range of FRAP, TEAC, and TPC values of 50 fruit peel extracts are 0.74–155.73 μmol Fe (II)/g, 1.72–93.10 μmol Trolox/g, and 0.38–22.95 mg GAE/g, respectively. The antioxidant capacities and TPC of the fruit seeds are approximate to those of peels, and the ranges are 0.34–181.39 μmol Fe(II)/g, 2.45–92.62 μmol Trolox/g, and 0.30–22.95 mg GAE/g, respectively. Among them, the antioxidant property of grape residues is higher, and the antioxidant activity and TPC of 30 grape peels and seeds are also evaluated in another study [28].

Table 9.1 The FRAP, TEAC, and TPC values of several plants.

Plant	Scientific name	Antioxidant activity		Total phenolic content (mg GAE/g)	References
		FRAP value (μmol Fe (II)/g)	TEAC value (μmol Trolox/g)		
Vegetable					
Chinese toon bud	*Toona sinensis*	60.90 ± 4.54	33.63 ± 1.59	23.27 ± 1.59	[22]
Cowpea	*Vigna unguiculata* (Linn.) Walp	17.01 ± 0.74	21.01 ± 1.11	8.28 ± 0.14	
Ginseng leaf	*Panax ginseng* C. A. Mey.	16.80 ± 1.41	15.35 ± 0.44	11.78 ± 0.44	
Loosestrife	*Lysimachia mauritiana* Lam.	25.22 ± 0.70	25.44 ± 1.30	13.13 ± 0.70	
Perilla leaf	*Perilla frutescens*	44.50 ± 3.15	23.94 ± 0.48	14.37 ± 0.48	
Fruit					
Chinese date	*Ziziphus jujuba* Mill.	72.11 ± 2.19	8.33 ± 0.12	5.85 ± 0.19	[24]
Guava	*Psidium guava*	23.80 ± 1.44	15.18 ± 0.81	1.94.11 ± 0.07	
Pearl Black Grape	*Vitis vinifera* L.	11.77	4.84	1.40	
Persimmon	*Diospyros kaki* L.	16.97 ± 0.26	9.38 ± 0.25	1.12 ± 0.05	
Pomegranate	*Punica granatum* L.	25.57 ± 0.53	40.61 ± 0.11	1.47	
Wild fruit					[26]
Baiqianceng	*Melaleuca leucadendron*	214 ± 8.86	461 ± 6.88	25.6 ± 1.59	
Maoren	*Melastoma sanguineum*	288 ± 10.4	404 ± 21.5	23.3 ± 0.49	
Swamp mahogany	*Eucalyptus robusta*	502 ± 6.41	1140 ± 7.93	54.8 ± 3.05	
Xichiyeling	*Eurya nitida*	428 ± 12.0	478 ± 14.1	35.0 ± 0.96	
Ziwei	*Lagerstroemia indica*	254 ± 5.75	109 ± 5.13	12.1 ± 0.30	
Fruit peel					[25]
Blueberry	*Vaccinium corymbosum* L.	104.55 ± 0.77	59.10 ± 3.85	8.92 ± 0.24	

Chinese olive	*Canarium album* L.	96.17 ± 6.69	77.12 ± 2.43	13.16 ± 0.32	
Hawthorn	*Crataegus pinnatifida*	89.57 ± 6.54	66.76 ± 2.77	12.66 ± 1.09	
Starfruit	*Averrhoa carambola*	84.52 ± 3.19	70.15 ± 4.91	10.45 ± 0.97	
Sweetsop	*Annona squamosa* L.	155.73 ± 8.46	84.14 ± 2.91	17.77 ± 0.16	
Fruit seed					[25]
Avocado	*Persea americana* Mill.	45.79 ± 3.01	42.63 ± 1.94	8.39 ± 0.27	
Grape (USA)	*Vitis vinifera* L.	181.39 ± 6.93	92.62 ± 0.37	22.95 ± 1.01	
Longan	*Dimocarpus longan* Lour.	86.39 ± 2.03	75.33 ± 2.76	2.37 ± 0.32	
Mango	*Mangifera indica* L.	31.17 ± 3.62	50.55 ± 2.52	7.54 ± 0.24	
Jujube	*Ziziphus jujube*	45.53 ± 1.21	56.03 ± 2.04	9.00 ± 0.94	
Medicinal plants					[29]
Chinese peony	*Paeonia lactiflora* Pall. (red)	481.36 ± 7.42	365.72 ± 5.08	31.48 ± 1.52	
Coltsfoot	*Tussilago farfara* L.	529.48 ± 10.40	225.61 ± 1.25	34.50 ± 0.52	
Danshen	*Salvia miltiorrhiza* Bge.	601.37 ± 27.50	1544.38	98.88	
Diyu	*Sanguisorba officinalis* L.	1844.85	20.84	9.37 ± 0.22	
Hongjingtian	*Rhodiola sacra* Fu	541.22 ± 3.61	471.53 ± 4.63	51.06 ± 1.75	
Peach	*Prunus persica (Linn)* Batsch.	5.07 ± 0.68	793.13 ± 35.89	55.23 ± 0.61	
Qinpi	*Fraxinus rhynchophylla* Hance.	548.25 ± 3.68	401.35 ± 9.28	52.31 ± 3.26	
Rhizoma cimicifugae	*Cimicifuga foetida* L.	707.02 ± 23.19	349.48 ± 3.12	24.97 ± 0.09	
Tuber fleeceflower root	*Polygonum multiflorum* Thunb. (Stem)	506.83 ± 2.35	538.75 ± 5.39	45.24 ± 1.32	
Xueteng	*Sargentodoxa cuneata* Rehd. et Wils.	521.62 ± 7.53	458.91 ± 8.74	65.28 ± 2.93	
Cereal grains					
Black rice	*Oryza sativa* L. indica	126.19 ± 2.91	30.03 ± 1.10	9.47 ± 0.48	[30]
Buckwheat	*Fagopyrum esculentum* Moench	31.20 ± 0.93	9.43 ± 0.35	4.48 ± 0.46	

(*Continued*)

Table 9.1 (Continued)

Plant	Scientific name	Antioxidant activity		Total phenolic content (mg GAE/g)	References
		FRAP value (μmol Fe (II)/g)	TEAC value (μmol Trolox/g)		
Corn	*Zea mays* L.	11.66 ± 0.40	4.52 ± 0.10	1.97 ± 0.06	
Millet	*Setaria italica* (L.) Beauv. var. germanica (Mill.) Schred.	11.29 ± 1.19	1.88 ± 0.11	2.05 ± 0.13	
Oat	*Avena sativa* Linn.	16.15 ± 1.06	1.57 ± 0.20	2.83 ± 0.16	
Flower					
Chinese rose	*Rosa hybrida*	629.64 ± 24.61	175.39 ± 1.74	35.84 ± 1.67	[31]
Geranium	*Pelargonium hortorum*	212.84 ± 3.94	132.18 ± 5.46	25.68 ± 1.02	
Qinyeshanhu	*Jatropha integerrima*	219.95 ± 10.08	115.01 ± 4.20	17.22 ± 0.77	
Statice	*Limonium sinuatum*	500.04 ± 70.78	157.42 ± 2.68	34.17 ± 1.17	
Sweet osmanthus	*Osmanthus fragrans*	163.57 ± 15.77	71.98 ± 3.13	16.00 ± 0.57	
Microalgae					
	Synechococcus sp. FACHB 283		29.56 ± 1.24	10.56 ± 0.11	[32]
	Nostoc ellipsosporum CCAP 1453/17		21.09 ± 1.83	60.35 ± 2.27	
	Chlamydomonas nivalis		24.13 ± 0.47	15.07 ± 0.26	
	Chlorella pyrenoidosa #1		17.32 ± 0.66	10.46 ± 0.20	
	Chlorella protothecoides #7		9.22 ± 0.35	19.03 ± 0.14	
Tea					
Black teas	*Camellia sinensis* (L.) O. Ktze.	1141.58 ± 13.92	724.28 ± 12.63	107.54 ± 1.80	[33]
Dark teas		1124.96 ± 23.87	589.43 ± 8.35	78.16 ± 1.33	
Green teas		3621.75 ± 81.44	1964.50 ± 38.29	195.79 ± 5.45	
Oolong teas		2013.37 ± 26.17	1460.46 ± 22.23	172.11 ± 2.09	
Yellow teas		3182.34 ± 31.31	2087.81 ± 42.89	198.44 ± 5.39	

9.3.3 Medicinal Plants

Antioxidants in food not only exist in fruits and vegetables, but also in medicinal plants. The antioxidant capacity of infusions from 223 medicinal plants in China are demonstrated by TEAC and FRAP assays, and the TPC values are tested by Folin–Ciocalteu method [29]. The range of FRAP values is 0.14–1844.85 μmol Fe(II)/g DW, the range of TEAC values is 0.99–1544.38 μmol Trolox/g DW, while the range of TPC values is 0.19–101.33 mg GAE/g DW. Ten medicinal plants infusions with high antioxidant activity and TPC are *Salvia miltiorrhiza* Bge., *Polygonum multiflorum* Thunb. (Stem), *Rhodiola sacra* Fu, *Sargentodoxa cuneata* Rehd. et Wils., *Fraxinus rhynchophylla* Hance, *Prunus persica* (Linn) Batsch., *Cimicifuga foetida* L., *Paeonia lactiflora* Pall. (red), *Tussilago farfara* L., and *Sanguisorba officinalis* L. Considering their high antioxidant activity and low toxicity, some of medicinal plants are expected to be developed as natural antioxidant products.

9.3.4 Cereal Grains

Cereal grains are an essential component of daily diet, and their antioxidant activities is also a concern. A finding shows that the antioxidant activity and the TPC of 24 hydrophilic and lipophilic fractions of cereal grains are determined [30]. The total FRAP values vary from 5.23 to 126.19 μmol Fe(II)/g, the total TEAC values vary from 0.62 to 30.03 μmol Trolox/g, and the TPC values vary from 1.35 to 9.47 mg GAE/g. The first five cereals with the highest antioxidant activity and TPC value are black rice, organic black rice, red rice, purple rice, and buckwheat. The results show that the antioxidant activity of pigmented rice is stronger than that of normal white rice, indicating that the pigmented rice may have better health benefits. Besides, gallic acid, kaempferol, quercetin, galangin, and cyanidin-3-glucoside are the main phenolic compounds in those cereal grains.

9.3.5 Flowers

Flowers are not only ornamental plants, but also a special kind of food. The history of adding flowers to the diet in Europe can be traced back to the sixteenth century, while some flowers can also be used as traditional Chinese medicine in China. The antioxidant capacities of 51 edible and wild flowers are evaluated, and the results show that the FRAP values ranged from 0.17 to 629.64 μmol Fe(II)/g, the TEAC values ranged from 0.23 to 175.39 μmol Trolox/g [31]. The TPC values of water-soluble and fat-soluble fractions range from 0.63 to 35.84 mg GAE/g. Combining the free radical scavenging capacity and ferric ion reducing capacity, it is concluded that the edible and wild flowers with strongest antioxidant capacity and highest TPC values are *Rosa hybrid*, *Limonium sinuatum*, *Pelargonium hortorum*, *Jatropha integerrima*, and *Osmanthus fragrans*. In addition, the antioxidant capacity is highly correlated with TPC, suggesting that phenols such as homogentisic acid, cyanidin-3-glucoside, protocatechuic acid, catechin, gallic acid, and epicatechin may be the main antioxidants in these flowers.

9.3.6 Microalgae

Microalgae are a kind of autotrophic plants with abundant nutrition and high photosynthetic utilization, which are widely distributed in river, lake, and sea. Li et al. measure the antioxidant activity of 23 kinds of microalgae by TEAC essay and determine the TPC values by Folin–Ciocalteu method [32]. The different species of microalgae have quite different radical scavenging capacity, and the TEAC values vary from 1.33 to 29.56 μmol Trolox/g. Several microalgae such as *Synechococcus* sp. FACHB 283, *Chlamydomonas nivalis*, and *Nostoc ellipsosporum* CCAP 1453/17 are found to have the strongest antioxidant capacity. The TPC values of these 23 microalgae range from 3.59 to 60.35 mg GAE/g. However, the correlation between the antioxidant capacity and the TPC of microalgae is very low, so it can be inferred that phenolic compounds are not the main antioxidants in microalgae, while carotenoids, polyunsaturated fatty acids, and polysaccharides might be the main antioxidants in microalgae.

9.3.7 Teas

Tea is a popular natural beverage with pleasant taste and health benefits. In a study, the antioxidant activity and TPC values of the fat-soluble, water-soluble, and bound-insoluble fractions of 30 teas from six categories (green, black, oolong, white, yellow, and dark teas) were examined [33]. The range of total FRAP values is 611.18–5375.18 μmol Fe (II)/g DW, the range of total TEAC values is 326.32 ± 0.48 to 3004.40 ± 112.89 μmol Trolox/g DW, and the range of TPC values is 37.25 ± 0.16 to 254.29 ± 15.51 mg GAE/g DW. Among them, several kinds of teas exhibit the highest antioxidant capacity, including Dianqing Tea, Lushan Yunwu Tea, and Xihu Longjing Tea. In addition, the antioxidant activity of green tea, yellow tea, oolong tea is higher than those of black tea, dark tea, and white tea. Green tea has a stronger antioxidant activity than other types of teas, and it also contains more phenolic compounds. Moreover, several phenolic compounds are isolated and identified from 30 teas, including catechins (epicatechin, epigallocatechin, epicatechingallate, and epigallocatechingallate), gallic acid, chlorogenic acid, ellagic acid, and kaempferol-3-*O*-glucoside.

Furthermore, the antioxidant activity and TPC values of the infusions of these 30 teas are measured as well [34]. The antioxidant activities of green tea, yellow tea, oolong tea, black tea, dark tea, and white tea decreased successively, and several phenolic compounds are found in the infusions, including eight catechins, three phenolic acids, two flavonols, two flavonol glycosides, and theaflavin. Overall, teas may be a promising source of natural antioxidants, with phenolic compounds as major contributors.

9.4 Applications of Plant Antioxidants

9.4.1 Food Additives

Most natural foods and some processed foods tend to become unpalatable, browning, and/or rotten during preservation, especially the oxidation of oils and fats. Oxidative rancidity is an important factor in food deterioration, mainly due to the formation of potentially toxic secondary compounds [35, 36]. Some plant antioxidants, such as phenolic compounds, are excellent participants in food antioxidant additives because they are relatively safe and efficient. For example, betanin is approved for use as a natural red colorant in food and pharmaceutical production. It also has the capacity of scavenging peroxy-radicals in pork, which can be used to maintain food quality [37]. The extracts of grape and olive pomaces can effectively replace the synthetic sodium ascorbate in raw lamb products [38].

Antioxidant dietary fiber from red grape pomace is rich in polyphenols, which can be added to cooked chicken hamburgers to markedly improve its oxidative stability. Also, the antioxidants in white grape pomace retard lipid oxidation in minced fish muscle [39]. Moreover, maclurin, a phenolic compound existing in white mulberry, is able to inactivate anti-polyphenol oxidase, inhibit the enzymatic browning of in potato supernatant, and improve its antioxidant capacity [40]. In addition, many other plants and their active ingredients have the potential to be applied as antioxidant additives in foods, such as *Carex distachya* roots, eucalyptus leaf, pomelo peels, oregano essential oil, dihydromyricetin, Andean berrypolyphenols, and olive biophenols [41–46]. Natural plant antioxidants are effective in preventing food from rapid deterioration, while being relatively safe for humans and environmentally friendly. Hence, the substitution of synthetic antioxidants with natural plant antioxidants is a promising direction for the research and development of food additives.

9.4.2 Dietary Supplements

Oxidative stress is associated with the pathogenesis of many diseases, such as neurodegenerative diseases, cardiovascular diseases, cancers, liver diseases, and several metabolic diseases [47]. Some plants and active compounds inhibit the progression of these diseases by alleviating oxidative stress. Several antioxidants found in fruits, vegetables, legumes, and medicinal plants, including flavone, apigenin, luteolin, kaempferol, piceatannol, esculetin, vanillin, and ellagic acid, may contribute to the prevention and treatment of neurodegenerative diseases such as Alzheimer's disease, Parkinson's disease, and Huntington's disease [48]. According to studies, natural plant antioxidant polyphenols have a protective effect on cardiovascular diseases, including quercetin, stilbenoids, blueberry phenolic acid, phlorizin, anthocyanins, etc. [49–53]. In addition, some plant-derived dietary antioxidants, including tea polyphenols, curcumin, genistein, resveratrol, lycopene, pomegranate, and lupeol, can help prevent and manage several cancers without significant side effects [54]. Some antioxidant flavonoids, alkaloids and terpenoids in plants have been shown to prevent alcoholic fatty liver disease, including anthocyanins, catechins, isorhamnetin, narirutin, hesperidin, chlorogenic acid, berberine, curcumin, and lignans [55]. Moreover, plant antioxidants are able to prevent and improve diabetes, obesity, and gout as well [56-58]. The superior preventive and therapeutic effects of natural plant antioxidants on diseases are worthy of further study. It will also be the development trend of modern medicine and health care industry to search for antioxidant substances from plants.

9.5 Conclusions

The moderate amount of antioxidants is beneficial for human health, and it could be used in the prevention and treatment of some diseases. The antioxidants are also important food additives, which can prevent or delay the oxidation and spoilage of food ingredients, and guarantee the quality and flavor of food. Plants are the treasure trove of natural antioxidants, including vegetables, fruits, medicinal plants, cereals, flowers, microalgae, and teas. According to the studies, the antioxidant capacity of teas is relatively high. However, considering the limited consumption of tea, it is recommended to have a balanced diet to ensure a reasonable intake of antioxidants. In addition to daily consumption, antioxidants from plants can also be developed into the dietary supplements and food additives. The antioxidants in the nonedible parts of plants, such as peels and seeds, also have potential to be exploited. Furthermore, given the high antioxidant activity of wild fruits and wild flowers, some unusual wild plants are also considered as an important source of natural antioxidants worth developing.

It is reported that phenolic compounds are the main contributors to the antioxidant activity of most plants. At present, several common phenolic compounds are isolated and identified, such as gallic acid, kaempferol, and quercetin, etc. Meanwhile, more other antioxidant compounds such as polysaccharides and carotenoids need further investigation. It should be noted that the contents of antioxidants in plants can be changed due to the environmental factors such as harvest, storage, and transportation conditions. With these issues solved, the application of plant-derived antioxidants has a promising prospect on the development of food additives to delay food oxidation and promote the production of functional food, which can prevent oxidative-stress-related diseases.

References

1 Valko, M., Leibfritz, D., Moncol, J. et al. (2007). Free radicals and antioxidants in normal physiological functions and human disease. *International Journal of Biochemistry and Cell Biology* 39: 44–84.

2 Bhat, A.H., Dar, K.B., Anees, S. et al. (2015). Oxidative stress, mitochondrial dysfunction and neurodegenerative diseases; a mechanistic insight. *Biomedicine and Pharmacotherapy* 74: 101–110.

3 Siti, H.N., Kamisah, Y., and Kamsiah, J. (2015). The role of oxidative stress, antioxidants and vascular inflammation in cardiovascular disease (a review). *Vascular Pharmacology* 71: 40–56.

4 Valko, M., Rhodes, C.J., Moncol, J. et al. (2006). Free radicals, metals and antioxidants in oxidative stress-induced cancer. *Chemico-Biological Interactions* 160: 1–40.

5 Li, S., Tan, H.Y., Wang, N. et al. (2015). The role of oxidative stress and antioxidants in liver diseases. *International Journal of Molecular Sciences* 16: 26087–26124.

6 Incalza, M.A., D'Oria, R., Natalicchio, A. et al. (2018). Oxidative stress and reactive oxygen species in endothelial dysfunction associated with cardiovascular and metabolic diseases. *Vascular Pharmacology* 100: 1–19.

7 Coronado, M., Leon, S.V.Y., Gutierrez, R. et al. (2015). Antioxidants: present perspective for the human health. *RevistaChilena De Nutricion* 42: 206–212.

8 Silva, M.M. and Lidon, F.C. (2016). An overview on applications and side effects of antioxidant food additives. *Emirates Journal of Food and Agriculture* 28: 823–832.

9 Ayala-Zavala, J.F., Vega-Vega, V., Rosas-Dominguez, C. et al. (2011). Agro-industrial potential of exotic fruit byproducts as a source of food additives. *Food Research International* 44: 1866–1874.

10 Sur, P., Chandhuri, T., Vedasiromoni, J.R. et al. (2001). Antiinflammatory and antioxidant property of saponins of tea *Camellia sinensis* (L) O. Kuntze root extract. *Phytotherapy Research* 15: 174–176.

11 Wang, J.Q., Hu, S.Z., Nie, S.P. et al. (2016). Reviews on mechanisms of in vitro antioxidant activity of polysaccharides. *Oxidative Medicine and Cellular Longevity* 2016: 5692852.

12 Pandey, K.B. and Rizvi, S.I. (2009). Plant polyphenols as dietary antioxidants in human health and disease. *Oxidative Medicine and Cellular Longevity* 2: 270–278.

13 Kaur, C. and Kapoor, H.C. (2001). Antioxidants in fruits and vegetables – the millennium's health. *International Journal of Food Science and Technology* 36: 703–725.

14 Halvorsen, B.L., Holte, K., Myhrstad, M.C.W. et al. (2002). A systematic screening of total antioxidants in dietary plants. *Journal of Nutrition* 132: 461–471.

15 Katalinic, V., Milos, M., Kulisic, T., and Jukic, M. (2006). Screening of 70 medicinal plant extracts for antioxidant capacity and total phenols. *Food Chemistry* 94: 550–557.

16 Li, H.B., Wong, C.C., Cheng, K.W., and Chen, F. (2008). Antioxidant properties *in vitro* and total phenolic contents in methanol extracts from medicinal plants. *Lwt-Food Science and Technology* 41: 385–390.

17 Moon, J.K. and Shibamoto, T. (2009). Antioxidant assays for plant and food components. *Journal of Agricultural and Food Chemistry* 57: 1655–1666.

18 Benzie, I.F.F. and Strain, J.J. (1996). The ferric reducing ability of plasma (FRAP) as a measure of "Antioxidant power": the FRAP assay. *Analytical Biochemistry* 239: 70–76.

19 Balasundram, N., Sundram, K., and Samman, S. (2006). Phenolic compounds in plants and agri-industrial by-products: antioxidant activity, occurrence, and potential uses. *Food Chemistry* 99: 191–203.

20 Sanchez-Rangel, J.C., Benavides, J., Heredia, J.B. et al. (2013). The Folin-Ciocalteu assay revisited: improvement of its specificity for total phenolic content determination. *Analytical Methods* 5: 5990–5999.

21 Li, Y., Cao, S.Y., Lin, S.J. et al. (2019). Polyphenolic profile and antioxidant capacity of extracts from *Gordonia axillaris* fruits. *Antioxidants* 8: 150.

22 Deng, G.F., Lin, X., Xu, X.R. et al. (2013). Antioxidant capacities and total phenolic contents of 56 vegetables. *Journal of Functional Foods* 5: 260–266.

23 Kalt, W. (2005). Effects of production and processing factors on major fruit and vegetable antioxidants. *Journal of Food Science* 70: R11–R19.

24 Fu, L., Xu, B.T., Xu, X.R. et al. (2011). Antioxidant capacities and total phenolic contents of 62 fruits. *Food Chemistry* 129: 345–350.

25 Liu, Q., Tang, G.Y., Zhao, C.N. et al. (2018). Comparison of antioxidant activities of different grape varieties. *Molecules* 23: 2432.

26 Fu, L., Xu, B.T., Xu, X.R. et al. (2010). Antioxidant capacities and total phenolic contents of 56 wild fruits from south China. *Molecules* 15: 8602–8617.

27 Deng, G.F., Shen, C., Xu, X.R. et al. (2012). Potential of fruit wastes as natural resources of bioactive compounds. *International Journal of Molecular Sciences* 13: 8308–8323.

28 Tang, G.Y., Zhao, C.N., Liu, Q. et al. (2018). Potential of grape wastes as a natural source of bioactive compounds. *Molecules* 23: 2598.

29 Li, S., Li, S.K., Gan, R.Y. et al. (2013). Antioxidant capacities and total phenolic contents of infusions from 223 medicinal plants. *Industrial Crops and Products* 51: 289–298.

30 Deng, G.F., Xu, X.R., Guo, Y.J. et al. (2012). Determination of antioxidant property and their lipophilic and hydrophilic phenolic contents in cereal grains. *Journal of Functional Foods* 4: 906–914.

31 Li, A.N., Li, S., Li, H.B. et al. (2014). Total phenolic contents and antioxidant capacities of 51 edible and wild flowers. *Journal of Functional Foods* 6: 319–330.

32 Li, H.B., Cheng, K.W., Wong, C.C. et al. (2007). Evaluation of antioxidant capacity and total phenolic content of different fractions of selected microalgae. *Food Chemistry* 102: 771–776.

33 Tang, G.Y., Zhao, C.N., Xu, X.Y. et al. (2019). Phytochemical composition and antioxidant capacity of 30 Chinese teas. *Antioxidants* 8: 180.

34 Zhao, C.N., Tang, G.Y., Cao, S.Y. et al. (2019). Phenolic profiles and antioxidant activities of 30 tea infusions from green, black, oolong, white, yellow and dark teas. *Antioxidants* 8: 215.

35 Franco, R., Navarro, G., and Martinez-Pinilla, E. (2019). Antioxidants versus food antioxidant additives and food preservatives. *Antioxidants* 8: 542.

36 Kumar, S., Singh, I., Kohli, D. et al. (2019). Waste pomelo (*Citrus maxima*) peels – a natural source of antioxidant and its utilization in peanut oil for suppressing the development of rancidity. *Current Research in Nutrition and Food Science* 7: 800–806.

37 da Silva, D.V.T., Baiao, D.D., Silva, F.D. et al. (2019). Betanin, a natural food additive: stability, bioavailability, antioxidant and preservative ability assessments. *Molecules* 24: 458.

38 Moon, K.M., Kim, C.Y., Ma, J.Y., and Lee, B. (2019). Xanthone-related compounds as an anti-browning and antioxidant food additive. *Food Chemistry* 274: 345–350.

39 Perez-Jimenez, J. and Sayago-Ayerdi, S.G. (2009). Grape antioxidant dietary fibre. A natural food additive and a dietary supplement. *Agro Food Industry Hi-Tech* 20: 17–19.

40 Moon, K.M., Kim, C.Y., Ma, J.Y., and Lee, B. (2019). Xanthone-related compounds as an anti-browning and antioxidant food additive. *Food Chemistry* 274: 345–350.

41 Fiorentino, A., Ricci, A., D'Abrosca, B. et al. (2008). Potential food additives from *Carexdistachya* roots: identification and *in vitro* antioxidant properties. *Journal of Agricultural and Food Chemistry* 56: 8218–8225.

42 Rodriguez-Garcia, I., Silva-Espinoza, B.A., Ortega-Ramirez, L.A. et al. (2016). Oregano essential oil as an antimicrobial and antioxidant additive in food products. *Critical Reviews in Food Science and Nutrition* 56: 1717–1727.

43 Teng, J., Liu, X.D., Hu, X.Q. et al. (2018). Dihydromyricetin as a functional additive to enhance antioxidant capacity and inhibit the formation of thermally induced food toxicants in a cookie model. *Molecules* 23: 2184.

44 Ciriminna, R., Meneguzzo, F., Delisi, R., and Pagliaro, M. (2017). Olive biophenols as new antioxidant additives in food and beverage. *Chemistryselect* 2: 1360–1365.

45 Amakura, Y., Yoshimura, M., Sugimoto, N. et al. (2009). Marker constituents of the natural antioxidant eucalyptus leaf extract for the evaluation of food additives. *Bioscience Biotechnology and Biochemistry* 73: 1060–1065.

46 Alzate-Arbelaez, A.F., Dorta, E., Lopez-Alarcon, C. et al. (2019). Immobilization of Andean berry (*Vacciniummeridionale*) polyphenols on nanocellulose isolated from banana residues: a natural food additive with antioxidant properties. *Food Chemistry* 294: 503–517.

47 Zhang, Y.J., Gan, R.Y., Li, S. et al. (2015). Antioxidant phytochemicals for the prevention and treatment of chronic diseases. *Molecules* 20: 21138–21156.

48 Pohl, F. and Lin, P.K.T. (2018). The potential use of plant natural products and plant extracts with antioxidant properties for the prevention/treatment of neurodegenerative diseases: *in vitro*, *in vivo;* and clinical trials. *Molecules* 23: 3283.

49 Lara-Guzman, O.J., Tabares-Guevara, J.H., Leon-Varela, Y.M. et al. (2012). Proatherogenicmacrophage activities are targeted by the flavonoid quercetin. *Journal of Pharmacology and Experimental Therapeutics* 343: 296–306.

50 Xie, C., Kang, J., Chen, J.-R. et al. (2011). Phenolic acids are *in vivo;* atheroprotective compounds appearing in the serum of rats after blueberry consumption. *Journal of Agricultural and Food Chemistry* 59: 10381–10387.

51 Kloypan, C., Jeenapongsa, R., Sri-in, P. et al. (2012). Stilbenoids from *Gnetummacrostachyum* attenuate human platelet aggregation and adhesion. *Phytotherapy Research* 26: 1564–1568.

52 Shen, L., You, B.-A., Gao, H.-Q. et al. (2012). Effects of phlorizin on vascular complications in diabetes *db/db* mice. *Chinese Medical Journal* 125: 3692–3696.

53 Kruger, M.J., Davies, N., Myburgh, K.H., and Lecour, S. (2014). Proanthocyanidins, anthocyanins and cardiovascular diseases. *Food Research International* 59: 41–52.

54 Khan, N., Afaq, F., and Mukhtar, H. (2008). Cancer chemoprevention through dietary antioxidants: progress and promise. *Antioxidants and Redox Signaling* 10: 475–510.

55 Han, K.H., Hashimoto, N., and Fukushima, M. (2016). Relationships among alcoholic liver disease, antioxidants, and antioxidant enzymes. *World Journal of Gastroenterology* 22: 37–49.

56 Nasri, H., Shirzad, H., Baradaran, A., and Rafieian-Kopaei, M. (2015). Antioxidant plants and diabetes mellitus. *Journal of Research in Medical Sciences* 20: 491–502.

57 Rahman, H.A., Saari, N., Abas, F. et al. (2017). Anti-obesity and antioxidant activities of selected medicinal plants and phytochemical profiling of bioactive compounds. *International Journal of Food Properties* 20: 2616–2629.

58 Teimoori-Boghsani, Y., Bagherieh-Najjar, M.B., and Mianabadi, M. (2018). Investigation of phytochemical and antioxidant capacity of fennel (*Foeniculumvulgare* Mill.) against gout. *Journal of Medicinal Plants and By-products* 7: 59–65.

10

Biodiversity and Importance of Plant Bioprospecting in Cosmetics

K. Sri Manjari[1], *Debarati Chakraborty*[2], *Aakanksha Kumar*[3], *and Sakshi Singh*[4]

[1] *University College for Women, Osmania University, Hyderabad, Telangana, India*
[2] *Department of Molecular Biology and Biotechnology, University of Kalyani, Kalyani, West Bengal, India*
[3] *Bioclues, Hyderabad, Telangana, India*
[4] *Department of Molecular Biology and Human Genetics, Banaras Hindu University, Varanasi, Uttar Pradesh, India*

10.1 Biodiversity, Bioprospecting, and Cosmetics – A Harmony of Triad

Biodiversity, a natural repository of all living things on earth encompassing microbes to humans has three levels – species, habitat, and gene. Biodiversity serves as the natural currency that contribute immensely toward the development of human civilization. Therefore, it is extremely important that biodiversity be used in a sustainable manner [1]. India is a land with diverse ethnic, cultural, and climate zones along with four biodiversity hot spots making it for a mega biodiverse country. Traditionally, Indians are aware of this distinguished biodiversity and have inculcated several practices to conserve and economically utilized these resources for medicine, food, agriculture, and other commercial products, growing its market both nationally and internationally [2].

Bioprospecting, also known as biodiversity prospecting, is a dynamic and methodical search for natural ingredients and genes in wildlife that have the potential to be turned into industrial products by biological, genetic, and chemical manipulation and without harming nature. Bioprospecting is applied in fields of cosmetics, pharmaceutical, agriculture biotechnology, enzymes, and proteins and sensor technology [3]. New methods involve high-throughput screening and other automated programs to identify, screen, and isolate novel wild bioactive compounds (plants, microbes, polyphenols, fungi, and animals, etc.) [1]. There is an ocean of untapped potential in unknown bioresources that can provide industries with novel developments. While bioprospecting leads to discovery and utilization of natural products, biotechnology is needed to turn the natural compound into a more accessible and potent product with higher efficacy. Biotechnology

Bioprospecting of Plant Biodiversity for Industrial Molecules, First Edition.
Edited by Santosh Kumar Upadhyay and Sudhir P. Singh.

can also contribute toward conserving and sustaining the original resources by utilizing the genetics of active ingredients thus preventing the exploitation or damage to the environment [4].

Cosmetics and make-up have been around for many centuries around the world applied equivalently among all the genders. Anatomically modern humans have used various substances to alter their appearance or adorn their features for at least 10,000 years and possibly a lot longer. The etymology of the Greek word "Kosmeticos" means "related to beautifying or decorative substance." From a consumer point of view, cosmetics are products applied locally on external body parts for beautifying mainly the complexion, skin, nails, teeth, and hair [5]. The term cosmetics differs slightly throughout the world. "The US Federal Food, Drug and Cosmetic Act defines cosmetics more precisely in two ways 1) as products intended to be rubbed, poured, sprinkled, or sprayed, introduced into or otherwise applied to the human body, or any part to cleanse or beautify or promote attractiveness or alter the appearance (for instance, shampoos and lipstick) and 2) products intended for use as a component of any such articles (the term soap is excluded)" [6]. Article 2 of the EU Cosmetics Regulation (Regulation (EC) No. 1223/2009) incorporating the following definition of a cosmetic product: "A 'cosmetic product' shall mean any substance or mixture intended to be placed in contact with the various external parts of the human body (epidermis, hair system, nails, lips, and external genital organs) or with the teeth and the mucous membranes of the oral cavity with a view exclusively or mainly to cleaning them, perfuming them, changing their appearance and/or correcting body odors" [7]. The Indian Drugs and Cosmetics Act, 1940, and the Drugs and Cosmetic Rules 1948 defines cosmetics almost similar to the US Federal Food, Drug and Cosmetic Act [8]. The word cosmeceutical a hybrid category of the words "cosmetics and pharmaceutical," was first coined by Raymound Reed, a member of U.S. society of cosmetic chemist. Furthermore, the term was popularized by Dr. Albert Kligman in 1984 describing it as a cosmetic with pharmaceutical action [9]. However, the term is still not officially recognized in US, EU, or India. Nutricosmetics is another closely related term used which conceptualize that orally ingestible dietary products can act as nutritional supplements to care skin, nails, and hair natural beauty [10].

Cosmetics can be synthesized artificially or naturally occurring and a combination of both [11]. The basic ingredients in a majority of cosmetics include water, thickeners, moisturizers, emulsifiers, stabilizers of pH, colors, fragrances, preservatives, and other active formulations specific to skin, hair, nails, or hygiene [12]. Cosmetics essentially require interaction and penetration to multilayers of skin and different cell types; hence, compounds having properties like biosurfactants, antioxidants, antiaging, etc. produced by various sources serve as the best replacement to chemical entities available in marker [5]. The cosmetic industry is constantly seeking new natural products and has explored the untapped bioresources in the past decade and is one of the frontiers in bioprospecting. Bioprospecting in cosmetics has made leaps from utilizing plant seeds to fungi, from traditional and herbal cosmetics to vegan and natural products made from seaweed and oats. A changing preference in the cosmetics industry from synthetic to natural and biologically obtained sources has undoubtedly changed the cosmetics market. Use of herbal and natural beauty products reduces the chance of side effects or any other harm that may be caused by chemical-based products. The natural genetic resources have an enormous capacity to produce novel compounds for the medical and food industry. The undiscovered potential of our ecosystem and diverse species has never-ending possibilities of biological value. The chapter attempts to recapitulate the bioprospecting of plant biodiversity in cosmetic industry.

10.2 The Fury of Synthetic Chemicals in Cosmetics on Health

Synthetic chemicals in cosmetics and skincare products often unleash their fury on users causing several mild to severe health problems. With the increasing use of cosmetics and the consequent exposure to the chemicals for a lengthy period and repeated frequency, the adverse biological effects of these products are becoming more prevalent globally [13, 14]. Both men and women are being affected by these harmful chemicals. Although there exist several stringent regulations and quality control tests for cosmetic manufacturing, these standards are often not met, as evidenced by the harmful effects still occurring in the cosmetic consumers [15]. The cosmetic industry is creative and always striving for improved products, which is in the never-ending process of finding and using new ingredients. Many of these new ingredients are not enlisted in the restriction's list and may serve as the latest potential allergens. Contrary to medicine manufacturing, there neither exists any specific agency to critically assess the standards of cosmetics and skincare products nor any marketing authorization is required [16]. Moreover, there is no analysis of the risk–benefit ratio involved in the production and no guarantee of maintenance of consistent quality from one batch of products to the other. It is, therefore, the practice of "cosmetovigilance," a kind of quality surveillance devoted to ensuring that the safety of cosmetics has been emerging in public health sciences [17]. In the past nine years, almost 12% of global cosmetic users had experienced toxic effects caused by cosmetics and skincare products [17, 18]. Thus, the increasing use of cosmetic products is emerging as a public health concern. Table 10.1 enlists several public health risks caused by chemicals used commonly in a wide variety of cosmetics.

10.3 India's Biodiversity and Its Traditional Knowledge/Medicine in Cosmetics

World Health Organization defines "Traditional medicine" as a collection of indigenous and long-established medicinal practices seen all around the world, among various countries, cultures, and civilizations, which has its roots from the humankind's knowledge and experience. India's rich biodiversity has been preserved and ethnically shared by people practicing traditional medicine like Yoga, Homeopathy, Siddha, Amchi, Unani, and Ayurveda in conjunction with modern medicine [19]. The recent few decades have seen an increase to search for new natural products in drugs and cosmetics industry from the traditional medicine systems as they bear distinct similarities to contemporary pharmacognosy along with experiential wisdom, holistic approach, better tolerance, and safety [20, 21]. Ethnobotany is again a study of traditional folklore knowledge and customs of mankind

Table 10.1 Chemical cosmetics and their toxic effects.

Causative chemical/s	Chemical cosmetics	Harmful effects
Dibutyl phthalate (DBP), formaldehyde, toluene (Toxic trio)	Nail polish	Allergic reactions and skin irritation, headache, eye irritation, dizziness. If ingested acts as endocrine disruptor, worsen infertility, causes liver and kidney failure
Benzophenones, debenzoylmethanes	Sun-screening agents	Irritant, allergic, phototoxic or photo-allergic dermatitis
Para-aminobenzoicacid (PABA)	Sun-screening agents	Irritant, allergic, phototoxic or photo-allergic dermatitis
Cinnamates	Sun-screening agents	Irritant, allergic, phototoxic or photo-allergic dermatitis
Para Phenylenediamine (PPD)	Commercial hair dye	Skin inflammation and irritation (eczema), blisters and surface oozing, contact dermatitis, erythematous rash, carcinogenic
Hydroquinone (HQ)	Skin lightening cream	Severe skin redness, burning sensation, ochronosis (gradual darkening of the skin)
Coumarin	Deodorants, anti-perspirants, fragrances	Potential carcinogen
Phethleugenol	Deodorants, anti-perspirants, fragrances	Potential carcinogen
Phthalates	Deodorants, antiperspirants, fragrances	Potential hormone disruptors
Hydrogen peroxide solutions	Hair bleaching products	Allergic contact dermatitis (Type I and IV)
Ammonium persulfate	Hair bleaching products	Allergic contact dermatitis (Type I and IV)
Butylated hydroxyanisole (BHA), Butylated Hydroxytoluene (BHT)	Moisturizers, lipsticks	Skin allergies, carcinogenic
Parabens (methylparaben, propylparaben, butylparaben, and ethylparaben)	Shampoo, toothpastes moisturizers, sunscreens, face and body washes, deodorants, antiperspirants, shaving gels	Skin irritation, contact dermatitis, endocrine disruptors
Diazolidinyl Urea	Child care products, skin care products, hair, nails, eye and face make-up	Mutagenic and carcinogenic agent, allergic contact dermatitis
Imidazolidinyl urea	Lotions, creams, shampoos, hair conditioners, deodorants	Mutagenic and carcinogenic agent, allergic contact dermatitis
1,4-dioxane	Toothpaste, mouthwash, shampoo	Potential carcinogen, can trigger breast, skin and liver cancer

Table 10.1 (Continued)

Causative chemical/s	Chemical cosmetics	Harmful effects
Heavy metals [lead (Pb), cadmium (Cd), nickel (Ni), arsenic (As), mercury (Hg), antimony (Sb), cobalt (Co)]	Lipstick, eyeliner, eye shadow, whitening toothpaste, nail color, face powder, hair cream, hair dyes (contaminants)	Contact dermatitis, systemic allergic dermatitis, circulatory and peripheral nervous disorders can serve as potential carcinogen, pneumoconiosis, emphysema, bronchitis, bone fragility, kidney damage
Methylchloroisothiazolinone-methylisothiazolinone (MCI-MI)	Personal hygiene products	Allergen, may cause hypersensitivity on long term exposure
Methyldibromoglutaronitrile-phenoxyethanol (MDBGN-PE)	Moisturizers	Contact dermatitis
Quaternium-15	Soaps, shampoos, hair conditioners, hair styling products, shaving creams, moisturizers	Allergen, may cause systemic anaphylaxis
Thimerosal	Tattoo inks, mascara	Respiratory and nervous system toxicant
Glyceryl monothioglycolate	Hair styling chemicals (straightening, perming)	Allergen
Myroxylon Pereirae (Balsam of Peru)	Fragrances	Allergen
Cocamidopropyl betaine	Shampoo, toothpaste, hand soaps, shower gels, liquid soaps, skin cleansers, deodorants	Allergen
Toluenesulfonamide-formaldehyde resin	Nail lacquers, polishes	Allergen
Propyl gallate, octyl gallate, dodecyl gallate	Moisturizing cream, lipstick	Allergen

concerning plants and its varied uses not just in medicine but also religious and other anthropological uses [22]. Herbal products have been used all over the world either as home remedies or on large-scale industrial beauty and health products with plant sources as the main ingredient. Ayurveda, a well-documented traditional medicine of India, is accepted by modern medicine and western countries for its rich and established history of effectiveness. These findings are published in ancient scriptures like Charaka Samhita, Sushruta Samhita, Ashtanga Hridayam, and Ashtanga Sangraha. It is still widely used in rural areas as their foremost health care. Bioproducts include health foods and medicines, botanicals, and supplements. In India, Ayurveda, Unani, and Siddha systems are based on a holistic lifestyle that incorporates everything from diet to the drug, from physical to spiritual healing of mind and body. Traditional and herbal medicine is also widely used around the globe, such as traditional Chinese medicine and Kampo, Japanese traditional medicine [23].

India's biodiversity in plants provides sources not only to medicine and drugs but also cosmetics, and hence incorporated in clinical practices backed by ages of results and ongoing evidence of being safe and natural. India has four biodiversity hotspots, The western Ghats, home to 6000 vascular plants with 2500 genus. Three thousand plants are endemic, and most spices of black pepper and cardamom in the world are all believed to have originated in the Western Ghats. The Himalayas are another biodiversity hotspot, where 10,000 species of plants, a third of which are endemic and cannot be located anywhere else in the world can be found. Similarly, the Indo-Burma Region and Sundaland are the other hotspots that are home to thousands of endemic species of fauna, half of which cannot be found anywhere else in the world. Due to the rich biodiversity and established benefits of traditional and herbal medicine, public interest in plant-based products has increased over the years, leading to a surge in the development of plant-based nutraceuticals and cosmetics. Thus, a multidisciplinary approach, including traditional knowledge-based bioprospecting, may offer promising leads in the cosmetic industry with natural and safe use.

10.3.1 Herbal Cosmetics

The plethora of plants and herbs, which have been traditionally used for ages in India for healing, led to the development of modern drugs and are now making a comeback for being a safe and effective substitute for chemical and synthetic medicine and drugs. Herbal cosmetics have one or more herb and shrubs as base ingredients that provide cosmetic benefits. Herbal cosmetics tend to suit all skin types and are available in a wide range and at a lower price than synthetic cosmetics and cosmeceuticals. They are natural and are mostly gentle on the skin and are cruelty-free. These features make herbal cosmetics a safer and more attractive option for the new generation of more conscious buyers. Few examples of the most popular and developed herbs and bioingredients of natural cosmetics are shown in Table 10.2.

10.4 Use of Plant-Based Products in the Cosmetic Industry

Earlier botanicals were source of a small percentage of cosmetics. However, the consumers worldwide are looking for cosmetic products devoid of any synthetic chemicals. Many factors attribute in interest in natural products research: First, unmet therapeutic needs; second, the remarkable diversity of both chemical structures and biological activities of naturally occurring secondary metabolites; third, the utility of bioactive natural products as biochemical and molecular probes; fourth, the development of novel and sensitive techniques to detect biologically active natural products; fifth, improved techniques to isolate, purify, and structurally characterize these active constituents, and last, advances in solving the demand for the supply of complex natural products.

Cosmetics typically contain around 15–50 ingredients. Many products differ from company to company and vary with ingredients. However, most cosmetics contain certain core components such as emulsifier, thickener, preservative, emollient, pH stabilizers, fragrances, and colors. Emulsifiers are ingredients used to produce homogenous and even texture usually to bind the water–oil mixtures in the cosmetics [25]. Some of the

Table 10.2 Popular herbs and their ingredients in natural cosmetics [24].

Skin type	Natural compound/ingredient	Feature and benefit
Dry skin	Coconut oil	Liquid or solid Directly applied on skin and hair as a moisturizer; can be used as cooking oil,
	Sunflower oil	Smoothing properties and considered non-comedogenic
Antiaging treatment	Golden Root (Rhodiola rosea)	Increases resistance to physical stress and has antioxidant properties.
	Carrot seed oil	Many essential vitamins like Vit A, promotes new skin cell formation, antiwrinkle, and natural toner.
	Ginkgo (*Ginkgo biloba)*	Circulatory tonic, strengthening capillaries, protects the nervous system, and fights oxidation.
All-round skincare and protection	Neem	Antiseptic, reduce dark spots, antibacterial
	Aloe	Moisturizer, sunscreen, emollient
	Green tea	Antioxidant that can be consumed as a drink or applied on the skin, lessen inflammation, and cell damage protection
	Calendula	Topically used for acne, reducing inflammation, controlling bleeding, and soothing irritated tissue
Haircare	Amla	It contains vitamin C, tannins, phosphorus, iron, and calcium that give nutrition to the hair, prevents graying, and stimulates thicker hair growth.
	Kapoor, Heena, Amalaki, Bhringraj, Hirda, Rosary pea, Brahmi, Lemon and Shikakai	Used in moisturizers, shampoos, hair growth, darkening of hair, hair oil, and preventing dandruff.
Essential Oils	Eucalyptus oil, Rose oil, Citronella oil	Copreservatives with antibacterial activity, for fragrance, in hair care and skincare for having sustained effects in the skin

commercially used natural emulsifiers are glyceryl monostearate (GMS) and polysorbate 20 derived from natural oils. Thickeners [26] are agents that help in increasing the viscosity of the final cosmetic products. Some of the natural ingredients have both emulsifying and thickening properties. The use of natural preservative from plants has both antioxidant and antibacterial effect. Many essential and aromatic oils of plant origin exhibit preservative properties [27]. Emollients are skin softeners by preventing dry skin formation [28]. Stabilizing agents maintain pH of the cometic preparation. Most of the stabilizers are of microbial origin than plant source. The fragrances used in cosmetics usually have their origin from a wide range of plants. Natural pigments of plant origin are gradually used in

Table 10.3 Core components of plant origin used in cosmetics.

Core component	Plant name	References
Emulsifier	*Helianthus annuus* L. - sunflower oil *Corylus americana* Marshall - hazelnut oil *Acacia senegal* - gum arabic *Cyamopsis tetragonoloba* - guar gum	[29–31]
	Elaeis guineensis - palm kernel oil *Glycine max* - soy oil *Cocos nucifera* - coconut oil	[32]
Thickeners	*Gossypium arboretum* - hydroxyethyl cellulose, *Cyamopsis tetragonoloba* - guar gum,	[33]
Preservatives	*Melaleuca alternifolia* - tea tree oil *Thymus vulgaris* - thyme oil	[34, 35]
Emollient	*Olea europaea* - olive oil, *Cocos nucifera* - coconut oil	[32, 36]
pH stabilizers	*Triticum aestivum* - hydrolyzed wheat protein	[37]
Colors	*Curcuma longa* - bright yellow *Croccus sativa* - yellowish orange *Beta vulgaris* - red Lithospermum - red	[38]

recent years. Many of the plant-based components are used commercially by many of the noted brands all around the world. Some of the plant-based core components are listed in Table 10.3.

Apart from the core ingredients, many other bioactive materials are added depending on the type of cosmetic action required. This may include skin protection from harmful UV irradiation, skin whitening, antiaging, hair care, acne-protection, anticellulite activity, and hair coloring, etc. Some the plant bioactives used in the natural cosmetic industry are enlisted in Table 10.4.

10.5 Green Cosmetics – Significance and Current Status of the Global Market

The word "green" is interchangeably used with "organic," "environment-friendly," "natural," "sustainable," or "chemical-free." Thus, "Green cosmetics" are those cosmetics which are not merely plant-based but manufactured from plants which are cultivated or growing naturally in organic and chemical-free condition. Green cosmetics "guarantees environmental conservation all along the production line, consumer's respect and utilization of natural matter with superior ecological quality" (Eco cert 2003). They are devoid of genetically modified ingredients and thus aim to achieve environmental conservation, minimal pollution, sustainable utilization of nonrenewable resources, and preservation species.

Table 10.4 Plant Bioactives and their mode of action in natural cosmetics.

Cosmetic property	Plant name	Bioactives	Action	References
Skin protection	Grape, *Vitis sp.*	Proanthocyanidins or procyanidins	High antioxidant activity, Skin whitening	[39]
	White mulberry, *Morus alba*	Phenolic compounds Eg. gallic acid; quercetin and the fatty acids e.g. linoleic acid, palmitic acid; Mulberroside F	Skin whitening property	[40]
	Liquorice/Licorice, *Glycyrrhiza glabra*	Glabridin, glabrene, isoliquiritigenin, licuraside, licochalcone a, liquiritin and its isoform	Skin whitening property	[41, 42]
	Aloe vera	Aloesin	Skin whitening, photoprotection property	[43]
	Citrus sinensis	Hesperidin (Flavonoid)	Skin whitening	[44]
	C. paradisi	Hesperidin (Flavonoid)	Skin whitening	[44]
	Cranberry, *Vaccinium oxycoccos, Vaccinium macrocarpon*	Proanthocyanidins or procyanidins	High antioxidant activity, skin whitening	[45]
Anti-aging	Cofee, *Coffea arabica*	Polyphenols, especially chlorogenic acid, condensed proanthocyanidins, quinic acid, and ferulic acid	Improves pigmentation, fine lines, wrinkles,	[46, 47]
Photo protection	*Moringa oleifera*	Rutin, quercetin, ellagic acid, chlorogenic acid, ferulic acid	Protects from UVA-induced oxidative stress, UV-induced erythema, photoaging, prevent collagen destruction, enhances expression of immune factors	[48]
Hair care activity	Chinarose, *Hibiscus rosasinensis*	Flavonoids, tannins, terpenoids, saponins, and alkaloids	Hair growth promotion and antigraying activity	[49]
	Clove basil, *Ocimum gratissum*	α-Pinene, camphene, *cis*-ocimene, β-ocimene, *trans*-caryophy-llene, eugenol	Enhance normal hair growth, promotes follicular proliferation in cyclophosphamide (used frequently in chemotherapy) – induced hair loss, antimicrobial activities	[50]

(*Continued*)

Table 10.4 (Continued)

Cosmetic property	Plant name	Bioactives	Action	References
Anti-acne	Bitter melon, *Momordica charantia*	Charantin (steroidal saponin), Momordicoside U, Geracrene	Antibacterial, anti-inflammatory efficacy, anti-acne activity, anti-psoriasis, anti- leprosy	[51]
	Neem, *Azadirachta indica*	Azadirachtin, nimbolinin, nimbin, nimbidin, nimbidol, sodium nimbinate, gedunin, salannin, quercetin, ß-sitosterol	Anti-inflammatory, anti-acne activity	[52]
	Eucalyptus globulus	Gamma-terpinene (Monoterpene)	Antibacterial, antifungal and anti-inflammatory activity, antiacne efficacy	[53]
Anti-cellulite activity	Coffee, *Cofffea robusta L., Coffea arabica L.*	Caffeine	Removes skin swellings and fatigue signs	[54]
	Coleus forskohlii	Forskolin	Dislodges localized fat	[55]
	Centella asiatica L.	Asiaticoside	Strengthens circulation, increases metabolism of proline and lysine	[56]
	Horse chestnut, *Aesculus hippocastanum L.*	Saponin	Anti-inflammatory and vasoprotective actions	[56]
	Ginkgo biloba L.	Proanthocyanidins	Prevent edema, anti-inflammatory activities	[56]
	Ivy, *Hedera helix L.*	Saponin	Have vasoconstrictor properties, prevents edema, improves body fluid drainage	[56]
Hair coloring	Henna, *Lawsonia inermis, Lawsonia alba, Lawsonia spinosa*	Lawsone/hennotannic acid	Red dye	[57]
	Juglans cinerea, J. Regia, J. Nigra	Naphthoquinones (Juglone and juglandin)	Yellow-brown dye	[58, 59]
	Indigofera tinctoria	Indigotin	Indigo dye	[60]

Green or environment-friendly cosmetics are manufactured from organically grown plant and fruit extracts and their concentrates [61].

The main reason for increasing consumer preferences toward these products is related to an increase in health awareness and environmental consciousness [62]. Our skin offers the first line of defense to our body in fighting an array of infections. It can wholly or partially soak the applied products, which thus can reach to our bloodstream. About 60% of the used cosmetics and skincare products reach our circulatory system. An average woman soaks up almost 30 pounds of the ingredients present in moisturizers in 60 years [63]. Hence, using chemical-free, green cosmetics is a safer avenue.

Moreover, the ability to perform physiological functions of the skin decreases with aging. Plant products can offer a solution in the fight against aging by performing hydration, antioxidant, antimicrobial, and UV-protection activity, reducing inflammation, irritation, and many other things. Some of the examples are coconut and argan oil, calendula, green tea, pomegranate, crocin occurring in crocus and gardenia flowers, soy, and feverfew [64].

The chemicals used in cosmetics also take a toll on the environment [65]. Examples are some preservatives (parabens, triclosan), UV-shields, plastics, and heavy metals [66]. The potential of environmentally safe green cosmetics toward solving this problem is also a reason for their gradually increasing popularity [67]. However, there is still a considerable lack of awareness among customers about these products [68], which requires unique marketing strategies. Strategies undertaken for Green marketing highlight a new edge of socioeconomic and environmental responsibility adopted by organizations [69].

The Indian market has also seen a surge in such Green companies in the past few decades. Brands like Khadi naturals, Ayur, Jovees, Himalaya, and Lotus herbals, Shahnaz Hussain, Biotique, Aroma essentials, Fab India, are successful ventures so far [70].

10.5.1 Sustainable Development Goals (Economic, Ecological Benefits) in Cosmetic Industry – How Bioprospecting and Green Cosmetics Can Help?

Intelligent innovations serve as the keystone business strategy driving not only economic growth but also plays a prominent role in solving an array of social and environmental issues. Sustainability related intelligent innovations or sustainable innovations, also called GreenTech solutions, act at production, service, strategy, process design level to achieve a win–win situation for both producer and consumer. The use of synthetic chemicals in the cosmetic industry causes several adverse effects, e.g., chemical abuse leading to mild to severe health issues (Table 10.1), environmental pollution, and unsustainable use of natural resources. The harmful effects of synthetic chemicals in cosmetics are raising health concerns among customers globally. Researchers have reported an increasing trend of consumer awareness toward sustainable cosmetic products in recent decades due to health issues caused by UV rays and customers becoming aware of the benefits of natural products [71–73].

Hence, several research efforts have been focused on the conceptualization and development of sustainably innovated green cosmetic products and their marketing models [73]. These products are characterized by the consumption of less water, material, and energy during production, least or zero environmental pollution [74, 75]. For reduction in environmental pollution caused by cosmetics and body care products, the use of an array of higher plants, macro, and micro algae-derived bioactive natural components, agro wastes, greener extraction methods is being envisaged [76–78]. Volatiles and nonvolatiles extracted from plants, mainly herbs, and spices serve as the primary sources of higher plant-derived bioactive ingredients [77]. Some examples of bioactive ingredients include gallotannins, caffeic acid, and that of agrowaste is the husk of quinoa (*Chenopodium quinoa*), which serves as a skin exfoliant. Consideration toward the materials used for packaging of the finished cosmetic is also of utmost importance. The use of eco-friendly products like paper, glass, wood is gradually increasing in cosmetic packaging.

"The United Nations defined 17 Sustainable Development Goals (SDGs) covers a range of environmental, economic, and societal problems required for sustainable, innovative actions." According to Reuters, the global cosmetic sector was valued to be 507.75 billion USD in 2018 and predicted to reach 758.45 billion USD by 2025. Hence, the activities undertaken in this sector can contribute directly toward achieving certain SDGs if implemented sustainably. Production of green cosmetics by using organically cultivated plant products, use of environment-friendly packaging, low energy, low water-intensive production techniques are such approaches that can directly contribute toward reducing the carbon footprint. Organically cultivated plants use natural fertilizer and are pesticide-free. Environment-friendly packaging is biodegradable and hence will contribute toward lesser or zero solid waste production. Hence, such actions can aid in achieving responsible production (SDG12) and climate action (SDG13). The use of natural bioactive compounds, ecofriendly packaging in cosmetics, and body care products will ensure customer health while ensuring economic benefits. These actions will contribute toward achieving good health and well-being (SDG3).

10.6 Ethical and Legal Implications of Bioprospecting and Cosmetics

Bioprospecting is defined as search for identification, collection, extracting, and screening of biological resources mainly of plants, animals, microorganisms, cells, or genes, for commercial purposes, such as for pharmaceutical products, cosmetics, dyes, and foods. Bioprospecting is hence governed by laws and has ethical implications and is connected and communicated by local populations, governments, other persons in the scientific community, and the international public.

From the first step of discovery of novel natural compounds, the process and activities of bioprospecting have direct implications to the environment, habitat, and natural surroundings and the indigenous community of that place. The search for bioresources and or traditional knowledge has gained tremendous momentum and has led to biopiracy, denial of rights, and exploitation of natural resources. Hence, both national and international laws as well as various private and regional measures have been put in place to regulate access to genetic resources and benefit-sharing.

There are various legal issues related to bioprospecting. Most prominent among them is the protection and conservation of traditional knowledge of biodiverse and indigenous lands, which gets exploited by developed countries. Other legal issues concerning bioprospecting revolve around pharmaceutical and other industry research based on traditional medicine and knowledge of the indigenous communities; Biopiracy, which is unauthorized and plagiarized use of biological resources or traditional knowledge or its patenting and lastly unequal benefit sharing between resources or knowledge provider and technology developer [79].

Apart from the laws pertaining bioprospecting, the ethical and legal guidelines applicable to cosmetics should also be taken into consideration. Bioprospecting of novel cosmetic compounds in given biota by the use of tools like bioinformatics and cheminformatics can also lead to quicker results with new botanical-based cosmetics. However, one major drawback would be the side effects of the use of such cosmetics in a long run. But the traditional knowledge has the advantage of the above limitation because they have observations over a long period of many generations. The future cosmetics can thus be based on time tested use of traditional herbal cosmetics developed by using advanced technologies bringing together the best use of knowledge, resources, science, and technology in an amicable manner.

10.6.1 International Laws Regulating Bioprospecting

1) Convention on Biological Diversity (CBD) was formed in 1992 to stop the loss of biodiversity and to regulate the inconsistency between the resource and knowledge giver and the technology and product developer [80]. The CBD acknowledges the state's sovereign rights upon its bioresources and builds the structure for equal access for the same. It recognizes the efforts and rights of indigenous communities for their knowledge and long traditional practices of sustainable use and conservation of the biological resources, all the while protecting and providing equal share and benefits leading from the use of those biological resources and traditional knowledge [79].
2) Rio Earth Summit formed the Statement of Forest Principles, which also oversees grants of equal share to indigenous communities for providing any traditional knowledge or resources from their environment and its ultimate use [81].
3) "Declaration on the Rights of Indigenous Peoples provides rights of traditional medicines (protecting plants, animals, and minerals) and health practices of Indigenous communities." It also supports the recognition of complete ownership, protection, and control of their traditional and intellectual property. This article clearly outlines the full rights of indigenous people to protect and create technologies and products and to control their cultural properties, which includes medicine, genetic resources, literature, humans and seeds, and their knowledge of flora and fauna [82].
4) Nagoya Protocol dictates the user of biological resources and traditional knowledge, to follow the benefit sharing and access protocols and regulations that are set as prerequisites of those resource countries. Bioprospectors must adhere to mutually agreed to conditions between the resource giver and user when trying to utilize their bioresources along with following the regulations of equal access and sharing. The Nagoya Protocol is relevant public international law, which makes it mandatory for associations which use and benefit (bioprospectors) from biological and genetic resources. The protocol helps the traditional knowledge to follow and adhere to jurisdictions and local Access and Benefit Sharing (ABS) governing laws of the provider country and, when appropriate, agree to ABS conditions with genuine providers of bioresources and knowledge [83].

10.6.2 Indian Law Regulating Bioprospecting

India is a hub for aromatic and medicinal plants, and its rich and diverse bio-history makes it extremely unique (from too hot and dry to cold regions with plenty of species variations) and an important country for bioprospecting. Thus, it requires special care and attention toward the safety, preservation, and conservation and sustainable utilization of the biological and traditional resources. The Biological Diversity Act 2002 (BD Act) and the Biological Diversity Rules 2004 (BD Rules) protect Indian biological and traditional resources, regulated by the center through the National Biodiversity Authority (NBA). The BD act delegate's responsibilities and influence the State Biodiversity Board and Biodiversity management committees at the local level [80].

As mentioned, India is a significant source for bioprospecting, considering that 80% of the medicinal plants patented in the US are of Indian origin, and 50% of the drugs reported in the British Pharmacopoeia have Western Himalayas medicinal plant origins. The Biodiversity Act and rule emphasis on Biosurvey and Bio Utilization and form the framework for safeguarding and providing approvals for administering bioprospecting in India. Among that is obtaining a beforehand approval from the NBA for bioprospecting activities such as intellectual property rights for any research or innovation based on biological resources or traditional knowledge from India; for collecting or procuring any biological resource originating in India or any traditional knowledge of India for any research or commercial use such as drugs, industrial enzymes, food flavors and extracts, fragrance, cosmetics, emulsifiers, colors, and genes used for improving crops and livestock through genetic intervention. It also covers the transfer of traditional knowledge or biological resources and biosurvey or bioutilization. With all these regulations and monitoring, the NBA must protect the terms and conditions of all approvals and secure equal sharing of benefits between the user and provider of the biological resources and traditional knowledge.

Another essential article of the BD act and rules is to provide the prior declaration and information of any Indian citizen or corporate body registered in India to the state biodiversity board before beginning any bioprospecting activity like biosurvey or procure biological resource for any commercial or research utilization [79].

10.6.3 Access and Benefit Sharing (ABS)

Access and benefit sharing (ABS) is the manner of accessing genetic resources between provider and user and the agreement among them for equal and fair sharing of the resultant benefits of the resource. CBD lays out rules that oversee the benefit sharing and access of bioresources and traditional knowledge that promote environment-friendly access of genetic resources and their purposes and to also protect the resultant benefits of their use, which should be fairly and equally shared between the provider and user.

Two key features in ensuring ABS are the prior informed consent (PIC), which is provided by the central governing body (NBA in India) in a legal and institutional framework and the mutually agreed terms (MAT) of fair and equal distribution of benefits between provider and user [80].

For centuries local communities and indigenous people have been using and benefiting and developing biological resources and lived off of their local biodiversity by using them from medicine, shelter, food, agriculture, clothing, and animal husbandry. Hence, traditional knowledge refers to practices and innovations and complete knowledge of local communities and indigenous people related to bioresources and genetic materials, which gets adapted and changed and developed based on needs and environment over centuries and continues to be passed down to new generations [81].

Traditional knowledge, including genetic resources, is relevant to indigenous and local communities (ILCs), which depend on these bioresources for their day to day activities and livelihood. The bioprospector or user benefits from traditional knowledge by seeking permission and accessing these resources to develop commercial and new products as well as for research in both commercial and academic fields. Competent National Authorities (CNAs) generally oversee the negotiations and agreements and validity between users and providers of traditional knowledge and maintain a balance in the bioprospecting area [82].

Bonn Guidelines – the CBD adapted Bonn Guidelines in 2002 to support governments in regulating ABS in their countries. It assists countries in developing and enforcing ABS effectively to make sure that both users and providers of bioresource share the benefits equally and fairly [81].

10.6.4 World Intellectual Property Organization (WIPO)

World Intellectual Property Organization (WIPO) – formed in 1967 as a global framework that promotes and protects Intellectual property (IP) policy under the United Nationals with 193 members. It is a forum to shape international IP rights and protects IP across borders and resolves conflicts. Through WIPO IP knowledge can be shared and can be used as a reference for world IP policies [84].

10.6.5 Intergovernmental Committee on Intellectual Property and Genetic Resources, Traditional Knowledge, and Folklore (IGC)

Intergovernmental Committee on Intellectual Property and Genetic Resources, Traditional Knowledge, and Folklore (IGC) was further formed in 2000 as a forum for WIPO associate states to examine IP concerns pertaining to accessing, sharing, and protecting bioresources and traditional knowledge. Formal negotiations are held by IGC to reach agreements regarding international laws that can establish adequate protection of bioresources and traditional and cultural practices and knowledge. Motivation to protect IP was taken from growing awareness of folklore creativity belonging to local people and is part of indigenous community identity and should hence be protected as intellectual property and should prevent misuse or exploitation by growing technologies [85].

10.7 Laws Regulating Cosmetics

Every country has laws governing access to provide safe and effective cosmetics. Cosmetic laws are regulated by Central Drugs Standards Control Organization (CDSCO) headed by drugs controller general of India and Bureau of Indian Standards (BIS) [86]. The manufacturer of cosmetics has to obtain licensing authority from the respective state governments and manufacture the product under the strict guidelines. The United States (U.S.) and European Union (EU) share similar goals of warranting the safety of cosmetics for consumers through critical and proven science-based regulation. The means by which each regulates the safety of cosmetics are quite similar. EU also has a Cosmetic ingredient Review that reviews and declares a given constituent of a cosmetic to be safe or not safe.

Proof of safety of cosmetic product, especially each ingredient, has to be provided by the manufacturer. Unlike pharmaceutical drugs, no pre-market approval is needed for cosmetics except color additives, active ingredients, sunscreen, and preservatives. The International Fragrance Association (IFRA) also establishes safety standards for fragrance materials. Safety of the cosmetic products is tested based on toxicity and irritancy potential, the rules which are similar to that of screening drugs for treatment of skin diseases. However, the strict guidelines for testing depend on the ingredients used in cosmetics. The cosmetics once manufactured have to abide by the labeling, import and export rules, and advertising rules of the country of origin [5].

10.8 Role of Biotechnology in Bioprospecting and Cosmetics

Although organic or green cosmetics are on a rise, they still remain a small part of the overall cosmetics industry. The main reason would be faster depletion of the natural plants than the time required for the plants to grow. Another reason is that the active ingredient which gives the magical effects still remains confined to certain rarer and endangered species. Hence, there is a simultaneous need for faster cultivation of such rarer plants without compromising on the plant properties. Advances in biotechnology and use of tissue culture of plants to produce secondary metabolites can be exploited for commercial processing of even rare plants and the metabolites they produce. Examples for such commercially available products are Shikonine a red pigment used especially for lipsticks and nail color [87] is produced from *Lithospermum erythrorhizon*. Arbutin is a whitening ingredient produced from *Catharanthus roseus* [88]. Both these ingredients are commercially produced via tissue culture by Mitsui Petrochemical Industries, Japan. However, the use of plant cell culture or their metabolite derived cosmetic active ingredients is still in its infant stage [89].

White biotechnology or industrial biotechnology, a component of "green chemistry," plays a key role in producing the cosmetics in the present world. The use of renewable substrates and biocatalyst results in an environmentally friendly valuable cosmetic product. This not only reduces the use of animal killing thereby protecting the animals from being extinct. Blue biotechnology exploits the ocean and sea resources to generate products of industrial interest. There is rise in the use of materials, to generate natural cosmetics with better efficacy and minimum side effects.

Ideally, a government and industry partnership should focus on multidisciplinary research. The research joins the forces starting from bioprospecting of plant from its natural habitat. This begins with the consultation of local people for their ethnic knowledge of plants for cosmetic uses. Once a right plant source with potential benefits is selected, further study to know the natural products chemistry, the underlying molecular and cellular biology, using the combined knowledge of biochemistry, synthetic, and analytical chemistry, and pharmacology which ends with a consumer product profitable to all.

Box 10.1 A Good and Successful Initiative

The International Cooperative Biodiversity Groups (ICBG) Panama, 1998, was started to establish shared benefits of Panama's bioresources with the country as a provider of genetic resources. This has resulted in benefits from drug discovery projects, scientific infrastructures, training of scientists, and forming many research programs. One of the objectives is to include local researchers in the center of commercially viable programs that use local biodiversity for research. The ICBG has led to a higher incentive for conservation helping advance the Coiba National Park to its UNESCO world heritage tag [83].

References

1 Pushpangadan, P., George, V., Ijinu, T.P., and Chithra, M.A. (2018). Biodiversity, bioprospecting, traditional knowledge, sustainable development and value added products: a review. *Journal of Traditional Medicine and Clinical Naturopathy* 7: 256.

2 Mishra, R.K. (2019). Medicinal and aromatic plants biodiversity in india and their future prospects. *Academic Journal of Botanical Sciences* 1: 11–15.

3 Sahu, T.R. and Sahu, P.K. (2019). *Biodiversity and Sustainable Utilization of Biological Resources*. Scientific Publishers.

4 Levin, S.A. (2013). *Encyclopedia of Biodiversity*. Elsevier Inc.

5 Larsen, W.G., Jackson, E.M., Barker, M.O. et al. (1992). A primer on cosmetics. *Journal of the American Academy of Dermatology* 27: 469–484.

6 US Department of Health and Human Services (2018). US Food and Drug Administration (2018). https://www.fda.gov/cosmetics/guidanceregulation/lawsregulations/ucm074162htm (accessed 01 October 2018).

7 UNION, P (2009). Regulation (EC) No 1223/2009 of the European Parliament and of the Council. *Official Journal of the European Union L* 342: 59.

8 Drugs and Cosmetics Act and Rules 1945 (2012). Definition as per Indian guidance document. http://www.slideshare.net/rakshitthumar/drugs-cosmetics-act-1940 (accessed 1 May 2015).

9 Pandey, A., Jatana, G.K., and Sonthalia, S. (2020). Cosmeceuticals. In: *StatPearls*. Treasure Island, FL: StatPearls Publishing.

10 Tabor, A. and Blair, R.M. (eds.) (2009). *Nutritional Cosmetics: Beauty from Within*. William Andrew.

11 Sawarkar, A., Sharma, R.K., and Gautam, V. (2019). Bioprospecting: creating value for biodiversity. *The Pharma Innovation* 8 (4): 256–265.

12 Augustine, R., Rajendran, R., Cvelbar, U. et al. (2013). Biopolymers for health food and cosmetic applications. In: *Handbook of Biopolymer-Based Materials: From Blends and Composites to Gels and Complex Networks*, 1ee (eds. S. Thomas, C. Chassenieux and P. Jyotishkumar), 801–849 Singapore: Wiley-VCH.

13 Pereira, J.X. and Pereira, T.C. (2018). Cosmetics and its health risks. *Global Journal of Medical Research* 18: 63–70.
14 Mohiuddin, A.K. (2019). Cosmetics safety: gray areas with darker inside. *Journal of Medical Research and Health Sciences* 2: 688–693.
15 Alani, J.I., Davis, M.D.P., and Yiannias, J.A. (2013). Allergy to cosmetics: a literature review. *Dermatitis* 24: 283–290.
16 Vigan, M. and Castelain, F. (2014). Fragrance and cosmetic contact allergy in children. *Current Treatment Options in Allergy* 1: 310–316.
17 Vigan, M. and Castelain, F. (2014). Cosmetovigilance: definition, regulation and use "in practice". *European Journal of Dermatology* 24: 643–649.
18 Maipas, S. and Nicolopoulou-Stamati, P. (2015). Sun lotion chemicals as endocrine disruptors. *Hormones* 14: 32–46.
19 World Health Organization (2000). *General Guidelines for Methodologies on Research and Evaluation of Traditional Medicine (No. WHO/EDM/TRM/20001)*. World Health Organization.
20 Patwardhan, B. and Mashelkar, R.A. (2009). Traditional medicine-inspired approaches to drug discovery: can Ayurveda show the way forward? *Drug Discovery Today* 14: 804–811.
21 Joshi, K., Ghodke, Y., and Shintre, P. (2010). Traditional medicine and genomics. *Journal of Ayurveda and Integrative Medicine* 1: 26–32.
22 Sharma, H. and Kumar, A. (2011). Ethnobotanical studies on medicinal plants of Rajasthan (India): a review. *Journal of Medicinal Plants Research* 5: 1107–1112.
23 Yuan, H., Ma, Q., Ye, L., and Piao, G. (2016). The traditional medicine and modern medicine from natural products. *Molecules* 21: 559.
24 Gediya, S.K., Mistry, R.B., Patel, U.K. et al. (2011). Herbal plants: used as a cosmetics. *Journal of Natural Product and Plant Resources* 1: 24–32.
25 Epstein, H. (2009). Skin care products, Chapter 11 In: *Handbook of Cosmetic Science and Technology*, 3e. *Edisi kedua* (eds. A.O. Barel, M. Paye and H.I. Maibach), 626–629 New York: *John Willy and Son Inc.*
26 Savary, G., Grisel, M., and Picard, C. (2016). Cosmetics and personal care products. In: *Natural Polymers* (ed. O. Olatunji), 219–261 Cham: Springer International Publishing.
27 Ku, J.E., Han, H.S., and Song, J.H. (2013). The recent trend of the natural preservative used in cosmetics. *Korean Journal of Aesthetics and Cosmetology* 11: 835–844.
28 Parente, M.E., Gambaro, A., and Ares, G. (2008). Sensory characterization of emollients. *Journal of Sensory Studies* 23: 149–161.
29 Athar, M. and Nasir, S.M. (2005). Taxonomic perspective of plant species yielding vegetable oils used in cosmetics and skin care products. *African Journal of Biotechnology* 4: 36–44.
30 Srivastava, S. and Ray, D.P. (2015). Natural gum resources in india and their commercial importance. *International Journal of Bioresource Science* 2: 151–155.
31 Thombare, N., Jha, U., Mishra, S., and Siddiqui, M.Z. (2016). Guar gum as a promising starting material for diverse applications: a review. *International Journal of Biological Macromolecules* 88: 361–372.
32 López, O., García, M.A., and Zaritzky, N.E. (2010). Novel sources of edible films and coatings. *Stewart Postharvest Review* 6: 1–8.
33 Harika, K., Sunitha, K., Kumar, P.P. et al. (2012). Basic concepts of cellulose polymers – a comprehensive review. *Archives of Pharmacy Practice* 3: 202.

34 Weseler, A., Geiss, H.K., Saller, R., and Reichling, J. (2002). Antifungal effect of Australian tea tree oil on Malassezia pachydermatis isolated from canines suffering from cutaneous skin disease. *Schweiz Arch Tierheilkd* 144: 215–221.

35 Varvaresou, A., Papageorgiou, S., Tsirivas, E. et al. (2009). Self-preserving cosmetics. *International Journal of Cosmetic Science* 31: 163–175.

36 Khan, Y., Panchal, S., Vyas, N. et al. (2007). Olea europaea: a phyto-pharmacological review. *Pharmacognosy Reviews* 1: 114–118.

37 Augustine, R., Rajendran, R., Cvelbar, U. et al. (2013). Biopolymers for health, food, and cosmetic applications. In: *Handbook of Biopolymer-Based Materials*, 1ee (eds. S. Thomas, D. Durand, C. Chassenieux and Jyotishkumar), 801–849 Wiley-VCH.

38 Vandamme, E.J. (2011). Natural colors... but of course. *SIM-News* 61: 121–128.

39 Rauf, A., Imran, M., Abu-Izneid, T. et al. (2019). Proanthocyanidins: a comprehensive review. *Biomedicine and Pharmacotherapy* 116: 108999.

40 Lee, S.H., Choi, S.Y., Kim, H. et al. (2002). Mulberroside F isolated from the leaves of Morus alba inhibits melanin biosynthesis. *Biological and Pharmaceutical Bulletin* 25: 1045–1048.

41 Fu, B., Li, H., Wang, X. et al. (2005). Isolation and identification of flavonoids in licorice and a study of their inhibitory effects on tyrosinase. *Journal of Agricultural and Food Chemistry* 53: 7408–7414.

42 Nerya, O., Vaya, J., Musa, R. et al. (2003). Glabrene and isoliquiritigenin as tyrosinase inhibitors from licorice roots. *Journal of Agricultural and Food Chemistry* 51: 1201–1207.

43 Choi, S., Park, Y.I., Lee, S.K. et al. (2002). Aloesin inhibits hyperpigmentation induced by UV radiation. *Clinical and Experimental Dermatology* 27: 513–515.

44 Razavi, B.M. and Hosseinzadeh, H. (2019). A review of the effects of citrus paradisi (grapefruit) and its flavonoids, naringin, and naringenin in metabolic syndrome, in bioactive food as dietary interventions for diabetes. *Phytotherapy Research*: 515–543.

45 Zhu, W. and Gao, J. (2008). The use of botanical extracts as topical skin-lightening agents for the improvement of skin pigmentation disorders. *Journal of Investigative Dermatology Symposium Proceedings* 13 (1): 20–24.

46 Baumann, L.S. (2007). Less-known botanical cosmeceuticals. *Dermatologic Therapy* 20: 330–342.

47 Farris, P. (2007). Idebenone, green tea, and Coffeeberry® extract: new and innovative antioxidants. *Dermatologic Therapy* 20: 322–329.

48 Baldisserotto, A., Buso, P., Radice, M. et al. (2018). Moringa oleifera leaf extracts as multifunctional ingredients for "natural and organic" sunscreens and photoprotective preparations. *Molecules* 23: 664.

49 Adhirajan, N., Kumar, T.R., Shanmugasundaram, N., and Babu, M. (2003). In vivo and in vitro evaluation of hair growth potential of Hibiscus rosa-sinensis Linn. *Journal of Ethnopharmacology* 88: 235–239.

50 Anand, A.K., Mohan, M., Haider, S.Z., and Sharma, A. (2011). Essential oil composition and antimicrobial activity of three Ocimum species from Uttarakhand (India). *International Journal of Pharmacy and Pharmaceutical Sciences* 3: 223–225.

51 Saeed, F., Afzaal, M., Niaz, B. et al. (2018). Bitter melon (Momordica charantia): a natural healthy vegetable. *International Journal of Food Properties* 21: 1270–1290.

52 Alzohairy, M.A. (2016). Therapeutics role of Azadirachta indica (Neem) and their active constituents in diseases prevention and treatment. *Evidence-Based Complementary and Alternative Medicine* 2016: 7382506.

53 Athikomkulchai, S., Watthanachaiyingcharoen, R., Tunvichien, S. et al. (2008). The development of anti-acne products from Eucalyptus globulus and Psidium guajava oil. *Journal of Health Research* 22: 109–113.

54 Bessada, S.M., Alves, R.C., and Oliveira, M.B.P.P. (2018). Coffee silverskin: a review on potential cosmetic applications. *Cosmetics* 5: 5.

55 Loftus, H.L., Astell, K.J., Mathai, M.L., and Su, X.Q. (2015). Coleus forskohlii extract supplementation in conjunction with a hypocaloric diet reduces the risk factors of metabolic syndrome in overweight and obese subjects: a randomized controlled trial. *Nutrients* 7: 9508–9522.

56 Abreu, E.N. and Moraes, C.A.P. (2019). Comparative analysis of anti-cellulite market products containing vegetal cosmetic assets. *Journal of Dermatology and Cosmetology* 3: 26–30.

57 Bhuiyan, M.R., Islam, A., Ali, A., and Islam, M.N. (2017). Color and chemical constitution of natural dye henna (Lawsonia inermis L) and its application in the coloration of textiles. *Journal of Cleaner Production* 167: 14–22.

58 Beiki, T., Najafpour, G.D., and Hosseini, M. (2018). Evaluation of antimicrobial and dyeing properties of walnut (Juglans regia L.) green husk extract for cosmetics. *Coloration Technology* 134: 71–81.

59 Jahanban-Esfahlan, A., Ostadrahimi, A., Tabibiazar, M., and Amarowicz, R. (2019). A comprehensive review on the chemical constituents and functional uses of walnut (Juglans spp.) husk. *International Journal of Molecular Sciences* 20: 3920.

60 Wahyuningsih, S., Ramelan, A.H., Wardani, D.K. et al. (2017). Indigo dye derived from indigofera tinctoria as natural food colorant. *Materials Science and Engineering, IOP Publishing* 193: 012048.

61 Lin, Y., Yang, S., Hanifah, H., and Iqbal, Q. (2018). An exploratory study of consumer attitudes toward green cosmetics in the UK market. *Administrative Sciences* 8: 71.

62 Asgharian, R. and Saleki, Z. (2012). Green product quality, green customer satisfaction, and green customer loyalty. *IRACST-International Journal of Research in Management and Technology* 2: 499–501.

63 Sethi, A., Kaur, T., Malhotra, S.K., and Gambhir, M.L. (2016). Moisturizers: the slippery road. *Indian Journal of Dermatology* 61: 279–287.

64 Campa, M. and Baron, E. (2018). Anti-aging effects of select botanicals: scientific evidence and current trends. *Cosmetics* 5: 54.

65 Gan, C., Wee, H.Y., Ozanne, L., and Kao, T. (2008). Consumers purchasing behaviour towards green products in New Zealand. *Journal of Innovative Marketing* 1: 93–102.

66 Juliano, C. and Magrini, G.A. (2017). Cosmetic ingredients as emerging pollutants of environmental and health concern. A mini-review. *Cosmetics* 4: 11.

67 Kumar, P. (2014). Greening retail: an Indian experience. *International Journal of Retail and Distribution Management* 42: 613–625.

68 Khan, A.F. and Khan, M.F. (2013). A study on the awareness of product awareness among skincare users in the state of Madhya Pradesh. *Journal of Business Management* 14: 65–72.

69 Kardos, M., Gabor, M.R., and Cristache, N. (2019). Green marketing's roles in sustainability and ecopreneurship. Case study: green packaging's impact on Romanian young consumers' environmental responsibility. *Sustainability* 11: 873.
70 Kaur, T. (2018). Forbes, April 2018 https://www.forbes.com/sites/tarandipkaur/2018/04/20/leading-herbal-beauty-brands-to-know-from-india/#4d1ecd652f5c. www.forbes.com.
71 Matić, M. and Puh, B. (2016). Consumers' purchase intentions towards natural cosmetics. *Econviews (Ekonomski Vjesnik)* 29: 53–64.
72 Tungmunnithum, D., Garros, L., Drouet, S. et al. (2019). Green ultrasound assisted extraction of trans rosmarinic acid from Plectranthus scutellarioides (L.) R.Br. Leaves. *Plants* 8: 50.
73 Amberg, N. and Fogarassy, C. (2019). Green consumer behavior in the cosmetics market. *Resources* 8: 137.
74 Zappelli, C., Barbulova, A., Apone, F., and Colucci, G. (2016). Effective active ingredients obtained through biotechnology. *Cosmetics* 3: 39.
75 Chin, J., Jiang, B.C., Mufidah, I. et al. (2018). The investigation of consumers' behavior intention in using green skincare products: a pro-environmental behavior model approach. *Sustainability* 10: 3922.
76 Pereira, L. (2018). Seaweeds as source of bioactive substances and skin care therapy – cosmeceuticals, algotheraphy, and thalassotherapy. *Cosmetics* 5: 68.
77 Li, Y., Bundeesomchok, K., Rakotomanomana, N. et al. (2019). Towards a zero-waste biorefinery using edible oils as solvents for the green extraction of volatile and non-volatile bioactive compounds form rosemary. *Antioxidants* 8: 140.
78 Panda, D. and Manickam, S. (2019). Cavitation technology – the future of greener extraction method: a review on the extraction of natural products and process intensification mechanism and perspectives. *Applied Sciences* 9: 766.
79 Gupta, P.L., Rajput, M., Oza, T. et al. (2019). Eminence of microbial products in cosmetic industry. *Natural Products and Bioprospecting* 9: 1–12.
80 Purkayastha, J. (ed.) (2016). *Bioprospecting of Indigenous Bioresources of North-East India*. New York: Springer.
81 Report of the United Nations Conference on Environment and Development Annex I, A.CONF.151/26 (Vol I) (Rio de Janeiro, 3–14 June 1992).
82 UN General Assembly. United Nations Declaration on the Rights of Indigenous Peoples resolution/adopted by the General Assembly, A/RES/61/295 (2 October 2007), 6.
83 Buck, M. and Hamilton, C. (2011). The Nagoya Protocol on access to genetic resources and the fair and equitable sharing of benefits arising from their utilization to the Convention on Biological Diversity. *Review of European Community & International Environmental Law* 20 (1): 47–61.
84 World Intellectual Property Organization (2001). Report of Traditional Knowledge Task Force.
85 Goffe, M. (2011). Recent developments in the WIPO Intergovernmental Committee on Intellectual Property and Genetic Resources, Traditional Knowledge and Folklore. *Queen Mary Journal of Intellectual Property* 1 (1): 90–98.
86 The Drug and Cosmetic Act, 1940 and The Drugs and Cosmetic Rules, 1945 as amended time to time. Ministry of Health and Family Welfare, Government of India. http://www.cdsco.nic.in/Drugs&CosmeticAct.pdf (accessed 10 June 2020).

87 Fujita, Y. and Tabata, M. (1986). Secondary metabolites from plant cells. *Proceedings of 6th International Congress of Plant Tissue and Cell Culture, IAPTC*, University of Minnesota, Minneapolis, MN (3–8 August 1986).

88 Misawa, M. (1994). *Plant Tissue Culture: An Alternative for Production of Useful Metabolites*, FAO Agricultural Services Bulletin, 1–87 Food & Agriculture Organization of the United Nations (FAO).

89 Schürch, C., Blum, P., and Zülli, F. (2008). Potential of plant cells in culture for cosmetic application. *Phytochemistry Reviews* 7: 599–605.

11

Therapeutic Lead Secondary Metabolites Production Using Plant *In Vitro* Cultures

Vikas Srivastava[1], Aksar Ali Chowdhary[1#], Skalzang Lhamo[1#], Sonal Mishra[2], and Shakti Mehrotra[3]

[1] *Department of Botany, Central University of Jammu, Samba, Jammu and Kashmir, India*
[2] *School of Biotechnology, University of Jammu, Jammu, Jammu and Kashmir, India*
[3] *Department of Biotechnology, Institute of Engineering and Technology, Lucknow, Uttar Pradesh, India*

11.1 Introduction

Plants being sessile in nature face adverse environment conditions and manage them by stimulating various physiological, biochemical and molecular processes which in turn alter the ion homeostasis, gene expression and also endogenous level of several metabolites [1, 2]. The plant metabolites have been identified as the basis for their nutritional and medicinal properties, and therefore plants are used by human being since ancient time for such purposes [3–5]. The metabolites are the product of metabolism, and based on their utility in governing plant life, it can be categorized into primary (amino acids, fats, carbohydrate, etc.) or secondary (terpene, alkaloid, phenolics, etc.) metabolites. However, there are no clear cut criteria, which distinguish between the primary and secondary metabolism. For example, both diterpenes (C_{20}) and triterpenes (C_{30}) represent primary and secondary metabolites (SMs). Similarly, proline is considered as the primary metabolite, whereas C_6 analog (pipecolic acid) is considered as an alkaloid [6, 7].

The significance of SMs has not only considered them as an effective stress mitigation entity for plant itself but also as an effective molecule to alleviate many human ailments. The research on phytochemistry, biochemistry and bio-prospection (biological activity testing) has revealed the establishment of many SMs with authorized medicinal values, which have also been qualified for commercialization in terms of medicines in various health care systems. Production and accumulation of SMs in plants tissues are associated with a variety of accessory functions, but primarily it is for the defence against bacteria, fungi, herbivores and as signals in plants [8]. SMs are highly inducible in response to stresses [9] and include active molecule responsible for pharmacological or biological activity. Further, in

[#] Equal Contribution

Bioprospecting of Plant Biodiversity for Industrial Molecules, First Edition.
Edited by Santosh Kumar Upadhyay and Sudhir P. Singh.

the last 50 years due to their medicinal properties, research on plant SMs has increased significantly [10]. The enormous opportunity offered by SMs leads to the distortion of the natures' sustainability through direct harvesting of medicinal plants from their natural habitat and therefore the alternative production system is highly desired.

Plant *in vitro* cultures though emerged as a system of plant culturing under controlled environment condition since a century ago. But, with the advancement of research in this area, the focus has slightly changed and now the researchers' effort to offer some value application out of it. Since previous half century or so, the focus on raising plant *in vitro* cultures of medicinal plants involves the analysis of their potential to produce the metabolite of interest. The range of plant *in vitro* cultures, viz., shoot, adventitious root, callus and cell suspension and hairy root cultures (HRCs) are explored for this purpose. Based on growth and the way to manage (easy handling) the *in vitro* cultures, callus and cell suspension culture and HRCs are emerged as the imperative choice. Further, based on genetic and biochemical stability along with culture cost criteria, HRCs [11, 12] currently supersede the entire range of *in vitro* cultures. Besides the production of SMs, nowadays, plant *in vitro* cultures are also used for many other biotechnological interventions such as elicitation, metabolic engineering, phytoremediation, molecular farming, plant–microbe interaction, bioreactor up-scaling, etc. [11–16]. In spite of all types of plant *in vitro* cultures, irrespective of their pace of utility today, have been explored sometimes for SMs production with pharmaceutical value, which is largely the subject of the current chapter. The below text will highlight some of the excellent approaches to enhance the metabolite production using plant *in vitro* cultures to further meet the desired production. Further, the emphasis will also be on some of the instances of utilization of plant *in vitro* cultures for therapeutically important plant metabolites.

11.2 Secondary Metabolites and Pharmaceutical Significance

The SMs are also known as plant constituents, natural products or phytochemicals and have medicinal properties [17]. These are produced by almost all plants, and plant has genetic information for the biosynthesis of these metabolites. Further, on the basis of their biosynthetic pathways and precursor, they are classified into three main categories as terpenes (or terpenoid, terpene derivatives), phenolics and alkaloids [7, 17, 18], and each category represents pharmaceutical lead molecules (Table 11.1). The terpenoids (both primary and SMs) and alkaloids are synthesized from C_5 precursor isopentenyl diphosphate and amino acids, respectively. However, the phenolics biosynthesis involves either shikimic acid pathway or the malonate/acetate pathway [7].

Some SMs such as terpenes and phenolic compounds are almost produced by major lineage of land plants, whereas alkaloids are restricted to a particular taxon, but they may also occur in other groups due to convergence [19]. These metabolites also serve as the basis for chemotaxonomy [20]. For instance, the plants have been demarcated into groups with specific metabolites such as tropane alkaloid bearing plants (*Atropa, Hyoscyamus, Scopolia,* etc.), terpene indole alkaloid bearing plants (*Catharanthus, Rauwolfia etc.*), benzyl isoquinoline alkaloid bearing plants (*Papaver*), etc. In plants, the functions of SMs are not crucial for vital processes such as growth, development and reproduction [21]. However, they

Table 11.1 Major classes of plant secondary metabolites (SMs) and biological activity of some of the relevant metabolites produced through plant *in vitro* system.

Major class of SMs	Metabolite	Plant system	Biological activities
Terpenoids	Artemisinin	*Artemisia annua* L.	Antimalarial and Anticancerous
	Astragaloside	*Astragalus membranaceus*	Cardioprotective, antitumor, antihypertensive, antiinflammatory, antidiabetic, neuroprotective, etc
	Tanshinone	*Salvia miltiorrhiza*	Used for the treatment of cardiovascular disease
	Eclalbatin	*Eclipta alba* L.	Antioxidant
	Abietane	*Cephalotaxus fortune*	Antitumor
Alkaloids	Codeine	*Papaver somniferum*	Antitussive and analgesic
	Morphine	*P. somniferum*	Narcotic and analgesic
	Atropine	*Hyoscyamus niger*	Anticholinergic
	Caffeine	*Coffea arabica*	Stimulant
	Quinine	*Cinchona officinalis*	Antimalarial
	Scopolamine & Hyoscyamine	*Hyoscyamus reticulatus* L.	Anticholinergic
	Reserpine and Ajmalicine	*Rauwolfia serpentina*	Antihypertensive and Antiarrhythmic
	Paclitaxel	*Corylus avellana*	Anticancerous
	Vinblastine	*Catharanthus roseus*	Antileukemic
Phenolics	Reseveratrol	*Vitis vinifera*	Anticancerous, anti-cardiovascular and antineurodegenerative
	Rosmarinic acid	*Rosmarinus officinalis L.*	Antioxidant
	Caffeic acid	*R. officinalis L.*	Antioxidant
	Daidzein	*Glycine max*	Anticancerous
	Genistein	*G. max*	Anticancerous

protect plants from any harmful situation imposed by the environment [21–24]. Further, the significance of SMs as therapeutics has also been identified in many studies [25].

The SMs are widely presented as an imperative molecule of high industrial significance and function as therapeutic, flavour, fragrance, nutraceutical, etc. [26]. Among them, the pharmaceutical significance of SMs is primarily targeted in many medicinal systems, and many commercial products utilizing these metabolites are available in the market. The current understanding of them suggested their role in curing many human diseases as mentioned in Table 11.2. Some of the important medicinal values they offer include their utility as anticancer, antidiabetic, antidepressant, antiinflammatory, antiviral, antioxidant, anticholinergic, antimalarial, antimicrobial, hypotensive, hepatoprotective, neuroprotective, etc. For instance, the metabolites like catharanthine, camptothecin, podophyllotoxin are known for their contribution in cancer therapy. Gymnemic acid and

Table 11.2 Some examples of pharmaceutical lead molecules produced through plant *in vitro* cultures.

Pharmaceutical value	Secondary metabolites
Anticancerous	Aloe-emodin, Andrographolide, Artemisinin, Camptothecin, Emodin, Ginsenoside, Glycyrrhizin, Paclitaxel, Plumbagin, Podophyllotoxin, Resveratrol, Rutamarin, Vinblastine, Vincristine
Anticholinergic	Atropine, Hyoscyamine, Scopolamine
Antidepressent	Chrysophanol, Harmine
Antidiabetic	Gymnemic acid, Serpentine
Antihypertensive	Ajmalicine, Reserpine
Antiinflammatory	Aloe-emodin, Aloin, Glycyrrhizin, Hypericin, Kaempferol, Verbascoside, Withanolides
Antimalarial	Artemisinin, Quinine
Antioxidant	Arbutin, Daidzin, Ginsenoside, Kaempferol, Plumbagin
Antiviral	Aloe-emodin, Hypericin, Glycyrrhizin, Rosmarinic acid
Hepatoprotective	Glycyrrhizin, Picroside I
Neuroprotective	Ginsenoside, Withanolide A
Sedative	Codeine

serpentine are known for the antidiabetic function. The reserpine and ajmalicine are good hypo-tensive metabolites. Picroside I and Withaferin A are hepatoprotective and neuroprotective, respectively.

11.3 Plant *In Vitro* Cultures and Strategies for Secondary Metabolite Production

Though being a mimic of *in vivo* plant, it is obvious that *in vitro* cultures may generate the biochemical reactions, similar to the parent plant, which is also true for SMs production. Therefore, many investigations during generation/induction of *in vitro* culture and subsequent maintenance are accompanied with secondary metabolite analysis. Successes of the *in vitro* production of various SMs of high medicinal value are reported such as scopolamine, reserpine, camptothecin, withanolides, serpentine, catharanthine, etc. (Table 11.2. Section 11.4). But, as independent system, these cultures also offer the route to further enhance the metabolite content, the knowledge of which can also be translated at bench scale. Some of the approaches that meet higher production of SMs are mentioned in the below text.

11.3.1 Precursor Feeding

This approach is used to supply the immediate or distant precursors of the targeted biosynthetic pathway. The genesis is the presumption that the targeted pathway is available, but due to scarcity of their immediate or distant precursors, the desired production of metabolite of interest could not be achieved. Precursor feeding is one of the types of biotransformation, where the efforts are done to make the substrate (precursor) available to obtain desired product (SMs), the other two are co-culture and exogenous/non-native molecule biotransformation [16, 27, 28]. Chao and Lan [29] have reported the production of scopolamine using 6-hydroxyhyoscyamine as biotransformation substrate in *in vitro* cultures of *Anisodus tanguticus*. The exogenous supplementation of tropic acid, tropinone and tropanol as precursors lead to increased production of hyoscyamine and scopolamine in *Atropa belladonna* suspension cultures [30]. Anitha and Ranjitha [31] have reported the stimulation of reserpine biosynthesis in the callus cultures of *Rauwolfia tetraphylla* L. using tryptophan as precursor. In *Solenostemon scutellarioides in vitro* culture, improved rosmarinic acid was noticed after feeding of phenylalanine and l-tyrosine [32]. Besides, the exogenous molecule biotransformation and co-culture approach also provide ways to develop molecule with therapeutic significance [16]. For instance, in an attempt of biotransformation of betuligenol in *A. belladonna* HRCs, the bioactive molecule raspberry ketone (antiobese) and betuloside (hepatoprotective) are generated [33]. The co-culture of *Atropa belladonna* hairy root with its shooty teratomas results in hyoscyamine production by one system (HRCs) which is translocated by the medium to the system (shooty teratoma) that convert hyoscyamine into scopolamine [34].

11.3.2 Metabolic Engineering

The production of SM requires a specific biosynthetic pathway. The substrate (precursor) enters into the pathway and leads to the production of desired metabolites. Besides, as the metabolic pathways are mostly complex in nature, some diversions are also possible that may divert the pathway, thereby reducing the expected outcome. The metabolic engineering (genetic engineering to regulate metabolism) offer the opportunity to strengthen the biochemical activity through over-expression of genes responsible for the production of desired metabolite or through suppression of any possible diversion of the pathway of interest [15]. For instance, the expression of *Catharanthus roseus tryptophan decarboxylase (CrTDC)* gene leads to enhanced accumulation of reserpine and ajmalicine in transgenic HRCs of *R. serpentina* [35]. Sun and Peebles [36] have reported increase alkaloid (ajmalicine, catharanthine, tabersonine, etc.) production in *Catharanthus roseus* HRCs overexpressed with *ORCA3* and *Strictosidine glucosidase (SGD)*. The expression of *AaPMT (Putrescine methyltransferase)* and *AaTRI (Tropinone reductase I)* in *Anisodus acutangulus* HRCs led to enhanced biosynthesis of tropane alkaloids [37]. Enhanced scopolamine production (17mg/L) which is 100 times more than the control clones was also reported in transgenic HRCs of *H. muticus* overexpressed with *H6H (Hyoscyamine 6β-hydroxylase)* [38]. Similarly, the overexpression of *PMT* and *H6H* simultaneously led to the improvement of tropane alkaloids in transgenic *A. belladonna* HRCs [39]. The expression of *Arabidopsis thaliana* squalene synthase gene (*AtSQS1*) in *W. somnifera* transformed HRCs demonstrated increased withaferin A content [40]. Besides, nowadays using heterologous expression approaches, the desired SMs have also been produced in non-native (plant with no report of desired SM production) plant or microbial cultures.

11.3.3 Elicitation

It is evident that the environmental cues may influence the plant growth and development; this is also true for the metabolism including secondary metabolism. Under such conditions (biotic or abiotic stress or media/nutrient stress), the additional SMs thus produced provide additional advantage to the plant and thus the plants better manage the stress impacts. This feature of induction of secondary metabolism leading to the production of desired metabolites is being used to enhance the production profile of plant *in vitro* cultures by application of certain physical, chemical or biological factors, an application known as elicitation. The approach has been widely used in *in vitro* cultures and offer enough opportunity to optimize factors, which can dramatically increase the metabolite content [13, 15, 41]. In *Hypericum perforatum* cell suspensions, the elicitation with MeJA (methyl jasmonate) exhibited stimulation of flavonoid production, which was 2.7 times as compared to control [42]. Similarly, elicitation with salicyclic acid enhanced production of chrysophanol and aloe-emodin more than 5–13 and 10–11 fold as compared to control in the adventitious root culture of *Aloe vera* [43]. The elicitation effect of *Piriformospora indica* on withaferin A biosynthesis was reported in cell suspension cultures of *Withania somnifera* [44]. The improved gymnemic acid accumulation was also achieved through endophytic fungi in the cell suspension cultures of *Gymnema sylvestre* [45]. Besides, the elicitation-based metabolite enhancement is also reported in HRCs [15]. For instance, Kai et al. [46] have reported the improved accumulation of tropane alklaoids by 1.51 (with ethanol), 1.13 (with methyl jasmonate) and 1.08 (with Ag^{+}) times in *Anisodus acutangulus* HRCs. In *Catharanthus roseus* HRCs, the improvement of ajmalicine content was noticed after elicitation with optimized mixture of elicitor composed of jasmonic acid, methyl jasmonate and KCl [47]. Zaheer et al. [48] have reported the enhanced daidzin production in *Psoralea corylifolia* L. HRCs after elicitation with jasmonic acid and acetyl salicylic acid.

11.3.4 Bioreactor Up-scaling

This should be the ideal interventions, when everything is set (*in vitro* cultures are producing maximum production) and one need to up-scale the entire process. Bioreactor is a physical instrument used for the proportionate increase in the culture volume, which is also proportional to the metabolites it contains. Several successful attempts for plant *in vitro* culture-based SMs production utilizing various type of bioreactor are reported [49, 50]. For instance, in the cultivated shoot cultures of *Artemisia annua* L. in bioreactors (a modified airlift bioreactor, a multi-plate radius-flow bioreactor and an ultrasonic nutrient mist bioreactor), artemisinin production was reported. Out of three bioreactors used, the nutrient mist bioreactor exhibited 1.4 and 3.3 fold higher production as compared to multi-plate radius-flow bioreactor and the modified airlift bioreactor, respectively [51]. In *R. serpentina* HRCs, the comparison was made in 11-week old HRCs grown in shake flask and bioreactor, which suggested an increase in the dry matter and reserpine content in later [52]. Baque et al. [53] have reviewed the utilization of bioreactor for the production of biomass and useful compounds from adventitious roots of many high value medicinal plants (*Morinda citrifolia, Echinacea purpurea* and *E.angustifolia, Hypericum perforatum* and *Panax ginseng*).

Besides, the mentioned interventions, the combinatorial approaches utilizing two or more strategies of them further provide better results.

11.4 Exemplification of the Utilization of Different Types of Plant *In Vitro* Cultures for SMs Production

The plant *in vitro* cultures have been presented as an important system for the production of plant SMs. Based on the site of *in planta* biosynthesis of SMs, the different types of *in vitro* cultures are used for this purpose (Table 11.3). For instance, if the SMs of interest are synthesized in natural roots, then the *in vitro* alternative would be adventitious roots or hairy roots. Similarly, if the choice of SMs belongs to aerial plant parts, the *in vitro* alternative would be either shoot or callus/cell suspension cultures. However, this does not make an error-free categorization and several exceptions do exist. The below text describe some of the instance of success in the therapeutic lead SMs production through different types of *in vitro* cultures.

11.4.1 Shoot Culture

The exploitation of shoot cultures was reported for the production of many pharmaceutical lead compounds such as artemisinin, scopolamine, kaempferol, digitoxin and digoxin, cocaine, vindoline, withanolide, etc. The shoot cultures of *Artemisia annua* were reported to produce artemisinin [51, 75]. Khanam et al. [76] have demonstrated the presence of hyoscymaine and scopolamine in the shoot cultures of *Duboisia myoporoides* R. Br. In *Scopolia parviflora* Nakai (a rare medicinal plant native to Korea), the shoot cultures were generated from adventitious shoots of rhizome [77]. The analysis of tropane alkaloids (hyoscyamine and scopolamine) was also studied, which demonstrated their occurrence in root, stem and leaves of *in vitro* propagated plants [77]. Another tropane alkaloid cocaine was reported in the shoot cultures of *Erythroxylum coca* var. coca, though its level was 50% as compared to the parent plant. Further, maximum accumulation was achieved in the leaves of shoot cultures [78].

The *Catharanthus roseus* is a well-known plant producing natural antineoplastic drugs (vinblastine and vincristine). Due to the complexity and spatiotemporal regulation of terpenoid indole alkaloids (TIAs) biosynthetic pathway responsible to their biosynthesis, the callus culture, cell suspension and hairy roots of *C. roseus* do not provide adequate system to produce them [79]. Yet, the multiple shoot cultures are explored to accomplish TIA biosynthetic pathway. The multiple shoot cultures of *C. roseus* (Madagascar periwinkle) induced directly from seedlings in the presence of 1.0mg/l benzyladenine, additionally it also composed of an unorganised tissue. The HPLC assay demonstrated the presence of vindoline and catharanthine in the shoot (especially leaf tissue), whereas ajmalicine was reported in the unorganized tissue. Comparison with parent demonstrated several fold higher catharathine content in leaf tissue; however, the vindoline content was comparable with the parent [80]. Miura et al. [81] have studied the formation of vinblastine in *C. roseus* multiple shoot cultures. Satdive et al. [82] have reported the production of ajmalicine in the

Table 11.3 Few examples of production of secondary metabolites using plant *in vitro* cultures.

S. No	Plant System	*In vitro* culture	Family	Class of SMs	Secondary metabolite	References
1	*Alstonia scholaris*	Callus culture	Apocynaceae	Alkaloid	Echitamine Picrinine Tubotaiwine	Jeet et al. [54]
2	*Anisodus luridus*	Hairy root culture	Solanaceae	Alkaloid	Scopolamine and hyoscyamine	Qin et al. [55]
3	*Artemisia annua* L.	Cell suspension culture	Asteraceae	Sesquiterpene	Artemisinin	Zebarjadi et al. [56]
4	*A. annua* L.	Hairy root culture	Asteraceae	Sesquiterpene	Artemisinin	Ahlawat et al. [57]
5	*Astragalus membranaceus*	Hairy root culture	Leguminosae	Triterpene saponin	Astragaloside	Jiao et al. [58]
6	*Catharanthus roseus*	Callus culture	Apocynaceae	Alkaloid	Vinblastine and vincristine	McGehee et al. [59]
7	*Calotropis gigantea*	Hairy root culture	Asclepiadaceae	Steroid	Cardenolides	Sun et al. [60]
8	*Corylus avellana*	Cell suspension culture	Betulaceae	Alkaloid	Paclitaxel	Salehi et al. [61]
9	*Datura metel*	*Hairy root culture*	Solanaceae	Alkaloid	Atropine	Shakeran et al. [62]
10	*Eclipta alba L.*	*Callus culture*	*Asteraceae*	Coumestan	Wedelolactone and demethylwedelol actone	Khurshid et al. [63]
				Triterpene saponin	Eclalbatin	
11	*Hyoscyamus reticulatus* L.	Hairy root culture	Solanaceae	Alkaloid	Hyoscyamine, scopolamine	Moharrami et al. [64]
12	*Ophiorrhiza mungos var. angustifolia*	Callus and cell suspension culture	Rubiaceae	Alkaloid	Camptothecin	Krishnan et al. [65]

13	*Papaver orientale* L	Hairy root culture	Papaveraceae	Alkaloid	Morphine, thebaine and codeine	Hashemi et al. [66]
14	*Rauwolfia serpentina*	Hairy root culture	Apocynaceae	Alkaloid	Ajmaline Ajmalicine	Srivastava et al. [67]
15	*R. serpentina*	Transgenic hairy roots	Apocynaceae	Alkaloid	Reserpine Ajmalicine	Mehrotra et al. [35]
16	*R. serpentina*	Callus, leaf. Root, and stem culture	Apocynaceae	Alkaloid	Reserpine Ajmalicine	Zafar et al. [68]
17	*Rosmarinus officinalis L.*	Callus culture	Lamiaceae	Phenolic	Rosmarinic acid and caffeic acid	Coskun et al. [69]
18	*Salvia miltiorrhiza*	Transgenic hairy roots	Lamiaceae	Diterpenes	Tanshinone	Hao et al. [70]
19	*Solanum khasianum*	Hairy root culture	Solanaceae	Alkaloid	Solasodine and α-solanine	Srivastava et al. [67]
20	*Tinospora cordifolia*	Callus culture	Menispermaceae	Alkaloid	Berberine	Pillai et al. [71]
21	*Vitis vinifera*	Hairy root culture	Vitaceae	Phenolic	Reseveratrol	Hosseini et al. [72]
22	*Withania somnifera*	Transformed Hairy Root Culture	Solanaceae	Steroid	Withaferin A	Yousefian et al. [40]
23	*W. somnifera*	Adventitious root culture	Solanaceae	Steroid	Withanolides, Withaferin A	Rangaraju et al. [73]
24	*W. somnifera*	Hairy Root Culture	Solanaceae	Steroid	Withanolide A, withaferin A and withanone	Sivanandhan et al. [74]

multiple shoot cultures of *C. roseus*. In a study of precursor feeding, Sharma et al. [79] have demonstrated the feeding of tryptamine and tryptophan was found effective, with better results in former.

The production of hypericin and pseudohypericin was achieved in the shoot cultures of *Hypericum hirsutum* and *H. maculatum*, which was highly stimulated by salicylic acid elicitor [83]. The multiple shoot cultures of *Rauvolfia serpentina* were reported to produce 3-epi-α-yohimbine [84]. The production of pharmaceutical important compounds such as casticin, rutin, neochlorogenic and *p*-coumaric acids was achieved in *Vitex agnus-castus* agitated shoot cultures [85]. The effect of plant growth regulators (PGR) on shoot development and biologically active indolizidine alkaloids (securinine, and allosecurinine) were assessed in *Phyllanthus glaucus* (Euphorbiaceae) *in vitro* shoot cultures. Among tested PGR, the highest number of shoot per explant was achieved on MS medium supplemented with IBA 0.5 mg/L and BAP 1.0 mg/L, and the optimal rooting response was obtained in MS + IBA 1.0 mg/L [86]. In the stationary liquid phase, the shoot cultures of *Ruta graveolens* L. were demonstrated to produce psoralen, bergapten, xanthotoxin, isopimpinellin (linear furanocoumarins), rutamarin (linear dihydrofuranocoumarin), kokusaginine and skimmianine (furanoquinoline alkaloids) [87]. Kilby et al. [88] have reported the production of quinoline alkaloid by juvenile shoot cultures of *Cinchona ledgeriana*. The production of two withanolides (Withaferin A and Withanolide D) was investigated in the multiple shoot cultures of *Withania somnifera* [89]. Quinoline alkaloids (camptothecin and 9-methoxycamptothecin) production from shoot cultures of *Nothapodytes foetida* was reported by Roja and Heble [90]. Gopalakrishnan and Shankar [91] have reported the *Camptotheca* alkaloids (Camptothecin) from multiple shoot cultures of *O. decumbens* (0.056% dry weight). Remarkable production of cardiotonic glycosides digitoxin and digoxin were reporetd in the shoot culture of *Digitalis purpurea* [92, 93].

11.4.2 Adventitious Root Culture

Adventitious root culture is another important choice for secondary metabolite production and used successfully in many plant species for the production of many metabolites of high pharmaceutical, nutraceutical and industrial values. It usually requires auxin (IBA, NAA, IAA, etc.) supplementation to the basal salt solution for the induction and maintenance as reported in the adventitious roots of many medicinal plants [94]. Production of several important SMs of pharmaceutical significance was carried out using the adventitious root cultures of diverse medicinal plants, viz., Withanolide (*Withania somnifera,* [95]; [96], Silymarin (*Silybum marianum,* [97], Scopolamine (*Scopolia parviflora,* [98], Psoralen (*Psoralea coryfolia,* [99]), Podophyllotoxin (*Podophyllum hexandrum,* [100]), Gingenoside (*Panax ginseng,* [101]), Rosmarinic acid (*Orthisipon stamineus,* [102]), Hypericin, (*Hypericum perforatum,* [103]), Andrographolide (*Andrographis paniculata,* [104]), etc. The improvement of SMs production was also observed after adoption of elicitation and bioreactor-based methods [105].

11.4.3 Callus and Cell Suspension Culture

The callus and cell suspension systems are simple, cost-effective method for secondary metabolite production. The callus culture represents a mass of mostly undifferentiated cells grown on semi-solid medium; however, the suspension cultures are free cell or small group of cells grown in the form of suspension in continuously agitated liquid culture medium. These cultures provide an alternative option to whole plant cultivation for the production of SMs of pharmaceutical interest [106]. For instance, the accumulation of reserpine in NaCl-treated calli was reported in *R. tetraphylla* [31]. Zafar et al. [107] have reported the improvement of reserpine content in the *R. serpentina* callus after alumium chloride elicitation. The callus and adventitious roots were developed from leaf segments of shoot cultures of *Cephalis ipecacuanha* A. Richard in phytohormone-supplemented MS media. The content of emetic alkaloids (emetine and cephaeline) was analysed by HPLC in calli, root and root suspension cultures, which demonstrated their presence; however, in calli, emetic alkaloids were present in traces [108]. Similarly, the cell suspension cultures of medicinal plants are largely utilized for production of various therapeutic metabolites. Some of relevant examples of such important metabolites produced through cell suspension cultures are ajmalicine (*C. roseus,* [109]), artemisinin (*A. annua,* [110]), camptothecin (*C. acuminata,* [111]), crocin (*Crocus sativus,* [112]), forskolin (*Coleus forskohlii,* [113]), gymnemic acid (*G. sylvestre,* [114]), hypericin (*Hypericum perforatum,* [115]), reserpine (*R. serpentina,* [116]), etc.

11.4.4 Hairy Root Cultures

Currently, the HRCs are considered as one of the effective method for the production of imperative SMs and also offer several other allied applications in biotechnology [11, 12, 15, 117]. In HRCs, the amalgamation of several other biotechnological interventions such as elicitation [13], metabolic engineering [14], precursor feeding [16], bioreactor up-scaling [49], ploidy alteration [15] are further reported to enhance the production profile of HRCs. The HRCs have also proved to be very effective source for long-term production of pharmaceutical lead molecules because of its genetic and biochemical stability and very low induction and maintenance associated cost. Using HRCs, the important pharmaceutical lead molecules produced are very rich in number, some of imperative examples are Andrographolide (*A. paniculata,* [118]), Artemisinin (*A. annua,* [57]), Ajmalicine (*C. roseus,* [47]), Atropine (*Datura metel,* [62]), Podophyllotoxin (*Linum album,* [119]), Reserpine (*R. serpentina,* [35]), withanolide (*W. somnifera,* [120]), Dopamine (*Portulaca oleracea,* [121]), Camptothecin (*Ophiorrhiza pumila,* [122]), Tanshinone (*Salvia miltiorrhiza,* [123]), etc.

11.5 Conclusion

In a nut shell, the plant *in vitro* cultures are widely used for the production of a variety of SMs having therapeutic values for the treatment of many serious human ailments, viz., cancer, diabetes, hypertension, etc. The diverse biotechnological interventions have also utilized to make the optimum yield of metabolite of interest, and up-scaling process through mediation of bioreactor is also well documented. The journey so far is very exciting, which provide sustainable and long-lasting alternative method for the production of therapeutic lead plant natural metabolites that will be available for generous supply to meet the desired pharmaceutical and commercial demand.

References

1 Ashraf, M.A., Iqbal, M., Rasheed, R. et al. (2018). Environmental stress and secondary metabolites in plants: an overview. In: Plant Metabolites and Regulation Under Environmental Stress (eds. P. Ahmad, M.A. Ahanger, V.P. Singh, et al.), 153–167. Academic Press.

2 Rodrigues-Corrêa, K.C.D.S. and Fett-Neto, A.G. (2019). Abiotic stresses and non-protein amino acids in plants. *Critical Reviews in Plant Sciences* 38 (5–6): 411–430.

3 Christodoulakis, N.S., Georgoudi, M., and Fasseas, C. (2014). Leaf structure of Cistus creticus L.(rock rose), a medicinal plant widely used in folk remedies since ancient times. *Journal of Herbs, Spices and Medicinal Plants* 20 (2): 103–114.

4 Mushtaq, S., Abbasi, B.H., Uzair, B., and Abbasi, R. (2018). Natural products as reservoirs of novel therapeutic agents. *EXCLI Journal* 17: 420.

5 Petrovska, B.B. (2012). Historical review of medicinal plants usage. *Pharmacognosy Reviews* 6 (11): 1.

6 Ahmed, E., Arshad, M., Khan, M.Z. et al. (2017). Secondary metabolites and their multidimensional prospective in plant life. *Journal of Pharmacognosy and Phytochemistry* 6: 205–214.

7 Croteau, R., Kutchan, T.M., and Lewis, N.G. (2000). Natural products (secondary metabolites). *Biochemistry and Molecular Biology of Plants* 24: 1250–1319.

8 Schoonhoven, L.M., Van Loon, B., van Loon, J.J., and Dicke, M. (2005). Insect-Plant Biology. Oxford University Press on Demand.

9 Pagare, S., Bhatia, M., Tripathi, N. et al. (2015). Secondary metabolites of plants and their role: Overview. *Current Trends in Biotechnology and Pharmacy* 9 (3): 293–304.

10 Mulabagal, V. and Tsay, H.S. (2004). Plant cell cultures-an alternative and efficient source for the production of biologically important secondary metabolites. *International Journal of Applied Science and Engineering* 2 (1): 29–48.

11 Srivastava, V., Mehrotra, S., and Mishra, S. (2018). Hairy Roots: An Effective Tool of Plant Biotechnology. Springer.

12 Srivastava, V., Mehrotra, S., and Mishra, S. (2020). Hairy Root Cultures Based Applications. Springer.

13 Goel, M.K., Mehrotra, S., and Kukreja, A.K. (2011). Elicitor-induced cellular and molecular events are responsible for productivity enhancement in hairy root cultures: an insight study. *Applied Biochemistry and Biotechnology* 165: 1342–1355.

14 Mehrotra, S., Rahman, L.U., and Kukreja, A.K. (2010). An extensive case study of hairy-root cultures for enhanced secondary-metabolite production through metabolic-pathway engineering. *Biotechnology and Applied Biochemistry* 56: 161–172.

15 Srivastava, V., Mehrotra, S., and Verma, P.K. (2017). Biotechnological interventions for production of therapeutic secondary metabolites using hairy root cultures of medicinal plants. In: Current Developments in Biotechnology and Bioengineering; Crop Modification, Nutrition, and Food Production (eds. S.K. Dubey, A. Pandey and R.S. Sangwan), 259–282. Amsterdam: Elsevier Science.

16 Srivastava, V., Mehrotra, S., and Mishra, S. (2017). Biotransformation through hairy roots: perspectives, outcomes, and major challenges. In: Transgenesis and Secondary Metabolism (ed. S. Jha), 1–24. Cham: Springer.

17 Kabera, J.N., Semana, E., Mussa, A.R., and He, X. (2014). Plant secondary metabolites: biosynthesis, classification, function and pharmacological properties. *Journal of Pharmacy and Pharmacology* 2: 377–392.

18 Zaynab, M., Fatima, M., Abbas, S. et al. (2018). Role of secondary metabolites in plant defense against pathogens. *Microbial Pathogenesis* 124: 198–202.

19 Wink, M. (2008). Plant secondary metabolism: diversity, function and its evolution. *Natural Product Communications* 3 (8) https://doi.org/10.1177/1934578X0800300801.

20 Wink, M. and Waterman, P.G. (1999). Chemotaxonomy in relation to molecular phylogeny of plants. In: Biochemistry of Plant Secondary Metabolism (ed. M. Wink), 300–341. Sheffield: Sheffield Academic Press and CRC Press.

21 Tiwari, R. and Rana, C.S. (2015). Plant secondary metabolites: a review. *International Journal of Engineering Research and General Science* 3 (5): 661–670.

22 Mishra, S., Phukan, U.J., Tripathi, V. et al. (2015). PsAP2 an AP2/ERF family transcription factor from Papaver somniferum enhances abiotic and biotic stress tolerance in transgenic tobacco. *Plant Molecular Biology* 89 (1–2): 173–186.

23 Mishra, S., Triptahi, V., Singh, S. et al. (2013). Wound induced tanscriptional regulation of benzylisoquinoline pathway and characterization of wound inducible PsWRKY transcription factor from Papaver somniferum. *PLoS One* 8 (1): 52784.

24 Samuni-Blank, M., Izhaki, I., Dearing, M.D. et al. (2012). Intraspecific directed deterrence by the mustard oil bomb in a desert plant. *Current Biology* 22 (13): 1218–1220.

25 Siddiqui, M.W. and Prasad, K. (eds.) (2017). Plant Secondary Metabolites, Volume One: Biological and Therapeutic Significance. CRC Press.

26 Jamwal, K., Bhattacharya, S., and Puri, S. (2018). Plant growth regulator mediated consequences of secondary metabolites in medicinal plants. *Journal of Applied Research on Medicinal and Aromatic Plants* 9: 26–38.

27 Srivastava, V., Negi, A.S., Ajayakumar, P.V. et al. (2012). Atropa belladonna hairy roots: orchestration of concurrent oxidation and reduction reactions or biotransformation of carbonyl compounds. *Applied Biochemistry and Biotechnology* 166 (6): 1401–1408.

28 Srivastava, V. (2015). Studies on hairy roots as bioreactors for normal and elicited biotransformation of exogenous molecules. Ph.D. thesis. University of Lucknow, India.

29 Chao, M. and Lan, L. (1989). The biotransformation of 6-hydroxyhyoscyamine to scopolamine by suspension and immobilized cells of Anisodus tanguticus. *Planta Medica* 55 (04): 391–392.

30 Simola, L.K., Parviainen, R., Martinsen, A. et al. (1990). Feeding experiments using suspension cultures of Atropa belladonna; limiting steps in the biosynthesis of tropane alkaloids. *Phytochemistry* 29 (11): 3517–3523.

31 Anitha, S. and Ranjitha, B.D. (2006). Stimulation of reserpine biosynthesis in the callus of Rauvolfia tetraphylla L. by precursor feeding. *African Journal of Biotechnology* 5 (8): 659–661.

32 Dewanjee, S., Gangopadhyay, M., Das, U. et al. (2014). Enhanced rosmarinic acid biosynthesis in Solenostemon scutellarioides culture: a precursor-feeding strategy. *Natural Product Research* 28 (20): 1691–1698.

33 Srivastava, V., Kaur, R., Chattopadhyay, S.K., and Banerjee, S. (2013). Production of industrially important cosmaceutical and pharmaceutical derivatives of betuligenol by Atropa belladonna hairy root mediated biotransformation. *Industrial Crops and Products* 44: 171–175.

34 Subroto, M.A., Kwok, K.H., Hamill, J.D., and Doran, P.M. (1996). Coculture of genetically transformed roots and shoots for synthesis, translocation, and biotransformation of secondary metabolites. *Biotechnology and Bioengineering* 49 (5): 481–494.

35 Mehrotra, S., Srivastava, V., Rahman, L.U., and Kukreja, A.K. (2013). Overexpression of a Catharanthus tryptophan decarboxylase (tdc) gene leads to enhanced terpenoid indole alkaloid (TIA) production in transgenic hairy root lines of Rauwolfia serpentina. *Plant Cell, Tissue and Organ Culture (PCTOC)* 115 (3): 377–384.

36 Sun, J. and Peebles, C.A. (2016). Engineering overexpression of ORCA3 and strictosidine glucosidase in Catharanthus roseus hairy roots increases alkaloid production. *Protoplasma* 253 (5): 1255–1264.

37 Kai, G., Yang, S., Luo, X. et al. (2011). Co-expression of AaPMT and AaTRI effectively enhances the yields of tropane alkaloids in Anisodus acutangulus hairy roots. *BMC Biotechnology* 11 (1): 43.

38 Jouhikainen, K., Lindgren, L., Jokelainen, T. et al. (1999). Enhancement of scopolamine production in Hyoscyamus muticus L. hairy root cultures by genetic engineering. *Planta* 208 (4): 545–551.

39 Yang, C., Chen, M., Zeng, L. et al. (2011). Improvement of Tropane alkaloids production in hairy root cultures of'Atropabelladonna'by overexpressing pmt and h6h genes. *Plant Omics* 4 (1): 29.

40 Yousefian, Z., Hosseini, B., Rezadoost, H. et al. (2018). Production of the anticancer compound withaferin A from genetically transformed hairy root cultures of Withania somnifera. *Natural Product Communications* 13 (8): 943–948.

41 Namdeo, A.G. (2007). Plant cell elicitation for production of secondary metabolites: a review. *Pharmacognosy Reviews* 1 (1): 69–79.

42 Wang, J., Qian, J., Yao, L., and Lu, Y. (2015). Enhanced production of flavonoids by methyl jasmonate elicitation in cell suspension culture of Hypericum perforatum. *Bioresources and Bioprocessing* 2 (1): 5.

43 Lee, Y.S., Ju, H.K., Kim, Y.J. et al. (2013). Enhancement of anti-inflammatory activity of Aloe vera adventitious root extracts through the alteration of primary and secondary metabolites via salicylic acid elicitation. *PLoS One* 8 (12): 82479.

44 Ahlawat, S., Saxena, P., Ali, A., and Abdin, M.Z. (2016). Piriformospora indica elicitation of withaferin A biosynthesis and biomass accumulation in cell suspension cultures of Withania somnifera. *Symbiosis* 69 (1): 37–46.

45 Netala, V.R., Kotakadi, V.S., Gaddam, S.A. et al. (2016). Elicitation of gymnemic acid production in cell suspension cultures of Gymnema sylvestre R. Br. through endophytic fungi. *3 Biotech* 6 (2): 232.

46 Kai, G., Yang, S., Zhang, Y. et al. (2012). Effects of different elicitors on yield of tropane alkaloids in hairy roots of Anisodus acutangulus. *Molecular Biology Reports* 39 (2): 1721–1729.

47 Thakore, D., Srivastava, A.K., and Sinha, A.K. (2015). Model based fed batch cultivation and elicitation for the overproduction of ajmalicine from hairy roots of Catharanthus roseus. *Biochemical Engineering Journal* 97: 73–80.

48 Zaheer, M., Reddy, V.D., and Giri, C.C. (2016). Enhanced daidzin production from jasmonic and acetyl salicylic acid elicited hairy root cultures of Psoralea corylifolia L.(Fabaceae). *Natural Product Research* 30 (13): 1542–1547.

49 Mehrotra, S., Mishra, S., and Srivastava, V. (2016). Bioreactor technology for hairy roots cultivation. In: Bioprocessing of Plant in vitro; Systems, Reference Series in Phytochemistry (eds. A. Pavlov and T. Bley). Cham: Springer.

50 Srikantan, C. and Srivastava, S. (2018). Bioreactor design and analysis for large-scale plant cell and hairy root cultivation. In: Hairy Roots (eds. V. Srivastava, S. Mehrotra and S. Mishra), 147–182. Singapore: Springer.

51 Liu, C.Z., Guo, C., Wang, Y.C., and Ouyang, F. (2003). Comparison of various bioreactors on growth and artemisinin biosynthesis of Artemisia annua L. shoot cultures. *Process Biochemistry* 39 (1): 45–49.

52 Mehrotra, S., Goel, M.K., Srivastava, V., and Rahman, L.U. (2015a). Hairy root biotechnology of Rauwolfia serpentina: a potent approach for the production of pharmaceutically important terpenoid indole alkaloids. *Biotechnology Letters* 37 (2): 253–263.

53 Baque, M.A., Moh, S.H., Lee, E.J. et al. (2012). Production of biomass and useful compounds from adventitious roots of high-value added medicinal plants using bioreactor. *Biotechnology Advances* 30 (6): 1255–1267.

54 Jeet, A., Singh, Y., Singh, P. et al. (2020). Strategies for indole alkaloids enrichment through callus culture from Alstonia scholaris (L.) R. Br. *Plant Growth Regulation* 90 (2): 383–392.

55 Qin, B., Ma, L., Wang, Y. et al. (2014). Effects of acetylsalicylic acid and UV-B on gene expression and tropane alkaloid biosynthesis in hairy root cultures of Anisodus luridus. *Plant Cell, Tissue and Organ Culture (PCTOC)* 117 (3): 483–490.

56 Zebarjadi, A., Dianatkhah, S., Mohammadi, P.P., and Qaderi, A. (2018). Influence of abiotic elicitors on improvement production of artemisinin in cell culture of Artemisia annua L. *Cellular and Molecular Biology* 64 (9): 1–5.

57 Ahlawat, S., Saxena, P., Alam, P. et al. (2014). Modulation of artemisinin biosynthesis by elicitors, inhibitor, and precursor in hairy root cultures of Artemisia annua L. *Journal of Plant Interactions* 9 (1): 811–824.

58 Jiao, J., Gai, Q.Y., Wang, W. et al. (2016). Enhanced astragaloside production and transcriptional responses of biosynthetic genes in Astragalus membranaceus hairy root cultures by elicitation with methyl jasmonate. *Biochemical Engineering Journal* 105: 339–346.

59 McGehee, D.L., Alimohammadi, M., and Khodakovskaya, M.V. (2019). Carbon-based nanomaterials as stimulators of production of pharmaceutically active alkaloids in cell culture of Catharanthus roseus. *Nanotechnology* 30 (27): 275102.

60 Sun, J., Xiao, J., Wang, X. et al. (2012). Improved cardenolide production in Calotropis gigantea hairy roots using mechanical wounding and elicitation. *Biotechnology Letters* 34 (3): 563–569.

61 Salehi, M., Farhadi, S., Moieni, A. et al. (2020). Mathematical modeling of growth and paclitaxel biosynthesis in Corylus avellana cell culture responding to fungal elicitors using multilayer perceptron-genetic algorithm. *Frontiers in Plant Science* 11.

62 Shakeran, Z., Keyhanfar, M., Asghari, G., and Ghanadian, M. (2015). Improvement of atropine production by different biotic and abiotic elicitors in hairy root cultures of Datura metel. *Turkish Journal of Biology* 39 (1): 111–118.

63 Khurshid, R., Ullah, M.A., Tungmunnithum, D. et al. (2020). Lights triggered differential accumulation of antioxidant and antidiabetic secondary metabolites in callus culture of Eclipta alba L. *PLOS One* 15 (6): e0233963.

64 Moharrami, F., Hosseini, B., Sharafi, A., and Farjaminezhad, M. (2017). Enhanced production of hyoscyamine and scopolamine from genetically transformed root culture of Hyoscyamus reticulatus L. elicited by iron oxide nanoparticles. *in vitro; Cellular & Developmental Biology-Plant* 53 (2): 104–111.

65 Krishnan, J.J., Gangaprasad, A., and Satheeshkumar, K. (2019). Biosynthesis of camptothecin from callus and cell suspension cultures of ophiorrhiza mungos L. var. angustifolia (Thw.) Hook. f. *Proceedings of the National Academy of Sciences, India Section B: Biological Sciences* 89 (3): 893–902.

66 Hashemi, S.M. and Naghavi, M.R. (2016). Production and gene expression of morphinan alkaloids in hairy root culture of Papaver orientale L. using abiotic elicitors. *Plant Cell, Tissue and Organ Culture (PCTOC)* 125 (1): 31–41.

67 Srivastava, M., Sharma, S., and Misra, P. (2016). Elicitation based enhancement of secondary metabolites in Rauwolfia serpentina and Solanum khasianum hairy root cultures. *Pharmacognosy Magazine* 12 (Suppl 3): S315.

68 Zafar, N., Mujib, A., Ali, M. et al. (2020). Cadmium chloride (CdCl 2) elicitation improves reserpine and ajmalicine yield in Rauvolfia serpentina as revealed by high-performance thin-layer chromatography (HPTLC). *3 Biotech* 10 (8): 1–14.

69 Coskun, Y., Duran, R.E., and Kilic, S. (2019). Striking effects of melatonin on secondary metabolites produced by callus culture of rosemary (Rosmarinus officinalis L.). *Plant Cell, Tissue and Organ Culture (PCTOC)* 138 (1): 89–95.

70 Hao, X., Shi, M., Cui, L. et al. (2015). Effects of methyl jasmonate and salicylic acid on tanshinone production and biosynthetic gene expression in transgenic Salvia miltiorrhiza hairy roots. *Biotechnology and Applied Biochemistry* 62 (1): 24–31.

71 Pillai, S.K. and Siril, E.A. (2019). Enhanced production of Berberine through callus culture of Tinospora cordifolia (Willd.) Miers ex Hook F. and Thoms. *Proceedings of the National Academy of Sciences, India Section B: Biological Sciences*: 1–9.

72 Hosseini, S.M., Bahramnejad, B., Baneh, H.D. et al. (2017). Hairy root culture optimization and resveratrol production from Vitis vinifera subsp. sylvesteris. *World Journal of Microbiology and Biotechnology* 33 (4): 67.

73 Rangaraju, S., Lokesha, A.N., and Aswath, C.R. (2019). Improved production of withanolides in adventitious root cultures of Withania somnifera by suspension culture method.

74 Sivanandhan, G., Dev, G.K., Jeyaraj, M. et al. (2013). Increased production of withanolide A, withanone, and withaferin A in hairy root cultures of Withania somnifera (L.) Dunal elicited with methyl jasmonate and salicylic acid. *Plant Cell, Tissue and Organ Culture (PCTOC)* 114 (1): 121–129.

75 Woerdenbag, H.J., Lüers, J.F., van Uden, W. et al. (1993). Production of the new antimalarial drug artemisinin in shoot cultures of Artemisia annua L. *Plant Cell, Tissue and Organ Culture* 32 (2): 247–257.

76 Khanam, N., Khoo, C., Close, R., and Khan, A.G. (2001). Tropane alkaloid production by shoot culture of Duboisia myoporoides R. Br. *Phytochemistry* 56 (1): 59–65.

77 Kang, Y.M., Min, J.Y., Moon, H.S. et al. (2004). Rapid in vitro; adventitious shoot propagation of Scopolia parviflora through rhizome cultures for enhanced production of tropane alkaloids. *Plant Cell Reports* 23 (3): 128–133.

78 Lydon, J., Zimmerman, R.H., Fordham, I.M., and Lusby, W.R. (1993). Tissue culture and alkaloid production of Erthroxylum coca var. coca. *Journal of Herbs, Spices and Medicinal Plants* 2 (1): 3–14.

79 Sharma, A., Mathur, A.K., Ganpathy, J. et al. (2019). Effect of abiotic elicitation and pathway precursors feeding over terpenoid indole alkaloids production in multiple shoot and callus cultures of Catharanthus roseus. *Biologia* 74 (5): 543–553.

80 Hirata, K., Yamanaka, A., Kurano, N. et al. (1987). Production of indole alkaloids in multiple shoot culture of Catharanthus roseus (L). G. Don. *Agricultural and Biological Chemistry* 51 (5): 1311–1317.

81 Miura, Y., Hirata, K., Kurano, N. et al. (1988). Formation of vinblastine in multiple shoot culture of Catharanthus roseus. *Planta Medica* 54 (01): 18–20.

82 Satdive, R.K., Fulzele, D.P., and Eapen, S. (2003). Studies on production of ajmalicine in shake flasks by multiple shoot cultures of Catharanthus roseus. *Biotechnology Progress* 19 (3): 1071–1075.

83 Coste, A., Vlase, L., Halmagyi, A. et al. (2011). Effects of plant growth regulators and elicitors on production of secondary metabolites in shoot cultures of Hypericum hirsutum and Hypericum maculatum. *Plant Cell, Tissue and Organ Culture (PCTOC)* 106 (2): 279–288.

84 Roja, P.C., Sipahimalani, A.T., Heble, M.R., and Chadha, M.S. (1987). Multiple shoot cultures of Rauvolfia serpentina: growth and alkaloid production. *Journal of Natural Products* 50 (5): 872–875.

85 Skrzypczak-Pietraszek, E., Piska, K., and Pietraszek, J. (2018). Enhanced production of the pharmaceutically important polyphenolic compounds in Vitex agnus castus L. shoot cultures by precursor feeding strategy. *Engineering in Life Sciences* 18 (5): 287–297.

86 Sparzak-Stefanowska, B., Krauze-Baranowska, M., and Hałasa, R. (2019). Influence of plant growth regulators on the shoot culture of Phyllanthus glaucus and accumulation of indolizidine alkaloids with evaluation of antimicrobial activity. *Acta Physiologiae Plantarum* 41 (1): 6.

87 Ekiert, H. and Kisiel, W. (1997). Coumarins and alkaloids in shoot culture of Ruta graveolens L. *Acta Societatis Botanicorum Poloniae* 66 (3–4): 329–332.

88 Kilby, N.J., Griggs, D.L., and Berry, S.F. (1992). Assessment of inter-and intra-population variation in quinoline alkaloid profile of juvenile shoot cultures of Cinchona ledgeriana: effect of method of nutrient delivery on alkaloid profile. *Plant Cell, Tissue and Organ Culture* 28 (3): 275–280.

89 Ray, S. and Jha, S. (2001). Production of withaferin A in shoot cultures of Withania somnifera. *Planta Medica* 67 (05): 432–436.

90 Roja, G. and Heble, M.R. (1994). The quinoline alkaloids camptothecin and 9-methoxycamptothecin from tissue cultures and mature trees of Nothapodytes foetida. *Phytochemistry* 36 (1): 65–66.

91 Gopalakrishnan, R. and Shankar, B. (2014). Multiple shoot cultures of Ophiorrhiza rugosa var. decumbens Deb and Mondal–a viable renewable source for the continuous production of bioactive Camptotheca alkaloids apart from stems of the parent plant of Nothapodytes foetida (Wight) Sleumer. *Phytomedicine* 21 (3): 383–389.

92 Patil, J.G., Ahire, M.L., Nitnaware, K.M. et al. (2013). in vitro; propagation and production of cardiotonic glycosides in shoot cultures of Digitalis purpurea L. by elicitation and precursor feeding. *Applied Microbiology and Biotechnology* 97 (6): 2379–2393.

93 Pérez-Alonso, N., Wilken, D., Gerth, A. et al. (2009). Cardiotonic glycosides from biomass of Digitalis purpurea L. cultured in temporary immersion systems. *Plant Cell, Tissue and Organ Culture (PCTOC)* 99 (2): 151–156.

94 Rahmat, E. and Kang, Y. (2019). Adventitious root culture for secondary metabolite production in medicinal plants: a review. *Journal of Plant Biotechnology* 46 (3): 143–157.

95 Praveen, N. and Murthy, H.N. (2010). Production of withanolide-A from adventitious root cultures of Withania somnifera. *Acta physiologiae plantarum* 32 (5): 1017–1022.

96 Thilip, C., Raju, C.S., Arutharaju, K. et al. (2015). Establishment of adventitious root culture fom cell suspension of Withania somnifora (L.) Dunal: an in vitro; approach for production of whitanolides. *International Journal of Pharma and Bio Sciences* 6 (1): 1030–1037.

97 Riasat, R., Riasat, Z., Abbasi, B.H. et al. (2015). Silybum marianum:adventitious roots induction along with free radical scavenging activity. *Journal of Plant Biology Research* 4 (1): 12–21.

98 Min, J.Y., Jung, H.Y., Kang, S.M. et al. (2007). Production of tropane alkaloids by small-scale bubble column bioreactor cultures of Scopolia parviflora adventitious roots. *Bioresource Technology* 98: 1748–1753.

99 Baskaran, P. and Jayabalan, N. (2009). Psoralen production in hairy roots and adventitious roots cultures of Psoralea coryfolia. *Biotechnology Letters* 31 (7): 1073–1077.

100 Rajesh, M., Jeyaraj, M., Sivanandhan, G. et al. (2012). Adventitious root culture in Podophyllum hexandrum Royle (syn. P. emodi Wall. ex Hook.f. &Thomas) – an important medicinal plant. *Journal of Biotechnology and Biomaterials* 2: 6.

101 Wu, C.H., Murthy, H.N., Hahn, E.J., and Paek, K.Y. (2008). Establishment of adventitious root co-culture of Ginseng and Echinacea for the production of secondary metabolites. *Acta Physiologiae Plantarum* 30: 891–896.

102 Ling, A.P.K., Kok, K.M., Hussein, S., and Ong, S.L. (2009). Effects of plant growth regulators on adventitious roots induction from different explants of Orthosiphon Stamineus. *American-Eurasian Journal of Sustainable Agriculture* 3 (3): 493–501.

103 Cui, X.H., Murthy, H.N., Wu, C.H., and Paek, K.Y. (2010). Sucrose-induced osmotic stress affects biomass, metabolite, and antioxidant levels in root suspension cultures of Hypericum perforatum L. *Plant Cell, Tissue and Organ Culture (PCTOC)* 103: 7–14.

104 Sharma, S.N., Jha, Z., and Sinha, R.K. (2013). Establishment of in vitro; adventitious root cultures and analysis of andrographolide in Andrographis paniculata. *Natural Product Communications* 8 (8): 1045–1047.

105 Murthy, H.N., Hahn, E.J., and Paek, K.Y. (2008). Adventitious roots and secondary metabolism. *Chinese Journal of Biotechnology* 24 (5): 711–716.

106 Yue, W., Ming, Q.L., Lin, B. et al. (2016). Medicinal plant cell suspension cultures: pharmaceutical applications and high-yielding strategies for the desired secondary metabolites. *Critical Reviews in Biotechnology* 36 (2): 215–232.

107 Zafar, N., Mujib, A., Ali, M. et al. (2017). Aluminum chloride elicitation (amendment) improves callus biomass growth and reserpine yield in Rauvolfia serpentina leaf callus. *Plant Cell, Tissue and Organ Culture (PCTOC)* 130 (2): 357–368.

108 Teshima, D., Ikeda, K., Satake, M. et al. (1988). Production of emetic alkaloid by in vitro; culture of Cephaelis ipecacuanha A. Richard. *Plant Cell Reports* 7 (4): 278–280.

109 Ten Hoopen, H.J.G., Vinke, J.L., and Moreno, P.R.H. (2002). Influence of temperature on growth and ajmalicine production by Catharantus roseus suspension cultures. *Enzyme and Microbial Technology* 30: 56–65.

110 Baldi, A. and Dixit, V.K. (2008). Yield enhancement strategies for artemisinin production by suspension cultures of Artemisia annua. *Bioresource Technology* 99: 4609–4614.

111 Pasqua, G., Silvestrini, A., and Monacelli, B. (2006). Triterpenoids and ellagic acid derivatives from in vitro; cultures of Camptotheca acuminata Decaisne. *Plant Physiology and Biochemistry* 44: 220–225.

112 Chen, S., Wang, X., and Zhao, B. (2003). Production of crocin using Crocus sativus callus by two-stage culture system. *Biotechnology Letters* 25: 1235–1238.

113 Mukherjee, S., Ghosh, B., and Jha, S. (2000). Establishment of forskolin yielding transformed cell suspension cultures of Coleus forskohlii as controlled by different factors. *Journal of Biotechnology* 76: 73–81.

114 Praveen, N., Murthy, H.N., and Chung, I.M. (2011). Improvement of growth and gymnemic acid production by altering the macro elements concentration and nitrogen source supply in cell suspension cultures of Gymnema sylvestre R. Br. *Industrial Crops and Products* 33: 282–286.

115 Walker, T.S., Pal Bais, H., and Vivanco, J.M. (2002). Jasmonic acid-induced hypericin production in cell suspension cultures of Hypericum perforatum L. (St. John's wort). *Phytochemistry* 60: 289–293.

116 Yamamoto, O. and Yamada, Y. (1986). Production of reserpine and its optimization in cultured Rauwolfia serpentina Benth cells. *Plant Cell Reports* 5: 50–53.

117 Mehrotra, S., Srivastava, V., Ur, R.L., and Kukreja, A.K. (2015b). Hairy root biotechnology-indicative timeline to understand missing links and future outlook. *Protoplasma* 252 (5): 1189–1201.

118 Sharmila, R. and Subburathinam, K.M. (2013). Effect of signal compounds on andrographolide in the hairy root cultures of Andrographis paniculata. *International Journal of Pharmaceutical Sciences and Research* 4 (5): 1773.

119 Bahabadi, S.E., Sharifi, M., Chashmi, N.A. et al. (2014). Significant enhancement of lignan accumulation in hairy root cultures of Linum album using biotic elicitors. *Acta Physiologiae Plantarum* 36 (12): 3325–3331.

120 Sivanandhan, G., Arunachalam, C., Selvaraj, N. et al. (2015). Expression of important pathway genes involved in withanolides biosynthesis in hairy root culture of Withania somnifera upon treatment with Gracilaria edulis and Sargassum wightii. *Plant Physiology and Biochemistry* 91: 61–64.

121 Ahmadi Moghadam, Y., Piri, K., Bahramnejad, B., and Ghiasvand, T. (2014). Dopamine production in hairy root cultures of Portulaca oleracea (Purslane) using Agrobacterium rhizogenes. *Journal of Agricultural Science and Technology* 16 (2): 409–420.

122 Cui, L., Ni, X., Ji, Q. et al. (2015). Co-overexpression of geraniol-10-hydroxylase and strictosidine synthase improves anti-cancer drug camptothecin accumulation in Ophiorrhiza pumila. *Scientific Reports* 5 (1): 1–9.

123 Deng, C., Hao, X., Shi, M. et al. (2019). Tanshinone production could be increased by the expression of SmWRKY2 in Salvia miltiorrhiza hairy roots. *Plant Science* 284: 1–8.

12

Plant Diversity and Ethnobotanical Knowledge of Spices and Condiments

Thakku R. Ramkumar[1] *and Subbiah Karuppusamy*[2]

[1] *Agronomy Department, IFAS, University of Florida, Gainesville, FL, USA*
[2] *Department of Botany, Botanical Research Center, The Madura College, Madurai, Tamil Nadu, India*

12.1 Introduction

Spices and condiments are used for flavoring the food and beverages. They may be defined variously that "Spices are plant parts like seed, bark, berries, buds, floral parts, fruits, kernels, leaves, rhizome, latex etc., which impart strong flavor or aroma to food or drinks" [1]. These aromatic vegetable products are used in the whole, broken, or in ground form to season food rather than for nutrients [2]. The meaning of spice derived from Latin term *spec*, the noun alluding to appearance or kind. Traditional medicinal practices are using herbs including spices, in Indian subcontinent and Chinese proximities dating back to at least 4000 BCE. In popular culture, spices usage was also connected with magic, religion, tradition, and cultural rituals. Condiments on the other hand are usually a combination of herbs and spices blended in a liquid form. Condiments are prepared food, mainly preserved or fermented and used in invariable quantities depending on diner's taste [3]. However, International Organization for Standardization [4] defines spices and condiments, do not show any clear-cut differences, and can be clubbed together under one term "spices and condiments," whichever used for flavoring and seasoning, including meat preparation, baking, confectioning, and other food processing and preserving.

India is bestowed with rich ethnic and cultural diversity. There are about 440 ethnic communities in this region [5]. India is also one of the mega biodiversity countries that has the one of the world's richest plant diversity reservoir supporting about 50% of India's biodiversity [6]. Most of the ethnic communities residing in hilly tracts and slopes form small, isolated remote villages. The ethnic people mostly depend on natural resources from the nearby forest for their food, livelihood, and ailments. They are repository of indigenous knowledge system belonging to agriculture, food, medicine, etc. As they have very good knowledge of their natural resources, the local crops, wild plants, ethnic vegetables, and indigenous fruits are mainly used in their local diet for food [7]. They use varieties of plants

Bioprospecting of Plant Biodiversity for Industrial Molecules, First Edition.
Edited by Santosh Kumar Upadhyay and Sudhir P. Singh.

in ethnic food for flavoring, seasoning, coloring, and food preservation as well. About 60 varieties of spices are grown in India [8], which are well-documented from ancient period onward [9].

The important spices traditionally traded throughout the world are mostly products of tropical environments. The major exceptions to this group are the capsicums (chilli, paprika) and coriander, which are grown over a much wider range of tropical and nontropical environments as well. Production of spices and essential oils in these wet and humid environments brings special difficulties for crop and product management. Drying the crop to ensure a stable stored product is of particular importance, and in wet humid environments, this creates the need for efficient and effective drying systems. In the passage of time, the uses of the spices have steadily increased; in culinary art, beverages, confectionery, liquors and cordials, perfumery, cosmetic industry, and even notably as medicines. Today, the principal uses of spices in medicine lie with their adjuvant and alleviative qualities [10].

Spices are even more important today all around the world for improving taste and clamor of food items. Most of the spices are rich sources of essential oils. Essential oils are volatile liquid products of steam or water distillation of plant parts such as leaves, stems, bark, seeds, fruits, roots, and plant exudates. An essential oil may contain up to several hundred chemical compounds, and this complex mixture of compounds gives the oil its characteristic fragrance and flavor. Usually spices are used directly or indirectly in several industries such as cosmetics, perfumery, medicine, and pharmaceutical. The main properties of spices are antioxidants, antimicrobial, antibiotic, and preservatives. Spices are also used for coloring the foodstuff [11]. Many leafy spices are sometimes used as "herbal" teas for improving general health. Apart from enhancing the taste and flavor of food, spices have been widely believed to exert digestive stimulant action. A few medicinal properties of spices such as tonic, carminative, stomachic, diuretic, and antispasmodic have long been recognized [12].

12.2 Habitat and Diversity of Major Spices and Condiments in India

India is historically known as the "The home of spices." India has produced the largest quantity spices and condiments and also consumed and exported highest level significantly when compared to other countries. The diverse agroclimatic regions across the subcontinent offer tropical, subtropical, and temperate conditions, suitable to grow almost all such species [13]. Among the 109 spices listed by International Organization for Standardization [4], India grows about 60 species and Indian spices flavor foods in over 130 countries of the world [14]. In India, at least 52 spices are being cultivated including tree spices, seed spices, and herbal spices. Most of these spices can be grown alone or in combination with other crops as a system. The choice of crops depends on the physiography, topography, soil, climate, and the market demands. Major spices, namely black pepper, cardamom, vanilla, ginger, turmeric; tree spices nutmeg, clove, cinnamon, allspice (dried unripe berry of *Pimenta dioica*) and garcinia; and seed spices, coriander, fennel, mustard are ideal crops for inter or mixed cropping [15]. Some of the important Indian spices and condiments with scientific name, family name, useful part, common name, distribution in Indian states, and their cultivation status are summarized family-wise in Table 12.1. Phenotypic images of selected spice plants also provided in Figure 12.1.

Table 12.1 Diversity and distribution of spices and condiments in India.

Family	Botanical name	Part(s) used	Common name(s)	Distribution	Cultivation status
Acoraceae	*Acorus calamus* L.	Rhizome	Sweet flag, myrtie flag, calamus, flag root	Throughout India	Cultivated
Amaryllidaceae	*Allium ascalonicum* L.	Bulb	Shallot	Assam, Tripura	Cultivated
Amaryllidaceae	*Allium cepa* L.	Bulb	Onion	Throughout India	Cultivated
Amaryllidaceae	*Allium cepa* var. *aggregatum*	Bulb	Potato onion	Assam, Maharashtra	Cultivated
Amaryllidaceae	*Allium fistulosum* L.	Bulb, leaves	Japanese bunching onion, stony leek	North eastern India	Cultivated
Amaryllidaceae	*Allium porrum* L.	Bulb, Leaves	Leek, winter leek	Northeastern States	Cultivated
Amaryllidaceae	*Allium sativum* L.	Bulb	Garlic	Throughout India	Cultivated
Amaryllidaceae	*Allium tuberosum* Rottler.	Bulb	Indian leek	Assam	Cultivated
Anacardiaceae	*Buchanania lanzan* Spreng.	Seed	Cuddapah almond	Andhra Pradesh, Tamil Nadu	Wild
Anacardiaceae	*Mangifera indica* L.	Immature fruit	Mango	Throughout India	Cultivated
Anacardiaceae	*Schinus molle* L.	Fruit	Californian pepper tree	Karnataka, Kerala, Tamil Nadu	Wild
Anacardiaceae	*Schinus terebinthifolia* Radde.	Fruit	Brazilien pepper tree	Maharashtra	Cultivated
Annonaceae	*Xylopia aethiopica* (Dunal) A. Rich.	Fruit	Negro pepper	Arunachal Pradesh, Assam	Cultivated
Apiaceae	*Anethum graveolens* L.	Leaves	Dill	Andhra Pradesh, Madhya Pradesh, Kerala, Tamil Nadu	Cultivated
Apiaceae	*Anethum sowa* Kurz.	Fruit	Indian Dill	Rajasthan, Gujarat, Odisha	Cultivated
Apiaceae	*Angelica archangelica* L.	Fruit, petiole, root	Garden angelica	Assam, Meghalaya, Tripura	Cultivated
Apiaceae	*Anthriscus cerasifolium* (L.) Hoffm.	Leaf	Chervil	Assam, Arunachal Pradesh	Wild

(*Continued*)

Table 12.1 (Continued)

Family	Botanical name	Part(s) used	Common name(s)	Distribution	Cultivation status
Apiaceae	*Apium graveolens* L.	Seeds, root, leaves	Celery	Assam, Nagaland, Maharashtra, Gujarat	Cultivated
Apiaceae	*Bunium persicum* (Boiss) Fedtsch.	Fruit, bulb	Black caraway	Rajasthan, Assam, West Bengal	Cultivated
Apiaceae	*Carum bulbocastanum* L.	Fruit, Bulb	Black caraway	Assam, Tripura, Meghalaya	Wild
Apiaceae	*Coriandrum sativum* L.	Leaves and fruits	Coriander	Andhra Pradesh, Madhya Pradesh Telangana, Tamil Nadu	Cultivated
Apiaceae	*Carum copticum* L.	Fruits	Ajowan caraway, blond caraway	Tamil Nadu, Kerala, Karnataka	Cultivated
Apiaceae	*Cuminum cyminum* L.	Fruits	Cumin	Throughout India	Cultivated
Apiaceae	*Eryngium foetidum* L.	Fruits	Elephant pepper	Kerala, Arunachal Pradesh	Wild
Apiaceae	*Ferula asa foetida* L.	Rhizome exudates	Asafoetida	Rajasthan	Cultivated
Apiaceae	*Foeniculum vulgare* Miller.	Seeds	Fennel	Kerala, Madhya Pradesh	Cultivated
Apiaceae	*Levisticum officinale* Koch.	Leaves	Lovage	Northern India	Cultivated
Apiaceae	*Pimpinella anisum* L.	Fruits	Aniseed	Assam, Tripura, Rajasthan, Gujarat, Odisha	Cultivated
Asteraceae	*Artemisia dracunculus* L.	Leaves	Tarragon	Arunachal Pradesh	Wild
Brassicaceae	*Armoracia rusticana* Gaertn.	Root	Horse radish	Assam, Arunachal Pradesh	Cultivated
Brassicaceae	*Brassica juncea* (L.) Czernj. et Cosson.	Seeds	Mustard	Andhra Pradesh, Bihar, Madhya Pradesh, Uttar Pradesh	Cultivated
Brassicaceae	*Brassica nigra* (L.) Koch.	Seeds	Black mustard	Bihar, Uttar Pradesh	Cultivated
Brassicaceae	*Brassica rapa* (L.) Hanelt.	Seed	Field mustard	Uttar Pradesh, Haryana, Chhattisgarh	Cultivated
Brassicaceae	*Sinapis alba* L.	Seeds	White mustard	Northern India	Cultivated
Caesalpiniaceae	*Tamarindus indica* L.	Fruit	Tamarind	Throughout India	Cultivated

Capparidaceae	*Capparis spinosa* L.	Flower bud	Caper bud	Tamil Nadu	Wild, rarely cultivated
Cleomaceae	*Cleome viscosa* L.	Seeds	Jakhiya	Throughout in India	Wild
Clusiaceae	*Garcinia gummi-gutta* (L.) Robson.	Fruit	Malabar tamarind, camboge	Kerala, Karnataka	Cultivated and wild
Clusiaceae	*Garcinia indica* (Thouars) Choisy.	Fruit	Kokam	Kerala	Wild
Cupressaceae	*Juniperus communis* L.	Fruit	Juniper berry (technically seed cone)	Assam, Meghalaya	Cultivated
Fabaceae	*Abrus precatorius* L.	Root	Indian liquorice, rosary pea	Throughout India	Wild
Fabaceae	*Glycyrrhiza glabra* L.	Root	Liquorice	Odisha, Andhra Pradesh, Tamil Nadu	Cultivated
Fabaceae	*Trigonella foenum-graecum* L.	Leaves, seeds	Fenugreek	Throughout India	Cultivated
Illiciaceae	*Illicium verum* Hook.f.	Fruit	Star anise	Sikkim, Meghalaya	Cultivated
Iridaceae	*Crocus sativus* L.	Anther and filament	Saffron	Jammu and Kashmir	Cultivated
Lamiaceae	*Hyssopus officinalis* L.	Leaves	Hysopp	Northeastern India	Cultivated
Lamiaceae	*Lippia graveolens* Kunth.	Leaves, terminal shoot	Mexican oregano	Northern India	Cultivated
Lamiaceae	*Melissa officinalis* L.	Leaves	Balm, Lemon balm	Kerala, Maharashtra, Karnataka	Cultivated
Lamiaceae	*Mentha arvensis* L.	Leaves	Mint, field mint	Throughout India	Wild
Lamiaceae	*Mentha citrata* L.	Leaves	Bergamol	Northeastern India	Cultivated
Lamiaceae	*Mentha piperita* L.	Leaves	Pepper mint	Throughout India	Cultivated

(*Continued*)

Table 12.1 (Continued)

Family	Botanical name	Part(s) used	Common name(s)	Distribution	Cultivation status
Lamiaceae	*Mentha spicata* L.	Leaves	Spear mint, garden mint	Throughout India	Cultivated
Lamiaceae	*Ocimum bascilicum* L.	Leaves	Sweet basil	Throughout India	Cultivated
Lamiaceae	*Ocimum sanctum* L.	Leaves	Holy basil	Throughout India	Cultivated
Lamiaceae	*Origanum hortensis* L.	Leafy shoots	Marjoram	Andhra Pradesh, Tamil Nadu	Cultivated
Lamiaceae	*Origanum vulgare* L.	Leaves	Origano	Throughout India	Cultivated
Lamiaceae	*Petroselinum crispum* (Miller) Nyman.	Leaves	Parsley	Kerala, Karnataka, Maharashtra	Cultivated
Lamiaceae	*Rosmarinus officinalis* L.	Leaves	Rosemary	Kerala, Tamil Nadu, Maharashtra, Assam	Cultivated
Lamiaceae	*Salvia officinalis* L.	Leaves	Garden sage, Dalmatian sage	Kerala, Tamil Nadu, Maharashtra	Cultivated
Lamiaceae	*Satureja hortensis* L.	Leaves	Summer savory	Northern India	Cultivated
Lamiaceae	*Thymus vulgaris* L.	Leaves	Thyme	Throughout India	Cultivated
Lauraceae	*Cinnamomum burmanii* Blume.	Bark	Indonesian Cassia	Meghalaya, West Bengal	Cultivated
Lauraceae	*Cinnamomum cassia* L.	Bark	Cinnamon cassia	Kerala	Cultivated
Lauraceae	*Cinnamomum loureiroi* Nees.	Bark	Vietnamese cassia bark	Kerala, Karnataka	Cultivated
Lauraceae	*Cinnamomum tamala* Nees.	Leaves	Indian cassia bark	Kerala, Karnataka, Tamil Nadu	Cultivated
Lauraceae	*Cinnamomum verum* Blume.	Leaves	Sri Lankan cinnamon, common cinnamon	Kerala, Tamil Nadu	Wild
Lauraceae	*Laurus nobilis* L.	Leaves	Bay leaf, true laurel	Meghalaya, Nagaland	Cultivated
Lauraceae	*Phoebe cooperiana* Kanjlal& Das	Leaves	Assamese mekhali	Arunachal Pradesh, Assam	Wild
Malvaceae	*Bombax ceiba* L.	Fruit	Maratimokka, cotton tree	Kerala, Tamil Nadu	Wild

Moringaceae	*Moringa pterygosperma* Gaertn.	Leaves	Drumstick leaf	Throughout India	Cultivated
Myristicaceae	*Myristica argentea*Warb.	Kernal and mace	Papuan nutmeg	Gujarat, Odisha, Assam	Cultivated
Myristicaceae	*Myristica fragrans* Houtt.	Kernal and mace	Nut meg	Throughout India	Cultivated
Myrtaceae	*Syzygium aromaticum* Thunb.	Flower bud	Clove	Kerala, Tamil Nadu	Cultivated
Myrtaceae	*Pimenta dioica* (L.) Merr.	Leaves	Allspice	Tamil Nadu	Cultivated
Myrtaceae	*Pimenta racemosa* (Miller) J. Moore.	Leaves	West Indian bay	Arunachal Pradesh	Cultivated
Orchidaceae	*Vanilla planifolia* Andrews.	Fruit	Vanilla	Kerala, Tamil Nadu	Cultivated
Orchidaceae	*Vanilla tahitensis* J. Moore.	Fruit	Tahitian vanilla	West Bengal, Assam	Cultivated
Orchidaceae	*Vanilla pompona* Schiede.	Fruit	Pompona vanilla	Arunachal Pradesh, Nagaland	Cultivated
Pandanaceae	*Pandanus amaryllifolius* Roxb.	Leaves	Pandanwangi	Maharashtra	Wild
Papaveraceae	*Papaver somniferum* L.	Seeds	Poppy, blue maw	Jammu and Kashmir, Sikkim	Cultivated
Parmeliaceae	*Parmotrema perlatum* (Huds.) Choisy.	Thallus	Kalpasi, black stone flower, lichens	Northeastern India	Wild
Pedaliaceae	*Sesamum indicum* L.	Seeds	Sesame	Throughout India	Cultivated
Piperaceae	*Piper guineense* Schum.	Fruits	West African pepper	Assam, Arunachal Pradesh	Cultivated
Piperaceae	*Piper longum* L.	Fruits	Long pepper	Kerala, Tamil Nadu	Cultivated
Piperaceae	*Piper mullesua* Buch.-Ham. ex D.Don.	Fruits	Wild pepper	Maharashtra, Karnataka, Kerala, Tamil Nadu	Wild
Piperaceae	*Piper nigrum* L.	Fruits	Black pepper	Kerala, Tamil Nadu	Cultivated

(*Continued*)

Table 12.1 (Continued)

Family	Botanical name	Part(s) used	Common name(s)	Distribution	Cultivation status
Poaceae	*Cymbopogon citratus* (DC.) Stapf.	Leaves	Lemon grass, West Indian lemon grass	Throughout India	Wild
Poaceae	*Cymbopogon nardus* L.	Leaves	Sri Lankan citronella	Throughout India	Wild
Punicaceae	*Punica granatum* L.	Seeds	Pomegranate	Throughout India	Cultivated
Ranunculaceae	*Nigella damascena* L.	Seeds	Damas black cumin	Rajasthan	Cultivated
Ranunculaceae	*Nigella sativa* L.	Seeds	Black cumin	Jammu and Kashmir, Rajasthan	Cultivated
Rutaceae	*Murraya koenigii* (L.) Sprengel.	Leaves	Curry leaf	Throughout India	Cultivated
Rutaceae	*Zanthoxylum armatum* DC.	Fruit	Mountain prickle pepper	Arunachal Pradesh	Wild
Rutaceae	*Zanthoxylum bungee* Planch.	Fruit	Chinese prickly ash pepper	Arunachal Pradesh	Wild
Rutaceae	*Zanthoxylum acanthopodium* DC.	Fruit	Chinese pepper	Arunachal Pradesh,	Wild
Rutaceae	*Zanthoxylum oxyphyllum* Edgew.	Fruit	Wild thorn pepper	Arunachal Pradesh	Wild
Rutaceae	*Zanthoxylumrhetsa*DC.	Fruit	Small thorn pepper	Karnataka, Odisha, Kerala, Arunachal Pradesh	Wild
Solanaceae	*Capsicum annuum* L.	Fruit	Wild chilli, capsicum	Throughout India	Cultivated
Solanaceae	*Capsicum frutescens* L.	Fruit	Bird eye chilli	Throughout India	Cultivated
Zigiberaceae	*Kaempferia galanga* L.	Rhizome	Galanga	Kerala	Cultivated
Zingiberaceae	*Aframomum angustifolium* Schum.	Fruit, seeds	Madagascar cardamom	Assam, Meghalaya, Tripura	Cultivated
Zingiberaceae	*Aframomum hanburyi* Schum.	Fruit, seeds	Cameroon cardamom	Assam, Tripura, Nagaland, Maharashtra	Cultivated
Zingiberaceae	*Aframomum corrorim* Engl.	Fruit, Seeds	Korarima cardamom	Assam, Meghalaya, Nagaland	Cultivated

Zingiberaceae	*Aframomum melegueta* Schum.	Fruit, seeds	Grain of paradise, guinea grains	Assam, Manipur, Mizoram, Nagaland	Cultivated
Zingiberaceae	*Alpinia calcarata* Roscoe.	Rhizome	Galanga	Southern states, Assam	Wild
Zingiberaceae	*Alpinia galanga* (L.) Willd.	Rhizome	Greater galangal	Throughout India	Cultivated
Zingiberaceae	*Alpinia officinarum* Hance.	Rhizome	Lesser galangal	Assam, Nagaland, Tripura	Cultivated
Zingiberaceae	*Amomum aromaticum* Roxb.	Fruit, seeds	Bengal cardamom	Throughout India	Cultivated
Zingiberaceae	*Amomum kepulaga* Sparague.	Fruit, seeds	Round cardamom, chester cardamom, Siamese cardamom	Arunachal Pradesh	Cultivated
Zingiberaceae	*Amomum subulatum* Roxb.	Fruit	Large cardamom	Arunachal Pradesh, Kerala, Sikkim, Tamil Nadu	Cultivated
Zingiberaceae	*Curcuma amada* Roxb.	Rhizome	Mango ginger	Arunachal, Maharashtra, Kerala	Cultivated
Zingiberaceae	*Curcuma longa* L.	Rhizome	Turmeric	Telangana, Tamil Nadu	Cultivated
Zingiberaceae	*Curcuma zedoaria* Roscoe.	Rhizome	Zedoary	Kerala	Cultivated
Zingiberaceae	*Elettaria cardamomum* (L.) Maton.	Fruits	Cardamom, Green cardamom, small cardamom	Kerala, Karnataka, Tamil Nadu	Cultivated
Zingiberaceae	*Zingiber montanum* (Koenig.) Link.	Rhizome	Wild ginger, mountain ginger	Arunachal Pradesh	Wild
Zingiberaceae	*Zingiber officinale* Roscoe.	Rhizome	Ginger	Throughout India	Cultivated

Figure 12.1 Selected Indian spice and condiment plants. (a) *Alipnia calcarata*, (b) *Cinnamomum zeylanicum*, (c) *Costus speciosus*, (d) *Curucuma neilgherrensis*, (e) *Elettaria cardamomum*, (f) *Melissa officinalis*, (g) *Pimenta dioica*, (h) *Piper longum*, (i) *Piper nigrum*, (j) *Piper mileusa*, (k) *Zanthoxylum armatum*. *Source:* Santosh Kumar Upadhyay.

Ethnobotanical evidences indicate that the use of turmeric in India has begun in ancient time in connection with Goddess Sakthi worship of divine mother and later as a commodity of trade as coloring material and as a condiment [16]. The genus originated in

Indo-Malayan region and has spread from Southeast Asia to tropical West Africa and East Africa and later introduced to Caribbean islands and to Americas. Turmeric is popular among rice-eating people of South Asia, Southeast Asia, and Indochina (continental portion of Southeast Asia) as a spice and condiment. Turmeric is the main cash crop in the tribal dominated districts of Kandhamal, Gajapati, Ganjam, Mayurbhanj, and Koraput in Orissa since time immemorial [17].

The cardamoms are the capsules of dried fruits in different genera of the Zingiberaceae family, primarily *Elettaria, Amomum*, and *Aframomum*. Among them, *Elettaria cardamomum* (cardamom/green cardamom/small cardamom) is the most important and is grown predominantly in southern India [18]. The false cardamom, large cardamom, and black cardamom from the allied genus *Amomum* are native to Nepal, Sikkim, Bengal, and Southeast Asian countries. African cardamom, which is botanically known as *Aframomum danielli*, is native to Southeast Africa especially in Tanzania, Cameroon, Madagascar, and Guinea. Small cardamom is extensively cultivated in Nepal and Sikkim and to a limited extent with the large cardamom (*Amomum subulatum*). However, international trade is now limited to Asian countries as far as small and large cardamoms are concerned because of high prices. Worldwide, cardamom is recognized as the "queen of spices" for its pleasant aroma and taste and is the third most expensive spice after saffron and vanilla [19, 20].

Spices produced from tree crops are called "Tree spices." Tamil Nadu, Kerala, Andhra Pradesh, Maharashtra, and Karnataka are the major states cultivating tree spices in larger areas of India. There are 17 tree spices, commonly grown in India, which are clove, nutmeg, cinnamon, tamarind, garcinia, curry leaf, and allspice [12]. The commercially important products have been obtained from whole spice, ground spice oil, and oleoresin. The Western Ghats belt of Kanyakumari and Nilgiris districts of Tamil Nadu and in a few areas in Karnataka and Kerala are cultivating the clove crop. Nutmeg cultivation is more common in Kerala and limited areas of Tamil Nadu and Karnataka. All spice was cultivated rarely in the gardens of Kerala, Tamil Nadu, Karnataka, West Bengal, and Odisha [21].

In India, Kerala is highly suitable for the cultivation of spices. Important spices cultivated are black pepper, cardamom, ginger, turmeric, clove, garcinia, nutmeg, and mace. Out of this 95% of the total production is contributed by pepper, cardamom, ginger, and turmeric. Pepper and cardamom are cultivated in the high ranges, while ginger and turmeric predominate in plains. The major production centers are Idukki and Wayanad districts [12].

12.3 Ethnobotanical Context of Spices and Condiments in India

The earliest literary record in India on spices is the Rig Veda (at least or before 4000 BCE), one of the ancient Hindu scriptures, listing more than a thousand healing plants including spices and condiments. Thus, the story of Indian spices dates back to at least 6000 years into the past. In the modem world, major trade is related to eating, and the spices provide the major thrust – traditionally a country of agriculture [15, 22]. Domestication of turmeric is probably much earlier to Indus valley civilization [23]. In the current scenario of ever-increasing human stress, ethnomedicinal plants having antioxidant properties are thought to be extremely fruitful to relieve stress and improve health and eventually economy. Many Indian spice plants have been investigated for their beneficial use as antioxidants or source of antioxidants using presently available experimental techniques [24]. This plant-based traditional knowledge has become a recognized tool in the search for new sources of drugs and nutraceuticals [25].

Tribals of Eastern Ghats in Andhra Pradesh make use of some rare plants for their condiments and culinary purposes that are specific to that region, including dried fruits of *Tacca leontopetaloides* and tubers of *Decalepis hamiltonii*, which are rare spices [26]. Arunachal Pradesh was inhabited by Adi, Apatani, Nyishi, Galo, Khampti, Khowa, Mishmi, Idu, Taroan, Momba, Sherdukpen, Singpho, Hrusso, Tagin, Khamba [27]. Some of the plant species are recorded in tribal belts of Arunachal Pradesh as new source of spice plants such as *Acmella paniculata*, *Dendrocalamus hamiltonii*, *Etlingera linguiformis*, *Gynura prcoumbens*, *Hydrocotyle sibthorpioides*, *Magnolia oblonga*, *Oenanthe javanica*, *Paederia foetida*, *Perilla frutescens*, *Persicaria nepalensis*, *Phyllostachys bambusoides*, and *Spilanthes acmella*. Some ethnic tribal communities of Assam, namely Ahom, Deori, Mishing, Sonowal-kachari use 51 species as spices and condiments, including *Garcinia pedunculata*, *Hibiscus sabdariffa*, *Houttuynia cordata*, *Neocinnamomum caudatum* and *Peperomia pellucida*, which are new spice species [28]. Tribes in Maharashtra such as Pawara, Bhil, Mavachi, Kokani, and Tadavi use 26 common spices and added a new spice species of *Guizotia abyssinica* as used in food flavoring [29].

Owing to their aromatic properties, three species, namely *Artemisia scoparia* (seeds, flowers, and leaves), *Carum carvi* (fruits, seeds), and *Murraya koenigii* (leaves), are added to pulses and vegetables as condiment and spice in Jammu and Kashmir region. *Ar. scoparia* is a locally grown wild spice [30]. The Tangkhul people of Manipur region are mainly dependent on the forest, as forest plants are gathered for food, medicinal, spices, fuel, etc. They use at least 30 plant species as spice belonging to 17 genera of 8 families. Several plant species used by them are recorded as new spice species such as *Allium hookeri*, *Allium tuberosum*, *Costus speciosus*, *Curcuma cassia*, *Elshotzia blanda*, *Elshotzia communis*, *Hedychium coronarium*, *Hedychium marginatum*, *Persicaria posumbu*, *Zingiber cassumunar* and *Zingiber zerumbet* [31]. Babu et al. [8] assessed the status of traditional practices of farmers and women, in particular for storage of spices and condiments in Odisha especially for ginger (*Zingiber officinale*), turmeric (*Curcuma longa*), chilli (*Capsicum annuum*), onion (*Allium cepa*), garlic (*Allium sativum*), and coriander (*Coriandrum sativum*). Observations revealed that a large number of farmers still practice the traditional storage system. Ginger and turmeric are stored in pit method, heap method and in situ method while chilli, onion, and garlic are stored in mesh bags and hanging method. In traditional method of storage, farmers primarily rely on traditional practices using local resources [8].

The Sikkim state was inhibited by various ethnic tribes such as Lepcha, Bhutis, Limboo, and Nepalese [32]. The tribes live mostly on hilltops and on slopes, forming small and isolated villages. It is important to save this traditional knowledge of biological heritage and explore new resources. Traditional and ethnic knowledge has played a significant role in the discovery of novel ideas about conservation of natural resources. Spices have good antioxidant and preservative properties as well as antimicrobial and antibiotic properties, and therefore, are also used for medicinal purposes, 14 such species are used by tribes of

Sikkim [7]. Spices and condiments used by traditional communities of Tripura were also record [33]. In West Bengal, local communities using 32 spices distributed in 27 families including some wild plants such as *Zanthoxylum armatum* were recorded [34].

Western Ghats are tribal-dominated areas in southern India where Kadar, Malasars, Mudhuvan, Kurumbans, Kaatunaickens, Thodar, Kanis, and Paliyars reside. Still they follow traditional way of life with close association of forest vegetation for their livelihood materials including food. Several local spice and condiments species are recorded as they are used for preservation, flavoring, and coloring material of the food stuffs. Some of the plants such as *Costus speciosus*, *Garcinia gummi-gutta*, *Curcuma neilgherrensis*, *Syzygium palghatense*, *Curculigo orchioides*, *Cleome monophylla* are exclusively used as spice plants in this region [35]. In Kerala and Tamil Nadu, crushed cardamom capsules are boiled with tea and water to impart a pleasant aroma to tea, which is popularly called "*Elakkai tea*," and which has been used to relieve tiredness and depression. Cardamom capsules contain significant concentration of β-carotene [36]. In traditional medicine, consumption of cardamom daily with a table spoon of honey improves the eyesight [37]. However, some people believe that excessive uses of cardamom capsules may cause impotency [20]. A total of 11 spices out of 109 ISO approved spices are most frequently used in South Indian cuisines, particularly in Kerala and Tamil Nadu [38].

12.4 Major Spices and Condiments in India

12.4.1 Black Pepper

Piper nigrum is a perennial climbing vine grown for its berries. The matured and dried berries are used in flavoring and medicinal purposes. India is a main producer, consumer, and exporter of black pepper in the world. Annually it produces 17,563 tonnes of black pepper worth Rs. 150.95 crore that were exported to various other countries mainly from Kerala and Karnataka and Tamil Nadu [11]. It is used in different medicines from very ancient periods. Long pepper (*Piper longum*) is a slender aromatic climber whose spike is widely used in Siddha, Ayurvedic, and Unani systems of medicines particularly for diseases of respiratory tract. Pipalarishta, Pippalyasava, Panchakola, Pippalayadilauha, and Lavanabhaskarchuran, which include contributions from both *P. nigrum* and *P. longum*, are common ayurvedic preparations [10]. Combination of *sukku* (dried *Z. officinale*) *milagu* (dried fruits of *P. nigrum*), and *thippili* (dried fruits of *P. longum*), generally referred as *sukku-milagu-thippili*, is a common home medicine for cold in South India, especially Tamil Nadu.

12.4.2 Capsicums

Capsicums are dried and processed fruits of annual peppers (*Ca. annuum* var. *annuum; Capsicum chinense; Capsicum frutescens*) commonly referred as chilli or chili pepper or Chile pepper or bell pepper. A rainfall of 600–1250 mm is desirable. Rainfall is needed over the growing season but is not needed as the fruits ripen. Heavy rain during flowering adversely affects pollination and wetness at ripening encourages fungal spoilage. Capsicums flourish in warm sunny conditions, require 3–5 months with a temperature range of 18–30 °C; growth retards at below 5 °C, and frost kills the plants at any growth stage. A seedbed temperature of 20–28 °C is optimum for germination. Red chili powder, a common spice, is dried and powdered red chilli fruits. Being the hottest part of the chili, the powder is exceptionally strong and used in small quantities. Originating in the Americas, chillies were then introduced to India by the Portuguese and since then became an integral part of Indian cooking. Chillies are used as powder, crushed, and in its whole form in various South Indian curries.

12.4.3 Cinnamomum

Cinnamomum verum bark is widely used in aromatic rice preparations such as Biriyani and Pulao in India. It is a sweet-tasting spice with a warm and woody aroma. These properties make it a great ingredient to be used in cakes and desserts. Cinnamomum bark powder is highly favored for flavoring processed drinks in western hemisphere. Apart from adding flavor to food, cinnamon also has various health benefits; thought to help prevent cancer and lower blood pressure. It is predominantly grown along the Western Ghats of Kerala and Tamil Nadu [39]. While the origins of this spice can be traced back to India, it is also native to Sri Lanka. *Ci. verum* is an evergreen tree of Lauraceae family whose bark and leaves are strongly aromatic. It is useful in bronchitis, asthma, cephalalgia odontalgia, cardiac diseases, diarrhea, uropathy, nausea and vomiting, flatulence, fever, halitosis, and skin protection [40].

12.4.4 Coriander

Coriandrum sativum, is a member of the parsley family, and its seeds are oval-shaped, ridged, and turn from bright green to beige when ripe. This spice tastes sweet and tangy, with a slightly citrusy flavor. This omnipresent spice is probably the oldest in the world and is widely grown in the states of Rajasthan, Madhya Pradesh, and Tamil Nadu. Coriander seeds are also used as an alternative to salt. Seeds and leaves contain essential oils ranging from 0.1 to 1.0%, which is used for preparation of soaps of pleasant odor and good leathering property. The dried fruit powder is used to flavor foods, such as pickles, sauces, and confectionery. The fresh leaves are added in almost all-sauce-like preparations in Tamil Nadu. The distilled essential oils from the fruits are used in perfumes, soaps, candy, chocolates, tobacco, meat products, baked food, canned foods, liquors and alcoholic beverages, and also in pharmaceutical preparations.

12.4.5 Cumin

Cuminum cyminum L., another Apiaceae member, is used to add a smoky note and a robust aroma to most Indian curries and vegetables. Fried in its dry form or roasted before use, cumin seed is usually the first spice added while cooking Indian dishes. It is also dry roasted and converted to powder before being added to dishes such as pudding and buttermilk. It is used to flavor rice, stuffed vegetables, many savory dishes, and curries. Since it burns easily and can become overpowering, it is used sparingly. Cuminol or cuminaldehyde is the typical volatile principle present in the cumin seeds, which is the reason for characteristic aroma of the cumin. The cumin seeds are used for flavoring vegetables, pickles, soups, sausages, cheese and for seasoning breads, cakes, and biscuits. The seeds are also used for the preparation of traditional and homeopathy medicines prescribed for several common ailments. The seeds have major properties such as carminative, stimulant, astringent, and emmenagogue.

12.4.6 Cardamom

Elettaria cardamomum (green/small cardamom) is the concerned species, a tall growing perennial herb from Zingiberaceae. The fruits borne on panicles at the base of the plants have trilocular capsule with 15–20 seeds. The natural altitudinal growing range is between 750 and 1500 m while the most productive cultivated zone is 1000–1200 m. The annual rainfall required is usually 2500–4000 mm in the monsoon belt. A temperature range of 10–35 °C occurs over the production areas with a lower limit of about 17 °C, and an optimum temperature between 22 and 24 °C is favored. Cardamom grows naturally in shade but will produce good yields in only partial shade if well-watered. It is an indigenous spice of the land of the Malabar Coast in India. This is the third most expensive spice in the world, mainly because it is hand-harvested and requires a lot of manual work [41]. While the green cardamom has a mild and light eucalyptus tone to it, the black/large cardamom (*A. subulatum*) is spicy, smoky, and generally used only for its seeds. The most common use for cardamom is to enhance the flavor of tea and puddings [11].

12.4.7 Fennel

Foeniculum vulgare is an annual crop from Apiaceae, which is commonly cultivated in frost-free area of Rajasthan and Gujarat. The fennels are aromatic fruits and used in most kinds of Indian foods. Fennel seeds for their fragrant odor and pleasant aromatic taste due to the presence of aromatic volatile oil principles are widely used in soaps, pickles, meat dishes, sauces, pastries, confectioneries, etc. Besides fruits, the tender leaves are used in salads. The seeds are used in several Allopathic, Siddha, and Ayurvedic medicines, which are administered in diseases such as cholera, piles, gripping, constipation, dysentery, and diarrhea.

12.4.8 Ginger

Zingiber officinale is an herbaceous perennial from Zingiberaceae, the rhizome of which is used as a spice, and it is also one important spice native of South and Southeast Asia. It is recommended as a medicine in Siddha and Ayurveda for curing liver complaints, flatulence, anemia, rheumatism, piles, and jaundice. Kerala and Meghalaya are major ginger-growing states in India. India exported 7,250 tonnes of ginger to the value of the Rs. 40.755 cores annually [11]. Dried ginger in locally called as sukku, and the decoction is the first home remedy for illness in much of southern parts of the country. Sukku decoction is also taken with milk and jaggery (a locally processed sugar), also a traditional greeting drink for guests.

12.4.9 Mustard Seed

In Indian cooking, brown mustard (*Brassica juncea* L.) seeds are more commonly used than the black mustard (*Brassica nigra* L.) seeds. They belong to Brassicaceae. These seeds can be fried whole in order to flavor oil that is then used for cooking raw food. This favored oil can also be used as a garnish. While the seeds are native to Rome, the earliest reference to their use is in stories of Buddha, where it refers the seeds to justify death as a natural phenomenon [42]. Rapeseed cultivation is confined only to northern India because of late maturity and shattering of pods owing to high temperature prevailing during harvest in February–March.

12.4.10 Nutmeg

Myristica fragrans of Myristicaceae is the common nutmeg, while *Myristica argentea*, the papuan nutmeg also is referred as nutmeg. Nutmeg is a perennial tree, reaching up to 20 m in height. The useful part of the nutmeg is kernel of the seed, while mace is the net-like crimson-colored leathery outer growth (aril) covering the shell of the seed. Nutmeg pickle is one of the favorites in southern parts of India. The tree requires an optimal growing temperature between 20 and 30 °C and the annual rainfall should be between1500 and 2500 mm. The common nutmeg cultivating areas of India are Kerala and parts of Tamil Nadu [11]. The seed and mace of nutmeg, *M. fragrans* are useful in vitiated conditions of kapa and vata, inflammations, cephalalgia, helminthiasis, dyspepsia, flatulence, cough, asthma, diarrhea, vomiting, ulcer, hepatopathy, skin diseases, cardiac disorders, fever, and generally debility. Thriphaladichoorna, Karppooradichoorna, Athisaragrahanichoorna, jeerakadichoorna, etc., in Siddha and Ayurveda contain nutmeg kernel as well as mace.

12.4.11 Saffron

Crocus sativus is the most expensive spice in the world. It is cultivated predominantly in the regions of Jammu and Kashmir and to a lesser extent in Himachal Pradesh and Ladakh regions. Saffron is derived from the stigma of crocus flowers. Saffron is believed to be more valuable than gold. The most striking feature of this spice is its pungent, honey-like aroma. The deeper the color of saffron, the purer it is. It is often used after being soaked in water or milk, which softens its strong aroma and taste. The glycosides crocin and picrocrocin are the coloring principles and the bitter substance of saffron. Saffron is an important ingredient of the Ayurvedic and Unani systems of medicine in India other than food additives and coloring agent. The major properties of saffron are nervous stimulant, helping in urinary, digestive, and uterine troubles.

12.4.12 Turmeric

Curcuma longa, another spice belonging to the ginger family, turmeric is probably the most commonly used spice in India. Turmeric was predominantly used as a dye and in Siddha medicine for thousands of years. Derived from the roots of *C. longa*, a leafy plant native to India, turmeric has an earthy consistency and a warm aroma and taste. Mainly used for its flavor and color, turmeric also has antiseptic qualities and is therefore used for its health benefits as well. Turmeric forms an integral part of the rituals, ceremonies, and cuisine. Due to the strong antiseptic properties, turmeric has been used as a remedy for all kinds of poisonous affections, ulcers, and wounds. It gives good complexion to the skin. The rhizomes of mango ginger (*Curcuma amada*) are useful in vitiated conditions of pitta, anorexia, dyspepsia, flatulence, colic, bruises, wounds, chronic ulcers, skin diseases, pruritus, fever, constipations, strangury, hiccough, cough, bronchitis, sprains, gout, halitosis, otalgia, and inflammations [16]. *C. aromatica* is another Zingiberaceous perennial tuberous herb with aromatic yellow rhizome, which is internally creamy in color and the fresh root has a camphoraceous odor. Rhizomes are used in combination with astringents and aromatics for bruises, sprains, hiccough, bronchitis, cough, leukoderma, and skin eruptions. Maximum area under turmeric is in Andhra Pradesh, followed by Maharashtra, Tamil Nadu, Odisha, Karnataka, and Kerala [23].

12.4.13 Vanilla

Vanilla belongs to the member of Orchidaceae with potential economic values due to its unique flavor and pleasant aroma of its fruits. Vanillin is an aromatic substance obtained from the unripe fruits of *Vanilla planifolia*. This plant is native to South-Eastern Mexico and Central America. *Vanilla tahitensis* (Tahitian vanilla) and *Vanilla pompon* are also cultivated in some parts of the South Pacific Islands for vanillin extraction. Vanilla was introduced in India during 1990s for covering an area of 90 hectares. At present, it is widely cultivated in Kerala and Tamil Nadu as an intercrop of coconut, areca nut, and black pepper plantations.

12.5 Importance of Indian Spices

Indian spices possess varieties of phytochemicals, for example, in curry leaves: α-Amorphene, γ-Eudesmol, β-Pinene, α-Terpinolene, Limonene, (Z)-β-Ocimene. Terpenes, Sesquiterpenes, Eucalyptol, Terpinyl acetate are found in bay leaves. In hemp seeds, Cannabinol, Tetrahydrocannabinol, Cannabidiol, Tetrahydrocannabivarin, Cannabivarin, Cannabichromene are found. Black pepper seeds contain Piperine, Guineensine, Piperamide, and PiperoleinB [43]. Some pharmaceutically important compounds are Eugenol (allspice, cinnamon, cassia, clove), Piperine (black pepper), Gingerol (ginger), Myristin (nutmeg), Curcumin (turmeric), and Vanillin (vanilla) (Table 12.2).

These are phytochemical compounds that are accumulated in the usable spice and condiments and are essential for the quality of the spices [44]. Almost all the spices and condiments are aromatic in nature, and it is by their respective volatile phytochemical usually essential oils with composition of several terpenoids and oleoresins. The essential oils and oleoresins help to impart specific aroma to the food, act as antioxidants, are active against microbes, preserve from chemical deterioration, and increase shelf-life of the food. They act as natural preservative. Quality of the spice is determined by the quality of respective phytochemicals present in them. Indian spices and condiments possess quality aromatic chemicals with superior quality [45]. Flavonoids and essential oils present in spice and

Table 12.2 Important flavor compounds in selected Indian spices and condiments.

Spice	Important flavor compound(s)
Allspice	Eugenol, β-caryophyllene
Anise	(E)-Anethole, Methyl chavicol
Sweet basil	Methyl chavicol, Linalool, Methyl eugenol
Bay laurel	1,8-Cineole
Black pepper	Piperine, S-3-Carene, β-caryophyllene
Caraway	d-Carvone, Carvone derivatives
Cardamom	α-Terpinyl acetate, 1-8-Cineole, Linalool
Chilli	Capsaicin, Dihydrocapsaicin
Cinnamon cassia	Cinnamaldehyde, Eugenol
Clove	Eugenol, Eugenyl acetate
Coriander	d-Linalool, C10-C14-2-alkenals
Cumin	Cuminaldehyde, p-1,3-Mentha-dienal
Dill	d-Carvone
Fennel	(E)-Anethole, Fenchone
Ginger	Gingerol, Shogaol, Neral, Geranial
Mace	α-Pinene, Sabinene, 1-Terpinen-4-ol
Marjoram	e- and t-Sabinene hydrates, 1-Terpinen-4-ol
Mustard	Allyl isothiocyanate
Nutmeg	Sabinene, α-Pinene, Myristicin
Oregano	Carvacrol, Thymol
Origanum	Thymol, Carvacrol
Parsley	Apiole
Peppermint	1-menthol, Menthone, Menthofuran
Rosemary	Verbenone, 1-8-cineole, Camphor, linanool
Saffron	Safranal
Sage, Clary	Salvial-4 (14)-en-1-one, Linalool
Savory	Carvacrol
Spear mint	1-Carvone, Carvone derivatives
Tarragon	Methyl chavicol, Anethole
Thyme	Thymol, Carvacrol
Turmeric	Turmerone, Zingiberene, 1,8-Cineole
Vanilla	Vanillin, p-OH-Benzyl-methyl ether

condiments act against a range of human illness including cancer, cardiovascular disease, nervous diseases, and digestive problems [46]. Gingerol in ginger is also an intestinal stimulant and promoter of the bioactivity of drugs. Capsaicin in chilli is an effective counterirritant used in both pharmaceuticals and cosmetics. Fenugreek, onion, and garlic help to lower cholesterol levels. A number of spices have also been identified as having antimicrobial properties. All these compounds are rich in Indian spices [47].

Selection of germplasm for cultivation is an important step for maintaining the quality of spice species. The classic sage is *Salvia officinalis* but widely traded *Salvia triloba* and *Salvia tomentosa* are commercial sages. Similarly, thyme is usually referred to as *Thymus vulgaris*, but most thymes traded are a mixture of *Thymus capitatus*, *Thymus serpyllum*, and *T. vulgaris*. The well-known turmeric is botanically defined as *C. longa*, but there are many subspecies such as Alleppy turmeric and Cuddapah turmeric used to market with varied qualities. The qualities of spices determined the trade niche. Hence, several spices are gathered from their natural habitat where mixing of their coped one and wild varieties occurred in that location [48].

Currently, biomedical efforts are focused on their scientific merits, to provide science-based evidence for the traditional uses and to develop either functional foods or nutraceuticals. The Indian traditional medical systems use turmeric for wound healing, rheumatic disorders, gastrointestinal symptoms, deworming, rhinitis, and as a cosmetic. Studies in India have explored its anti-inflammatory, cholekinetic, and antioxidant potentials with the recent investigations focusing on its preventive effect on pre-carcinogenic, anti-inflammatory, and anti-atherosclerotic effects in biological systems. Both turmeric and curcumin were found to increase detoxifying enzymes, prevent DNA damage, improve DNA repair, decrease mutations rate and tumor formation, and exhibit antioxidative potential [49]. Recently several molecular targets have been identified for therapeutic or preventive effects of turmeric. Fenugreek seeds, a rich source of soluble fiber used in Indian cuisine, reduce blood glucose and lipids and can be used as a food adjuvant in diabetes. Similarly garlic, onions, and ginger have been found to modulate favorably the process of anti-carcinogenesis [50].

12.6 Spice Plantation and Cultivation in India

India produces 75 varieties of spices of the 109 listed by ISO and is also the world's largest producer. According to the Spices Board of India, every state of India is cultivating at least 10 varieties of spice plants, which are utilized in domestic use and export. The black pepper production in Kerala (94.1%) and followed by Karnataka (3.6%) contributed major production in India [3]. India is particularly suitable for cultivation of many spices due to existing various soil-related parameters such as soil moisture, water content, soil temperature, and soil nutrients, environmental parameters such as climate and rainfall [51].

As a result of diverse agroclimatic conditions, India produces more than 20 seed spices. Cumin, coriander, dill seeds, fenugreek, and fennel are the major seed spices cultivated in the country. Different states are known for different spices, but seed spices are mostly grown in Rajasthan and Gujarat with more than 80% contribution. [52]. India has been producing 7.07 million tonnes of spices, coming from 3.52 mha area during 2015–2016. Seed spices play a crucial role in Indian spice economy contributing 50.31 and 21.30% area and production share to nation's total spices. Individually chillies, cumin, coriander, garlic, and fenugreek are the largest grown spices in the India with

23.05, 22.79, 17.48, 8.39, and 6.42% area share to total spice and 21.88, 7.25, 8.06, 23.07, and 3.56 % production share to total spices production, respectively [53]. Tamil Nadu state occupies the second position in both production and area under turmeric among all the states in India after Odisha [54]. Small cardamom is cultivated in the states of Kerala, Karnataka, and Tami Nadu in India about 50,000 hectares intercropping with black pepper plantations [55]. In India, cardamom is being cultivated majorly in Cardamom hill range of Western Ghats covering an area of 1050 km^2 designated as Cardamom hill reserves [41].

Mustard cultivation is also being extended to nontraditional areas of southern states such as Karnataka, Tamil Nadu, and Andhra Pradesh. It is cultivated in 26 states in the northern and eastern plains of the country, about 6.8 mha is occupied under these crops. Rajasthan is the giant mustard-growing state and alone contributes 43% of the total mustard seed production in India. The cultivation of brown sarson, which once dominated the entire rapeseed–mustard-growing region is now shadowed by Indian mustard (*B. juncea*). There are two different ecotypes of brown sarson (*Brassica rapa* var. brown sarson) and lotni (self-incompatible). The lotni is predominantly cultivated in colder regions of the country particularly in Kashmir and Himachal valley. Yellow sarson (*Brassica rapa* var yellow sarson) is now mainly grown in Assam, Bihar, north-eastern states, Odisha, eastern Uttar Pradesh, and West Bengal [56].

A recent analysis showed that growth and instability are there in spice cultivation in India. For example, chilli registered a higher production and yield growth in Andhra Pradesh, Karnataka, and at all India level. The area growth was 1.32% per annum in Andhra Pradesh and negative in Karnataka (−1.05%), Maharashtra (−1.85%), and at all India level (−0.06%). However, West Bengal registered a positive and significant growth in area, production, and yield of chilli during 2008–2009. Black pepper has registered a substantial growth in both area and production in Karnataka and Kerala states and also at all India level. Interestingly, yield growth was found to be negative in Karnataka. Turmeric production growth was found to be the highest in Andhra Pradesh (7.58%) and about 4–5% in Karnataka, Odisha, and at the national level, while it is Odisha that registered highest growth in turmeric yield [57].

12.7 Cultivation Technology of Caper Bud in India

Capers (*Capparis spinosa*) of commerce are immature flower buds, which have been pickled in vinegar or preserved in granular salt, and it is a new spice crop for India. Semi-mature fruits (caperberries) and young shoots with small leaves may also be pickled for use as a condiment. Capers are said to reduce flatulence and to be antirheumatic in effect. In Ayurvedic system, capers are recorded as hepatic stimulants and protectors, improving liver function. Capers have reported uses for arteriosclerosis, as diuretics, kidney disinfectants, vermifuges, and tonics. Infusions and decoctions from caper root bark have been traditionally used for dropsy, anemia, arthritis, and gout. Capers contain considerable amounts of the antioxidant bioflavonoid rutin [58]. Capers probably originated from dry regions in west or central Asia. Known to be used millennia, capers were mentioned by Dioscorides as being a marketable product of the ancient Greeks. In India, the plant is growing in wild almost in all the states of dry and arid regions. Indians are not aware of the use of caper buds as spice except some local medicinal uses [59]. Recently the plant is under cultivation in Tamil Nadu after recognizing its trade importance.

Ecological conditions are needed for the caper plants in dry heat and intense sunlight that make the preferred environment for caper plants. Plants are productive in zones having 350 mm annual precipitation (precipitating mostly in winter and spring months) and easily survive summer temperatures higher than 37 °C (100 °F). Caper plants are small shrubs and may reach about 1 m upright. However, uncultivated caper plants are more often seen hanging, draped, and sprawling as they scramble over soil and rocks (Figure 12.2a and b). The caper's vegetative canopy covers soil surfaces, which help to conserve soil water reserves. Leaf stipules may be transformed into spines. Flowers are born on year-old branches. The cultivated variety is thornless and developed through in vitro technique by selecting somaclonal variants. In Tamil Nadu, the stem cuttings were procured from the Argentina and the variety is known as "inermis."

Caper plants propagated usually through the stem cuttings. Cuttings are collected in February, March, or April. Basal stem portions, greater than 1 cm diameter and 8 cm in length with 6–10 buds, are used. A dip in IBA solution of 1.5–3.0 ppm is recommended (15 seconds). A 70% rooting percentage would be considered good. Transplanting is carried out during the wet winter and spring periods, and first-year plants are mulched with stones. In India, plants are spaced 1–1.5 m apart (depending on the roughness of the topography; about 2000 plants per hectare). A full yield is expected from second or three year. Plants are pruned back in winter to remove dead wood and water sprouts. Pruning is crucial to high yield. Heavy branch pruning is necessary, as flower buds arise on one-year-old branches. Three-year-old plants will yield 1–3 kg of caper flower buds per plant. The unopened flower buds should be picked on a dry days. Harvesting is carried out regularly throughout the growing season. Caper flower buds (Figure 12.2c) are collected by hand mechanically about every 8–12 days, resulting in 9–12 harvest times per season. Immature flower buds are preserved either in vinegar or under layers of salt in a jar (Figure 12.2d). Raw capers are bland flavored and need to be cured to develop their piquant flavor. Mechanized screens are used to sort out the various-sized capers after being hand-picked from the cultivation. After processing the dried flower buds are preserved in air-tight packs and are ready for exports.

12.8 Export of Indian Spices

Spices trade is a big business from time immemorial. Spices from India and far Eastern Asia were in demand from ancient times. Caravan carried them from India and China to the ports of the Mediterranean Sea or the Persian Gulf and then to the marketplaces of Athens, Rome, and other cities, where they were sold at exorbitant Prices. When the Mongols and Turks cut off overland trade routes from Asia, the European demand for spices was a major factor in motivating a search for new trade routes around Africa and across the Atlantic and Pacific oceans. The high price obtainable for spices was partially responsible for the bitter rivalry of European powers for the control of spice-producing areas and of trade routes. Even after adequate supplies of spices were found and means of

Figure 12.2 Caper bud form in Tamil Nadu. (a) Farm field, (b) *Capparis spinosa* – caper plant, (c) fresh caper buds, (d) processed Caper buds in US market (Whole Foods Market, Gainesville, FL, dt. 24th May 2020). *Source:* Santosh Kumar Upadhyay.

transportation made available, the cost long remained very high in Europe and America. This was largely because of the transportation costs, expenses incident to attempts to retain monopoly of markets and to limit crops in order to secure high prices. The Indian spices are having high demand in the global market due to their geographically indicative quality principles [3].

India as "the land of spices" plays a significant role in the global spices as quality spices come from Kerala. Some of the spices such as chillies, coriander seeds, cumin, fenugreek, fennels are coming from other parts of India. At present, India produces around 2.75 million tonnes of different spices valued at approximately 4.2 billion US$ and holds the premier position in the world spices market. Climatic diversity of India is from tropical to subtropical with varied temperature gradient (45–0 °C), suitable for growing several spice crops in this country. In almost all of the 28 states and eight union territories of India, at least one spice is grown in abundance. About 4,00,000 tonnes equivalent to $1.5 billion worth of spices are traded worldwide annually [12]. In 2009–2010, the export of spices from India has been 502750 tonnes valued Rs. 5560 crores [60]. Though every state of the country grows at least a few spice crops, the states of peninsular India and bordering region, Kerala, Andhra Pradesh, Odisha, Gujarat, Maharashtra, West Bengal, Karnataka, Tamil Nadu, and Madhya Pradesh are the major states in spice production. Odisha grows several spices such as chillies, ginger, turmeric, coriander, garlic in the area of 2.37 lakh ha with a production of 2.17 lakh tonnes. Turmeric and ginger are two most important spice crops grown in Odisha and more than 50% of these crop growers are tribals [8].

Initially, the trade of export was mostly confined with black peeper, cumin, and cardamom, but later covered a number of spices. Out of 109 spices listed by ISO, about 75 are produced by India. The average export touches around 550 thousand tonnes of spices annually. Major share is contributed by chilly, ginger, cumin, and turmeric in terms of volume. Other important products are cardamom, black pepper, and nutmeg. In terms of value, mint and mint products stand first followed by chilly and oleoresins. In the past 10 years, the Indian spice exports increased substantially in terms of volume and value [61].

Among the agricultural commodities, total export contributes around 7% and spices exports contribute around 3% in India [11]. In world trade of spices, India is at number three with 8.8% of the share. The major spices exported by India are chillies (40%), turmeric (10%), cumin (10%), coriander (9.5%), fenugreek (4.2%), black pepper (4%), and others constitute up to 19%. Though these spices provide innumerable benefits, they should be used sparingly [3]. India is a leading producer, consumer, and exporter of black pepper in the world. Andhra Pradesh leads in chilly and turmeric production in the country with 49 and 57%. In coriander, cumin, and fenugreek production, Rajasthan stands as the largest producer in the country with 63, 56, and 87% of domestic production [62].

Turmeric of India has tremendous demand in the market of foreign countries. India was the largest producer (4.35 lakh tonnes) and exporter (24900 tonnes) of turmeric during 1996–1997. The yellow coloring chemical 'Curcumin' is gaining wider use in the food industry, pharmaceuticals, in preservatives, and in health and body care. The Alleppey Finger Turmeric (AFT) with a curcumin content of more than 5.5% is in great demand abroad. India exports turmeric to the foreign countries such as UAE, Iran, Japan, Bangladesh, and South Africa. India exported about 6580 tonnes of turmeric worth Rs. 2296 crores during2009–2010. Though, growth and instability fluctuated in area, production, productivity, and export of major seed spices, namely cumin, coriander, and fenugreek, grown along with total spices from 1985 to 2015 [53].

The main export centers in India are Spice Board, Cochin, Kerala; Agricultural and Processed Food Products Export Development Authority, New Delhi; All India Spices

Exporters Forum (AISEF), Cochin, Kerala; Directorate of Marketing and Inspection, Faridabad, Haryana.

12.9 Conservation Efforts Against Selected Uncultivated Wild Spices and Condiments

India has a good amount of diversity in spices such as black pepper, cardamom, ginger, turmeric, cinnamon, tamarind and, garcinia. The other important spices relevant to India are coriander, fennel, fenugreek, cumin, nutmeg, clove, and vanilla. Apart from these common spices and condiments, some of the plants and plant products are used for preservation and flavoring of foods by traditional knowledge or folklore system. These spice or condiment plants are available locally and use of the plants in such a way for purpose for spices in addition to medicinal and local therapeutic applications. India is a land of spice, where several spice genera originated in the diversity hotspots of either The Himalayas or The Western Ghats. Considering gingers (Zingiberaceae), reports indicate that around eight tuber-bearing, one stoloniferous, and 16 non-tuberous species of gingers are found in The Western Ghats and most of the genera are indigenous to India [63].

Nearly 25% of spice and condiment plants are growing in wild and collected from local vegetation. Several local spices are wild relatives of the known valuable spice, for example, *Allium*, *Alpinia*, *Cinnamomum*, *Curcuma*, *Hedychium*, *Zanthoxylum*, etc. Wild germplasms of cultivated spice plants and wild relatives of the usable spice species are much important for the improvement and development of quality spice varieties [62]. Many research centers are actively engaged in collection, cultivation, and agronomic development of wild relatives of the usable spice plants. Collection and conservation of valuable wild spice species in Indian region have become imperative due to increasing interest in herbal spices for health care across the globe. India alone contributes 140 species and condiments plant species used locally as well export to other countries from which 60 spices are used in industries for making drugs [27, 28].

12.10 Institutions and Organization Dedicated for Research and Development in Spices and Condiments in India

The Indian Institute of Spices Research (IISR), Kozhikode (Calicut), a constituent body of Indian Council of Agricultural Research (ICAR), is a major institute devoted to research on spices and condiments. Regional station of the Central Plantation Crops Research Institute (CPCRI), Kasaragod, engaged in research on spices since 1976. A National Research Centre for Spices was established in 1986 with its headquarters at Kozhikode, Kerala, by merging the erstwhile Regional Station of CPCRI at Kozhikode and Cardamom Research Centre at Appangala, Karnataka. Realizing the importance of Spices Research in India, this Research Centre was upgraded to Indian Institute of Spices Research in 1995. The National Research Centre on Seed Spices, at Ajmer, Rajasthan, is an apex center of ICAR working on improvement of seed spices and betterment of their stakeholders since its inception in 2000. Government of India has notified the formation of 10 spice development agencies (SDAS) in the main spice-growing regions for the overall development of spices grown in the region. These agencies will be chaired by the Chief Secretary of the concerned State Government and consist of members from Ministry of Commerce and Industry, the State/Central Agriculture/Horticulture Ministry, other related Central/State organizations, Agri university, Member of Spices Board from the region, and various stakeholders of the industry, namely growers, traders, and exporters of spices. Indian government has also formed the "Saffron Production and Export Development Agency" (SPEDA) for the overall development in the saffron industry in the state of Jammu and Kashmir in 2015. SPEDA acts as a subordinate agency under Ministry of Commerce.

Other institutions and research organizations in India for development of spice and condiments crops are Indian Cardamom Research in Myladumpara, Kerala; Jawaharlal Nehru Tropical Botanical Garden and Research Institute, Palode, Kerala; Cardamom Research Centre (ICAR), Appangala, Karnataka; Central Institute of Medicinal and Aromatic Plants (CSIR), Lucknow, Uttar Pradesh; Central Food Technological Research Institute (CSIR), Mysore, Karnataka; Several state- and central-funded agriculture, science, and technical universities are also engaged in spices research, with focus on localized species concerned with localized problems and development.

12.11 Recent Researches on Spices and Condiments

Plants used as spices and condiments are usually aromatic and pungent. Most of the spice plants have been reported that they owe these properties to the presence of varying types of essential oils. Several studies indicated that the rich presence of essential oils and oleoresins determines the aromatic, flavoring, coloring, and pungent properties of spices and condiments [2] (Table 12.2). Indian spices are "part and parcel" of traditional health care for thousands of years. Over the couple of decades, largely due to the growth in popularity of complementary and alternative medicine, spices have regained attention due to their physiological and functional benefits. By applying modern research methods to traditional remedies, it is possible to discover what made these spices such effective ailment treatments [64].

Low productivity in the spice sector is one of the serious problems facing the Indian spice industry, resulting in low competitiveness in the international markets. The rapid disappearance of some indigenous varieties of spices due to mixing of planting material results in loss of genetic purity; examples are varieties contributing to the production of Cochin ginger (namely Kuruppampady, Ellackal), Alleppey finger turmeric (namely Elanji), and Byadagichilli [65]. Major challenges facing seed spices production in India are low productivity, lack of high-yielding varieties. Such challenges should be overcome by developing fertilizer-responsive varieties and developing pests and disease resistance. Introduction of germplasm of spices from their centers of diversity both in general and for specific characters will help in broadening the genetic base and for attempting meaningful crop improvement programs [66]. India being the primary and secondary center of diversity for many spices, the genetic variability in major spices such as black pepper, cardamom, cumin, coriander, nigella can be addresses by assessing diverse germplasm and integrating beneficial agronomic trait into cultivable varieties. Genetic resource enhancement, its evaluation, and valuation for effective use to meet the challenges of biotic and abiotic stresses to sustain the impact of climate change are also the need of the hour besides yield quality and nutritional value.

12.12 Conclusion and Future Perspectives

Spices are rich source for phytochemicals having precise health benefits. Spices are used individually or in combination as food adjuncts to impart flavor, color, and aroma. Traditional knowledge prevailing in countries such as India has shown the medicinal properties of many spices for treating wounds, cough and cold, fever, hyperglycemic, and hyperlipidemic conditions. Some of the important bioactive compounds of spices that are shown to possess medicinal value include curcumin from turmeric, capsaicin from red pepper, piperine from black pepper, eugenol from cloves, allyl sulfides from garlic and onion. These compounds are shown to possess antioxidant, anti-inflammatory, antimicrobial, hypolipidemic, and anti-lithogenic activities and pro-anti-cancer properties as well. There is need to increase the productivity to fulfill the domestic requirement and for export of valued spices and condiments. It has been observed that technological interventions such as rhizome treatment, soil application of biocontrol agents, manures, fertilizers, micronutrients, crop rotation, mulching, and correct form of plant protection measures increase the yield performance. Reintroducing the germplasm from the centers of diversity in crops such as paprika, nutmeg, clove, allspice, and vanilla will improve our genetic stocks for future utilization. Popular spices such as lavender, anise, dill, oregano, marjoram etc. need to be introduced to help in diversifying Indian spices production. Introduction of spices species to nontraditional areas must be done in order to conserve the germplasm. To recognize the climate-dependent high-yielding spice varieties, new improved spice crops for Indian climates are to be developed through modern research tools. Introduction of suitable modern hybrids, considering the market demand and more emphasis on utilization of exotic wild species in breeding programs, is to be considered. To enhance the productivity of spices and condiments, integration of advanced farming practices with novel technologies and coupling it with ecofriendly traditional production technologies among the farming community is the need of the hour.

Acknowledgments

No funding has been received to conduct this work.

Authors' Contribution

SK and TRR designed the work. SK did major contributions, compiled the information, and written the manuscript. TRR did minor contributions and formatted the manuscript.

References

1 Rathore, M.S. and Shekhawat, N.S. (2008). Incredible spices of India: from traditions to cuisine. *Am.-Eurasian. J. Bot.* 1 (3): 85–89.
2 UNIDO, FAO (2005). Herbs, Spices and Essential Oils. Post-harvest Operations in Developing Countries. United Nations Industrial Development Organization (UNIDO), Food and Agriculture Organization (FAO).
3 Dubey, S. (2017). Indian spices and their medicinal value. *Indian J. Pharm. Educ. Res.* 51: 5330–5332. https://doi.org/10.5530/ijper.51.3s.41.
4 ISO, International Standards Organization (1993). 'Saffron (Crocus sativus L.)', ISO 3632-1 and 3632-2, 1e. Geneva, Switzerland: International Standards Organization.
5 Rai, P.K. and Lalramnghinglova, H. (2011). Ethnomedicinal plants of India with special reference to an Indo-Burma hotspot region: an overview. *Ethnobot. Res. Appl.* 9: 379–420. https://doi.org/10.17348/era.9.0.379-420.
6 Mao, A., Hynniewta, T., and Sanjappa, M. (2009). Plant wealth of Northeast India with reference to ethnobotany. *Indian J. Tradit. Knowl.* 8: 96–103.
7 Gudade, B.A., Babu, S., Deka, T.N. et al. (2015). Spices biodiversity and their ethnomedicinal uses by tribal community of Sikkim. *India Vegetos* 28: 141–145. https://doi.org/10.5958/2229-4473.2015.00019.1.
8 Babu, N., Srivastava, S.K., and Agarwal, S. (2013). Traditional storage practices of spices and condiments in Odisha. *Indian J. Tradit. Knowl.* 12: 518–523.
9 Mathew, J. (2013). indigenous aromatic and spice plants described in Van Rheed's Hortus Indici Malabarici. *Indian J. Appl. Res.* 3: 30–33. https://doi.org/10.15373/2249555X/NOV2013/10.
10 Skaria, B.P., Joy, P.P., Mathew, G., and Mathew, S. (2005). Spices of Kerala in Taraditional medicine. *National Seminar on Recent Advances in Medicinal and Aromatic Plants*, Meerut, India (15–17 February 2005).
11 Parthasarathy, V.A. and Kandiannan, K. (2007). Horticulture: Vegetable Science: Spices and Condiments. Calicut, India: Indian Institute of Spice Research (ICAR).
12 Sugasini, D., Yalagala, C.R.P., Kavitha, B. et al. (2018). Indian culinary ethnic spices uses in foods and palate of paradise. *Acta Sci. Nutr. Health* 2: 22–28.
13 Singh, A.K., Verma, N., Tyagi, V., and Dimree, S. (2010). Indian needs of crop genetic resources setting priorities. *Prog. Agric.* 10: 1–16.
14 Mohan, S.S., Rajan, S.S., and Unnikrishnan, G. (2013). Marketing of Indian spices as a challenge in India. *Int. J. Bus. Manage. Invent.* 2: 26–31.
15 Patil, D.A. and Dhale, D.A. (2013). Spices and Condiments Origin, History and Applications. New Delhi: Astral International Pvt. Ltd.
16 Sopher, D.E. (1964). Indigenous uses of turmeric (Curcuma domestica) in Asia and Oceania. *Anthropos* 59: 93–27.
17 Velayudhan, K.C., Muralidhran, V.K., Amalraj, V.A. et al. (1999). Curcuma genetic resources, Monograph No.4. National Bureau of Plant Genetic Resources Regional Station, Thrissur, Kerala, India, pp. 1–27.
18 Govindarajan, V.S., Narasimhan, S., Raghuveer, K.G. et al. (1982). Cardamom – production, technology, chemistry, and quality. *CRC Crit. Rev. Food Sci. Nutr.* 16: 229–326. https://doi.org/10.1080/10408398209527337.

19 Adegoke, G.O., Rao, L.J.M., and Shankaracharya, N.B. (1998). A comparison of the essential oils of *Aframomumdaniellii* (Hook. F.) K. Schum. and *Amomum subulatum* Roxb. *Flavour Fragr. J.* 13: 349–352.

20 Nair, K.P.P. (2011). The agronomy and economy of cardamom (*Elettaria cardamomum* M.): the "Queen of Spices". In: Agronomy and Economy of Black Pepper and Cardamom (ed. K.P.P. Nair), 109–366. Elsevier https://doi.org/10.1016/B978-0-12-391865-9.00002-5.

21 Divakaran, M., Jayasree, E., Nirmal Babu, K., and Peter, K.V. (2018). Legacy of Indian spices: its production and processing. In: Indian Spices (ed. A.B. Sharangi), 13–30. Cham: Springer International Publishing https://doi.org/10.1007/978-3-319-75016-3_2.

22 Shah, N.C. (2008). Ethnobotany in India. In: Encyclopaedia of the History of Science, Technology, and Medicine in Non-Western Cultures (ed. H. Selin). Dordrecht: Springer https://doi.org/10.1007/978-1-4020-4425-0.

23 Velayudhan, K.C., Dikshit, N., and Nizar, M.A. (2012). Ethnobotany of turmeric (*Curcuma longa* L.). *Ind. J. Trad. Know.* 11: 607–614.

24 Ali, S.S., Kasoju, N., Luthra, A. et al. (2008). Indian medicinal herbs as sources of antioxidants. *Food Res. Int.* 41: 1–15. https://doi.org/10.1016/j.foodres.2007.10.001.

25 Sharma, P. and Mujumdar, A.M. (2003). Traditional knowledge on plants from Toranmal plateau of Maharashtra. *Indian J. Tradit. Knowl.* 2: 292–296.

26 Reddy, K.N., Pattanaik, C., Reddy, C.S., and Raju, V.S. (2007). Traditional knowledge on wild food plants in Andhra Pradesh. *Ind. J. Trad. Know.* 6: 223–229.

27 Bharali, P., Sharma, M., Sharma, C.L., and Singh, B. (2017). Ethnobotanical survey of spices and condiments used by some tribes of Arunachal Pradesh. *J. Med. Plants Stud.* 5: 101–109.

28 Bharali, P., Sharma, C.L., Singh, B., and Sharma, M. (2017). Ethnobotanical studies of spice and condiment plants used by some communities of Assam. *Int. J. Adv. Sci. Res.* 3: 01–11.

29 Tayade, S.K. and Patil, D.A. (2010). Ethnomedicinal applications of spices and condiments in Nandurbar district (Maharashtra). *J. Ecobiotechnol.* 2: 8–10.

30 Bhatia, H., Sharma, Y.P., Manhas, R.K., and Kumar, K. (2018). Traditionally used wild edible plants of district Udhampur, J&K, India. *J. Ethnobiol. Ethnomed.* 14: 73. https://doi.org/10.1186/s13002-018-0272-1.

31 Salam, S. and Jamir, N.S. (2016). Common spices plant used as medicine by the Tangkhul Tribe of Ukhrul District, Manipur. *Indian Int. J. Sci. Res. Pub.* 6: 22–25.

32 Rai, P.C., Sarkar, A., Bhiyet, R.B., and Das, A.P. (1998). Ethnobotanical studies in some fringe areas of Sikkim and Darjeeling Himalayas. *J. Hill Res.* 11: 12–21.

33 Chakraborty, S. and Chaturvedi, H.P. (2015). Some wild edible genetic resources of vegetables and spices of Tripura. *Indian Res. J. Gent. Biotechnol.* 7: 132–137.

34 Singh, A.G. (2017). Ethnomedicinally important plants used as spice and condiments in the Rupandehi district, West Bengal. *Life Sci. Leafl.* 85: 64–71.

35 Yesodharan, K. and Sujana, K.A. (2007). Wild edible plants traditionally used by the tribes in the Parambikulam Wildlife Sanctuary, Kerala, India. *Nat. Prod. Radiance* 6: 74–80.

36 Ashokkumar, K., Pandian, A., Murugan, M. et al. (2020). Profiling bioactive flavonoids and carotenoids in select south Indian spices and nuts. *Nat. Prod. Res.* 34: 1306–1310. https://doi.org/10.1080/14786419.2018.1557179.

37 Singh, V.B. and Singh, K. (1996). Spices. New Delhi. India: New Age International Ltd.

38 Salma, S. and Ramakrishanan, L. (2017). Study on most frequently consumed spices in households and the consumer's knowledge and preference to organic spices in Coimbatore district, Tamil Nadu, India. *Int. J. Herb. Med.* 5: 49–52.

39 Maridass, M. and Victor, B. (n.d.). Ethnobotanical uses of Cinnamomum species, Tamil Nadu, India. *Ethnobot. Leafl.* 12: 150–155.

40 Baruah, A. and Nath, S.C. (2006). Ethnobotanical evaluation of *Cinnamomum* species used as spice and condiments in northeastern India. *Ethnobotany* 18: 27–36.

41 Ashokkumar, K., Murugan, M., Dhanya, M.K., and Warkentin, T.D. (2020). Botany, traditional uses, phytochemistry and biological activities of cardamom [Elettariacardamomum (L.) Maton] – a critical review. *J. Ethnopharmacol.* 246: 112244. https://doi.org/10.1016/j.jep.2019.112244.

42 Madhusooshanan, K.J., Hrideek, T.K., Kuruvilla, K.M., and Thomas, J. (2004). Mustard – cultivation practices. *Ind. J. Ancient Spices Med. Plants* 5: 157–162.

43 Roy, L. (2019). Phytochemical profile of medicinal Indian spices (Indian medicinal spices). *Int. J. Pharm. Bio Sci.* 10: 105–111. https://doi.org/10.22376/ijpbs.2019.10.4.b105-111.

44 Nirmal, B.K., Ravindran, P.N., and Peter, K.V. (2000). Biotechnology of spices. In: Biotechnology in Horticulture and Plantation Crops (eds. K.L. Chadha, P.N. Ravindran and L. Sahijram). New Delhi: Malhotra Publishing House.

45 Peter, K.V. (1998). Spices research. *Indian J. Agric. Sci.* 68: 527–532.

46 Johnson, I.T. (2000). Anti-tumour properties. In: Functional Foods: Concept to Product (eds. G.R. Gibson and C.M. Williams), 141–166. Cambridge, UK: Woodhead Publishing Ltd.

47 Kumaravel, S. and Alagusundaram, K. (2014). Antimicrobial activity and phytochemical analysis of selected Indian spices. *J. Pure Appl. Microbiol.* 8: 4131–4136.

48 Muggeridge, M. and Clay, M. (2001). Quality specifications for herbs and spices. In: Handbook of Herbs and Spices (ed. K.V. Peter), 13–21. Elsevier https://doi.org/10.1533/9781855736450.13.

49 Manigandan, M. and Saranraj, P. (2018). Antimicrobial activity of secondary metabolites in an Indian spices – a review. *Life Sci. Arch.* 4 (1): 1248–1259.

50 Krishnaswamy, K. (2008). Traditional Indian spices and their health significance. *Asia Pac. J. Clin. Nutr.* 17: 265–268.

51 Kodali, R.K. and Muraleedhar, A. (2015). WSN in spice cultivation. *2015 International Conference on Green Computing and Internet of Things (ICGCIoT). Presented at the 2015 International Conference on Green Computing and Internet of Things (ICGCIoT)* (8–10 October 2015), pp. 1173–1177. Greater Noida, Delhi, India: IEEE. doi:https://doi.org/10.1109/ICGCIoT.2015.7380640.

52 Joshi, D. and Singh, H.P. (2015). An empirical analysis of growth and instability in major spices in India. *Int. J. Agric. Sci.* 7: 440–444.

53 Meena, M.D., Lal, G., Meena, S.S., and Meena, N.K. (2018). Production and export performances of major seed spices in India during pre and post-WTO period. *Int. J. Seed Spices* 8: 21–30.

54 Dhanalakshmi, K., Chitra, K., Menimekalai, R. et al. (2018). Production and economics of turmeric cultivation. *Int. J. Curr. Microbiol. App. Sci.* 7: 3496–3502. https://doi.org/10.20546/ijcmas.2018.711.399.

55 Vijayan, A.K., Pradip Kumar, K., and Remashree, A.P. (2018). Small cardamom production technology and future prospects. *Int. J. Agric. Sci.* 16: 6943–6948.

56 Kumrawat, M. and Yadav, D.M. (2018). Trends in area, production, and yield of mustard crop in Bharatpur region of Rajasthan. *Int. J. Eng. Dev. Res.* 6: 315–321.

57 Krishnadas, M., Mundinamani, S.M., and Guledagudda, S.S. (2009). Growth and instability in production of major spices in India. *Conference: National Workshop on Current Trends and Future Prospects in Production and Export of Spice Crops with Special Reference to Chillies*, Bangalore (1 January 2009).

58 Mishra, S.N., Tomar, P.C., and Lakra, N. (2007). Medicinal and food value of Capparis – a harsh terrain plant. *Ind. J. Trad. Know.* 6: 230–238.

59 Manikandaselvi, S., Vadivel, V., and Brindha, P. (2016). Review on ethnobotanical studies of neutraceutical plant *Capparis spinosa* L. (caper). *Asian J. Pharm. Clin. Res.* 9: 1–4.

60 Parthasarathy, V.A. and Kandiannan, K. (2010). Spices: bio-diversity, of different spices. *Hindu Survey Indian Agric.* 7: 37–39.

61 Tilak, B. (2017). Quality certification of spices. In: Spices Handbook (eds. Shivakumar and Gajendra), 20–25. Bangalore, India: Vidya Herbs. Pvt Ltd.

62 Pedapati, A., Tyagi, V., Singh, A.K. et al. (2014). Spices and condiments: status of genetic resources and setting priorities for introduction in India. *Environ. Ecol.* 32: 1051–1056.

63 Sabu, M. (2006). Zingiberaceae and Costaceae of South India. Calicut, Kerala, India: Indian Association for Angiosperm Taxonomy.

64 Mishra, N. (2020). Ethnopharmacological Investigation of Indian Spices. USA: IGI Global Publisher.

65 Bhardwaj, R.K., Sikka, B.K., Singh, A. et al. (2011). Challenges and constraints of marketing and export of Indian spices in India. *International Conference on Technology and Business Management*, Dubai, UAE (10–12 April 2011).

66 Peter, K.V. and Babu, K.N. (2006). Genetic resources of spices. *Indian J. Plant Genet. Resour.* 19: 376–387.

13

Plants as Source of Essential Oils and Perfumery Applications

Monica Butnariu

Chemistry & Biochemistry Discipline, Banat's University of Agricultural Sciences and Veterinary Medicine "King Michael I of Romania" from Timisoara, Timis, Romania

13.1 Background

The plants are a valuable source of secondary metabolites (SMs), which due to their biological activity are widely used in the food, pharmaceutical, and cosmetology industries. SMs are derivatives of primary metabolites (PMs) obtained as a result of methylation, hydroxylation, glycosylation, and other biochemical reactions. Although they are very important, MS have a much more limited distribution in nature compared to PMs and are not essential for the functioning of the whole plant. They are not continuously synthesized and are often produced during the non-growth phase of cells.

Among the SMs can be mentioned phenols, phenylpropanoids, flavonoids, essential oils or volatile oils (EOs), terpenes, alkaloids, steroids, lignins, resins, etc. Over time, views on the role of SM have changed significantly. This hypothesis was put forward by A. Kossel in 1891, a German physiologist, considered to be the first to belong to the concept of SMs [1].

According to another definition, SMs have the role of storage (accumulation). Subsequently, it was assumed that all SMs are in fact PMs, because the important processes in which SMs participate are not yet known. The starting point for the development of this idea was the discovery of the shikimic path of biosynthesis of phenolic compounds. Another determinant of SMs derived from the protective role of secondary compounds against various abiotic and biotic stressors. Thus, the definition of SMs has changed over the years.

The most acceptable definition of SMs is that they are naturally occurring substances and play an explicit role in the internal economy of the organism that produces them [2].

The functions of the SMs are diametrically different from those of the PMs, and their main function is to maintain the fundamental processes in plant cells and the survival of the species in unfavorable environmental conditions. The delimitation between primary and secondary metabolism is not absolute, as many of the intermediate PMs play similar roles in SMs. Thus, some amino acids, such as 3-β-alanine and γ-aminobutyric acid, are

Bioprospecting of Plant Biodiversity for Industrial Molecules, First Edition.
Edited by Santosh Kumar Upadhyay and Sudhir P. Singh.

certainly SMs, while sterols are essential structural compounds of many organisms and should therefore be considered PMs. Therefore, the overlap of the biological functions of several components of metabolism creates a complex interrelationship between primary and secondary metabolism. A more precise presentation for the description of SMs is the one proposed by Luckner. According to him, the following particularities are characteristic for SMs: the diversity of chemical structures; limited distribution in plants; formation of enzymes encoded by special genetic material; strict control of SMs biosynthesis by regulating the amount of enzymes and their activity; compartmentalization of enzymes, precursors, intermediates, and products involved in biosynthesis, storage, and breakdown; the expression of the secondary metabolism as an aspect of the cellular specialization or of the cells formed de novo individualized by the integration in the processes of differentiation and development of the producing organs; the lack of importance of SMs for the synthesized cell itself, but essential for the whole organism. Plant PMs and SMs of economic interest have several common characteristics: most of them are non-protein biochemical compounds, and they can be extracted from plant material by steam distillation or by extraction with organic or inorganic solvents. Over the last 20–30 years, SMs analysis has advanced greatly.

The use of modern analytical techniques by researchers, such as chromatography, electrophoresis, the labeled isotope method, and enzymology, allowed them to elucidate the exact structural formulas and biosynthesis pathways of SMs. The SMs are a diverse class of compounds produced by plants, which perform many important functions. Unlike PMs, these compounds are not essential for the basic metabolism of the plant but still perform significant functions that allow the plant to adapt and thrive in its living environment. Because plants do not have certain fundamental characteristics that we encounter in humans and animals, they must have the means to meet additional challenges. All plants taken together produce over 100 000 different varieties of secondary compounds that serve a variety of purposes.

The main categories of SMs in EOs are terpenes and terpenoids, phenols, and alkaloids. The compounds are grouped into one of these categories based on their chemical structure. Terpenes are volatile organic compounds, which are based on a 5-carbon hydrocarbon, known as isoprene. The smallest and most volatile compounds are monoterpenes, which are biosynthesized by joining two isoprene molecules. Larger and less volatile compounds are biosynthesized by joining three or more isoprene molecules. The monoterpenes are followed by sesquiterpenes, which are formed by the union of three isoprene molecules. Terpenes are SMs that give the plant its organoleptic characteristics (aroma and taste) and which make up most of the EOs produced by aromatic plants [3].

13.2 Biochemistry of Essential Oils

13.2.1 The Physiological Mechanism of Biosynthesis of Essential Oils

EOs have a still unknown biological role. It accumulates in various plant organs, but especially in petals, seeds, roots, bark, and wood. These substances are mainly composed of terpene hydrocarbons. Rose petal oil contains an acyclic terpene called citronellol. D-limonene and L-limonene are found in orange peels, lemons, cumin, and celery. In the mint leaves, there are menthol, pine, lemon, in those of basil (ocimen methyl chavicol), and wormwood thion + phellandrene. The scent of chrysanthemum flowers has the following chemical compounds: tetracolsan hexacosan; in lavender, there is lavandulol, ethylamyl ketone linalool, butyrate of isoamyl; chamomile contains a mixture of acetylene and furfuryl paraffins [4].

EOs produced by glandular cells accumulates, as in all *Lamiaceae* species, between the outer wall of glandular cells and the cuticle. This leads to the detachment of the cuticle from the glandular cells and the emergence of a common storage space for EOs.

When the subcuticular space is filled with the secretion product, the glandular brushes, entering the post-secretory phase, degenerate. Only a slight enlargement of the subcuticular space was observed in the capped brushes, in which case the secreted material accumulates in the lumen of the glandular cell. At maturity, the multicellular secretory gland has a large subcuticular space. This cavity will be filled with EOs at the beginning. The cell space that thus arises increases as the oil is secreted and released into this cavity. The secretions pass through the plasma membrane and the wall of the secretory cells and accumulate in this cavity to accumulate the secretion products. When the cuticle ruptures, the secretion products are released. In secretory brushes with a multicellular gland, this subcuticular space exceeds in height the glandular cells themselves. The EOs also exhibit optically active properties due to the content of compounds with asymmetric carbon atoms. Among these compounds are linalool, camphor, pinene, borneol, etc. Each of these compounds is characterized by its own angle of rotation of the plane of polarization, and the value of the angle of rotation of the volatile oil is dependent on the values of the angles of rotation of each component, as well as the percentage content of these compounds in the mixture.

Although EOs contain optically active compounds, the value of the angle of rotation for each oil falls within relatively well-defined limits and can serve to establish the identity of the EOs compounds. EOs are biosynthesized in various organs or tissues of plants, including leaves and stems (e.g. basil, bay, eucalyptus, fennel, lemon grass, mints, oregano, parsley, pine, rosemary, sage, spicebush, tarragon, and wintergreen), seeds (e.g. anise, almond, cumin, and celery), berries (e.g. allspice and juniper), bark (e.g. cinnamon and sassafras), fruit peel or rind (e.g. citron, grapefruit, lemon, lime and orange), roots (e.g. valerian), rhizomes (e.g. ginger), and flowers (e.g. chamomile, clove, geranium, jasmine, lavender, orange, and rose) [5].

13.2.2 The Role of Terpenes in Plants

The two main functions are protection against inspections and herbivores, as well as protection against high temperatures. Plants react by producing terpenes in areas affected by the action of insects and herbivores, and these terpenes acting as bitter compounds or, in some cases, even as pesticides that repel or "attract" (phenolic in nature [e.g. menthol from *Mentha piperita*], volatile alkaloids [e.g. Valerotropes from *Valeriana* off.], resins and gums [e.g. *Humulus lupulus*]). Plants in turn have evolved rather interesting strategies to "attract."

They include flowers that produce differently colored, often hairy "nectar guides" on their petals (e.g. *Iris*); plants that produce ultraviolet pigments that insects see as "bulls-eyes"; various colored petals and/or sepals whose flavonoid and anthocyanin pigments attract specific pollinators; flowers that open only at night when moth-type pollinators are active in flight (as with yucca flowers visited by hawkmoths); flowers that produce a rotten meat smell (due to indoles, skatole, or amines) that attract flies or beetles, as in the case of skunk cabbage and other aroids; flowers that produce pheromones/plant hormones (e.g. ethylene, methyl jasmonate, and methyl salicylate). The flowers produce essential oils as olfactory cues that "attract specific" because of their highly evolved sensing systems.

Monoterpenes (2 isoprene units, so 10 carbon atoms), which are more volatile, dominate the inflorescences (e.g. limonene and menthone from the youngest leaves of peppermint), can acting as a good spatial repellent. Sesquiterpenes (C15-terpenoids built from three isoprene units), which are more bitter, are more abundant on the leaves, acting against herbivores and can good contact repellent (e.g. α-bisabolene in black pepper [*Piper nigrum*] and the candeia tree [*Vanillosmopsis erythropappa*] and β-caryophyllene in ylang ylang [*Cananga odorata*], yomogin [*Artemisia princeps* Pamp.], and nerolidol in genus *Tanacetum* [Compositae]). When plants notice an increase in temperature, they begin to synthesize more terpenes and thus, at higher temperatures, day or night, more terpenes are released.

Terpenes evaporate at high temperatures, producing air currents that cool the plant and decrease perspiration, preventing the plant from drying out. In the case of hemp, terpenes are exuded in the resin and give it a sticky and viscous appearance that will catch and immobilize some insects, thus acting as a protection against insects and high temperatures. Thus, it is easy to notice that the hemp plant smells stronger during the first hours of the morning than during the hottest part of the day, when a larger amount of terpenes evaporates.

This is why it is recommended that mature plants should be harvested in the early hours of the morning in order to have a maximum production of EOs. An important feature of EOs and their compounds is hydrophobicity, which allows them to affect the lipid structure of the bacterial cell membrane and increase their permeability, with cells losing ions and other cellular components. The chemical structure of the individual compounds of EOs affects their specific mode of antibacterial action. The importance of presence of the hydroxyl group in phenolic compounds such as carvacrol and thymol has been confirmed [6].

13.2.3 The Prevalence Essential Oils in Plants

EOs are relatively widespread in the plant kingdom, with some families being very rich in such substances. Among these families, more representative is *Pinaceae, Labiatae, Umbelliferae, Myrtaceae, Lauraceae, Rutaceae, Caryophylaceae, Compositae,*and *Zingiberaceae*. Plants containing EOs are commonly called aromatic plants or hetero-oleaginous plants. Aromatic plants contain EOs in amounts from hundreds of percent to 20–25% relative to dry mass. For example, in the inflorescences of white acacia and lily of the valley, there is about 0.05% EOs, in the rose flowers 0.06–0.20%, in the inflorescences of lavender 1.0–2.0%, in the fruits of fennel 4.0–6.0%, in star anise fruits up to 11.5%, and in the flower buds of the clove tree up to 22%. About 2500 species of aromatic plants are known in the world, of which more than 40% grow in the tropics.

However, of the total number of aromatic plants, only about 200 are of industrial importance. The others are not recovered either due to a low content of EOs, or the unsatisfactory quality of the volatile oil they contain. EOs are spread unevenly in the plant's body, usually concentrating in one or more of the organs. For example, in the rose, the organ in which EOS accumulates is the flower, in mint – leaves and inflorescences, in vetiver – in roots, in bay and eucalyptus – in the leaves and young shoots.

Usually, only the parts of the plant where the content of EOs is maximum is used in industry. Sometimes, the raw material may contain parts of the plant (ballast) in which practically no EOs are contained (mint, geranium, lavender stems, etc., are the ballast in the process of extracting the respective EOs). The content of EOs in this type of raw material depends a lot on the ratio between the parts rich in EOs and ballast.

The quantity and quality of volatile oil in the plant vary depending on a number of endogenous (species, life cycle phase, etc.) and exogenous factors (temperature and humidity of the air and soil, light intensity, etc.). In the body of the aromatic plant, EOs are contained in the free and/or fixed state (in the form of glycosides).

EOs in the free state are located in such structures as: isolated secretory cells – idioblasts (family *Lauraceae*); secretory organs (schizogenous oil glands from the secretory pockets – characteristic of the family *Rutaceae*, *Myrtaceae*); secretory canals (*Pinaceae*, *Umbelliferae*); secretory hairs (*Lamiaceae*, *Compositae*). The form in which EOs are found in the plant raw material influences the storage method and its processing technology. EOs are common aromatic compounds from different parts of plants (flowers, buds, seeds, branches, bark, wood, fruit, or roots). They can be obtained by fermentation or extraction, but the most commercially used method is hydrodistillation. It is estimated that over 3000 EOs are known so far, of which about 300 are of practical importance in the perfume industry [7].

EOs contain a mixture of compounds that are chemically derived from terpenes or their oxygenated compounds. The widest use of EOs in the European Union is in the food (spices), cosmetics, and pharmaceutical industries. Terpenes are volatile organic compounds, which are based on a hydrocarbon with five carbon atoms, known as isoprene.

The smallest and most volatile compounds are monoterpenes, which are biosynthesized by joining two isoprene molecules. Larger and less volatile compounds are biosynthesized by joining three or more isoprene molecules. The monoterpenes are followed by sesquiterpenes, which are formed by the union of three isoprene molecules. Terpenes are SMs that give the plant its organoleptic characteristics (aroma and taste) and which make up most of the EOs produced by aromatic plants. The ratio of monoterpenes to sesquiterpenes in leaves and flowers has been shown to be different. This is due to the predominance of sessile trichomes in the leaves, which are much more specialized in the synthesis of terpenes, while capitate trichomes are much more abundant in flowers and are specialized in the synthesis of volatile terpenes (monoterpenes and sesquiterpenes). Normally, the proportion of terpenes in the plant is less than 1%, reaching of up to 10% of the resin composition [8].

13.2.4 Paths of Biosynthesis of Volatile Compounds in Plants

13.2.4.1 Metabolic Cycles Involved in the Biosynthesis of Different Groups of Secondary Metabolites

SM biosynthesis involves a relatively small number of chemical compounds, which are subsequently modified by various cycles into an unlimited number of secondary substances.

There are four cycles involved in the biosynthesis of SMs, each being responsible for the biosynthesis of different classes of compounds: the shikimic cycle: simple phenols, hydroxycinnamic acids, phenylpropene, phenylpropanoids, flavonoids, lignins, coumarins, naphthoquinones, anthraquinones, tannins; mevalonic cycle and methylerythritol-phosphate cycle: terpenes; acetate cycle: alkaloids. At the same time, according to their biosynthetic origin, SMs are classified into several classes, including the class of phenylpropanoids, terpenes, alkaloids, etc. These cycles are carried out with the involvement of the essential derivatives of acetyl-CoA intermediates, shikimic acid, mevalonic acid, and methylerythritol-phosphate, respectively [9].

13.2.4.2 Metabolic Cycles of Biosynthesis of Phenolic Compounds

13.2.4.2.1 The Shikimic Ccle of Biosynthesis of Phenolic Compounds The shikimic cycle takes place through the cyclization of some pseudo aromatic intermediate compounds (2-phosphoenol-pyruvic acid and 4-phosphoerythrosis), which lead to the formation of the structure of the benzoic ring.

The shikimic cycle includes a key intermediate, namely shikimic acid. It is the main precursor of the biosynthesis of aromatic amino acids, such as L-phenylalanine, L-tyrosine, and L-tryptophan, as well as other compounds such as alkaloids, phenols, and phenylpropanoids. That cycle is common for the biosynthesis of aromatic acids in bacteria, fungi, yeasts, microorganisms, and plants, but not for the animal kingdom. This cycle is essential because it is involved in the generation of structural blocks for the biosynthesis of proteins, vitamins, and electron-carrying compounds, such as co-factors and quinones.

Shikimic acid was first isolated from the fruits of a species native to Japan *Illicium religiosum* Sieb. It is currently the raw material for the synthesis of antiviral oseltamivir, used to treat avian influenza. The main source of extraction of shikimic acid is aniseed fruits (*Illicium verum* Hook F.). The precursors of the synthesis of shikimic acid in the respective cycle-2-phosphoenol-pyruvic acid are formed from 3-phosphoglyceric acid, and 4-phosphatitrose is formed by cleavage of the condensation product between 6-phosphofructose and 3-phosphoglyceraldehyde. Thus, upon condensation of the two precursors, 3-deoxy-D-arabino-2-heptulosonic-7-phosphate (DAHP) is formed [10].

The compound is a combination of carbohydrate residues, which although have a linear chain, the carbon atoms are oriented in space according to a pseudo cyclic structure, which allows the subsequent cyclization of the chain. The cyclization process takes place in several stages, through oxidation and reduction reactions, with the formation of 3-dehydroquinic acid, subsequently entrained in a series of transformations, with the formation of cyclohexane derivatives. In the next step, the dehydration of 3-dehydroquinic acid takes place with the formation of 3-dehydroschemic acid. The latter, under the influence of the oxidative-reducing enzyme shikimat-dehydrogenase, is converted to shikimic acid.

Shikimic acid, unlike aromatic compounds, contains only a double bond. Subsequently, as a result of a series of biochemical conversions, shikimic acid is transformed into horismic acid, another important intermediate, which already has two double bonds. At that stage, the branching of the shikimic cycle takes place. One of the biosynthetic pathways based on horismic acid leads to the formation of L-tryptophan (and other indole derivatives), the second pathway to the formation of L-phenylalanine and L-tyrosine. It is with the second biosynthetic pathway that subsequent transformations are associated, which ultimately lead to the formation of phenolic compounds in plant cells and tissues. This is the end point of the shikimic cycle, being the source of these amino acids, it is also one of the elements of primary cellular metabolism. Secondary specific transformations lead to phenol biosynthesis [11].

13.2.4.2.2 The Metabolic Cycle of Phenylpropanoid Biosynthesis L-phenylalanine and L-tyrosine are important precursors of many SM in higher plants, and the enzymes responsible for redirecting these essential MPs in secondary metabolic cycles are the enzymes L-phenylalanine ammoniacylase (PAL) and tyrosine decarboxylase (TirDC). These precursors form the basic units of C6C3 phenylpropanoids (FPs), which are found in many natural products, such as cinnamic acid, coumarins, lignins, and flavonoids.

Due to the commercial interest, regarding the use of FPs in pharmacological and other industrial fields, their functions and biosynthesis cycle have been intensively studied in many plant species. In the general biosynthesis scheme of FPs, the aromatic amino acid-L-phenylalanine derivative of the shikimic cycle is subsequently subjected to biochemical transformations, which consists in its deamination in the reaction catalyzed by the enzyme PAL (EC 4.3.1.5) in trans-cinnamic acid. The latter is exposed to the para-hydroxylation process with cinnamate-4-hydroxylase (C4H, EC 1.14.13.11), obtaining 4-hydroxycinnamic acid (p-coumaric acid). P-coumaric acid is the first, and from a biogenetic point of view, the simplest phenolic compound, being also the primary compound in the biosynthesis of other plant phenolic compounds. Following the enzymatic transformations of p-coumaric acid in plant cells, the biosynthesis of derivatives of different classes of phenolic compounds takes place. By oxidative reduction of the side chain of p-coumaric acid, acetophenones, phenylacetic acids, and phenylcarbonic acids are formed. The recovery of its side chain, together with the subsequent dimerization or polymerization of the reconstituted product, leads to the formation of lignin and phenolic resins-lignin type [12].

13.2.4.2.3 Metabolic Biosynthesis Cycles of Terpenes Terpenes, also called terpenoids or isoprenes, are the largest class of plant metabolites reported to date, which includes compounds with a wide structural variety.

They are produced by all living organisms. To date, more than 45 000 plant terpenes have been identified and this number is constantly growing. Of the tens or thousands of plant terpenes, only a few can be considered MPs, considered essential for the functioning of the whole plant and are therefore common to all plant species. These include molecules involved in the processes of respiration (ubiquinones), photosynthesis (chlorophylls, carotenoids, tocopherols, phylloquinones, and plastoquinones), regulating the growth and development of the plant (brassinosteroids, cytokinins, gibberellins and abscisic acid), strigol.

The rest of the terpenes are specialized metabolites, whose biosynthesis is usually limited to certain plant families or even certain species. Terpenes usually participate in protecting plants against herbivores and pathogens, in attracting pollinators and dispersing seeds and as allelochemicals, which influence competition between plant species [13].

A large number of terpene metabolites have commercial value and are used as pigments, flavors, polymers, or drugs. Typical terpene structures contain carbon skeletons represented by (C5)*n* units and are classified as hemiterpenes (C5), monoterpenes (C10), sesquiterpenes (C15), diterpenes (C20), sesterterpenes (C25), triterpenes (C30), and tetraterpenes (C40). According to the isoprene rule, all terpene compounds derive from two common precursors containing two isoprene units bound head-to-tail, namely from isopentenyl pyrophosphate (PPI) and dimethylallyl pyrophosphate (DMAPP) (allylic isomer of PPI). In plants, two independent cycles are responsible for the biosynthesis of isoprene units: the mevalonic acid cycle (MVA) and the methylerythritol-phosphate cycle (MEP) according Figure 13.1.

Abbreviations: *PPP*, pentose phosphate pathway; *EDP*, Entner-Doudoroff pathway; *G3P*, glyceraldehyde-3-phosphate; *DXP*, deoxy-D-xylulose 5-phosphate; *MEP*, 2-C-methyl-D-erythritol-4-phosphate; *HMBPP*, 4-hydroxy-3-methyl-but-2-enyl pyrophosphate; *MVA*, mevalonate pathway (mevalonic acid); *HMG*, 3-hydroxy-3-methylglutaryl; *IPP*, isopentenyl pyrophosphate; *DMAPP*, dimethylallyl pyrophosphate; *GPP*, geranyl diphosphate; *FPP*, farnesyl diphosphate; *GGPP*, geranylgeranyl diphosphate; *IDP*, isopentenyl diphosphate; *FDP*, farnesyl diphosphate; *GAP*, glyceraldehyde-3-phosphate; *GGDP*, geranylgeranyl diphosphate; *HMG-CoA*, 3-hydroxy-3-methylglutaryl CoA; *IDP*, isopentenyl diphosphate.

The normal biosynthesis of terpenes in plants depends essentially on the flow of precursors released by the nucleus of the isoprene metabolic cycles and, consequently, on the dynamic regulation of these cycles. The MVA cycle offers predominantly biosynthesis precursors of sesquiterpenes, triterpenes, and brassinosteroids. Respectively, in the biosynthesis of hemiterpenes, monoterpenes, diterpenes, tetraterpenes, carotenoids and their degradation products-cytokinins, gibberellins, chlorophyll, tocopherols, and plastoquinones participate in the MEP cycle derivatives. This cycle is considered exclusively plastid, because the set of enzymes involved in SM biosynthesis exists only in plastids, the

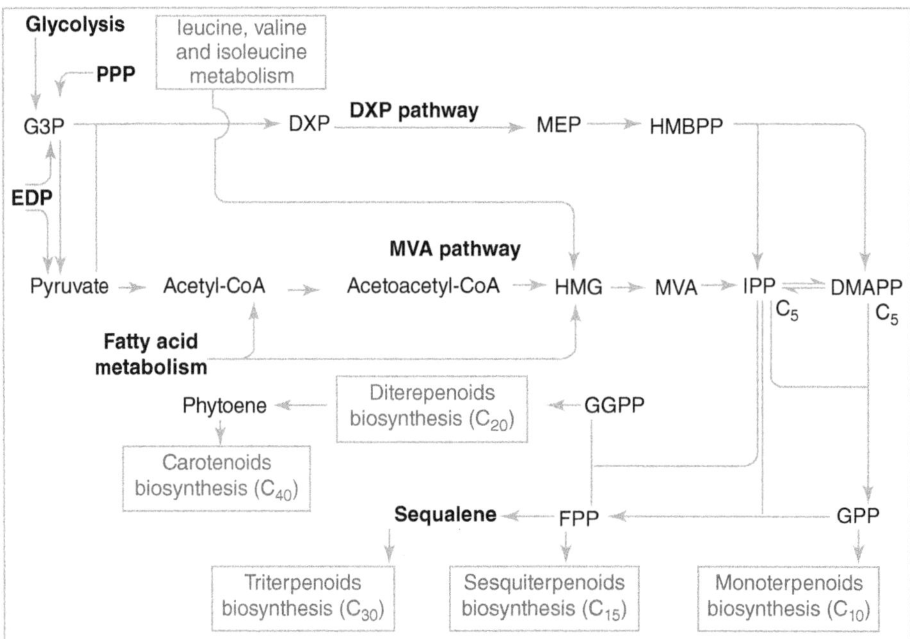

Figure 13.1 Schematic representation of a biosynthesis of terpenes in plants.

data being based on experimental research and predictions of their subcellular location. Unlike the previous one, the subcellular location of the MVA cycle is not as clear. Historically, this cycle has been referred to as the cytosolic cycle, and new claims suggest that the MVA cycle is distributed between the cytosol, the endoplasmic reticulum, and peroxisomes [14].

13.3 The Metabolism Terpenes

The metabolism of terpenes synthesized in the mevalonate cycle has as initial product acetyl CoA. In this cycle, monoterpenes, sesquiterpenes, diterpenes, triterpenes, and tetraterpenes are synthesized, among which we can mention: linalool, limonene, gibberellin, abscisic acid, ubiquinone, carotenoid pigments, etc. The biosynthesis of these substances is genetically coordinated, by coding the synthesis of enzymes that catalyze the reactions in this cycle: phytoen-desaturase, phytoen-synthase, hydroxymethylglutaryl-CoA reductase, geranylgeranyl-pyrophosphate synthase, lycopene cyclase, and capsanthin-capsorubin synthase.

In the first stage of terpene biosynthesis, the formation of 3-hydroxy-3-methylglutaryl-CoA takes place, from three molecules of acetyl CoA, a reaction catalyzed by an enzyme that cofactor with iron and quinone. In the second step, catalyzed by hydroxymethyl-glutaryl CoA reductase, mevalonate is formed. By decarboxylation and phosphorylation, mevalonate results in isopentyl pyrophosphate (C5), from which geranyl pyrophosphate (C10), farnesyl pyrophosphate (C15), and geranylgeranyl pyrophosphate (C20) are formed. These substances are precursors for different terpenes [15].

Thus, monoterpenes are formed from geranyl pyrophosphate: linalool and limonene. Sesquiterpenes and squalene are formed from farnesyl pyrophosphate. From geranylgeranyl-pyrophosphate are synthesized diterpenes, kauren which is the precursor of gibberellins and abscisic acid and phytoene, which is the precursor of carotenoid pigments. Locating at the cellular level, the various stages of the terpene synthesis process is difficult. Isopentyl pyrophosphate can be biosynthesized in all cellular structures where terpenes are synthesized.

The specific site for the synthesis of monoterpenes is plastids. The synthesis of farnesyl pyrophosphate and derived sesquiterpenes, as well as triterpenes including phytosterols, takes place in the cytoplasm and in the endoplasmic reticulum. The biosynthesis of diterpenes takes place in plastids, that is, where the activity of geranylgeranyl-pyrophosphate synthase has been identified. The activity of entkauren synthase, which catalyzes the synthesis of entkauren, the precursor of cytokinins, has been identified in chloroplasts. Also, in chloroplasts, carotenoid pigments and tocopherols are biosynthesized, and ubiquinone is biosynthesized in mitochondria and microsomes.

Terpene biosynthesis takes place in all plant cells. In the case of species that synthesize large amounts of terpenes, their synthesis takes place in specialized cells such as resinous channels in pine leaves or isolated resinous cells in *Thuja*.

In the case of angiosperm plants, monoterpenes are synthesized in glandular hairs on *Mentha* leaves, in flower petals or pistils (e.g. linalool), and latex in latex. Numerous terpenes with 10 or 15 carbon atoms, with a high degree of volatility, are known as EOs and give the characteristic aroma to some plant organs: flowers, fruits, seeds, etc. [16].

Sterols are made up of five isoprene units, such as cholesterol, sitosterol, stigmasterol, and campesterol, and are found in the composition of plasma membranes having a role in regulating their permeability. Some isoprenoid compounds secreted by plant roots are toxic to the roots of other plants, being considered allelopathic substances. The shikimic acid cycle has as substrate erythroso-4-phosphate, produced in the pentosophosphate cycle. The result is phenols, lignins, anthocyanins, and phenolic amino acids such as tryptophan from which auxin is biosynthesized. It is estimated that under normal conditions about 20% of the fixed carbon of plants is used in the shikimat cycle. In this cycle, phosphoenolpyruvate and erythroso-4-phosphate are condensed into a compound with seven carbon atoms: 3 deoxy-D-arabino-heptulosonate-7-phosphate (DHAP), which results in chorismate, as follows:

$$\text{DAHP} \rightarrow \text{3 dihydroquinate} \rightarrow \text{3 dihydroshikimat} \rightarrow \text{shikimat} \rightarrow \text{chorismat.}$$

Chorismate is the starting compound for three cycles: of the first results in phenylalanine, lignins, and flavonoids, of the second tryptophan, auxins, glucosinolates, phytoalexins, and alkaloids, and of the third tyrosine and melanin. The key enzyme in the shikimic acid cycle is phenylalanammonium lyase (PAL), which is bound to the membrane of the endoplasmic reticulum, chloroplasts, mitochondria, and plasma membranes. Mechanical damage and pathogen attack induce the formation of mRNA encoding the synthesis of DAHP synthase, stimulating the shikimat cycle. The enzymes that catalyse the reactions in this cycle are synthesized in the ribosomes in the cytoplasm. The activity of these enzymes has been shown in chloroplasts and its identification in the cytoplasm is uncertain [17].

Some of the amino acids that form in this cycle are in turn precursors of other substances. Tryptophan is the precursor of the growth hormone auxin, phenylalanine is the precursor of flavonoid pigments and lignins, tyrosine is the precursor of ubiquinone, an electron transporter in the process of respiration. In the shikimic acid cycle, which takes place in the endoplasmic reticulum and in the cytoplasm, phenolic acids are also biosynthesized: p-coumaric, cinnamic, caffeic, ferulic, chlorogenic, as well as galotanins. Phenolic substances can stimulate or inhibit the action of hormones, inhibit the synthesis of ATP in mitochondria, as well as the activity of enzymes, or cytoplasmic currents in the cells of absorbent hairs.

Some phenolic substances (ferulic, lunular, chlorogenic acid, and catechins) have an inhibitory effect on seed germination. By oxidizing substances such as tyrosine from the tubers of *Solanum tuberosum*, dopamine from bananas, phenolic acids from apples, in the presence of phenolase black melanin is formed, which gives the characteristic color for mechanically damaged or senescent fruits. Some phenols are allelopathic substances: juglone produced by *Juglans regia*, salicylic acid produced by *Quercus falcata*, and ferulic acid produced by *Adenostoma* etc. Cinnamic acid derivatives have been identified in vacuoles and chloroplasts. Enzymes involved in the biosynthesis of flavonoid substances: chalcone-flavone isomerase and flavonoid hydrolase have been identified in the endoplasmic reticulum and chloroplasts, and flavonoid substances have been identified in vacuoles, extraplasmatic space, and chloroplasts. The biosynthesis of flavonoid substances is genetically coordinated [18].

The synthesis of anthocyanins can be performed in any plant cell, being located in the cytoplasm for monomers, dimers, and trimers and in vesicles, improperly called anthocyanoplasts (intensely pigmented organelles), for the final products. The anthocyanins formed are transported from these vesicles, in vacuole, through a process of pinocytosis, and the membranes of the vesicles can be incorporated into the tonoplast. Organic acids result from the Krebs cycle, and phenolic substances from the shikimic acid cycle.

Alkaloids are nitrogen heterocyclic substances, found in over 13 000 species. Most plant alkaloids come from amines or amino acids and only some come from isoprenoid precursors in which nitrogen is incorporated into a late stage of the biosynthetic cycle. This is the case with solanine (glycoalkaloid) from *Solanum tuberosum* and tomatine and its aglycone, tomatidine from *Lycopersicon esculentum*. Alkaloids have been identified in vacuoles, chloroplasts, and extraplasmatic space. Enzymes involved in the synthesis of these substances have been identified in endoplasmic reticulum membranes, plasmalemma, and tonoplast.

Among alkaloids, nicotine has the best studied biosynthetic cycle [19].

The primary compounds are arginine and ornithine which are decarboxylated and metabolized to the putrescein ($C_4H_{12}N_2$) conjugate form. Nicotine biosynthesis takes place in small vesicles, which come from the endoplasmic reticulum or Golgi bodies and which contain the enzymes involved in this process. The membrane of these vesicles is permeable to tertiary compounds that can be synthesized at other sites. The formed quaternary compounds are released into the vacuole, following its fusion with vesicles. Cyanogenic glycosides such as amygdalin ($C_{20}H_{27}NO_{11}$) and prunasin ($C_{14}H_{17}NO_6$) have been identified in cell vacuoles, and enzymes involved in the synthesis of cyanogenic glycosides have been identified in the membrane of the endoplasmic reticulum. Cyanogenic glycosides have been identified in over 1000 species, 500 genera, and 100 plant families. Of these substances, the best known are amygdalin and prunasin from *Rosaceae* and sambunigrin from *Caprifoliaceae*.

Biogenic amines are widespread in plants both as simple amines (primary, secondary, and tertiary) and as amines with different functional groups (alcoholic, phenolic, carboxylic, etc.). Among the amines, putrescine is important, which is formed from ornithine (*Pisum*, *Nicotiana*), cadaverine is formed from lysine (*Lupinus* and *Pisum*), tryptamine is formed from tyrosine (*Hordeum* and *Lolium*), dopamine is formed from dihydroxyphenyl alanine.

Putrescein, spermine, and spermidine interact with nucleic acids and may be involved in protein biosynthesis. Putrescein, cadaverine, spermine, and spermidine in a concentration of 10^{-4}–10^{-6} M stimulate the growth process, and dopamine is the precursor for the formation of melanoid compounds in bananas. Volatile substances that impart fruit flavor are intermediate compounds of metabolism: alcohols, aldehydes, ketones, esters, ethers, etc., substances with a high degree of volatility [4].

13.3.1 Metabolic Cycle of Mevalonic Acid Biosynthesis

The first step in this cycle is the formation of coenzyme A-(*S*)-3-hydroxy-3-methylglutaryl (HMG-CoA) by the condensation of acetyl-CoA with acetoacetyl-CoA.

The next step involves the irreversible enzymatic reduction of HMG-CoA with hydrogen from nicotinamide-adenine-dinucleotide-phosphate (2 × NADPH) to obtain mevalonic acid (MVA). Subsequently MVA undergoes two successive phosphorylations with adenosine triphosphate (2 × ATP) to form 5-pyrophosphate.

The next step already involves the trans-elimination of the hydroxyltertiary group and the carboxyl group of 5-pyrophosphate, with the formation of PPI. This transformation is in balance with the formation of DMAPP. The key enzymes involved in the MVA cycle are acetoacetyl-CoA thiolase (AACT), HMG-CoA synthetase (HMGS), HMG-CoA reductase (HMGR), and mevalonate diphosphate decarboxylase (MPDC or MVD) [20].

13.3.2 Metabolic Cycle of Methylerythritol Phosphate Biosynthesis

Plasticid isoprenoids are derived from PPIs via the 2-*C*-methyl-D-erythritol 4-phosphate (MEP) ring. The first stage of this cycle involves the condensation of pyruvic acid (2-hydroxypropanoic acid) and D-glyceraldehyde 3-phosphate. The initial reaction is catalyzed by 1-deoxy-D-xylulose 5-phosphate (DXP) synthetase (DXS) to produce DXP. The second step is intramolecular rearrangement and reduction of DXP using the enzyme 1-deoxy-D-xylulose 5-phosphate reductisomerase (DXR) leading to the formation of 2-*C*-methyl-D-erythritol 4-P (MEP). It, in turn, continues to undergo a series of reactions to form PPIs.

The next step is the stereospecific and reversible isomerization of the PPI double bond that leads to the formation of DMAPP. This step is very important because it generates reactive pyrophosphate alkyl, which helps to combine two isoprene units to form geranyl pyrophosphate (GPP and C10). Farnesyl diphosphate (FPP and C15) is derived from two units of PPI and one unit of DMAPP, geranylgeranyl diphosphate (GGPP and C20) is from three PPIs and one DMAPP. The biosynthesis of monoterpenes uses GPP as a substrate to synthesize monoterpenes with linear and cyclic structure; the biosynthesis of sesquiterpenes uses FPP, and of diterpenes-GGPP. The key enzymes involved in the MEP cycle are 1-deoxy-D-xylulose 5-phosphate synthetase (DXS), 1-deoxy-D-xylulose reductoisomerase 5-phosphate (DXR), and (*E*)-4-hydroxy-3-methylbut-2-enyl diphosphate reductase (HDR) [21].

13.4 The Role of Essential Oils and the Specificity of Their Accumulation in Plants

Although only 20–30% of higher plants have been investigated so far, tens of thousands of SMs have already been isolated, and their structures have been determined by mass spectrometry (MS), nuclear magnetic resonance (1H NMR, ^{13}C NMR), or single crystal X-ray diffraction. The role of SM has long been overlooked but is now widely recognized. They are considered non-essential for sustaining life, but crucial for the survival of the producing organism. EOs are often accumulated by plants in smaller quantities than MP, about 1–3% of the dry mass, and can be identified only at a certain stage of growth and development of the species or within only one species, which makes the extraction and purification their difficult.

Although EOs have been known for hundreds of years and have been used as dyes (e.g. indigo, and shikonin), flavorings (vanillin, capsaicin, and mustard oil), flavors (rose oil, lavender oil, and other EOs), stimulants (caffeine, nicotine, and ephedrine), hallucinogens (morphine, cocaine, scopolamine, and tetrahydrocannabinol), insecticides (nicotine, piperine, pyrethrin, and rotenone), human and vertebrate toxins (coniine, strychnine, aconitine, colchicine, and glycoside), and therapeutic agents (atropine, quinine, and codeine), their supposed biological functions were a contradictory issue. In natural ecosystems, there is a regulation of population numbers, a natural control, in which the internal factors of the respective populations (endogenous factors) intervene, as well as the external factors (exogenous). In turn, exogenous factors can be abiotic and biotic. Among the abiotic factors, the most relevant are temperature, UV radiation (affects phenols and phenylalanine derivatives), soil composition (affects SMs containing nitrogen). Some EOs, for example flavonoids, are essential compounds that absorb UV rays and also show protective functions against those rays, thus preventing the photodegradation of chlorophylls, or even the destruction of the structural integrity of chloroplasts. However, biotic effects include plant pathogens, such as viruses and bacteria, as well as herbivorous animals. During the periods of stress, caused by the attack of microorganisms, the biosynthesis of SM is activated [22].

Among them were described SM with antibiotic, antifungal, antiviral action, which are able to protect plants from pathogens (phytoalexins). They are also known the EOs with anti-germinative properties, such as monoterpenes, in the following order of inhibitory power: geraniol > carvone > borneol > β-citronellol > α-terpineol > camphor > menthol > m entone > limonene > citral. At the same time, camphor also shows toxic (allelopathic) effects for plants, which have an important role in the interaction with pollinators and the seed coat.

In addition, plants also have mechanisms for responding to stressors, similar to the animal's immune system. They are based on the action of salicylic and jasmonic acid. Their biosynthesis seems without direct significance for the synthesizing cell but can be decisive for the development and functioning of the whole organism. Some SMs determine the taste and aroma of fruits and vegetables, and ethers, esters, terpenes, etc., through their pleasant smell, and they have an important role in plant pollination. Pigments, which are also products of secondary metabolism, play an important role in redox reactions. They determine the color of flowers, fruits, vegetables, and all plant organs and play an important role in the pollination process [23].

Phenylpropanoids (FPs) are one of the most important groups of natural compounds. They can be found in a wide variety of very valuable herbs as well *Echinacea purpurea*, *Rhodiola rosea*, *Silybum marianum*, *Melissa officinalis* etc. FPs play an essential role in plant integrity and have a protective effect against biotic and abiotic stresses. They are divided into several major classes, such as coumarins, flavonoids, hydroxycinnamic acids, and phenylpropenes. FPs have antioxidant activity and have beneficial effects on human health. FPs are also known for their antibacterial, antifungal, and anti-inflammatory properties.

Terpenes, being synthesized by all living organisms, represent the most numerous class of natural compounds. They are an essential component of human nutrition, and many of these substances are economically important as pharmaceuticals, aromatics, and potential biofuels of the latest generation. In the pharmaceutical industry, terpenes have a wide spectrum of use, being used both as excipients to improve skin penetration and as active medicinal principles with pharmacological properties anticancer, antimicrobial, antifungal, antiviral, antihyperglycemic, anti-inflammatory, etc. [24].

Data from the literature suggest that most of the materials studied describe the analgesic profile of monoterpenes and sesquiterpenes, alone or in combination with other biochemical components such as EOs extracted from medicinal plants. Examples of biologically derived drugs currently marketed as terpene products are artemisinin (sesquiterpenoid), taxol (diterpenoid), and vincristine (monoterpenoid). Hydroxycinnamic acids (chlorogenic, ferulic, caffeic, etc.) and other aromatic hydroxy acids (gallic and salicylic acids, etc.) are the most common phenolic acids in higher plants and act as biogenetic precursors in the biosynthesis of other plant phenolic compounds. Both acids and their derivatives are powerful antioxidants, which possess pronounced antiradical properties in in vitro; tests. The pronounced collagenic effect of ferulic, caffeic, and chlorogenic acids was established. All caffeic, chlorogenic, ferulic, coumaric, and other hydroxycinnamic acids are essential in the prevention and treatment of obesity, diabetes, and associated disorders, improve kidney function, stimulate antitoxic function of the liver, and have an antimicrobial and antimalignant effect.

The high content of hydroxycinnamic acids and their derivatives is characteristic of the species *Echinacea purpurea*, *Rhodiola rosea*, whose plant materials are widely used in medicine to correct immune system disorders. Therefore, the evaluation of the mechanisms involved in the accumulation of SM with beneficial effect for the improvement of health, as well as the highlighting of those exogenous factors with stimulating effect of active principles, represents a scientific issue of significant social and economic importance [25].

EOs are complex liquid mixtures consisting mainly of terpene and terpenoid hydrocarbons, extracted from parts of hetero-oil plants (by hydrodistillation, maceration, enfleurage, pressing, etc.) and used in the composition of perfumery, cosmetics, food, etc. in order to change their smell and/or taste. EOs are a by-product of plant metabolism that has a characteristic odor. The role of these compounds in the body of the plant is not well defined, but it is known that some substances behave as attractants (attract pollinating insects), others on the contrary-repellents-keep away potentially harmful organisms to the plant (show antifungal effects, herbicides, insecticides, etc.).

Also, these compounds, evaporating in hot weather, partially reduce the temperature of the plant. Some of them serve as intermediates in the biosynthesis of complex products necessary for the plant organism or participate directly in various biochemical processes. In the general monograph "Aetherolea" from the European Pharmacopoeia, tenth edition, EOs are defined as mixtures of volatile and lipophilic substances, with aromatic odor, belonging to different classes of organic compounds (especially terpenes and their oxygenated derivatives).

Named "oils," these mixtures were obtained due to their resemblance to the appearance of fatty oils. The full name "EOs" is appropriate, as it expresses one of the most characteristic properties of these compounds, namely the increased volatility at normal temperature.

Due to this property, they differ from fatty oils, which under normal conditions are virtually non-volatile. Other names, such as essential oil and essential oil, are less characteristic because neither these oils are composed of essential (and/or ester) compounds nor is the name essential. EOs contain different classes of chemical compounds, of which terpene compounds and terpenoids predominate [26].

Usually, non-terpenic compounds are found in smaller quantities, among them aromatic and aliphatic compounds (aldehydes, alcohols, amines, etc.) can be highlighted. In total, up to 500 compounds can be contained in an EOs, and their number vary depending on the plant from which the EOs were extracted. Thus, EOs from geranium and rock rose contain about 300 compounds, and those obtained from rose, bergamot, orange ofabout 500 compounds each.

Of this set of compounds, one or several are contained in high concentrations influencing the odor character of the volatile oil.

For example, in coriander EOs, the basic component is tertiary alcohol-linalool (38%); in rose EOs alcohols predominate: phenylethyl, citronellol, geraniol, and nerol (summary up to 95%); in lavender EOs-linalool (51%) and linalyl acetate (35%).

The class of terpene compounds in EOs predominate acyclic terpenes: monoterpenes (myrcene and ocimen), sesquiterpenes (farnesen); cyclic terpenes: monocyclic monoterpenes (limonene, terpinolines, and terpenes) and bicyclic monoterpenes (pinens, camphene, and hull); monocyclic sesquiterpenes (bisabolen), bicyclic sesquiterpenes (cadinen), etc. Oxygenated derivatives predominate in the class of terpenoids: terpenic alcohols (linalool and geraniol), terpenic aldehydes (citral and citronellal), and terpenic esters (linalylacetat and geranylbutyrat). Among the compounds of non-terpenic nature are phenols (anethole and eugenol), organic acids (acetic acid and benzoic acid), aliphatic alcohols and aldehydes, amines, thiocompounds, heterocyclic compounds, etc. [27].

EOs together with synthetic odor compounds are the basic raw material in the production of perfumery, cosmetics, but also food flavors.

Thus, EOs of sage, eucalyptus, sandalwood, and geranium are applied in various antiacne products due to their antiseptic properties. Many of these compounds are used as raw materials in various organic syntheses, for example, linalool in the synthesis of vitamin E, myrcene in the synthesis of cyclohexenes applied in perfumery, carvone in the synthesis of bravellin (antimalarial action), tetrahydrocannabinoids (anti-analgesic action, etc.), and some EOs (mint, lemon, rose, coriander, etc.) are used as taste and odor correctors in pharmaceuticals.

Terpenoids are one of the most common classes of natural compounds.

The aromas of many flowers and fruits are due to the mixtures of volatile compounds that they emit. These are often called terpenes.

Their molecule consists of $5n$ carbon atoms where n is a natural number. Many terpenes are used as food flavorings (clove extract, mint, etc.), as perfumes (rose, lavender, etc.), or as solvents (turpentine). Hydrocarbons included in the category of terpenoids have the general formula $(C_5H_8)n$, where n is a natural number greater than or equal to 2.

For this reason, they were considered isoprene oligomers in which the binding of monomers is performed in positions 1, 4 (head = tail).

Natural EOs are composed together with the mentioned hydrocarbons and a series of their functional derivatives containing oxygen (alcohols, carbonyl compounds, etc.). The most common classification of terpenoids is made according to the value of n in the mentioned molecular formula. In reality, plants do not synthesize terpenoid compounds from isoprene.

The source of these units is acetic acid (precursor) which with coenzyme A leads to acetyl-coenzyme A and then by trimolecular condensation to hydroxy-b-methyl-glutaryl-coenzyme A, which is reduced fermentatively to mevalonic acid, a compound which then transforms in isopentenyl pyrophosphate, the key element in terpenoid synthesis.

Examining the polycyclic structures of terpenes, it is obvious that they can be detailed by conformation and configuration formulas that approximate more correctly the molecules of these compounds, e.g. (−) menthol 2 has a chair configuration with equatorially oriented substituents.

Volatile terpenes are low molecular weight molecules derived from isopentenyl diphosphate (PPI) and play an important role in direct and indirect defense, being also used as a signal for neighboring plants. Table 13.1 shows the provenance, components, and physiological action of several plant species representative for the content of EOs.

Plants constantly secrete small amounts of volatile terpenes, but in some cases, such as the attack of herbivorous insects, terpenes are secreted in larger quantities in response to the negative action of biotic factors.

Table 13.1 Components of essential oils from some plant's species.

Chemical structure/origin	Oil	Plants species	Components
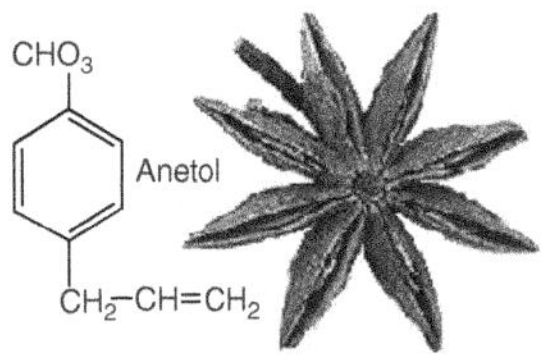	Anise	Fruit of anise (*Pimppinella anisum*)	Anethol (80–90%)
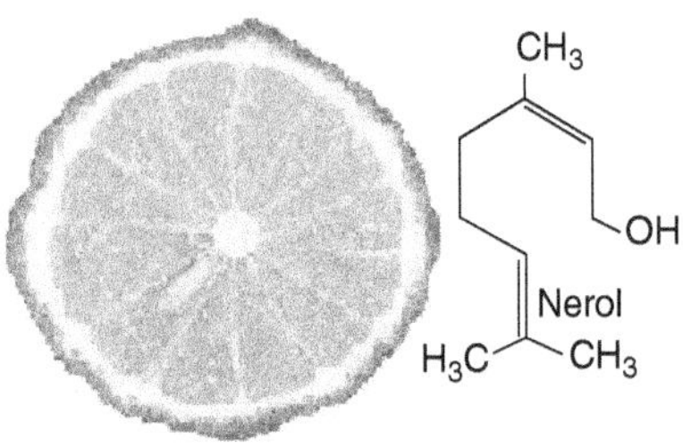	Bergamot	Orange peel (*Citrus aurantum, Bergamia subspecies*)	(−) Linalool acetate, nerol, terpineol, (−) pinene, (−) camphen, (+) lemon
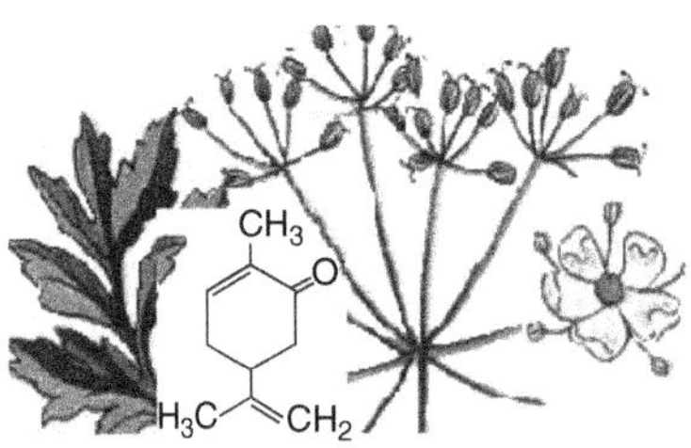	Caraway	Fruits *Carum carvi*	Carvone (50–60%), (+) limonen, dihydrocarbon,
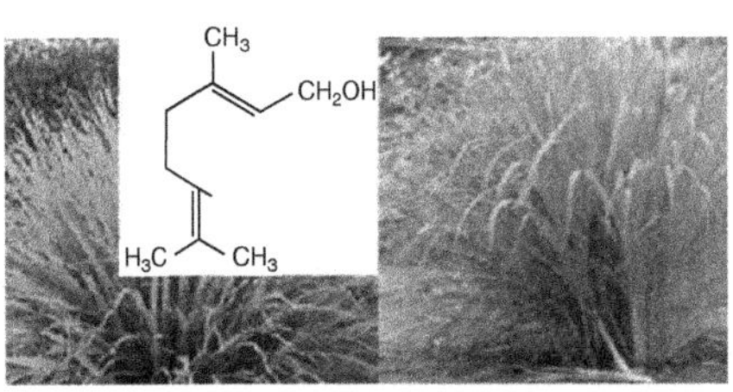	Lemon grass	*Cymbopogon flexuosus*	Citral (70–85%), methylheptenone, nerol, farnesol, geraniol, dipentene

Table 13.1 (Continued)

Chemical structure/origin	Oil	Plants species	Components
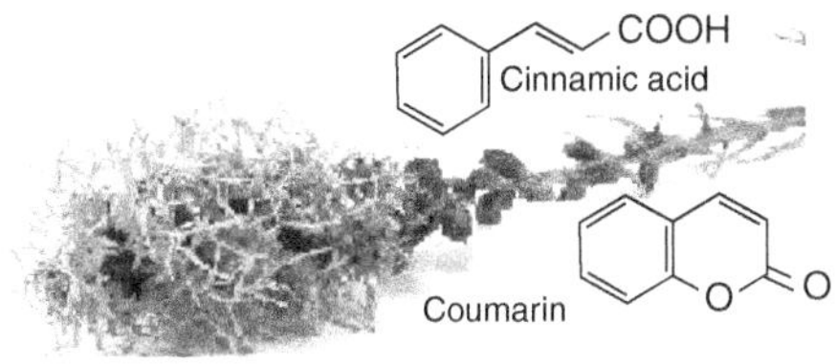	Lavender	Flowers *Lavendula officinalis*	Linolyl acetate (30–60%), linolyl butyrate, geraniol, coumarin, cineole
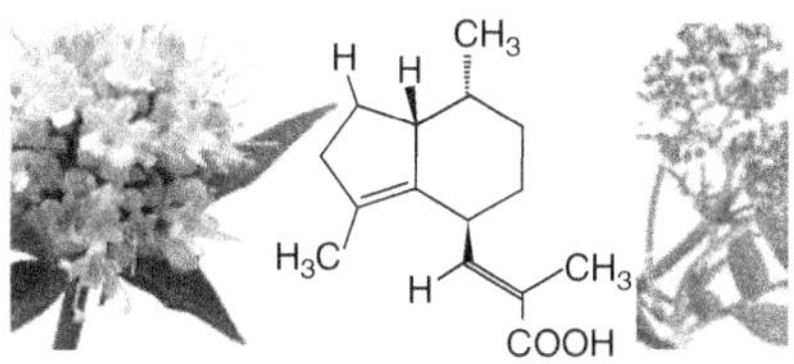	Valerian	Odolean (*Valeriana officinalis*)	Butyric acid, valerian acid, (−) camphene, (−) pinene, (−) borneol, bornyl valerianate
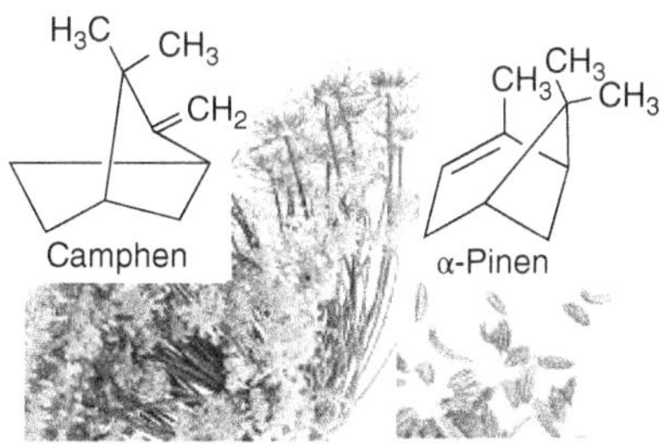	Fennel anethole	Seed of fennel (*Foeniculum vulgare*)	(50–60%), (+) pine, camphen, dipentene,
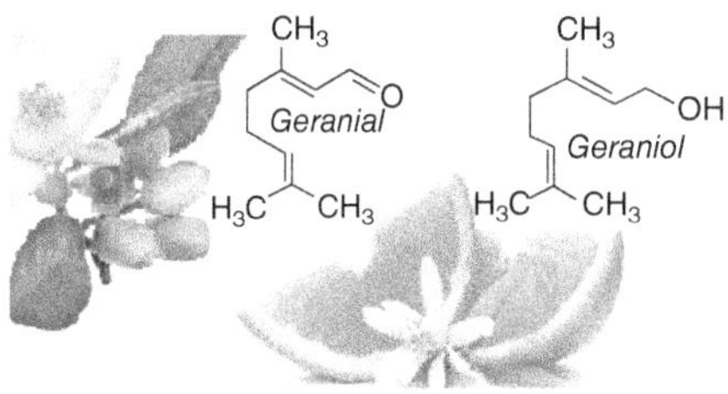	Neroli (orange blossom)	*Citrus bigaradiarisso*	27% (pine, camphen, dipenten), (−) linalool 30%, geraniol, nerol, anthranilic acid methyl ester 16%

Source: Santosh Kumar Upadhyay.

Terpenes are an important source of olefinic compounds (alkenes) that are involved in the formation of phytotoxic products.

Volatile terpenes combine with nitrogen oxides and form ozone-type photo-oxidants, thus increasing the stress around the plant.

The synthesis of terpenes is very energy-consuming, and in essence, the cost of producing volatile terpenes is higher than that of any primary or SMs.

Terpenes are generally accumulated in plants at the level of specialized secretory structures: glandular hairs (*Lamiaceae*, *Asteraceae*), secretory pockets and cavities (*Fabaceae, Rutaceae*), secretory canals (*Pinaceae*, *Apiaceae*), laticifers (*Euphorbiaceae*, *Asteraceae*), and idioblasts (*Magnoliacs*, *Lauraceae*) [28].

The compound that contributes to the scent depends on the plant species, and it can be seen from Table 13.2.

Table 13.2 The compound that contributes to the scent some spices plant.

Common name of the plant species	The compound/compounds responsible for odors
Carnations	Eugenol/beta-caryophyllene, and benzoic acid derivatives
Violets	Ionones
Lilies	(e)-Beta-ocimene and linalool, eucalyptol (also referred to as 1,8-cineole), (*E*)-ocimene, 8-oxolinalool, benzyl methyl ether, indole, lilac aldehyde, lilac alcohol, and hydroquinone dimethyl ether
Hyacinth	Hyacinth, ocimenol, cinnamon, and ethyl 2-methoxybenzoate
Chrysanthemums	Alpha-pinene, eucalyptol, camphor, and borneol, chrysanthenone and chrysanthenyl acetate, beta-caryophyllene
Lilacs	(e)-Beta-ocimene, benzyl methyl ether
Roses	(−)-Cis-rose oxide, beta-damascenone, geraniol, nerol, (−)-citronellol, farnesol, and linalool
Lily	1,8-Cineole, benzaldehyde, methyl benzoate, ethyl benzoate, creosol, and isoeugenol
Coffee	Linalool, nerol and geraniol (as major constituents 40% total), epoxygeraniols and epoxynerols
Jasmine	(−)-Jasmine lactone, (*Z*)-jasmone, (−)- and (−)-epi-methyl jasmonate
Lavender	(−)-(*R*)-linalool, (−)-(*R*)-lavandulol, 1-octen-3-yl acetate
Lemon	Geranial + neral, (+)-limonene
Lemongrass	Geranial, neral, myrcene, neomenthol, linalyl acetate, (*Z*)-beta-ocimene
Lily-of-the-valley	Floral-rosy-citrusy notes: citronellol, citronellyl acetate, geraniol, nerol, geranyl acetate, geranial+benzyl acetate, neral, benzyl acohol, phenethyl alcohol, phenylacetonitrile, farnesol and 2,3-dihydrofarnesol, green-grassy notes: (*Z*)-3-hexenal and (*E*)-2-hexenal, (*Z*)-3-hexenyl acetate, (*Z*)-3-hexen-1-ol, green pea and galbanum-like notes: 2-isobutyl-3-methoxypyrazine and 2-isopropyl-3-methoxypyrazine, fatty, waxy, aldehydic notes: octanal, nonanal, decanal and fruity, raspberry notes: beta-ionone
Magnolia	(+)-Verbenone, isopinocamphone, (*Z*)-jasmone. Magnolia (+)-verbenone, isopinocamphone, (*Z*)-jasmone
Mandarin	Methyl N-methyl-anthranilate, linalool, trans-4,5-epoxy-(E)-2-decenal, alpha-sinensal and (E,Z)-2,6-dodecadienal
Mango	Delta-3-carene, limonene, terpinolene, (*E*)-beta-ionone, a-phellandrene, (*E*,*Z*)-2,6-nonadienal, ethyl 2-methylpropanoate, (*E*)-2-nonenal, ethyl butanoate, methyl benzoate, decanal and 2,5-dimethyl-4-methoxy-3[2*H*]-furanone (mesifurane)
Neroli (orange flower)	(+)-Linalool, (+)-(*E*)-nerolidol and (*E*)(*E*)-farnesol, methyl anthranilate, indole, phenylacetonitrile and 1-nitro-2-phenylethane
Petunia	Benzaldehyde, phenylacetaldehyde, isoeugenol, methyl benzoate, phenethyl alcohol, benzyl benzoate
Peppermint	(−)-Menthol, (−)-menthyl acetate, (−)-menthone and (+)-menthofurane
Pink pepper	Alpha-cadinol

Table 13.2 (Continued)

Common name of the plant species	The compound/compounds responsible for odors
Pine	Alpha- and beta-pinene and two exotic macrolides from maritime pine, longifolene from long-leaved pine, and the two derivates isolongifolene and isolongifolanone
Raspberry	Raspberry ketone, (*R*)-(+)-(*E*)-alfa-ionone, mesifurane, beta-damascenone
Robinia	2-Aminobenzaldehyde and 3(*Z*)-hexen-1-ol, 3(*Z*)-hexen-1-ol
Rosemary	(+)-Borneol, (+)-bornyl acetate, (+)-camphor, (+)-alpha-pinene, (+)-verbenone and 1,8-cineole
Saffron	Safranal and 2-hydroxy-4,4,6-trimethyl-2,5-cyclohexadien-1-one
Tobacco	Beta-damascenone, megastigmatrienone, oxo-edulan, 4-oxo-beta-ionone
Thyme	Thymol, carvacrol
Vanilla	Vanillin and a vanilla vitispirane
Patchouli	(−)-Patchoulol and norpatchoulenol
Peach	Gamma-decalactone and delta-decalactone
Pear	Ethyl 2(*E*),4(*Z*)-decadienoate
Turmeric	Ar-turmerone, turmerone and curcumin
Yuzu	Branched aldehydes, yuzu lactone and 6-methyl-5-hepten-2-ol
Ylang-Ylang	p-Cresyl methyl ether, benzyl acetate, methyl salicylate, methyl benzoate and cinnamyl acetate

Both the effective synthesis of volatile terpenes and the formation of these specialized structures in their accumulation are very expensive for organisms. Terpenes serve as signals that can induce systemic defensive responses, in the unattacked areas of the plant, but also, through the air, can reach the neighboring plants where they can induce defensive responses to possible attacks. It is the volatile nature of terpenes that gives them the ability to behave as the highly efficient signaling molecules. Mircene is found in hops, geraniol has a "rose"-like (the scent of geraniums and roses) in rose flowers, limonene (the major scent of citrus flowers and fruits) in cumin, nerol (orange blossom), menthol in mint, camphor in *Laurus camphor* wood and wormwood, borneol in lavender, etc.

The origin and some physical properties of terpenoids are presented in Table 13.3.

The quality of EOs depends very much on the concentration of the basic components. But this criterion is not a determining one because some EOs that correspond to all physico-chemical quality parameters may receive a low grade from the perfumer due to the content of unpleasant-smelling compounds, which exceed the allowable limits.

As such compounds serve: for rose EOs-organic acids with low molecular weight; for lavender and coriander EOs-camphor; for EOs of mint and geranium-mint, etc. Also, the odor nuances of an EOs are influenced by a series of compounds that are contained in low concentrations (tenths or hundreds of percent). Thus, the presence or absence of rosenoxid, methylenugenol, eugenol, and acetic aldehyde influences the smell of volatile rose oil; menthylacetat and mentofuran – the smell of volatile peppermint oil, etc.

Table 13.3 Provenance and some physical properties of terpenoids.

Terpenoid	Provenance (essential oil)
Ocimen	Basil
Citronelol (R = CH_2OH)	(−) Of rose oil
Citronelal (R = CHO)	Essential oils of geranium and rose
4 Geraniol (R = CH_2OH)	Rose oil, geranium, animal tissues
Geranial (R = CHO) (citral a)	Lemon oil
Nerol (R = CH_2OH)	Rose oil, bitter orange blossom oil
Neral (R = CHO)	(citral b) Cis, trans, in lemon peel
Limonen (+)	In lemon peel (−) In turpentine (fir needles) (±) it is called dipentan
Terpineol	Peppermint oil
Menthol	Peppermint oil
Pulegonă	Peppermint oil
Alpha-pinene	Turpentine oil
Beta-pinene	Turpentine oil
Camphor	From camphor tree wood, from wormwood
Thujone	Thuja oil
Farnesol	From linden, rose, citrus oil, pearl
Nerolidol	Nerol oil
Juvenile hormone III	In insects
G-bisabolene	Lemon and bergamot oil
Abscisic (+) acid	Plant growth hormone
Caryophyllene	Clove oil
Drimenol	Drimys winteri tree (America)
Geranyl-geraniol	Conifer resin .
Geranyl-linalool	Conifer resin
Phytol	Component of chlorophyll and vitamins K and E
Vitamin A	Growth factor in milk, butter, eggs, fruits
Manool	Yellow pine oil
Sclareol	Sage oil, incense
Abietic acid	Conifer resin
Pimaric acid	Conifer resin
squalene	In the shark liver
Ambreina	In gray amber (sperm whale)
Cucurbitacina	In cucumbers, it gives a bitter taste

The quality parameters of EOs include odor, color, taste, acidity index, relative density, refractive index, iodine index, peroxide index, esterification index, saponification index, and the content of specific individual compounds of certain oils (content of linalool, menthol, charcoal, etc.). For a qualitative EOs, these parameters must correspond to those indicated in the corresponding normative documentation. EOs are complex liquid mixtures consisting of volatile organic compounds of plant origin. The composition of EOs is dominated by terpenic compounds and terpenoids (monoterpenes, sesquiterpenes, monoterpenic alcohols, etc.).

Classes of aliphatic compounds (alcohols, aldehydes, aliphatic esters, etc.), aromatics, macrocyclics, and their derivatives (amines, organic sulfides, heterocyclic compounds, etc.) are contained in small quantities [29].

13.5 Essential Oils from Plants in Perfume

Perfume is a mixture of aromatic EOs or aromatic compounds, fixatives and solvents, that used to give the human body, animals, food, objects, and living spaces a pleasant smell. The types of perfume reflect the concentration of aromatic compounds in a solvent, in which the fine perfume is usually ethanol or a mixture of water and ethanol. Various sources differ considerably in the definitions of perfume types. The intensity and longevity of a perfume are based on the concentration, intensity, and longevity of the aromatic compounds, or perfume EOs used. As the percentage of aromatic compounds increases, so does the intensity and longevity of the perfume. Specific terms are used to describe the approximate concentration of a perfume with the percentage of perfume oil in the volume of the final product.

The precise formulas of commercial perfumes are kept secret. Even if they were widely published, they would be dominated by ingredients and fragrances so complex that they would not be very useful in providing a guide to the general consumer in describing the experience of a perfume. However, perfume connoisseurs can become extremely skilled at identifying the components and origins of perfumes in the same way as wine experts.

The most practical way to start describing a perfume depends on the elements of the perfume notes of the perfume or "family" it belongs to, all affecting the overall impression of a perfume from the first application to the last persistent suggestion of the perfume.

The traces of the perfume left behind by a person wearing perfume are called his "*sillage*," after the French word for "wake," like the traces left by a boat in the water. The perfume is described in a musical metaphor as having three sets of notes, which makes the harmony of the perfume harmonious.

The EOs (aroma compounds) produced by plants can be systematized by functional groups. These groups involve alcohols (e.g. eugenol, furaneol, hexanol, and menthol), aldehydes (e.g. acetaldehyde [pungent], benzaldehyde [marzipan, almond], cinnamaldehyde [cinnamon], citral [lemon oil and lemon grass], furfural [burnt oats], hexanal [green and grassy], nonanal, octanal, and vanillin [vanilla]), amines (e.g. skatole and indole), esters (e.g. lutein fatty acid esters from marigold), ethers (nerolin = methyl β-naphthyl ether), terpenes (e.g. caryophyllene, citronellol in rose, geraniol, linalool in many flower species, nerol, and β-ionone).

The notes unfold over time, with the immediate impression of the top note leading to deeper middle notes, and the base notes gradually appear as the final stage [30].

The smells like, tobacco flavors (which are similar to cigarettes), fruit flavors (peach, blueberry, etc.), menthol flavors, sweet flavors (chocolate, candy, etc.), and other flavors (black tea, coffee, wine, etc.) are carefully created by knowing the process of evaporation of the perfume and the three categories listed below:

- *Top notes*: they are also called top notes. Perfumes that are perceived immediately when applying a perfume. Top notes consist of small, light molecules that evaporate quickly. They form the initial impression of a person's perfume and are therefore very important in selling a perfume, e.g. of top notes include mint (rich in menthol, menthone, menthofuran, 1,8-cineole, menthyl acetate, etc.), lavender (rich in linalool, perillyl alcohol, linalyl acetate, camphor, limonene, tannins, etc.), and coriander (rich in linalool, limonene, camphor, geraniol, etc.).
- *Middle notes*: also called heart notes. The scent of a perfume that appears just before the top note dissipates. Compounds with middle notes form the "heart" or main body of a perfume and act to mask the often-unpleasant initial impression of the base notes, which become more pleasant over time, e.g. of middle notes include sea water (oceanic, salty/seawater vibe, idea is smell "breezy," "outdoorsy,") sandalwood (sesquiterpenic alcohols/tricyclic α-santalol/β-santalol, etc.), and jasmine (ester benzyl acetate, nerolidol, cedrol, jasmone, etc.).
- *Base notes*: the smell of a perfume that appears close to the departure of the middle notes. Base and middle notes are together the main theme of a perfume. The base notes bring depth and solidity to a perfume. Compounds in this class of flavors are usually rich and "deep" and are usually not perceived until 30 minutes after application, e.g. of base notes include tobacco/tobacco smoke (e.g. cyclohexane, ionone, theaspirone, safanal, cyclocitral etc.), amber (dry woody amber ambergris musk sweet, etc.), and musk (*Angelica archangelica*, *Abelmoschus moschatus* produce musky-smelling macrocyclic lactone compounds, etc.).

The perfumes in the top and middle notes are influenced by the base notes; conversely, the smells of base notes will be altered by the types of scented materials used as middle notes. Manufacturers who publish perfume notes typically do so with perfume components presented as a perfume pyramid, using imaginative and abstract terms for the listed components. The various aroma constituents are produced in tobacco leaf via oxidative carotenoid degradation, according Table 13.4

Grouping perfumes can never be completely objective or definitive. Many perfumes contain aspects of different families. Even a fragrance designated as a "single flower" will have subtle hints of other aromatics. There are almost no real unique perfumes made from a single aromatic material [31]. Table 13.5 lists some of the chemical compounds in essential oils and their characteristic aroma in alphabetical order.

According to the fragrance classification chart using the disk of aromas, there are five main families Floral, Oriental, Woody, Fougère aromatic, and Fresh, the first four in classical terminology and the last in the modern oceanic category. Each of them is divided into subgroups and arranged around a wheel. In order to be able to describe the notion of smell, they were conceived to odor descriptor corresponding to flavor index.

Table 13.4 Aroma/smell/odor characteristics of tobacco carotenoid derivatives.

Name	Aroma/smell/odor characteristics	Structure
3-Oxo-alpha-Ionone	Sweet, floral	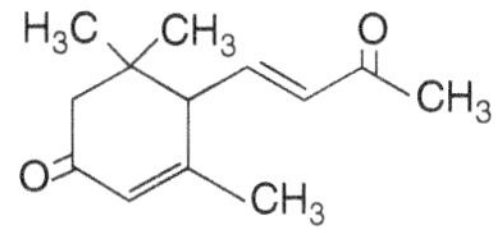
4-Oxo-beta-ionone	Sweet rich like virginia tobacco	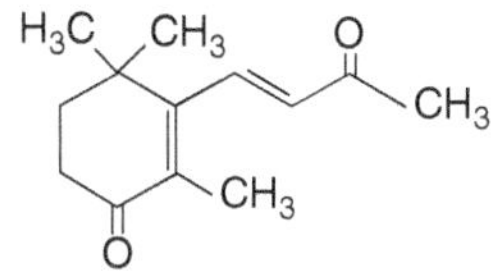
Alpha-Ionone	Woody balsamic, violet-raspberry in dilution	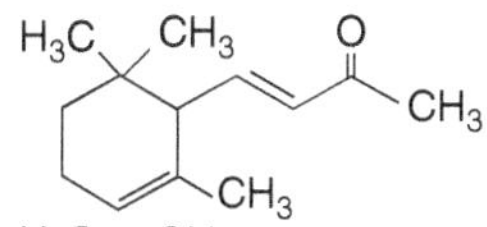
Beta-Cyclocitral	Green, grassy hay like odor	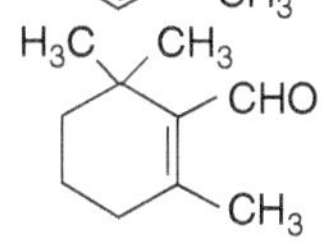
Beta-Damascenone	Fruity, floral with apple, plum-raisen, tea, rose, tobacco note	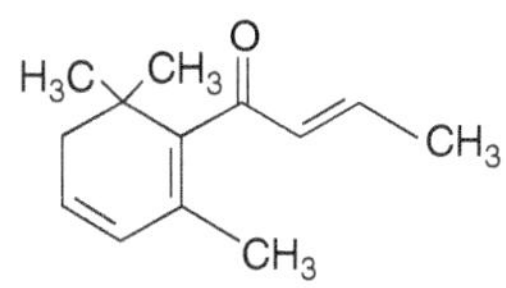
Beta-Damascone	Fruity (apple-citrus), tea-like with slight minty note	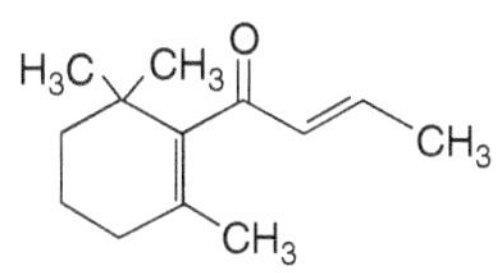
Beta-Ionone	Woody, violet, fruity; woody-raspberry on dilution	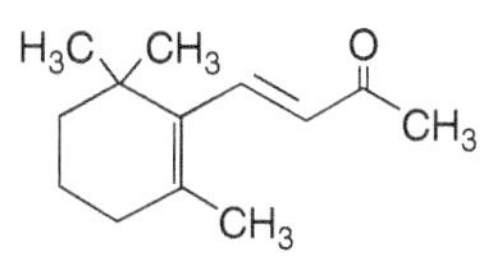
Dihydroactinodiolide	Weak, slightly cooling	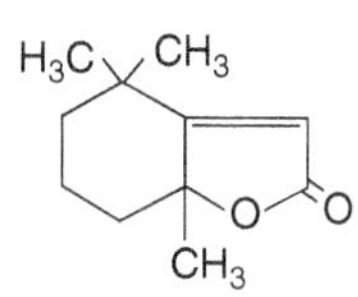
Oxo-Edulan I	Oriental tobacco like	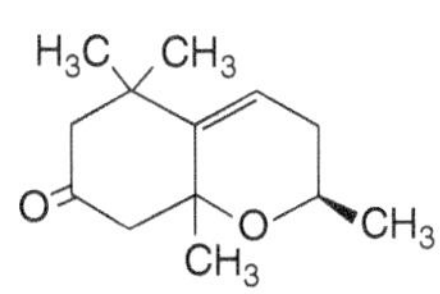

(*Continued*)

Table 13.4 (Continued)

Name	Aroma/smell/odor characteristics	Structure
Oxo-Edulan II	Oriental tobacco like	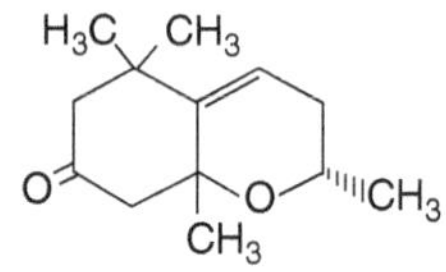
Safanal	Saffron, green, sweet, hay-like	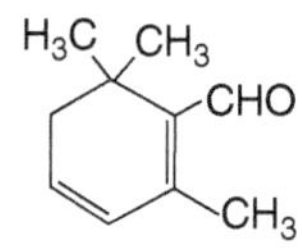
Theaspirone	Tea like	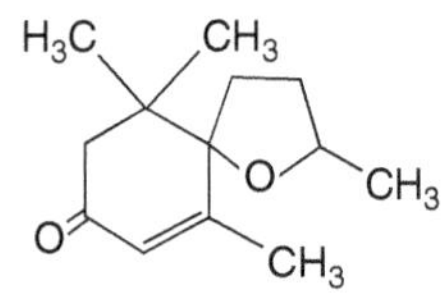

Table 13.5 Chemical components of essential oils and aroma/smell/odor characteristic.

Chemical components	The aroma/smell/odor characteristic
6-Methyl heptan-2-one	Green, herbaceous, weak oily
Benzyl acetate	Floral-fruity, fresh-sweet-notes
Benzyl alcohol	Faint aromatic, weak floral
Benzyl benzoate	Sweet-balsamic, faint-floral
Benzyl salicylate	Faint-sweet, weak herbaceous-medicinal
Chavicol	Sweet-phenolic, anise-like, green-minty-notes
cis-3-Hexenol	Green, fresh-grass
cis-3-Hexenyl acetate	Intense green, weak fruity, pungent
cis-3-Hexenyl benzoate	Green-herbal, weak woody
cis-Jasmone	Intense jasmine-like, warm-floral, weak fruity
cis-Linalool oxide	Fresh-floral, citric-notes
cis-Methyl jasmonate	Floral-sweet, herbaceous
Eugenol	Spicy, strong clove-cardamon-notes
Geraniol	Rose-notes, sweet-floral, weak fruity
Geranyl linalool	Floral, rose-geraniol-lavender-notes, weak fruity
Indole	Unpleasant-fecal, musty-cadaverous-floral in dilution
Isoeugenol	Spicy, intense clove-like, sweet and woody undertones
Isophytol	Weak floral-balsamic
Linalool	Fresh-floral, sweet-fruity-woody, lavender-notes
Methyl anthranilate	Fruity, grape-like

Table 13.5 (Continued)

Chemical components	The aroma/smell/odor characteristic
Methyl benzoate	Fragrant-fruity
Methyl linoleate	Fatty-oily
Methyl oleate	Fatty-oily
Methyl palmitate	Fatty-oily
Methyl salicylate	Fragrant-minty, sweet-spicy, wintergreen-notes
Nerol	Floral, weak rose-notes, sweet-fruity
p-Cresol	Medicinal-aromatic
Phenylethyl acetate	Sweet-fruity, rose-honey-notes
Phenylethyl alcohol	Floral, rose-like, fragrant-honey-notes
Phytol	Weak floral-balsamic, sticky side-notes
Phytyl acetate	Faint-floral-balsamic
trans-Jasmone	Floral-jasmine-like, warm-fruity, weak spicy
trans-Linalool oxide	Fresh-floral, weak sweet-citric
trans-Nerolidol	Floral-fruity, rose-apple- and green-citrus-notes, woody-waxy
α-(*E*,*E*)-Farnesene	Weak floral
α-Terpineol	Fragrant-floral, weak fruity with lilac-notes
δ-Jasmine lactone	Jasmine-like, warm-floral

Perfume compounds in perfumes will degrade or decompose if improperly stored in the presence of heat, light, oxygen, and foreign organic matter. Proper preservation of perfumes involves keeping them away from heat sources and storing them where they will not be exposed to light. An open bottle will keep its aroma intact for several years, as long as it is well preserved. However, the presence of oxygen in the space of the glass head and environmental factors will change the smell of the perfume in the long run.

Fragrances are best stored when stored in airtight aluminum bottles or in their original packaging when not in use and refrigerated at relatively low temperatures: between 3 and 7 °C (37–45 °F). Although it is difficult to completely remove oxygen from the headspace of a stored perfume bottle, opting for spray dispensers instead of "open" rollers and bottles will minimize oxygen exposure. Sprays also have the advantage of isolating the perfume inside a bottle and preventing it from mixing with dust, skin, and detritus, which would degrade and alter the quality of a perfume.

The family classification is a starting point for describing a perfume, but it does not fully characterize it. E.g., *Roses* spp. come in two types of perfume extracts: as an essential or absolute oil (deeper and sweeter than its oil counterpart). Rose was the test of time due to its ability to combine perfectly with other floral notes, wood and citrus. Rose extracts contain hundreds of molecules, which explains why its scent is so rich and multifaceted.

Rose perfume extracts have notes of citrus (lemongrass), green, fruity (peach, plum, wine), spicy (cloves), amber, and sweet veneers, all in one fragrance. *Pelargonium graveolens* (green, accentuated, and herbaceous scent) meaning strong smell, geranium is often confused with "the other" rose (but geranium has an aromatic quality [similar to lavender], which makes it smell more "masculine"), but with a less powdery and more lemony, herbaceous aroma with a soft but strong green scent. Known molecules found in geranium oil are citronellol, nerol, geraniol, and linalool. Nerol is found in lemongrass, contributing to the smell of lemony geraniums, and geraniol is one of the primary components in rose oil, contributing to the smell of rose geraniums [32].

Among the compounds contained in a relatively large number of EOs can be highlighted:

13.5.1 Linalool (3,7-dimethylocta-1,6-dien-3-ol), $C_{10}H_{18}O$

Smell and taste: it has a floral scent, free of terpenic and camphor shades, if it is pure enough, with fresh, woody, and citrus shades. The taste is spicy, similar to coriander EOs. Distribution in nature: it has been identified in over 200 EOs of different origins (leaves, flowers, wood, and fruit). Important sources: for the dextro form – coriander oil (60–70%); for the form levo – HoSho oil (80%), linaloe (80%), bois de rose (80–90%), ylang-ylang and bergamot (linalool 7–20%, linalyl acetate 20–40%). Field of use: it is used as an odorant compound in perfumery compositions and cosmetics. It serves as a raw material in the synthesis of other odorous compounds (such as citral, linalylacetate, etc.) and some biologically active compounds (vitamin E, cyclohexylgeranilacetic acid [Cygerolum]), etc. [33].

13.5.2 Camphor (1,7,7-trimethylbicyclo [2.2.1] heptan-2-one), $C_{10}H_{16}O$

Physico-chemical properties: chemical compound of the class of bicyclic terpenes. It has two asymmetric carbon atoms in positions C1 and C4, but only two enantiomeric forms (+) and (−) are possible. It is slightly soluble in water (0.12%), well soluble in ethanol, ether, chloroform, acetone, acetic acid, benzene. Smell and taste: after the smell, camphor can be considered as a prototype, and it shows a warm, mentholated smell, almost radiant, but a little tenacious. The taste is fresh, bitter-hot, and cool. Spread in nature: it has been identified in over 180 EOs. It is the main component of *Cinnamomum camphora* camphor tree oil in China. From this oil, camphor is separated by crystallization. Field of use: it is used as a component of perfumery compositions; as a plasticizer in the production of plastics; in medicine it is applied as an antiseptic, antirheumatoid, and analeptic remedy; as a raw material for the synthesis of other valuable compounds (such as 3-bromocamphor, borneol) etc. [34].

13.5.3 Cedrol (1*S*, 2*R*, 5*S*, 7*R*, 8*R*)-(2,6,6,8-tetramethyltricyclo [5.3.1.01,5] undecan-8-ol or cedran-8-ol), $C_{15}H_{26}O$

Physico-chemical properties: chemical compound of the class of sesquiterpene alcohols. It is in the form of colorless crystals, well soluble in benzyl benzoate, moderately soluble in glycols and mineral oils, insoluble in water. It contains five asymmetric carbon atoms. Smell and taste: it manifests a smell of moderate intensity, similar to the smell of cedar wood. It has weak and woody taste. Distribution in nature: it is contained in EOs from some conifers (cedar [*Cedrus atlantica*], juniper [*Juniperus virginiana*], cypress [*Cupressus sempervirens*] etc.), some species of oregano, etc. Field of use: it is used as an odorant and flavoring compound or in the synthesis of methyl ether and cedrilacetat, which have the same applications [35].

13.5.4 Eugenol (2-methoxy-4-allylphenol; 1-hydroxy-2-methoxy-4-allylbenzene), $C_{10}H_{12}O_2$

Physico-chemical properties: chemical compound of the phenol class. It appears as a colorless to slightly yellowish liquid. Smell and taste: it manifests a strong clove smell, characterized as hot, spicy, hot, less hot, and spicy than clove oil. It is hot and spicy taste. Distribution in nature: It contained in many EOs, including: EOs from cloves (85%), basil (60–70%), cinnamon leaves (even over 90%), in smaller quantities in obligatory EOs, ylang-ylang, Citronella, Sassafras, etc. Field of use: It is used as an odorant compound; as a raw material in the synthesis of other odorous compounds, such as dihydroeugenol, isoheugenol, etc. Esters are a class of chemical compounds derived from inorganic or organic acids in which at least one hydroxyl (−OH) group is substituted by the −O-alkyl group. In nature, the most common odorant aliphatic esters consist of the residues of formic, acetic, propionic, butyric acids, and the residues of methyl, ethyl, butyl, isoamyl alcohols, etc. Carboxylic acids and their esters are contained in various EOs, in fruits, flowers, fermented foods (cheeses, etc.). Uses: esters, due to their specific odor, represent the most numerous class of odorous and flavoring compounds. It is used in perfumery, but especially in the food industry to make food flavor compositions (food essences), which give the products aromas of flowers, fruits (the largest group), and berries [36].

13.5.5 Citral (3,7-dimethyl-2,6-octadien-1-al), $C_{10}H_{16}O$

Belongs to the class of monoterpenic aldehydes, it is a weak-yellow liquid with a strong lemon smell, with a bitter-bitter taste. It is a mixture of two geometric isomers: the trans-form (geranial or citral A) (55–70%) and the cis-form (neral or citral B) (35–45%). Distribution in nature: citral is found in various EOs, in large quantities in EOs of various species of *Cymbopogon* or *Lemongrass* (50–95%) and in EOs of *Litsea cubeba* (plant native to China, Indonesia) (60–80%). In smaller quantities, it is contained in EOs from: lemon (≈2%), eucalyptus, verbena (widespread in tropical and subtropical regions), ginger, rhubarb, etc. Uses: due to the pronounced smell of lemon, it is applied as an odorant or flavoring agent in cosmetics and food, and it is rarely used in perfumery (due to the tendency of polymerization and oxidation). It is added to chewing gum, pastries, sweets, ice cream, and various drinks. The concentration of citral in these products varies in the range of 9–170 ppm. To give the smell of lemon and verbena, it is added in the composition of perfumes, soaps, detergents, creams, and cosmetic lotions. Citral is also used as an intermediate in the synthesis of vitamin A, ionone, and methil ionone [37].

13.5.6 Vanillin (4-hydroxy-3-methoxybenzaldehyde) $C_8H_8O_3$

Vanillin comes in the form of white aciform crystals with a pleasant smell, vanilla, and sweet taste. In general, the toxicity of vanillin is relatively low (LD50 [rat, oral administration] = 1580 mg kg^{-1}; LD50 [rabbits, dermal administration] = 5010 mg kg^{-1}), but it can still induce dermal reactions in people allergic to compounds such as acid benzoic, *Peruvian balm*, cinnamon, and cloves. Natural vanillin is extracted from fermented vanilla pods. Three species of vanilla are known to be produced to obtain natural vanillin:

Vanilla planifolia A., *Vanilla pompona* S., and *Vanilla tahitensis* M. Uses: vanillin is used as a flavoring in various foods: in confectionery (ice cream [0.005–0.03%], chocolate [0.03–0.08%]-the ice cream and chocolate industry together consumes about 75% of all vanillin used as a flavoring), in confectionery (0.03–0.1%), in dairy products, and in various beverages. Vanillin is also used as an odorous compound in perfumery but also to mask the unpleasant smell and taste of medicines, animal feed and cleaning products. It is also used in the production of biologically active compounds. Other fields of application of vanillin are: in the composition of antifoams, in galvanostegy (gives gloss to zinc-coated surfaces), attractive in some insecticides, catalyst in the polymerization reaction of methyl methacrylate, yeast inhibitor in food, etc. [38].

13.5.7 Syringe Aldehyde (4-hydroxy-3,5-dimethoxybenzaldehyde) $C_9H_{10}O_4$

It comes in the form of colorless crystals soluble in ethanol (and other polar organic solvents), chloroform, etc. In nature, it is found in spruce and maple wood; it is also formed in alcoholic beverages kept in oak pots (the content of the compound increasing during the maturation of the drink, gives a spicy, smoky aroma with woody notes). At the same time, the syringe aldehyde together with vanillin participates in the chemical communication of some insect species. Uses: in the literature, it can be found studies that show properties: antioxidants, anti andfungals, moderate hypoglycemic antimalarials (in studies in rats), anticancer (due to inhibition of nitrosamines or antiproliferatives in colon cancer) for syringe aldehyde [39].

Some researcher, conducting research on a species of beetle (*Acanthoscelides obtectus*), identified insecticidal properties for syringe aldehyde. It is used as a flavoring compound in various alcoholic beverages, giving them an aroma reminiscent of chocolate, grapes, smoke, and wood. There are thousands of types of mono-, sesqui-, di-, tri-, and tetra-terpenic compounds (which differ from each other in molecular mass), and many of them have been studied extensively. These compounds are found in nature and in most plants.

Many monoterpenes are known for their strong aromas and high degree of volatility through which they attract insects to the flowers of plants for pollination [40].

Myrcene or beta-myrcene is a linear carbohydrate monoterpene and is the main component of wild thyme EOs, which comprises 40% myrcene. Myrtle is also found in high concentrations in other plants such as hops, mangoes, and citronella. *Pinene* is the common name for two isomeric bicyclic monoterpenoids, alpha-pinene, and beta-pinene, which are the main components of the pine resin that gives it its name, and of the resin of other conifers, being also the most widespread terpene in nature. In fact, the pine is not only found in the plant kingdom, the two compounds being part of the insect's communication system and also acting as an insecticide. Alpha-pinene is an acetylcholinesterase inhibitor. *Limonene* is a cyclic carbohydrate and the main component of lemon EOs, hence its name, and other citrus EOs. Limonene is also the second most widespread terpene in nature and is an intermediate in other terpene biosynthesis processes. In contrast to pine, limonene is not found in insects, but it still has some insecticidal effects. Limonene is widely used in the food and pharmaceutical industry as a flavoring. *Linalool* is a linear monoterpenic alcohol, resulting from the main substances of lavender EOs, but we also find it in many other plants. It is widely used as a flavoring in cleaning and hygiene products, as an intermediate in the chemical industry and as an insecticide against flies and beetles, but is not a broad-spectrum insecticide. Lavender EOs relieve skin burns and can even reduce the need for morphine when inhaled by patients with postoperative treatment. These effects are attributed to linalool because it is the main component of lavender EOs, because after ingestion, other substances, such as monoterpene linalyl acetate, hydrolyze into linalool. *Eucalyptol*, also known as 1,8-cineole, is a monoterpenic ester that is almost entirely eucalyptus oil, hence its name, but is also widespread in the plant kingdom. It acts as an insecticide, although it is produced by certain species of orchids to attract bees. Eucalyptol is used as a food additive to add a certain flavor. *Caryophyllene* is the generic name for a mixture of three compounds: alpha-caryophyllene or humulene, beta-caryophyllene, which is the main component of black pepper EOs and caryophyllene oxide, the result of oxidation of the rhubarb and eucalyptus plant [41].

The EOs isolated from rose, mint, lemon, and lavender contain numerous oxygenated monoterpenes, aliphatic, and aromatic compounds that give these oils their unique pleasant scent. Other non-aromatic terpenes and higher molecular weight terpenes can act as protective compounds, giving the plant a more bitter taste, a pungent aroma or a sticky texture, to repel potentially harmful predators. Another protective feature of terpenes is hydrophobicity (the water repellent characteristic), which allows them to easily pass through the membrane of invading cells. When crossing a cell membrane, these compounds can increase the fluidity of the membrane, so that the cell no longer has the ability to maintain a balanced internal environment. Because cell survival depends heavily on the balance of the internal environment, this can cause apoptosis (cell death) [42].

Although terpenes often do not endanger the lives of large organisms (such as animals and humans), they can be effective against many environmental threats.

13.6 Conclusions and Remarks

EOs consist largely of monoterpenes and a variable proportion of sesquiterpenes. These proportions, together with the extraction performance, will be mainly affected by the drying degree of the plant, when it is processed for the extraction of EOs. In fact, the performance of EOs extraction by steam distillation of fresh plant is less than 1%, with a composition of 80–90% monoterpenes and 10–20% sesquiterpenes. However, the extraction performance will be around 0.1% in the case of the dried plant and its composition will be lower in monoterpenes, reaching instead up to 50% sesquiterpenes, due to the fact that monoterpenes are very volatile and evaporate quickly during the process of drying plants. Some sesquiterpenes remain in the plant even after 15 minutes of decarboxylation at 120 °C. This is the case of caryophyllene, which has the characteristic aroma of moist earth of ripe or boiled hemp. Also, the evaporation of monoterpenes during the drying process is responsible for the transformation of the aroma from that of a fresh plant to the aroma of a well-dried plant, although the change in taste comes from the degradation of chlorophyll. Thus, fresh plants have menthol, citrus, fruity aromas, etc., which fade when the plants are dry. However, terpenes are not only responsible for the aroma of plants, but they also have an important biological and therapeutic activity. It has been proven that essential plant oils have therapeutic properties and are the pharmacological basis of aromatherapy. These pure oils and terpenes can also be used as flavorings in the food industry, being nontoxic compounds. The therapeutic properties will depend specifically on the respective terpene.

Plants need more attention in order to complete and complete studies on immunity, because acting at this level can ensure the food needs of the population on Earth. It is likely that the slow progress in the study of plant immunity can be attributed to the impossibility of knowing the mechanisms at the cellular level due to the lack of technology to allow this. In recent years, science and technology have advanced almost simultaneously, which gives us hope that perhaps in the coming years the food needed for the population of our planet can be provided. Nutritional support must provide all the necessities of human development, both quantitatively and qualitatively, and this poses quite serious problems. However, due to the fact that we have clear data about some plant immune mechanisms and the fact that at certain levels it has been possible to manipulate plants in order to reduce the activity of pathogens, we can hope that in the future, plant diseases will no longer be an economic catastrophe.

References

1 Ross, S.M. (2010). Aromatic plants, spirituality, and sacred traditions II. *Holistic Nursing Practice* 24 (6): 355–357.

2 Luo, X.N., Yuan, M., Li, B.J. et al. (2020). Variation of floral volatiles and fragrance reveals the phylogenetic relationship among nine wild tree peony species. *Flavour and Fragrance Journal* 35: 227–241.

3 Guo, X., Ho, C.T., Schwab, W. et al. (2019). Aroma compositions of large-leaf yellow tea and potential effect of theanine on volatile formation in tea. *Food Chemistry* 280: 73–82.

4 Chaintreau, A., Bicchi, C., and Rubiolo, P. (2018). Quantifying the constituents of flavours, fragrances and essential oils. *Flavour and Fragrance Journal* 33: 337–339.

5 Koopman, W., Wissemann, V., De Cock, K. et al. (2008). AFLP markers as a tool to reconstruct complex relationships, a case study in Rosa (Rosaceae). *American Journal of Botany* 95: 353–366.

6 Yazaki, K., Arimura, G.I., and Ohnishi, T. (2017). 'Hidden' terpenoids in plants: their biosynthesis, localization and ecological roles. *Plant and Cell Physiology* 58: 1615–1621.

7 Hung, K., Hu, X., and Maimone, T.J. (2018). Total synthesis of complex terpenoids employing radical cascade processes. *Natural Product Reports* 35: 174–202.

8 El Hadi, M.A., Zhang, F.J., Wu, F.F. et al. (2013). Advances in fruit aroma volatile research. *Molecules* 18: 8200–8229.

9 Kiran, G.D., Singh, B.B., Joshi, V.P., and Singh, V. (2002). Essential oil composition of Damask rose (*Rosa damascena* Mill) distilled under different pressures and temperatures. *Flavour and Fragrance Journal* 17: 136–140.

10 Rodríguez-Concepción, M. (2014). Plant isoprenoids: a general overview. *Methods in Molecular Biology* 1153: 1–5.

11 Truan, C., Peres, C., and Engel, E. (2020). Unraveling ingredients in complex mixtures by chromatographic spectrum recognition: application to perfume deformulation. *Flavour and Fragrance Journal* 35: 309–319.

12 Tetali, S.D. (2019). Terpenes and isoprenoids: a wealth of compounds for global use. *Planta* 249: 1–8.

13 Miyazawa, Y., Ohashi, T., Kawaguchi, K. et al. (2020). Synthesis and odour evaluation of double-bond isomers of DAMASCENOLIDE, 4-(4-methylpent-3-en-1-yl)-2(5*H*)-furanone, which has a citrus-like odour. *Flavour and Fragrance Journal* 35: 341–349.

14 Boelens, M.H. and Van Gemert, L.J. (1993). Volatile character-impact sulfur compounds and their sensory properties. *Perfumer and Flavorist* 18: 29–39.

15 Hoffmann, M., Sita, J.P.M.K., Kleider, C. et al. (2019). (*R*)-Tonkafuranone and related compounds: improved synthesis, stereochemical purity in nature, and bioactivities of the pure enantiomers. *Flavour and Fragrance Journal* 34: 329–338.

16 Chapuis, C., Skuy, D., and Richard, C.A. (2019). Syntheses of methyl jasmonate and analogues. *Chimia* 73: 194–204.

17 Bhavsar, K.V. and Yadav, G.D. (2019). Synthesis of geranyl acetate by transesterification of geraniol with ethyl acetate over *Candida antarctica* lipase as catalyst in solvent-free system. *Flavour and Fragrance Journal* 34: 288–293.

18 Tholl, D. (2015). Biosynthesis and biological functions of terpenoids in plants. *Advances in Biochemical Engineering/Biotechnology* 148: 63–106.

19 Holstein, S.A. and Hohl, R.J. (2004). Isoprenoids: remarkable diversity of form and function. *Lipids* 39: 293–309.

20 Cataldo, V.F., Lopez, J., Carcamo, M., and Agosin, E. (2016). Chemical vs. biotechnological synthesis of C-13-apocarotenoids: current methods, applications and perspectives. *Applied Microbiology and Biotechnology* 100: 5703–5718.

21 Bonikowski, R., Switakowska, P., Jablonska, A. et al. (2014). Characteristics of some synthetic terpenoids with a grapefruit odour discovered by serendipity. *Flavour and Fragrance Journal* 29: 380–387.

22 Berger, R.G. (2009). Biotechnology of flavours-the next generation. *Biotechnology Letters* 31: 1651–1659.

23 Merle, P. (2018). Clearwood((R)) & Patchouli: some clarifications. *Flavour and Fragrance Journal* 33: 367–367.

24 Brazinha, C. and Crespo, J.G. (2009). Aroma recovery from hydro alcoholic solutions by organophilic pervaporation: modelling of fractionation by condensation. *Journal of Membrane Science* 341: 109–121.

25 Chin, S.T. and Marriott, P.J. (2015). Review of the role and methodology of high resolution approaches in aroma analysis. *Analytica Chimica Acta* 854: 1–12.

26 Lim, B., Jung, H., Yoo, H. et al. (2020). Synthetic strategy for tetraphenyl-substituted all-E-carotenoids with improved molecular properties. *European Journal of Organic Chemistry* 2020: 1769–1777.

27 You, J.S., Jeon, S., Byun, Y.J. et al. (2015). Enhanced biological activity of carotenoids stabilized by phenyl groups. *Food Chemistry* 177: 339–345.

28 Frank, A. and Groll, M. (2017). The methylerythritol phosphate pathway to isoprenoids. *Chemical Reviews* 117: 5675–5703.

29 Lempenauer, L., Appleson, T., Lemiere, G., and Dunach, E. (2019). Synthesis and olfactory evaluation of allylic alpha-quaternary thioether ketones. *Flavour and Fragrance Journal* 34: 36–42.

30 Sobotková, M., Fialová, J., Roberts, S.C., and Havlíček, J. (2017). Effect of biological relatedness on perfume selection for others: preliminary evidence. *Perception* 46 (3–4): 498–515.

31 Sanchez-Prado, L., Llompart, M., Lamas, J.P. et al. (2011). Multicomponent analytical methodology to control phthalates, synthetic musks, fragrance allergens and preservatives in perfumes. *Talanta* 85: 370–379.

32 Plessis, C. (2008). The search for innovative fragrant molecules. *Chemistry & Biodiversity* 5: 1083–1098.

33 Yang, T., Stoopen, G., Thoen, M. et al. (2013). Chrysanthemum expressing a linalool synthase gene 'smells good', but 'tastes bad' to western flower thrips. *Plant Biotechnology Journal* 1 (7): 875–882.

34 Filiou, R.P., Lepore, F., Bryant, B. et al. (2015). Perception of trigeminal mixtures. *Chemical Senses* 40 (1): 61–69.

35 Welge-Lüssen, A., Drago, J., Wolfensberger, M., and Hummel, T. (2005). Gustatory stimulation influences the processing of intranasal stimuli. *Brain Research* 1038 (1): 69–75.

36 Stotz, S.C., Vriens, J., Martyn, D. et al. (2008). Citral sensing by transient [corrected] receptor potential channels in dorsal root ganglion neurons. *PLoS One* 3 (5): e2082.

37 Li, S., Crump, A.M., Grbin, P.R. et al. (2015). Aroma potential of oak battens prepared from decommissioned oak barrels. *Journal of Agricultural and Food Chemistry* 63: 3419–3425.

38 Schütze, M., Negoias, S., Olsson, M.J., and Hummel, T. (2014). Perceptual and processing differences between physical and dichorhinic odor mixtures. *Neuroscience* 258: 84–89.

39 Hummel, T., Rissom, K., Reden, J. et al. (2009). Effects of olfactory training in patients with olfactory loss. *Laryngoscope* 1193: 496–499.

40 Takahashi, T., Mizui, K., and Miyazawa, M. (2010). Volatile compounds with characteristic odour in moso-bamboo stems (*Phyllostachys pubescens* Mazel ex Houz. De ehaie). *Phytochemical Analysis* 21: 489–495.

41 De Vincenzi, M., Silano, M., De Vincenzi, A. et al. (2002). Constituents of aromatic plants: eucalyptol. *Fitoterapia* 73: 269–275.

42 Kiyama, R. (2017). Estrogenic terpenes and terpenoids: pathways, functions and applications. *European Journal of Pharmacology* 815: 405–415.

14

Bioprospection of Plants for Essential Mineral Micronutrients

Nikita Bisht[1,2] *and Puneet Singh Chauhan*[1,2]

[1] *CSIR-National Botanical Research Institute (CSIR-NBRI), Lucknow, India*
[2] *Academy of Scientific and Innovative Research (AcSIR), Ghaziabad, India*

14.1 Introduction

As a result of the green revolution, hunger/energy deficits have seen steady improvement in developing countries. According to the Food and Agriculture Organization (FAO), the undernourishment in the global population has declined from 18.6% in 1990–1992 to 10.9% in 2014–2016. But still, one in nine people worldwide suffers from hunger [1]. Regardless of the improvement, more than two billion people suffer from deficiency of micronutrients, also recognized as hidden hunger [2]. Micronutrients, as the name suggests, are the compounds that are required in small amounts to aid the growth, development, and maintenance of the body and are classified into two groups, i.e. vitamins and minerals [3]. Micronutrients are essential to human health and are significant for human body functions [4]. Despite their small requirement, micronutrient deficiencies, in particular of mineral micronutrients such as iron (Fe) and zinc (Zn), are common in developing countries, as well as in developed countries [5]. Micronutrient shortages thus create major public health problems worldwide, posing nutritional and serious health effects to approximately two billion people and causing 25 000 children to die every day [6, 7].

14.2 Plants as a Source of Mineral Micronutrients

The mineral elements that living beings require come into the food chain via plants. Therefore, mineral element concentrations in the edible plant tissues are essential for human nutrition. Up to two-thirds of the global population are estimated to be at risk of deficiencies in one or more important minerals with deficiencies of Fe and Zn being the most frequent [8, 9]. People require mineral elements for their well-being, and plants are the dietary source of most of these elements [10]. The diets widely eaten worldwide include a variety of plant products such as grains and vegetables. The combination of different food

Bioprospecting of Plant Biodiversity for Industrial Molecules, First Edition.
Edited by Santosh Kumar Upadhyay and Sudhir P. Singh.

ingredients and culinary practices such as heating, sprouting, fermentation, etc. can therefore have a significant impact on the availability of micronutrients from the plant-based diet [5]. However, edible plant tissues may contain low concentration of mineral elements for different reasons, or some plant species inherently have low concentration of particular mineral elements, for example, crops grown in areas with low phyto availability of minerals [11–15], or those edible plant parts, such as fruits, seeds, and tubers are used that are naturally low in mineral elements with restricted phloem mobility [8, 16].

14.3 Bioavailability of Micronutrients from Plants

Plants are packed with essential mineral micronutrients, but it is important to better understand their bioavailability to derive appropriate amount of nutrition from them. In a plant-based diet, the key concept for achieving balanced nutrition is food synergy that is defined as an additive or more than additive influences on health from dietary patterns, foods, and food constituents [17]. Maintaining the bioavailability of micronutrients as a cornerstone and developing intelligent food synergies have the benefit of being close to the population's psyche but involve efforts to build them as national missions [18]. Mineral micronutrients such as Fe and Zn have poor bioavailability from plant foods that are also affected by various dietary components including both inhibitors and enhancers of absorption. Phytic acid, tannins, dietary fibers, and calcium are among the most active inhibitors, while organic acids are known to help absorb Fe. Zn bioavailability from food grains has also been reported to be similarly influenced by the dietary factors [5]. Table 14.1 summarizes the micronutrient content of different food groups derived from plants based on the food composition database [18, 19].

14.3.1 Bioavailability of Fe and Zn

Nonheme Fe is the most important form of Fe present in plant foods. The absorption of nonheme Fe is inhibited by the phytic acid found in whole grains, lentils, and nuts. Moreover, in tea, coffee, red wines, and a number of cereals, vegetables, and spices, polyphenols impede Fe absorption [18, 20]. In India, a study was conducted on previous reports to estimate Fe absorption and was found to be 5% for adult male and 10% for adult female [21, 22] and was 12–15% low compared to the suggested Western diets [21, 22]. Another report on Fe bioavailability documented that the bioavailability of Fe in the cereals and

Table 14.1 Recommended dietary limits and dietary micronutrient content (per 100 mg) for adults.

			Plant-derived foods				
Micronutrient	**AR (M/F)**	**RDA**	**Vegetables**	**Fruits**	**Cereals, millets**	**Pulses, legumes**	**Nuts, oilseeds**
Iron (mg)	6/8.1	17/21	0.99	1.4	3.78	5.57	5.4
Zinc (mg)	9.4/6.8	12/10	0.34	0.3	1.71	3.31	5.18

AR (M/F) = average requirement (Male/Female); RDA = recommended dietary allowances.

millets ranged from 7.1% in pearl millets to 15% in rice [5, 23]. However, in the same study authors reported that in another study bioavailability of Fe differed greatly and suggested that the differences in the values could be due to regional differences in the food grains examined and the procedure employed for determination [5, 23].

Plant foods such as whole grains, legumes, nuts, and seeds are rich source of Zn, but they also contain high amounts of phytic acids (inhibitor of Zn absorption) due to which only 30–35% absorption of Zn takes place [24]. Thus, as compared to the nonvegetarian foods that are high in bioavailable Zn, the bioavailability of Zn from vegetarian foods is lower [24, 25]. Moreover, when the bioaccessibility of Fe and Zn was determined, the bioaccessibility of Zn from food grains was higher than that of Fe, and the difference in pulses was more prominent. While Fe's bioaccessibility from cereals ranged from 4% in sorghum to 8% in rice, Zn's bioaccessibility ranged from 5.5% in sorghum to 21% in rice. The availability of micronutrients from the diets was very poor, ranging between 3.3 and 4.4% for Fe, and between 7.8 and 8.7% for Zn. The low quality of minerals was due to the presence of high phytate and dietary fiber content in the diet [26]. Given comparable concentrations of Fe and Zn in rice and finger millet–based meals, the bioaccessibility of both of these minerals in finger millet–based meal was lower, and this was due to the higher tannin content of the finger millet, which was found to be the only difference in the two meals [27]. Furthermore, when the availability of Fe and Zn from 60 vegetarian diets eaten by infants, teenagers, adults, and older adults was evaluated in vitro, the abundance of these minerals was found to be very low in terms of diets, ranging from 3.3 to 4.4% for Fe and from 7.8 to 8.7% for Zn. [28].

14.3.2 Impact of Food Processing on Micronutrient Bioavailability from Plant Foods

It is generally observed that most of the plant-based food that is consumed normally undergoes some form of processing that leads to alterations in the food matrix and inherent components of foods. Hence, food processing techniques such as heat processing, sprouting, fermentation, etc. affect nutrient bioavailability from the plant-based foods (Figure 14.1).

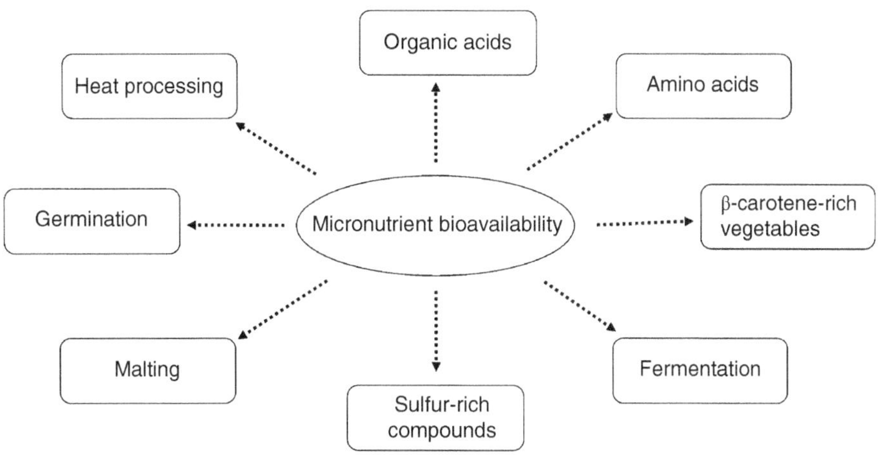

Figure 14.1 Different food processing techniques that influence nutrient bioavailability from the plant-based foods.

14.4 Manipulating Plant Micronutrients

The concentration of mineral elements in crops can be increased by the careful application of mineral fertilizers and/or by the use of genotypes with a higher mineral content. Besides, mineral element bioavailability may also be increased by crop husbandry, breeding, or genetic engineering [8]. For ensuring sufficient dietary intake of essential mineral nutrients, scientists have turned with renewed interest for the research and manipulation of plants. A major focus is to identify and isolate the genes needed to synthesize and accumulate a target compound in order to induce the desired dietary shift by increasing its levels in staple crops. A critical point for all micronutrient research, however, is that unlike macronutrients that can account for up to 30% of a tissue's dry weight, individual micronutrients are generally much less than 0.1% of a tissue's dry weight, and thus it is technically feasible to increase the micronutrient levels [29]. Moreover, due consideration should be given to identifying target compounds, their efficacy, and whether excessive dietary consumption may have unintended negative health consequences before attempting to alter nutrient components in food crops. Before addressing the potential of manipulating plant micronutrient content through emerging technologies, it is important to emphasize that there are more conventional approaches and that both separately and in conjunction with emerging technologies should be followed. Modern farming and breeding programs have aimed mainly to increase productivity and yields over the past 50 years, a task that will remain a major concern in supplying the caloric intake required by the world's increasing population. Nevertheless, the micronutrient composition and crop density are equally important but largely overlooked in breeding programs. In the rare cases where the content of micronutrients was assessed, significant genotypical variations were observed. Such variability can and will be used to grow nutritionally improved cultivars and to help establish the genetic and physiological basis for differences in nutrients [29]. Plant scientists have started to use genomic tools and the DNA technology's ability to research all areas of plant biology. The established biochemical knowledge for individual pathway steps, well-defined protein motifs common to different reaction mechanisms, and the presence or absence of subcellular targeting information in the primary amino acid sequence are combined for the identification of gene(s) by bioinformatics. These techniques complement biochemical and genetic approaches and can easily be integrated to add new dimensions to complex plant pathways elucidation.

14.4.1 Improving Bioavailability of Micronutrients from Plant Foods

Improving the micronutrient content of crop products through biotechnology is a promising technique for the worldwide fight against micronutrient malnutrition. Modifying the nutritional composition of plant foods is a critical global health issue, as essential nutritional requirements for most part of the world's population remain unfulfilled. Throughout most of the developing countries, there are large numbers of people on simple diets that consist mainly of a few staple foods (cassava, wheat, rice, maize, etc.) that are poor sources of several important micronutrients [29]. To fix mineral micronutrient shortages in human populations, plant scientists are working to establish methods for the application of fertilizers and/or to encourage plant breeding techniques to increase concentrations/bioavailability of mineral elements in plant products [2, 8, 13, 14, 30]. These approaches to

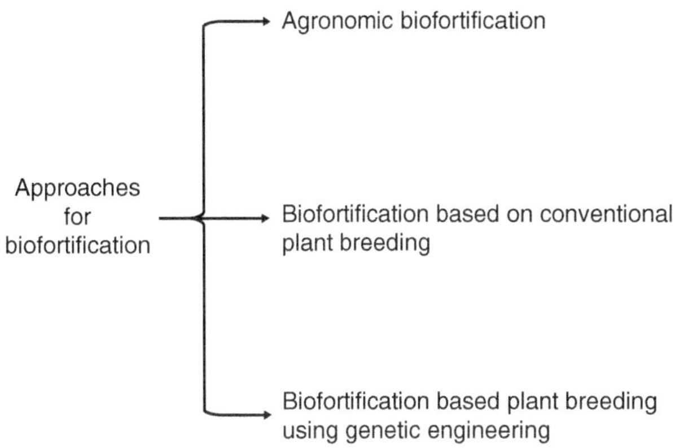

Figure 14.2 Different approaches that are used for the biofortification of mineral micronutrients in agricultural crops.

biofortification are classified according to genetic engineering as agronomic, conventional plant breeding, and plant breeding (Figure 14.2).

Agronomic biofortification through the application of fertilizers temporarily enriches micronutrients. This approach helps to increase micronutrients that can be ingested directly by the plant, such as Zn, but less so for micronutrients that are synthesized in the plant and cannot be absorbed directly [31]. In the light of sustainable economic growth and environmental health, various authors have recently examined the agronomic approaches for growing mineral nutrient concentrations in edible portions of major crop plants [2, 8, 13, 14, 30]. These included reviews of suitable methods, requirements for infrastructure, and practical benefits of agronomic biofortification of edible crops with Fe and Zn for economic sustainability, food production, and human health [8, 13, 32]. Conventional plant breeding recognizes and establishes high-mineral parent lines and crosses and segregates the generations to grow plants with the required nutrient and agronomic characteristics [30, 33]. Biofortification by genetic engineering aims to do the same and was primarily used in crops where the target nutrient naturally does not exist at the necessary levels [2]. Researchers have also tried to investigate genetic variation in mineral concentrations, the interaction between the environment and the genotype, and breeding potential for increased mineral element concentrations in the produce [8, 14, 30]. Previous studies provided a detailed overview of genetic factors influencing concentrations of mineral elements in edible tissues of popular crops and also established research under the Harvest Plus program to increase concentrations of Fe and Zn in dietary staples [8, 33].

14.4.2 Metabolic Engineering of Micronutrients in Crop Plants

Metabolic engineering can be used, as in molecular breeding, to produce crops with increased micronutrient content, each with its advantages and disadvantages. Metabolic engineering uses genetic modification by engineering metabolic pathways to enhance nutritional value in crop plants. This strategy involves modulating endogenous metabolic pathways or introducing one or the introduction of one or more heterologous performers to enhance the production of a target compound, decrease the amount of unwanted molecules, or alter the flux to accumulate a more bioavailable, stable, and active compound [31]. However, this requires detailed knowledge of the endogenous metabolic pathways involved in the regulation. Only then a successful engineering strategy can be developed for the overexpression/downregulation of key enzymes gene(s) responsible to enhance micronutrient biosynthesis/accumulation without affecting crop development and yield.

Fe and Zn are essential elements for many of the human metabolic processes [34, 35]. The biofortification of crop with these micronutrients involves many processes to be orchestrated, such from taking up by the roots to retaining in the edible parts. From the rhizosphere, Fe and Zn are absorbed via the root epidermis, taken to the xylem, and transferred to different tissues throughout the plant where they can be stored. The transgenic strategy for increasing the content of Fe and Zn in crops is to primarily improve the absorption and effectiveness of Fe and Zn by modulating transporter gene(s) expression [31, 36] and reducing antinutritional factors such as phytic acid [37]. The coexpression of LACTOFERRIN (Fe-chelating glycoprotein) and FERRITIN further enhances the content of Fe in crops [38–41]. The simultaneous expression of NAS (nicotianamine synthase) and Ferritin increases not only the content of Zn but also iron content in crops [42–44]. In transgenic rice, it has been documented that the combined expressions of four genes, NAS, FERRITIN, PSY, and CrtI, increase the Fe, Zn, and β-carotene contents significantly [45]. Therefore, a number of biofortified crops such as rice, maize, and wheat with enriched contents of Fe and Zn were produced to counter global human mineral deficiencies [46].

While the metabolic engineering has made tremendous progress in the field of crop biofortification, many challenges still remain [47]. The main issue is lack of knowledge of an organism's metabolic pathways and key regulators. Though at present the sequencing of plant genomes has become easy, but the lack of annotations for effective gene functions makes it difficult to determine the composition of genes that encode key enzymes in different metabolic pathways. The constitutive metabolite synthesis is likely to trigger abnormal cell development and plant growth. Metabolic processes may at the same time be affected by feedback regulation. Nevertheless, the expression of transgenes expressed in specific tissues must increase in order to achieve metabolite synthesis and accumulation in specific tissues and to avoid adverse effects on normal plant development [48]. Furthermore, most of the metabolic pathways require numerous regulatory factors and enzymes. The application of high-efficiency multigene expression vector systems (TGS II system) and the CRISPR gene editing tool would enable the expression and regulation of upstream and downstream genes of whole metabolic pathways in more versatile and precise ways [49–51]. Thus, with the advances in metabolic engineering technology, and deep understanding of the metabolic pathways, synthetic metabolic engineering can then accomplish more accurate reconstruction and regulation of complex multistage metabolic networks, producing more novel biofortified crop varieties.

14.5 Microbes in the Biofortification of Micronutrients in Crops

The concept of hidden hunger is well documented in the last two decades [52]. Different strategies have been used to grow biofortified crop varieties with improved micronutrient bioavailability, such as conventional and molecular plant breeding or the use of chemical supplements. The role of microorganisms in enhancing nutrient content in crops has also been documented [53–55]. There are vast amounts of Fe and Zn found in the earth's crust, but are not available to plants because they are present in the form of insoluble salts. Intrinsic plant-based strategies such as the production of organic acids or phytosiderophores/chelator secretions not always deliver sufficient micronutrients to plants [56–58]. The use of microorganisms to help the crop take up and translocate Zn and Fe more efficiently and effectively is a promising option that needs to be effectively integrated into agricultural or breeding approaches [58]. The mobilization of micronutrients by microorganisms highlighted the importance of (i) rhizospheric soil acidification, (ii) phenolic secretion stimulation, (iii) root morphology and architecture modifications, (iv) reduction of phytates, and (v) upregulation of transporters. The formulation(s) of these microbes can be explored as seed priming or soil dressing options for the biofortification of Zn and Fe [59].

14.6 Conclusions

Plants have been important for human health and wellbeing since ancient times and were used as a source of mineral micronutrients. But low concentration of mineral micronutrients and presence of their inhibitors in plants is the main problem. However, in recent years, advancement in the field of plant research have created new opportunities to extract maximum advantage from the plants. Plants are now considered as a rich resource for the production of mineral micro nutrients. With time as the genomic, transcriptomic, proteomic and metabolic data has rapidly escalated, the number of target molecules/pathways related to accumulation of micronutrients has also increased. Given that micronutrient nutrition is of global concern, it is imperative that the synergies between scientists, policy makers and educators concentrate on developing multi-pronged, environmentally friendly solutions and incorporating microbial options into the mainstream of integrated agricultural practices. Thus, convergence of all the above-mentioned factors with the potential to provide good quantities of mineral micronutrients from plants.

References

1 FAO, IFAD, and WFP (2015). *The State of Food Insecurity in the World*. Rome: FAO.

2 Saltzman, A., Birol, E., Oparinde, A. et al. (2017). Availability, production, and consumption of crops biofortified by plant breeding: current evidence and future potential. *Annals of the New York Academy of Sciences* 1390: 104–114.

3 Jones, D., Caballero, S., and Pardo, G.D. (2019). Bioavailability of nanotechnology-based bioactive and nutraceuticals. *Advances in Food and Nutrition Research* 1: 2–9.

4 Wang, K.M., Wu, J.G., Li, G. et al. (2011). Distribution of phytic acid and mineralelements in three Indica rice (*Oryza sativa* L.) cultivars. *Journal of Cereal Science* 54: 116–121.

5 Platel, K. and Srinivasan, K. (2016). Bioavailability of micronutrients from plant foods: an update. *Critical Reviews in Food Science and Nutrition* 56: 1608–1619.

6 Kirsten, G. (2010). *Global Challenges and Their Impact on International Humanitarian Action*. New York: United NationsOffice for the Coordination of Humanitarian Affairs (OCHA).

7 Anandan, A., Rajiv, G., Eswaran, R., and Prakash, M. (2011). Genotypic variation and relationships between qualitytraits and trace elements in traditional and improved rice (*Oryza sativa* L.) genotypes. *Journal of Food Science* 76: H122–H130.

8 White, P.J. and Broadley, M.R. (2009). Biofortification of crops with seven mineral elements often lacking in human diets—iron, zinc, copper, calcium, magnesium, selenium and iodine. *New Phytologist* 182: 49–84.

9 Stein, A.J. (2010). Global impacts of human mineral malnutrition. *Plant and Soil* 335: 133–154.

10 White, P.J. and Brown, P.H. (2010). Plant nutrition for sustainable development and global health. *Annals of Botany* 105: 1073–1080.

11 Frossard, E., Bucher, M., Mächler, F. et al. (2000). Potential for increasing the content and bioavailability of Fe, Zn and Ca in plants for human nutrition. *Journal of the Science of Food and Agriculture* 80: 861–879.

12 Rengel, Z. (2001). Genotypic differences in micronutrient use efficiency in crops. *Communications in Soil Science and Plant Analysis* 32: 1163–1186.

13 Cakmak, I. (2004). *Identification and Correction of Widespread Zinc Deficiency in Turkey - A Success Story*. York, UK: International Fertiliser Society.

14 Cakmak, I. (2008). Enrichment of cereal grains with zinc: agronomic or genetic biofortification? *Plant and Soil* 302: 1–17.

15 Broadley, M.R., White, P.J., Hammond, J.P. et al. (2007). Zinc in plants. *New Phytologist* 173: 677–702.

16 Karley, A.J. and White, P.J. (2009). Moving cationic minerals to edible tissues: potassium, magnesium, calcium. *Current Opinion in Plant Biology* 12: 291–298.

17 Jacobs, D.R. Jr., Tapsell, L.C., and Temple, N.J. (2009). Food synergy: the key tobalancing the nutrition research effort. *Public Health Reviews* 33: 507–529.

18 Nair, K.M. and Augustine, L.F. (2016). Food synergies for improving bioavailability of micronutrients fromplant foods. *Food Chemistry* 238: 180–185.

19 Gopalan, C., Rama Sastri, B.V., Balasubramanian, S.C. et al. (1999). *Nutritive Value of Indian Foods*. Hyderabad: National Instituteof Nutrition, ICMR.

20 Hunt, J.R. (2003). Bioavailability of iron, zinc, and other trace minerals fromvegetarian diets. *American Journal of Clinical Nutrition* 78: 633S–639S.

21 Nair, K.M. and Iyengar, V. (2009). Iron content, bioavailability and factors affecting ironstatus of Indians. *Indian Journal of Medical Research* 130: 634–645.

22 World Health Organization and Food and Agriculture Organization (2004). *Vitamin and Mineral Requirements in Human Nutrition: Report of a Joint FAO/WHO Expert Consultation, Bangkok, Thailand (21–30 September 1998)*, 2ee. Geneva: World Health Organization.

23 Rao, B.S.N. and Prabhavathi, T. (1978). An in vitro method to predict the bioavailabilityof iron from foods. *The American Journal of Clinical Nutrition* 31: 169–175.

24 Hunt, J.R. (2003). Bioavailability of iron, zinc, and other trace minerals from vegetarian diets. *The American Journal of Clinical Nutrition* 78: 633S–639S.

25 Krebs, N.F. (1998). Zinc supplementation during lactation. *The American Journal of Clinical Nutrition* 68: 509S–512S.

26 Hemalatha, S., Platel, K., and Srinivasan, K. (2007). Zinc and iron content andtheir bioaccessibility in cereals and pulses consumed in India. *Food Chemistry* 102: 1328–1336.

27 Bhavyashree, S.H., Prakash, J., Platel, K., and Srinivasan, K. (2009). Bioaccessibilityof minerals from cereal-based composite meals and ready-to-eatfoods. *Journal of Food Science and Technology* 46: 431–435.

28 Pushpanjali and Khokhar, S. (1996). In vitro availability of iron and zincfrom some Indian vegetarian diets: correlations with dietary fibre and phytate. *Food Chemistry* 56: 111–114.

29 DellaPenna, D. (1999). Nutritional genomics: manipulating plant micronutrients to improve human health. *Science* 285: 375–379.

30 Pfeiffer, W.H. and McClafferty, B. (2007). HarvestPlus: breedingcrops for better nutrition. *Crop Science* 47: S88–S105.

31 Blancquaert, D., Steur, H., Gellynck, X., and Van Der Straeten, D. (2017). Metabolic engineering of micronutrients in crop plants. *Annals of the New York Academy of Sciences* 1390: 59–73.

32 Graham, R.D., Welch, R.M., Saunders, D.A. et al. (2007). Nutritious subsistence food systems. *Advances in Agronomy* 92: 1–74.

33 Bouis, H.E. (2003). Micronutrient fortification of plantsthrough plant breeding: can it improve nutrition in manat low cost? *Proceedings of the Nutrition Society* 62: 403–411.

34 Underwood, E.J. (1977). *Trace Elements in Human Nutrition*, 4e, 545. New York: Academic Press.

35 Prasad, A.S. (1978). *Trace Elements and Iron in Human Metabolism*. New York/Chichester: Wiley.

36 Kerkeb, L., Mukherjee, I., Chatterjee, I. et al. (2008). Iron-induced turnover of the Arabidopsis iron-regulated transporter1 metal transporter requires lysine residues plant. *Physiology* 146: 1964–1973.

37 Aluru, M.R., Rodermel, S.R., and Reddy, M.B. (2011). Genetic modification of low phytic acid 1-1 maize to enhance iron content and bioavailability. *Journal of Agricultural and Food Chemistry* 59: 12954–12962.

38 Goto, F., Yoshihara, T., Shigemoto, N. et al. (1999). Iron fortification of rice seed by the soybean ferritin gene. *Nature Biotechnology* 17: 282–286.

39 Drakakaki, G., Christou, P., and Stoger, E. (2000). Constitutive expression of soybean ferritin cDNA in transgenic results in increased iron levels in vegetative tissues but not in seeds. *Transgenic Research* 9: 445–452.

40 Lucca, P., Hurrell, R., and Potrykus, I. (2001). Genetic engineering approaches to improve the bioavailability and the level of iron in rice grains. *Theoretical and Applied Genetics* 102: 392–397.

41 Borg, S., Brinch-Pedersen, H., Tauris, B. et al. (2012). Wheat ferritins, improving the iron content of the wheat grain. *Journal of Cereal Science* 56: 204–213.

42 Lee, S., Jeon, U.S., Lee, S.J. et al. (2009). Iron fortification of rice seeds through activation of the nicotianamine synthase gene. *Proceedings of the National Academy of Sciences of the United States of America* 106: 22014–22019.

43 Wirth, J., Poletti, S., Aeschlimann, B. et al. (2009). Rice endosperm iron biofortification by targeted and synergistic action of nicotianamine synthase and ferritin. *Plant Biotechnology Journal* 7: 631–644.

44 Zheng, L., Cheng, Z., Ai, C. et al. (2010). Nicotianamine, a novel enhancer of rice iron bioavailability to humans. *PLoS One* 5: e10190.

45 Singh, S.P., Gruissem, W., and Bhullar, N.K. (2017). Single genetic locus improvement of iron, zinc and β-carotene content in rice grains. *Scientific Reports* 7: 6883.

46 Kumar, S., Palve, A., Joshi, C. et al. (2019). Crop biofortification for iron (Fe), zinc (Zn) and vitamin A with transgenic approaches. *Heliyon* 5: e01914.

47 García-Granados, R., Lerma-Escalera, J.A., and andMorones-Ramírez, J.R. (2019). Metabolic engineering and synthetic biology: synergies, future, and challenges. *Frontiers in Bioengineering and Biotechnology* 7: 36.

48 Peremarti, A., Twyman, R.M., Gómez-Galera, S. et al. (2010). Promoter diversity in multigene transformation. *Plant Molecular Biology* 73: 363–378.

49 Zalatan, J.G., Lee, M.E., Almeida, R. et al. (2015). Engineering complex synthetic transcriptional programs with CRISPR RNA scaffolds. *Cell* 160: 339–350.

50 Li, Z., Zhang, D., Xiong, X. et al. (2017). A potent Cas9-derived gene activator for plant and mammalian cells. *Nature Plants* 3: 930–936.

51 Zhu, Q., Yu, S., Zeng, D. et al. (2017). Development of "purple endosperm rice" by engineering anthocyanin biosynthesis in the endosperm with a high-efficiency transgene stacking system. *Molecular Plant* 10: 918–929.

52 Nilson, A. and Piza, J. (1998). Food fortification: a tool for fighting hidden hunger. *Food and Nutrition Bulletin* 19: 49–60.

53 Zaidi, A., Khan, M.S., and Amil, M.D. (2003). Interactive effect of rhizotrophic microorganisms on yield and nutrient uptake of chickpea (*Cicer arietinum* L.). *European Journal of Agronomy* 19: 15–21.

54 Bisht, N. and Chauhan, P.S. (2020). Comparing the growth-promoting potential of *Paenibacillus lentimorbus* and *Bacillus amyloliquefaciens* in *Oryza sativa* L. var. Sarju-52 under suboptimal nutrient conditions. *Plant Physiology and Biochemistry* 146: 187–197.

55 Bisht, N., Mishra, S.K., and Chauhan, P.S. (2020). *Bacillus amyloliquefaciens* inoculation alters physiology of rice (*Oryza Sativa* L. Var. IR-36) through modulating carbohydrate metabolism to mitigate stress induced by nutrient starvation. *International Journal of Biological Macromolecules* 143: 937–951.

56 de Santiago, A., Quintero, J.M., Avilés, M., and Delgado, A. (2011). Effect of *Trichoderma asperellum* strain T34 on iron, copper, manganese, and zinc uptake by wheat grown on a calcareous medium. *Plant and Soil* 342: 97–104.

57 Mishra, P.K., Bisht, S.C., Ruwari, P. et al. (2011). Bioassociative effect of cold tolerant *Pseudomonas* spp. and *Rhizobium leguminosarum*-PR1 on iron acquisition, nutrient uptake and growth of lentil (*Lens culinaris* L.). *European Journal of Soil Biology* 47: 35–43.

58 Pii, Y., Mimmo, T., Tomasi, N. et al. (2015). Microbial interactions in the rhizosphere: beneficial influences of plant growth-promoting rhizobacteria on nutrient acquisition process. A review. *Biology and Fertility of Soils* 51: 403–415.

59 Singh, D. and Prasanna, R.P. (2020). Potential of microbes in the biofortification of Zn and Fe in dietary food grains. A review. *Agronomy for Sustainable Development* 40: 15.

15

Algal Biomass

A Natural Resource of High-Value Biomolecules

Dinesh Kumar Yadav, Ananya Singh, Variyata Agrawal, and Neelam Yadav

Molecular Biology and Genetic Engineering Laboratory, Department of Botany, University of Allahabad, Prayagraj, India

15.1 Introduction

Algae (microalgae and macroalgae) the extremely diverse group of photoautotrophs have almost ubiquitous inhabitation preferentially in aquatic environments. They successfully dwell in highly acidic and frozen soils, rocks, snow, in/on plants and animals as symbionts, hot spring, volcanic waters, and arid conditions [1]. They represent ecologically imperative group of organisms.

All microscopic unicellular prokaryotic (Cyanophytes) and eukaryotic (other algae phyla), and microscopic multicellular algae are commonly known as *microalgae*, whereas macroscopic, multicellular algae are called as *macroalgae*, are found in fresh and marine waters, and frequently referred as seaweeds. Algae constitute a fundamental part of aquatic food chains in various ecosystems. Microalgae contribute up to 80% of the biomass and primary productivity in marine ecosystems and perform >40% of global photosynthesis [2]. Representing their enormous ability to sequester carbon dioxide through photosynthesis and produce a huge biomass resource.

Algae have been classified based on their photosynthetic pigments, reserve food materials, and mode of reproduction. The additional criteria considered for their classification include morphological and cytological features, chemical composition of cell wall, and specific gene sequences along with photosynthetic pigments [3]. The latest seven kingdoms classification has divided living organisms into Archaebacteria, Eubacteria, Protozoa, Chromista, Fungi, Plantae, and Animalia. Seven-kingdom classification has distributed algae into eight phyla under four kingdoms [4]. Majority of algae have been clustered into seven phyla under domain Eukaryota. The Heterokontophyta lineage has c. 15 000 identified and about 10 million anonymous species remain. The bacillariophytes or diatoms are predominant form in phytoplankton populations and largest group of biomass producers on the Earth [5]. Rhodophtya lineage has ~5000 identified and ~15 000 new unidentified

Bioprospecting of Plant Biodiversity for Industrial Molecules, First Edition.
Edited by Santosh Kumar Upadhyay and Sudhir P. Singh.

species. Chlorophytes have ~16 000 known and ~100 000 undescribed extant species. The members of four minor lineages (Dinophyceae, Euglenophyceae, Cryptophyceae, and Glaucophyceae) are very ambivalent and under revision [2]. The domain Prokaryota has a single algal phylum, Cyanobacteria that represent another highly abundant group of algae next to Heterokontophyta and Chlorophyta [6, 7].

Algae are one of the fastest-growing life forms with short multiplication time, which may be due to their enormous strain diversity showing remarkable tolerance to seasonality leading to >3% greater annual photon-to-biomass conversion efficiencies [8]. Based on these properties, algae offer tremendous natural resource for commercial exploitation of high-value molecules like proteins, carbohydrates, lipids, triglycerides, pigments, polyphenols, phytosterols, hormones, vitamins, minerals for food and health, aquaculture feed, animal feed, dyes, cosmetics, and others. With the increasing awareness for advantages of using natural products over the synthetic ones, their utilization is expanding especially in food and pharmaceutical industries. Additionally, algae can be cultivated on nonagricultural landmass or wastewater bodies without pesticide so quality of food or produce is not compromised and atmospheric CO_2 sequestration reduces environmental issues. Improvement in bioprocessing strategies for biomolecules extraction, bioconversion (chemical catalysis maceration, maturation, and supercritical liquid extraction), and algal biotechnology has led to unexceptional utilization algal bio-resource to meet current food, energy, and environmental challenges [9–11]. This chapter focuses on algal bio-resources under four major categories, i.e., carbon dioxide capture and sequestration, high-value biomolecules, biofuel production, and feed for livestock and aquaculture (Figure 15.1).

15.2 Carbon Dioxide Capture and Sequestration

Carbon dioxide is the principal control button that governs the Earth's temperature. So, menace caused by global warming is worsening due to increasing atmospheric CO_2 concentration by anthropogenic activities. The global emission of CO_2 has increased from 2 to 36 billion tons in 115 years. The rate of increase in atmospheric CO_2 concentration in 2018 was recorded as 1.231 ppm year^{-1}. The rate CO_2 emission increased by eightfold in 2018 from that of 1960 [12]. At this rate of increment, CO_2 is the major component augmenting in total global warming, and climate change that may lead to substantial biological extinctions.

Algae have being considerably recognized as most productive biological systems for CO_2 capture and biomass generation. Microalgae have ~10 fold higher photosynthetic efficiency compared to land plants [13]. Tsai et al. [14] reported that algae can absorb 159 mg l^{-1} day^{-1} CO_2 at a consumption efficiency of 93% in high rate ponds (lab-scale algal niche), compared to continuous stirred tank reactor (178 mg^{-1} l^{-1} day^{-1} and 96%). The bicarbonate pumps arrayed in plasma- and chloroplast outer membrane of eukaryotic algae transport bicarbonate ions into the cells with ~90% efficiency in open water bodies [15]. The concentrated bicarbonate in the chloroplast is dehydrated spontaneously or by carbonic anhydrase, releasing CO_2 for photosynthesis and producing algal biomass. Some dominant algal species suitable for CO_2 capture are *Anabaena, Oscillatoria, Lyngbya, Spirulina, Chlorella, Chlamydomonas, Monoraphidium, Nannochloropsis, Oedogonium,*

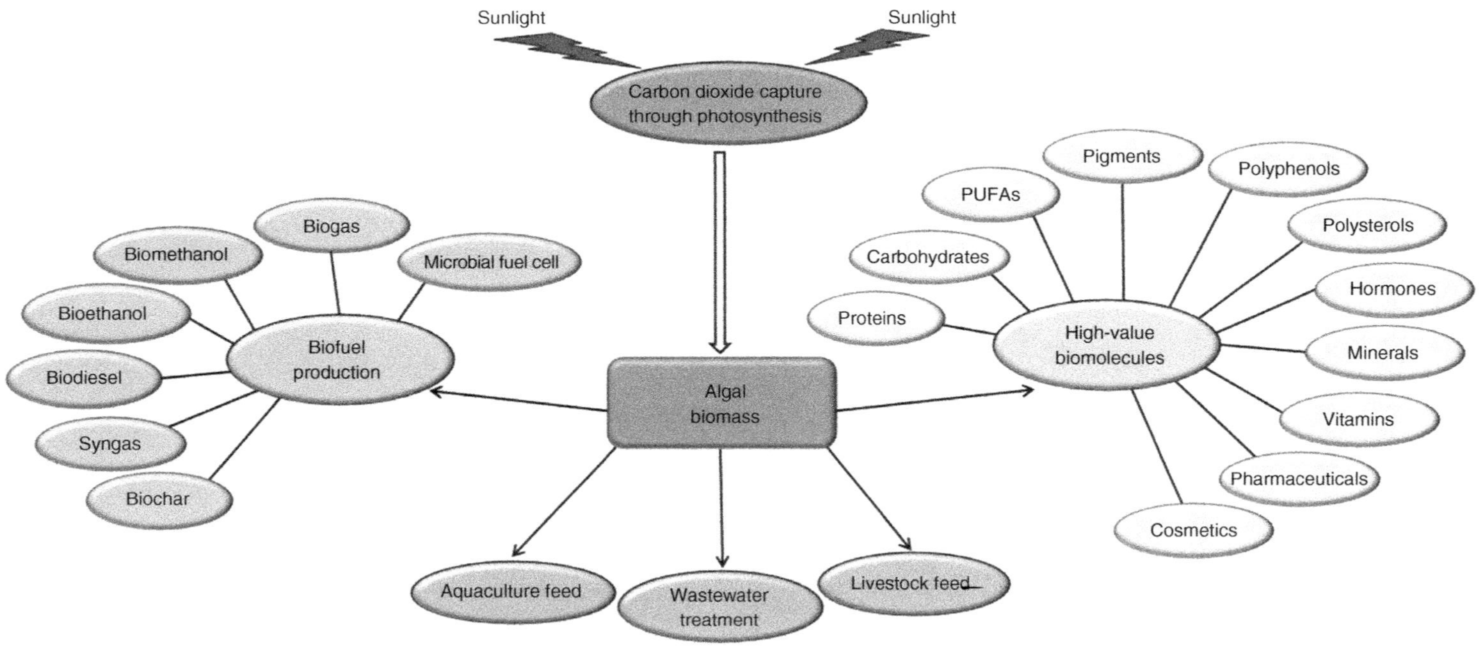

Figure 15.1 High-value natural products obtained from algal resources.

Microspora, Scenedesmus, etc. About 1.6–2.0 g of CO_2 is sequestered for each gram of algal biomass produced [16]. Complete harvesting and processing of the algal biomass are more efficient than terrestrial biomass production systems. It suggests that algae can accumulate considerable amount of carbon within short period of time and could be the best substitute to biologically sequester CO_2 to mitigate global warming. Flue gases released from fossil-fuel power plants typically contain high CO_2 concentrations (10–20%) blended with biologically significant concentrations of nitrous- and sulfuroxides. Douskova et al. [17] reported that use of flue gases as fertilizer in algal ponds increases biomass productivity by 30% in comparison to direct injection of an equivalent amount of pure CO_2. High CO_2 fixation efficiency of ~99% can be accomplished under optimal conditions with as short as two seconds of retention time of flue gases [18]. An algal pond of 3600 acres with the biomass productivity rate of 20 g dry weight m^{-2} day^{-1} can efficiently fix 80% of CO_2 in flue gas exhaust during day from a typical 200 MWh fossil-fuelled power plant. However, the efficiency of CO_2 mitigation by algae can be altered by pond chemistry, physiological condition of algae, temperature, and available nutrition.

The exploitation of algae for CO_2 sequestration offers several other advantages. Algae can be cultivated in open water bodies as well as in closed photobioreactors. The later system offers high productivity due to stringent control over physiochemical growth parameters, sterile conditions, and easy operations. Airlift bioreactors are provided with light period at regular intervals, controlled stress to cells, and even distribution of nutrients resulting in higher energy-to-mass transfer [19]. The huge revivable algal biomass is a rich source of high-value biomolecules and renewable green energy.

The CO_2 capture solely may not have the greatest impact on the accretion of greenhouse gases unless it is effectively sequestered over geological time intervals preventing its re-entry into carbon cycle. Multiple strategies may be adopted for algal carbon sequestration. It can be permanently deep buried in a geologic formation where ~7% of algal dry weight is not available for natural biochemical cycles. The disadvantage of burying algal biomass can be overcome by depositing only after the selective extraction of hydrocarbon or neutral lipid fraction, i.e. triacylglycerols (TAG), containing 75% carbon and accounting for more than 60% of the total dry weight. Additionally, burial of TAGs is safer as it prevents risk of escape of gaseous CO_2 from the geologic deposits. Carbon dioxide can also be chemically converted into stable solid or liquid carbonate salts as construction material. Although this approach has potentially lower risks of carbon escape, but is energy inefficient. Another method of geological storage for mitigated carbon in algal biomass is to transform it into biocharby pyrolysis at elevated temperature. Biochar contains >90% carbon and can endure for centuries under soil [20].

15.3 Algae in High-Value Biomolecules Production

Algal biomass represents a sustainable and cost-effective source of diverse essential nutrient supplements to meet the growing needs of undernourished population of developing countries. A large number of high-value, bioactive, novel compounds from algae have been recognized as viable source in nutraceuticals (protein, lipids carbohydrates, vitamins, and *other* important metabolites) with key role in nutritional food security of the world (Table 15.1).

Table 15.1 Major high-value health biomolecules in microalgae [6, 7].

High-value health products	Algae	Applications
Food	*Chlorella, Spirulina, Odontella auriata, Tetraselmis chuii, Aphanizomenon flosaquae, Nostoc, Ascophyllum sacrum, Spirogyra, Oedogonium, Caulerpa, Hydroclathrus, Enteromorpha, Ulva, Monostroma, Codium, Hizikia, Gracilaria, Cladosiphon, Sargassum, Laminaria, Porphyra, Gelidiella, Halymenia, Hypnea, Laurencia, Macrocystis, Ascophyllum, Laminaria, Undaria pinnatifida, Porphyra, Gelidium, Palmaria palmata, Chondrus crispus*	Nutraceuticals
Feed	*Chlorella, Arthrospira platensis, Pavlova, Tetraselmis, Isochrysis, Chaetoceros, Phaeodactylum, Thalassiosira, Nannochloropsis, Skeletonema, Porphyra, Kappaphycus, Gracilaria, Undaria, Laminaria, Hizikia fusiforme, Porphyridium valderianum, Hypnea cervicornis, Cryptonemia crenulata*	Livestock feed, aquaculture feed, heterocystous cyanobacteria used as biofertilizers
Proteins, peptides and amino acids	*Arthrospira platensis, Chlorella vulgaris, Chlorella ellipsiodea, Dunaliella salina, Nannochloropsis oculata, Porphyridium cruentum, Haematococcus pluvialis, Scenedesmus*	Nutraceuticals Cosmeceuticals, Pharmaceuticals,
PUFAs	*Nannochloropsis oculata, Chlorella vulgaris, Botryococcus braunii, Crypthecodinium cohnii, Ulkenia, Scenedesmus obliquus, Schizochytrium, Labyrinthula, Isochrysis galbana, Thraustochytrium, Phaeodactylum tricornutum, Nannochloropsis, Porphyridium cruentum, Monodus subterraneus, Pavlova salina, Chaetoceros calcitrans, Isochrysis galbana*	Food, Nutraceuticals, Pharmaceuticals
Carbohydrates	*Tetraselmis, Isochrysis galbana, Porphyridium cruentum, Odontella aurita, Porphyridium purpureum, Chlorella* sp., *Rhodella reticulate, Chlorella stigmatophora, Dunaliella salina, Haematococcus pluvialis, Nannochloropsis, Diacronema vlkianum, Pavlova lutheria, Phaeodactylum tricornutum, Scenedesmus dimorphus, Anabena cylindrica, Spirilina platensis, Ankistrodesmus angustus, Aphanizomenon flos-aquae, Laminaria hyperborea, Laminaria digitata, Laminaria japonica, Ascophyllum nodosum, Macrocystis pyrifera*	Food, Nutraceuticals, Pharmaceuticals

(*Continued*)

Table 15.1 (Continued)

High-value health products	Algae	Applications
Pigments chlorophyll	All algae, *Chlorella vulgaris, Monoraphidium, Aphanizomenon flos-aquae, Gloeothece membranacea*	Food, Colouringagents, Nutraceuticals, Pharmaceuticals, Cosmeceuticals,
Carotenoids	*Dunaliella salina, Dunaliella bardawil, Dunaliella tertiolecta, Chlorella zofingiensis, Chlorococcum, Scenedesmus almeriensis, Isoehrysis galbana, Haematococcus pluvalis, Codium fragile, Synechococcus, Nanocloropsis gaditana, Gracilaria chilensis, Porphyridium cruentum, Phordium autumnale, Anabaena cylindrica, Undaria pinnatifida, Spirulina platensis, Nostoc, Porphyridium marinum, Rhodella reticulata, Galdieria, Aphanizomenon flosaquae, Porphyra haitanensis*	Food, Colouring agents, Nutraceuticals, Pharmaceuticals, Cosmeceuticals,
Phycobilliproteins	*Arthrospira platensis, Pseudanabaena mucicola Porphyridium marinum, Rhodella reticulata, Galdieria, Aphanizomenon flosaquae, Porphyra haitanensis, Griffithsia pacifica, Kappaphycus alvarezii*	Fluorescent dyes used as fluorophores in Research &diagnostics
Vitamins and carotenoids		Food, Nutraceuticals,
Pro-vitamin A		Pharmaceuticals,
α-Carotene	*Chlorella, Tetraselmis suecica*	Cosmeceuticals
β-Carotene	*Dunaliella salina, Dunaliella bardawil, Dunaliella tertiolecta, Haematococcus pluvialis, Scenedesmus almeriensis, Codium fragile, Synechococcus, Nanocloropsis gaditana, Gracilaria chilensis, Porphyridium cruentum, Phordium autumnale, Anabaena cylindrica, Undaria pinnatifida, Ulva lactuca, Chlorella, Spirulina, Tetraselmis suecica, Isoehrysis galbana*	
Vitamin B complex	*Dunaliella salina, Chlorella stigmatophora, Chaetoceros gracilis, Thalassiosira pseudonana, Pavlova lutheri, Nannochloris atomus, Eisenia arborea, Nannochloropsis oculata, Anabaena cylindrica, Pleurochrysis carterae, Undaria pinnatifida, Ulva lactuca*	
Vitamin C	*Porphyra umbilicalis, Porphyridium cruentum, Himanthalia elongata, Gracilaria changii, Tetraselmis suecica, Isoehrysis galbana, Eisenia arborea, Dunaliella tertiolecta, Chlorella stigmatophora, Eisenia arborea, Ulva lactuca*	

Table 15.1 (Continued)

High-value health products	Algae	Applications
Pro-vitamin D	*Pediastrum, Scenedesmus, Crucigenia, Coelastrum, Chlorella, Cosmarium, Navicula, Cyclotella, Gomphosphania, Oscillatoria*	
Vitamin E	*Porphyridum cruentum, Tetraselmis suecica, Isoehrysis galbana, Dunaliella tertiolecta, Dunaliella salina, Chlorella stigmatophora, Nannochloropsis oculata, Macrocystis pyrifera, Haslea ostrearia, Chaetoceros calcitrans*	
Vitamin K_1	*Anabaena cylindrical*	
Polyphenols	*Phaeodactylum tricornutum, Nannochloropsis gaditana, Nannochloris* sp., *Haslea ostrearia, Dunaliella tertiolecta, Ankistrodesmus, Spirogyra, Caespitella pascheri, Tetraselmis suecica, Chlorella, Euglena cantabrica, Tolypothrix, Chlamydomonas, Synechocystis, Dunaliella salina, Nostoc* sp., *Nostoc commune, Leptolyngbya protospira, Anabaena, Nodularia spumigena, Arthrospira platensis, Phormidiochaete*	
Phytosterols (Sitosterol, Stigmasterol, Brassicasterol)	*Pavlova lutheri, Tetraselmis, Isochrysis galbana, Phaeodactylum tricornutum, Chattonella antique, Dunaliella tertiolecta, Dunaliella salina, Nannochloropsis, Chrysoderma, Chrysomeris, Chrysowaernella, Giraudyopsis, Peyssonnelia, Chlorella vulgaris, Glaucocystis nostochinearum*	
Phytohormones	*Nannochloropsis oceanica*	
Mineral nutrients		Food, Nutraceuticals
Inorganic salts	*Fucus vesiculosus, Tetraselmis suecica, Isochrysis galbana, Dunaliella tertiolecta, Chlorella stigmatophora, Undaria pinnatifida, Chondrus crispus, Porphyra tenera, Anabena cylindrica, Spirilina platensis*	
Iodine	*Alaria esculenta, Palmaria palmata, Ulva intestinalis, Ulva lactuca, Laminaria digitata, Saccharina japonica, Undaria pinnatifida*	
Iron	*Ulva lactuca, Sargassum, Porphyra, Gracilariopsis, Pyropia yezoensis*	

Use of seaweed and microalgae as food greatly depends upon the heterogeneity in biomass composition that can be varied by strain selection and altered growth conditions. Algae tremendously modulate their biochemical composition by synthesis and accumulation of secondary metabolites in response to variation in their biochemical niche. These features of algae have been exploited to develop metabolic imbalance-based strategies to divert algal cell to preferably synthesize desired metabolites [21]. The metabolic imbalance-based strategies involve stressing procedures such as high salt, light intensity and pH, unfavorable temperatures, or altered concentrations of metal salts.

15.3.1 Proteins, Peptides, and Amino Acids

Algae are unorthodox and protein rich source of protein containing balanced amount of essential and nonessential amino acids as compared to animal and plant protein sources like meat, egg, milk, or legumes [22]. Most of the microalgae (*Arthrospira platensis, Chlorella vulgaris, Nannochloropsis oculata, Porphyridium cruentum, Haematococcus pluvialis, Scenedesmus,* etc.) have proteins, i.e. ~50% their total dry weight, that can be scaled up to 70% by selecting suitable algal species and changing the physicochemical growth conditions [23]. Microalgal proteins are primarily used in the formulation of nutraceuticals or functional foods [24]. It is assumed that by the year 2054, 50% of protein market will be replaced by unconventional protein sources like algae or insects. In spite of having balanced amino acid profile, algal proteins have reduced digestibility in comparison to conventional sources [25] which varies from 50 to 82% species to species. The lower digestibility of algal proteins is attributed to thicker, more rigid, and high cellulose content in the cell wall. Their digestibility can be improved up to 78–84% on average by treating the cell with disruption processes [26, 27].

Furthermore, algal proteins are widely used in the skin-related cosmetic industry because of their elegant biological activities [28]. For example, oligopeptide extracted from *Chlorella vulgaris* has been used in the formulation of Dermochlorella DG® for skin rejuvenation. Application of Dermochlorella DG on skin induces collagen fiber production that increases the compactness of skin, lessens wrinkles and stretch marks, and increases vasculature [29]. The *Arthrospira* oligopeptide containing formulation is used as an antiaging ointment [30]. Micosporin-like amino acid containing formulation has antioxidant and photo-protective properties and is used in sunscreen lotion [31]. Three to twenty residue-long oligopeptides obtained after enzymatic hydrolysis of microalgal protein soup have shown antibacterial, antioxidant, antihypertensive, anticancer, and anti-inflammatory properties which can be exploited in pharmaceuticals for health benefits [32]. *In vitro* enzymatic hydrolysate of protein soup from *Chlorella ellipsiodea* generated a unique, antioxidant hexapeptide (LNGDVW) which scavenged peroxy 1,1,1-diphenyl-2-picrylhydrazyl (DPPH) and hydroxyl-free radicals. Also, under *in vitro* condition this purified hexapeptide protected monkey kidney cells from hydrochloride-induced apoptosis and necrosis [33]. The same lysate generated another unique tetrapeptide (VDGY) with an antihypertensive effect, i.e. angiotensin I-converting enzyme (ACE), inhibitory activity and systolic blood pressure-lowering, in mice on oral administration [34]. Various peptides purified from *Chlorella vulgaris* and *Spirulina platensis* hydrolysate showed antihypersensitive effect [35], and *Dunaliella salina* peptides exhibited cytotoxic effects [36]. These hydrolysate-originated peptides with 3 to >10 kDa molecular weight also exhibited significant antimicrobial activity against *Escherichia coli, Helicobacter pylori*, and *Staphylococcus aureus*. Smaller peptide of <3 kDa from *Dunaliella salina* hydrolysate showed antiproliferative activity even at lower concentrations. Studies have shown that enzymatic hydrolysate of microalgal biomass is safe for use in food additives nutraceuticals [37].

15.3.2 Polyunsaturated Fatty Acids (PUFAs)

Lipid is a distinct group of chemical compounds composed of predominantly nonpolar carbon–carbon or carbon–hydrogen bonds. It is an integral component of cell membrane and is crucial for the maintenance of structural and functional integrity of cell and organism. Lipids in most of organism can be broadly classified as *storage lipids* (primarily triglycerides) used for energy generation; and *structural lipids* (phospholipids, glycolipids, and sterols) are major constituent of cellular membranes. Specialized lipids serve a variety of cellular functions like cofactors (vitamin K), pigments (carotene, retinal pigment), detergents (bile juice), hormones (sex hormones, vitamin-D derivatives), transporters (dolichols), and different cellular signaling molecules (eicosanoids, phosphatidylinositol derivatives).

Fatty acids, a type of storage lipids, can be further categorized depending on the number of carbon–carbon (C–C) double bonds (unsaturated hydrocarbon) in their carbon chain as saturated, monounsaturated (MUFA), or polyunsaturated fatty acids (PUFAs). PUFAs have a long hydrocarbon chain possessing more than one double bond and have significant roles in metabolism. PUFAs can be further categorized into omega-3 and omega-6 (n–3 and n–6 families) fatty acids, depending on the position of first C–C double bond present in fatty acid chain with 18 or more carbon atoms present with respect to the terminal methyl end of the of PUFA molecule.

PUFAs include eicosapentaenoic acid (EPA), arachidonic acid (AA), γ-linolenic acid (GLA), and docosahexaenoic acid (DHA), as precursors for synthesis of a range of fatty acids like cofactors, pigments, hormones, and transporters. These fatty acids function as signaling molecules, usually of 20- or 22-CPUFA, i.e. very long-chain PUFAs (VLCPUFAs), and are essential for proper growth and development of human. Humans are deficient of endogenous synthesize PUFAs longer than 18-C [32], while fishes and terrestrial plants do not anabolize it de novo, but only accumulate EPA and DHA. Microalgae have emerged as promising exogenous dietary sources of long-chain PUFAs of $\omega 3$ (EPA and DHA) and $\omega 6$ (GLA and AA) families. The fatty acid profile and quantity of PUFAs in fishes are naturally low and they amass it by feeding on PUFA-rich microalgae. The fatty acid profile and content varies with algal species, growth conditions, and seasonal variations [38]. Various attempts have been made to understand role of environmental changes in lipid profile. It has been shown that factors such as nitrogen starvation, heterotrophic nutrition conditions, and high irradiation-induced elevated levels of reactive oxygen species; activated lipid, and DHA accumulation by ~57 and ~43% respectively, in microalgal biomass [39]. Although, stress factors increase lipid accumulation but significantly reduce algal growth rates and biomass production. Paes et al. [40] proposed a two-phase culture cultivation procedure to mitigate the deleterious effects of exposure to stresses, where *Nannochloropsis oculata* culture was initially cultivated under optimum conditions for maximum growth of algal biomass. Then biomass was exposed to nutrient deficient or excess irradiations for lipid accumulation. This showed similar biomass yields both under nitrogen supplemented and starved conditions. However, cell volume of nitrogen starved cells increased due 26–34% increased lipid accumulation of total dry mass.

The fish oil extracted from salmon and codfishes is key dietary sources of EPA and DHA. These oils have limited usage due to their low oxidative stability, unpalatable taste and smell, and nonvegetarian origin. As a result, EPA and DHA of microalgal origin have wider acceptability and market values. However, production of algal fatty acids at industry scale is limited to DHA, whereas that at laboratory scale is limited to EPA [11]. *Schizochytrium, Crypthecodinium*, and *Ulkenia* are microalgae that yield high amounts of DHA, i.e. >40% of total fatty acids [10]. Other common PUFA-producing microalgae are *Labyrinthula, Thraustochytrium, Phaeodactylum tricornutum, Nannochloropsis, Porphyridium cruentum, Monodus subterraneus, Pavlova salina, Chaetoceros calcitrans, Isochrysis galbana, and Crypthecodinium cohnii* [31] (Table 15.1).

15.3.3 Polysaccharides

Simple saccharides are primary by-products of photosynthesis in oxygenic photoautotrophs. Polysaccharides are glycosidic bond-linked hydrophilic sugar polymers and complex biomolecules used in numerous biochemical processes. Based on their biochemical role, polysaccharides can be categorized as energy-providing *storage polysaccharides* like starch; *structural polysaccharides* that make up the cell wall, and *polysaccharides associated in cellular signaling* as recognition molecules [41]. Five types of species-specific storage polysaccharides (starch, floridean starch, glycogen, chrysolaminarin, and paramylon) are found in microalgae and blue-green algae. Cyanobacteria exclusively store carbohydrates in the form of glycogen, sucrose, or glucosylglycerol [42]. The algal starch, floridean starch, and glycogen are polyglucans having α-1,4-/α-1,6-type of glycosidic linkages in distinctive proportions. Glucose residues in monomer unit of chrysolaminarin starch are β-1,3-/β-1,6-linked, whereas in paramylon are only β-1,3-linked. The starch granules are accumulated in chloroplasts; chrysolaminarin in vacuoles, whereas floridean starch, paramylon, and glycogen in cytoplasm [43]. Different types of algal polysaccharides vary in degree of polymerization, branching, and topology within the cell, imparting high degree of structural and compositional diversity.

The storage and structural polysaccharides have great industrial applications in food, pharmaceutical, and cosmetics industries. Microalgal species like *Pavlovalutheri, Porphyridium cruentum, Odontella aurita*, etc. can accumulate carbohydrates by >50% of total dry mass [25]; however, polysaccharide accumulation largely depends upon the genotype and growth conditions. Storage polysaccharides have been largely exploited in biofuel industry as starting material (in Section 15.5).

Alginates are complex polysaccharides made of linear copolymers of β-1-4-linked D-mannuronic acid and β-1-4-linked L-guluronic acid monomers. They are found in brown algae and considered as dietary fibers due to resistance to digestion in human gastrointestinal tract. Consumption of algae like *Codium rediae, Rhodella reticulate*, and *Gracilaria* containing 23–64% alginate of total dry weight supplement high amounts of dietary fiber but have nonnutritive physiological effects [44]. Fermentation of such dietary fibers in intestine by inhabitant microflora produces short-chain fatty acids like acetic acid, propionic acid, and butyric acids which foster both gut epithelia and exert a probiotic effect on its microbial consortia [45]. Hydrogels formed by alginates in the presence of divalent cations are used in wound healing, drug delivery, and tissue engineering.

Heavily sulfated heteroglycans, ulvans are obtained from members of Ulvales. Ulvans are unusually hydrophilic that possess anionic property and structural similarities with animal glucosaminoglycan regulators and other lectins. Ulvan polysaccharides show antibacterial, antiviral, antioxidant, anticoagulant, antihyperlipidemic, antitumor, and immunomodulatory properties [46].

Carrageenans are heavily sulfated galactans formed by disaccharide polymeric units of 3-linked β-D-galactopyranose or 4-linked α-D-galactopyranose or 4-linked 3,6-anhydro-α-D-galactopyranose. Primary sources of carrageenans are red algae orders Gelidiales, Gigartinales, and Gracilariales. They exhibit antiviral activities against herpes simplex, dengue-2, rabies virus, herpes zoster, and vesicular stomatitis virus. On the other hand, partially depolymerized carrageenan extracts show anticancer and immunostimulatory activities [46]. Carrageenan supplemented diet act as prebiotics in poultry and rat feeds [47, 48]. Polysaccharides from *Porphyridium cruentum* show antiproliferative activity [49]. Even carrageenans of low molecular mass have potential harmful effects. There high doses in feed caused ulceration in animals and modified cytokine profile innormal human colonic mucosal epithelial cell line culture [50]. Oral administration of λ-κ carrageenan to mice increased their blood glucose, and carrageenan-induced insulin tolerance might lead to diabetes [51].

Laminarin is a sulfated polysaccharide of phaeophytes and chemically made of β-1,3-D-glucan with β-1,6- branching with different reducing ends having either mannitol or glucose residues. The dietary effects of laminarins (β-1,3-D-glucan) depend on complexity of its primary structure. Oral doses of M-series β-1,3-D-glucan have antitumor and hepatoprotective effects [52, 53]. Fucoidans are sulfated fucose-containing polysaccharides found in edible phaeophytes like *Cladosiphon okamuranus, Undaria pinnatifida, Saccharina japonica,* etc., and used as antiaging biomolecules in cosmetics [54].

15.3.4 Pigments

Pigments are fundamental, essential, and one of the most studied biomolecules present in thylakoid membrane of chloroplast that selectively absorb visible spectrum of sunlight. Pigments are widely used as colorants, food additives, vitamin precursors, in pharmaceuticals as antioxidants, anticarcinogenic, antihypertensive, anti-inflammatory molecules, and feed of livestock and aquaculture [6, 10]. Algae produce pigments according to their genotype that are grouped as chlorophylls, carotenoids, and phycobiliproteins (PBPs), which impart green-, yellow/orange- and, red/blue-colored appearance to algae, respectively, as per their abundance.

15.3.4.1 Chlorophylls

Chlorophylls (*a, b,* and *c*) are lipid-soluble green pigments essentially present in virtually all photoautotrophs. Use of natural pigments and chlorophylls as colorants in nutraceutical, pharmaceutical, and cosmetics industry is growing rapidly due to intense green pigmentation and health and environmental concerns associated with synthetic dyes. Sodium and cupric derivatives of chlorophylls (chlorophyllin) are popularly used as food or beverage additives for rich green color [29]. Chlorophyll and its derivatives like phaeophytin, pyrophaeophytin, and chlorophyllin are strong protectants against mutagens like alkylating agents, heterocyclic amines, polycyclic aromatic hydrocarbons, aflatoxins, etc., and used as therapeutic biomolecules. Combinations of chlorophylls have shown strong antioxidant activity, increased glutathione *S*-transferase content, reduced cytochrome P450 enzyme activity, reduced cell differentiation, induced cell arrest, and programmed cell death [55, 56]. Chlorophyll content is dependent on the genotype and growth conditions. Low light intensity, red/green light spectrum, elevated temperature, and nitrogen and phosphorus-rich medium can significantly increase chlorophyll content. Chlorophyll c is found only in brown algae, and reports on its health benefits are elusive.

15.3.4.2 Carotenoids

Carotenoids form a family of isoprenoid structured lipophilic yellow to orange-red accessory light-harvesting pigments. These are found in certain bacteria, fungi, microalgae, and plants. Carotenoids harbor the membranes of chloroplasts, mitochondria, and endoplasmic reticulum. Sometimes are present in chloroplast matrix as plastoglobulin, or cytoplasmic lipid globules. Carotenoids are very important dietary components with therapeutic and cosmetic uses as precursor of vitamin A. Natural β-carotene is exclusive source of 9-*cis* β-carotene which is a strong quencher of oxygen-free radicals and is reported to improve retinal and visual dysfunctions in humans and animals [57]. Additionally, they are very strong antioxidants because of their free radical quenching properties and protect organisms from oxidative stresses. The β-carotene has been shown prevent advancement of atherosclerosis (plaque formation in arteries) in humans [58]. Oral doses of β-carotene prevent UV-induced erythema in humans [59]. There is an increasing demand for algae as source of carotenoids as they possess a variety of carotenoids and also microalgae have enormous capacity to produce them high amounts, i.e. ~0.1–0.2% of total dry mass [9]. However, in green microalgae, it can be increased up to 14% by exposure of algal cells to high irradiance, elevated temperature, high salinity, and reduced nutrition [60]. More than 600 different natural carotenoids have been reported and categorized as oxygen-free hydrocarbon carotenoids, carotenes, oxygenated derivatives of carotenes, and xanthophylls. Green microalgae produce a huge range carotenes (β-carotene, lycopene) and xanthophylls (astaxanthin, antheraxanthin, violaxanthin, lutein, zeaxanthin, neoxanthin, lutein, loroxanthin, canthaxanthin). Caretenoids of diatoms and brown algae are diatoxanthine, diadinoxanthin, and fucoxanthin [61]. The β-carotene or provitamin A is first commercially produced microalgal biomolecule used as an additive in pharmaceutical products and as colorant in food products like cheese, butter, and margarine [30, 62]. Common β-carotene producing microalgae are *Dunaliella bardawil, D. tertiolecta*, and *Scenedesmus almeriensis* [7]. *Dunaliella salina* and *Haematococcus pluvialis* are richest commercial producers of two most desirable caretenoids β-carotene and astaxanthin, respectively [61]. *D. salina* produces ~14% carotenoids of its total biomass and up to 98.5% β-carotene of its total carotenoid content [62]. *Phordium autmnale* produces 24 different all-trans-carotenoids with β-carotene as a major component [63].

Astaxanthin, a red xanthophyll pigment, is the second most desirable carotenoid and its microalgal sources include *Chlorella zofingiensis*, *Scenedesmus almeriensis*, and *Chlorococcum*. A freshwater green microalga, *Haematococcus pluvalis*, produces up to 7% carotenoids of its total dry mass and ~81% astaxanthin of total carotenoid content [62] and is a promising species for industrial scale astaxanthinproduction. Astaxanthin is a popular aquaculture feed that elicits pinkish-red color to the flesh of salmon, shrimp, lobsters, and shellfish [10, 29, 64]. Also, it is most potent antioxidant which is ~10 fold more active than other carotenoids and protects from oxidative stresses, cancer, diabetes, ophthalmic, and neurodegenerative diseases [65]. Consumption of natural astaxanthin reduces inflammatory reactions, improves health of patients with cardiovascular issues, prevents lipid peroxidation in heavy smokers [66], and increases serum high-density lipid and adiponectin [67].

Lutein and zeaxanthin (yellow carotenoids) are essential pigments in human retina and are obtained from microalgae. They protect photoreceptor cells against blue light-induced damage to lens and retinal cells [68]. Their regular dietary consumption protects from cataract, diabetic retinopathy, and age-related retinal degeneration by free radicals [69]. Common lutein-producing genotypes are *Chlorella zofingiensis*, *Chlorella protothecoides*, *Neospongiococcus gelatinosum*, *Scenedesmus almeriensis*, *Chlorococcum citriforme*, *Muriellopsis*, etc. [29]. Zeaxanthin producing strains are *Scenedesmus almeriensis* and *Nannochloropsis oculata*.

Other carotenoids are lycopene (red), violaxanthin (orange), canthaxanthin (reddish-orange), and fucoxanthin (olive green). Violaxanthin from *Dunaliella tertiolecta* and *Chlorella ellipsoidea* is anti-inflammatory, antiproliferative and anticancerous [70]. Lycopene is used in formulation of sunscreen and antiaging products in cosmetic industry. Fucoxanthin from brown algae and some marine diatomsis used as a strong antioxidant, anti-inflammatory, anticancerous, antidiabetic, tanning, and neuroprotective in pharmaceutical and cosmeceuticals.

15.3.4.3 Phycobilliproteins (PBPs)

PBPs functioning as accessory photosynthetic pigments are found in members of Cyanophyta, Rhodophyta, Cryptophyta, and Glaucophyta. PBPs are disc-shaped multi-subunit hydrophilic protein complexes with covalently linked linear tetra-pyrrole groups (bilins). They absorb light between 470 and 660 nm and transfer it to photosynthesis reaction center thereby enhance the photosynthetic efficiency. Based on spectral properties PBPs are sub-grouped asphycocyanin (PC), phycoerythrin (PE), allophycocyanin (APC), and phycoerythrocyanin (PEC). PBP's composition and content may vary with genotype and growth conditions and may reach up to 13% of total dry biomass. Being natural PBPs are extensively used in food, pharmaceutical, and cosmetic industries. In food industries, PBPs are used to make chewing gums, popsicles, wasabi, confectionaries, beverages, and dairy products [30]. Physiologically active PBPs are widely used as strong antioxidants, antiviral, anticarcinogenic, immunity booster, anti-inflammatory, hepatoprotective, and neuroprotective in pharmaceuticals and in cosmeceuticals for perfumes and eye makeup powder. Phycocyanin, a blue fluorescent dye from *Spirulina*, is used to prepare fluorescent-labeled probes for various research activities [30].

15.3.5 Vitamins

Vitamins are vital micronutrients that are precursors of many important enzyme cofactors, and antioxidants essentially required for normal metabolic processes. As majority of vitamins are neither synthesized de novo nor stored in body, so their regular dietary intake is needed. Microalgae are an excellent source of almost all vitamins (Table 15.1) such as provitamin A (α- and β-carotene), vitamin B complex (B1, B2, B3, B5, B6, B8, B9, and B12), vitamins C, D, E, and K_1 [71]. The quality and quantity of vitamins in microalgae are largely dependent on their genotype and physicochemical growth phase, culture conditions, and seasonal variations [32]. Fabregas and Herrero [72] reported that *Tetraselmis suecica*, *Isoehrysis galbana*, *Dunaliella tertiolecta*, and *Chlorella stigmatophora* produce high vitamin contents analogous to traditional sources. *D. tertiolecta* produces Provitamin A, vitamins B_{12}, B_2, E, and β-carotene. *T. suecica* is rich in vitamin B_1, B_3, B_5, B_6 and vitamin C, whereas *Chlorella* contains high levels of vitamin B_7 and five times more B_{12} than fruits and vegetables. So, microalgal supplemented vegan foods can fulfil vitamin B_{12} requirements, *viz.*, 1 g of *Anabaena cylindrica* powder supplements 64% of adult's vitamin B_{12} requirement [71]. Seaweeds are rich source of water- (vitamins B_1, B_2, B_{12}, and C) and fat-soluble (pro-vitamin A, β-carotene, and vitamin E) vitamins. Also they contain other vitamins of B-complex (B_3, B_6, B_9, and H) but in lesser amount. Sea vegetables like laver (*Porphyra umbilicalis*), sea spaghetti (*Himanthalia elongata*), and *Gracilaria changii* have vitamin C contents comparable to tomatoes and lettuce. Vitamin C content in a brown alga, *Eisenia arborea* is ~35 mg/100 g dry weight which is fairly close to mandarin oranges. *G. chilensis* and *Codium fragile* contain more β-carotene than carrots [73]. Common kelp, *Macrocystis pyrifera*, contains more of vitamin E (tocopherol) than soybean and sunflower oil.

Although microalgae are rich in vitamins, their bioavailability depends upon algal genotypes, seasons, and growth conditions [32]. Thus require high-throughput screening of promising algal genotypes, implementation of appropriate strategies for harvesting, drying, and processing of algal biomass to protect heat-labile vitamins.

15.3.6 Polyphenols

Polyphenols or phenolic compounds form a large group of secondary metabolites produced by plants and microorganisms and have gained attention as an important class of natural antioxidants. They possess antiallergic, antimicrobial, anticancer, anti-inflammatory, antidiabetes, antiaging activities. Polyphenols extracted from *Nannochloropsis* and *Spirulina* exhibitantifungal and antimycotoxigenic activities. Primary sources of polyphenols are fruits, vegetables, and beverages. Majority of phenolic compounds are derivatives of the aromatic amino acid metabolism. Polyphenols are the aromatic compounds consisting of one or more phenyl ring bearing one or more hydroxyl group that makes them polar. Based on their chemistry, polyphenols have been categorized into phenolic acids (hydroxycinnamic acids, hydroxy benzoic acids), flavonoids (anthocyanins flavonols, flavones, flavanonols, flavanones), isoflavonoids (coumestans, isoflavones), lignans, stilbenes, and phenolic polymers (hydrolyzable tannins, proanthocyanidins-condensed tannins) [74]. Phlorotannins are oligomers of phloroglucinol units and predominantly found in cell walls of phaeophycean genera. Halogenation of phlorotannins increases their structural complexities. Polyphenols protect organisms from oxidative damages caused by various environmental stresses and their profile and amounts are dependent on algal species and their growth conditions. Seaweeds (green-, red-, and brown-algae) are the primary source of polyphenols [75], however, microalgae (*Dunaliella tertiolecta, Haslea ostrearia Phaeodactylum tricornutum, Ankistrodesmus, Caespitella pascheri, Spirogyra, Euglena cantabrica*) and cyanophytes (*Nostoc* spp., *Nostoc commune, Leptolyngbya protospira Nodularia spumigena, Arthrospira platensis, Phormidiochaete* spp.) are also reported as their sources [76]. Polyphenol contents in *Haematococcus pluvialis, Tetraselmis, Neochloris oleoabundans, Arthrospira platensis,* and *Chlorella vulgaris* ranges from 54 to 375 mg Gallic acid Equivalent (GAE)/100 g dry weight is almost same as that from classical sources [77].

Microalgae have distinct flavonoid compositions from higher plants and a rich source of catechins and flavonols. Phaeophyceae genera have higher phlorotannin content than other algal groups. Eckols, a phlorotannin, isolated from brown alga, *Ecknonia cava* has antiadipogenic and neuroprotective effects, and *Eisenia bicyclis* has high free radical scavenging content that inhibits melanin formation [78, 79]. Bromophenols, commonly found in all algal groups, have significant pharmaceutical values due to their antioxidative, antimicrobial, anticarcinogenic, antithrombotic, and antidiabetic activities [80]. Thus, suitable methods producing significant amounts of polyphenols should be studied in detail.

15.3.7 Phytosterols

Sterols are ubiquitous amphipathic steroid alcohols, a form of lipid that are essential component of eukaryotic cellular membrane and regulate its fluidity, stability, and permeability by interacting with other membrane phospholipids and proteins [32]. Sterols function as hormones or their precursors and neurotransmitters. Sterols of plants, animals, and microorganisms are termed as phytosterols, zoosterols, and mycosterols, respectively. Sterols are triterpenes derived from isoprenoid biosynthesis with tetracyclic cyclopenta-α-phenanthrene (rings A, B, C, and D) including a long elastic aliphatic side chain attached to C-17 of ring D [81]. Most of the phytosterols have double bond-linked C-5/C-6, and methyl groups attached to C-10 and C-13. Also, aliphatic side chain length, position of double bond, availability of alkyl group, saturation condition, and stereochemistry of C-24 alkyl side chain of phytosterols have functional importance. Cholesterol is the primary zoosterol and about 200 structurally and functionally similar phytosterols are reported among which β-sitosterol is most abundant. Other common phytosterols are campesterol and stigmasterol. Phytosterols can lower the low-density lipids in blood and promote cardiovascular health, hence, have commercial values as nutraceuticals and pharmaceuticals. Additionally, they have antiatherogenicity, anticancer, anti-inflammatory, antioxidative activities and used for curing neural disorders, like Alzheimer's disease and autoimmune encephalomyelitis [6].

Microalgal phytosterols have 24-ethylcholesterolas dominant form with an extra alkyl group in side chain [32]. Extremely diverse algal phytosterols consist of cholesterol, fucosterol, isofucosterol, clionosterol, dihydroxysterols, etc., and are biosynthesized by mevalonate or methyl-D-erythritol 4-phosphate pathways. Algal strains and physicochemical growth conditions can modify phytosterol diversity and content. Microalgae may produce 5.1% phytosterols (brassicasterol, sitosterol, and stigmasterol) of their total biomass.

Phytosterol-producing microalgae are *Pavlova lutheri, Tetraselmis, Isochrysis galbana, Phaeodactylum tricornutum, Chattonella antique, Dunaliella tertiolecta, Dunaliella salina, Nannochloropsis, Chrysoderma, Chrysomeris, Chrysowaernella, Giraudyopsis, Peyssonnelia, Glaucocystis nostochinearum, Chlorella vulgaris*, etc. [81].

15.3.8 Phytohormones

Phytohormones are endogenously produced low molecular weight trace signal molecules that regulate diverse cellular processes in plants. Phytohormones like auxins, gibberellins (GAs), cytokinins (CKs), abscisic acid (ABA), ethylene (ET), salicylic acid (SA), jasmonic acid (JA), and brassinosteroids (BRs) function synergistically to initiate appropriate signaling network in response to various stimuli and maintain cellular homeostasis. Lu and Xu [82] have reported various essential and bioactive phytohormones (auxin, GA, ABA, CK, and ET) from diverse microalgal lineages. Despite controversial functional significance of algal phytohormones and related signaling pathways, endogenous ABA and CK in *Nannochloropsis oceanica* have shown physiological effects similar to higher plants [82]. Although, microalgal phytohormone profiles are similar to higher plants, but their nature and abundance, and biosynthetic intermediates seem to vary among species. Comprehensive understanding of phytohormones pathways in microalgae can be utilized to increase algal biomass, production of desired secondary metabolites, and many other biomolecules of industrial use.

15.3.9 Minerals

Mineral nutrients include essential inorganic dietary elements in addition to carbon, hydrogen, oxygen, and nitrogen needed for normal metabolism and growth of organisms. Important dietary elements are calcium, magnesium, phosphorus, copper, iron, manganese, molybdenum, potassium, sodium, iodine, and zinc. Based on their quantity stored in body minerals are classified into major minerals and trace minerals. Major minerals are recycled and stored in large quantities, *viz.* calcium, sodium, potassium, magnesium, phosphorus, chloride, and sulfur. However, trace minerals are equally vital, but required in very small amounts, *viz.* manganese, zinc, copper, iron, molybdenum, chromium, iodine, fluoride, and selenium. Algae are excellent source of mineral supplements such as potassium, iron, magnesium, calcium, and iodine are frequently used as sea vegetables and in aquaculture. Marine algae like *Fucus vesiculosus, Tetraselmis suecica, Isochrysis galbana, Dunaliella tertiolecta, Chlorella stigmatophora, Undaria pinnatifida, Chondrus crispus, Porphyra tenera*, etc. contain a significant amount of inorganic ions such as phosphorus, calcium, sodium, potassium, iron, magnesium, zinc, cobalt, and copper. They particularly have high concentration of Cl^- and Na^+ possibly due to their noteworthy osmotic potential [83, 84]. Algae have high bioadsorptive and bioaccumulative capacities leading to higher mineral content than terrestrial plants. Ash content of specific seaweeds may reach up to 40% of dry weight whereas mineral content in spinach can be 20% of dry weight [85]. However, the concentration of minerals may vary with genotypes, seasons, and geographic locations. Seaweeds like *Alaria esculenta, Palmaria palmata, Laminaria, Saccharina japonica, Undaria pinnatifida, Ulva intestinalis*, etc. are excellent sources of iodine. Iodine content is altered during preparation and storage as many iodine compounds are water-soluble or may vaporize under humid conditions [86]. Macroalgae like *Gracilariopsis, Sargassum, Ulva,* and *Porphyra* from the Venezuelan sea contain a potentially rich source of iron suitable for human consumption [46]. Thus, edible brown and red algae can be consumed as sea vegetables to supplement essential dietary mineral nutrition.

15.4 Algae in Biofuel Production/Generation

Steep population growth has driven enormous global industrialization where most of the energy demand is met with nonrenewable fossil fuels like petroleum oil, coal, and natural gas. Thus, conventional fossil fuel reserves are exhausting rapidly and global energy demand is anticipated to inflate by 1% per year by 2040 [87]. Environmental issues associated with fossil fuels and inflating prices have led the soaring energy sector to quest for alternative inexpensive, renewable, and green energy resources to ensure energy security. Solar, hydroelectric, wind, and biomass has represented potential and dynamic genesis of renewable energy. Biomass from agriculture and forestry has been used as feedstock for the production of biofuels like bioethanol, biodiesel, and biogases [88]. Exhaustion of agriculture and forestry biomass feedstock has serious concerns as their combustion expands carbon imbalance and their inappropriate harvesting has led to other crisis, *viz*. deforestation, soil erosion, and loss of biodiversity due to their slow growth rate.

Algal biomass has emerged an alternate feedstock for biofuel production over food-crop and non-food crop biomasses due to high growth rate, lipid, and carbohydrate contents as compared to terrestrial plants. Various biofuel products like biodiesel, bioethanol, biogas, biohydrogen, bio-oil, syngas, and biochar can be obtained from algal biomass according to the species type, growth conditions, and conversion processes. Members of Cyanophyceae, Chlorophyceae, and Pyrrophyceae contain higher amount of fatty acids and are most suitable for biofuel production [89], whereas those of Phaeophyceae are rich in carbohydrates and most suitable feedstock for bioethanol production [90]. Various algal biomass transformation pathways for biofuel production are summarized in Figure 15.2. Technologies for algal biomass conversion into biofuel involve four broad ways, i.e. thermochemical, chemical, biochemical, photosynthetic microbial fuel cell.

15.4.1 Thermochemical Conversion

In process of thermochemical conversion algal biomass is disintegrated followed by its chemical transformation into biofuels in the presence or absence of oxygen at different temperatures using combustion, gasification, pyrolysis, or hydrothermal liquefaction methods [91]. Direct combustion involves thermal conversion of biomass at >800 °C temperatures in excessive presence of an oxidant (typically oxygen) in a steam turbine or boiler to generate heat. Primary by-products produced are CO_2 and water vapor. Drying of algal biomass is preceded by its direct combustion for better efficiency. Gasification is the thermal conversion of algal biomass under-regulated oxygen at ~1000 °C into synthesis gas or syngas, which is primarily a mixture of hydrogen, carbon monoxide, and methane. Syngas can be directly used as fuel or can be further fractionated in a range of fuels, chemical

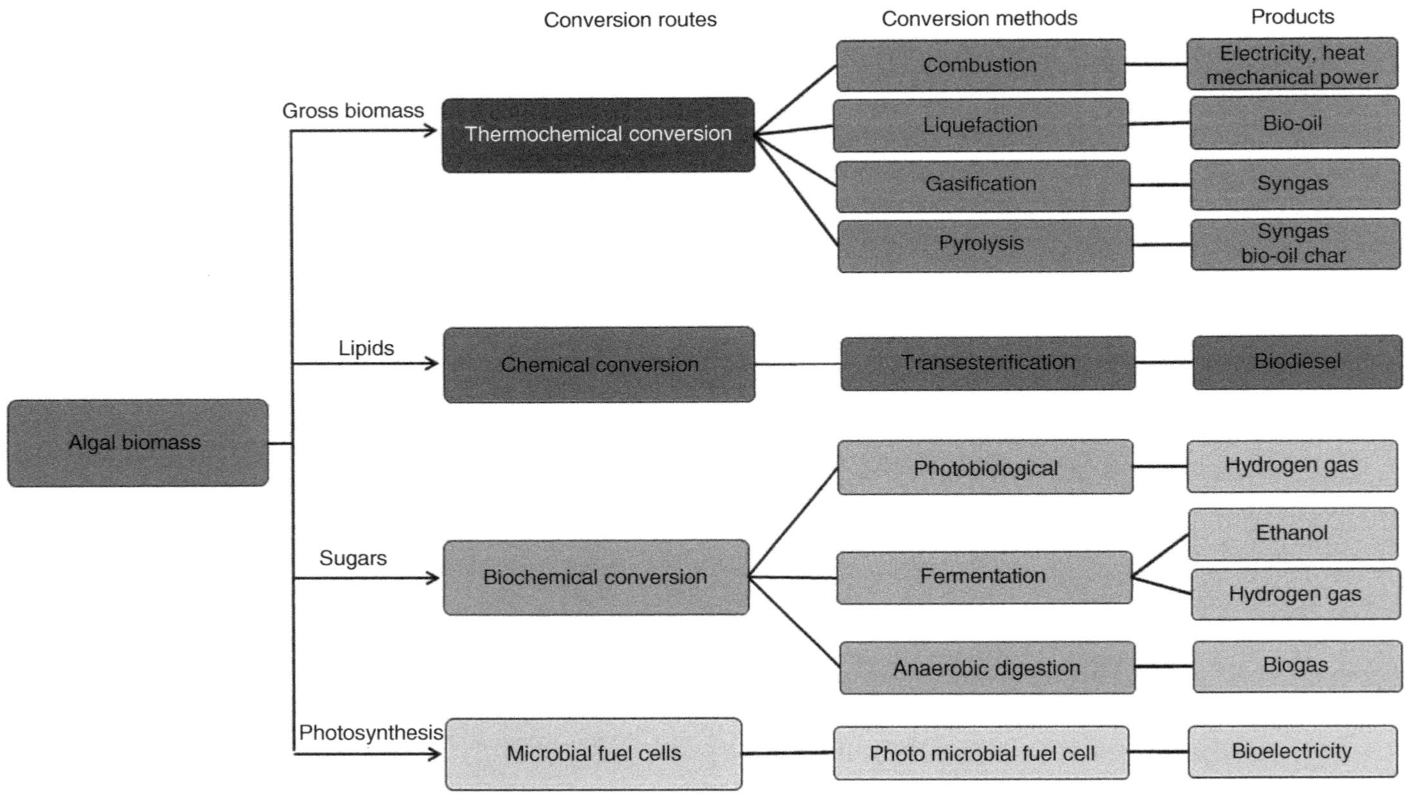

Figure 15.2 Different algal biomass modification pathways for biofuel production [7, 88].

intermediates, and liquid products. Pyrolysis is the thermal conversion of biomass by heating it under anoxic conditions producing charcoal (solid), bio-oil (liquid), and syngas [92]. Pyrolytic thermal conversion can be conventional (biomass heating rate is 0.1–1 $K\,s^{-1}$ for a long resident time of 45–550 s), fast (10–200 $K\,s^{-1}$ for a short resident time of 0.5–10 s), and flash (>1000 $K\,s^{-1}$ for a very short residence time of <0.5 s). Pyrolysis is the most preferred thermal conversion procedure due to high ash content in algal biomass. Depending upon the biomass feedstock, fast pyrolysis may yield 60–75% of bio-oil, 15–25% char, and 10–20% noncondensed gases. The flash pyrolysis procedure can transform biomass into crude oil with ~70% efficiency. Crude bio-oil produced through pyrolysis can be used as biofuel in engines and turbines. In hydrothermal liquefaction, the algal biomass slurry is deoxygenated with alkali solution at more than 400 °C and 2900 psi pressure in the absence of oxygen. Several compounds such as high viscosity bio-oil (~73%), gas (~20%), and char (0.5%) can be collected from algal biomass [93]. Average bio-oil yield with the catalytic hydrothermal liquefaction is nearly twice (63%) than with noncatalytic aqueous liquefaction [91].

15.4.2 Chemical Conversion by Transesterification

High lipid contents of algal biomass allow its chemical conversion in biofuel. Heterotrophic algal cells contain significantly high amount of total and neutral lipids than the photoautotrophics [94]. Although, lipid content of microalgae depends upon its ecological niche and species, they have significantly high lipid content (50–86% total dry weight) as compared to macroalgae except chlorophytes. Microalgal cells exhibit extensive lipid accumulation under high light intensity or nutritional (N or P) stresses [95]. High lipid content of algal cells can be converted into high efficiency biodiesel by the process of chemical transesterification.

Transesterification is an acid or base-catalysed chemical exchange of organic alkyl group of an ester with the organic alkyl group of an alcohol. It is utilized for the production of biodiesel by chemical reaction between algal fatty acids or triglycerides and alcohol in the presence of an acid, base, or lipase as a catalyst. Transesterification converts algal fatty acids or triglycerides to fatty acids methyl ester (FAME) or biodiesel and glycerol using alcohol in the presence of catalyst (Figure 15.3).

Although any alkyl alcohol (methyl, ethyl, propyl, or butyl, and amyl alcohol) can be used for transesterification but, methanol, and ethanol are preferred at commercial scale due to their physicochemical advantages and low cost [96]. This reaction highly depends on the molar ratio of reactants, alcohol, and catalyst used thus regulating the fluidity of biodiesel enabling its direct use in engines by blending with petroleum diesel [97].

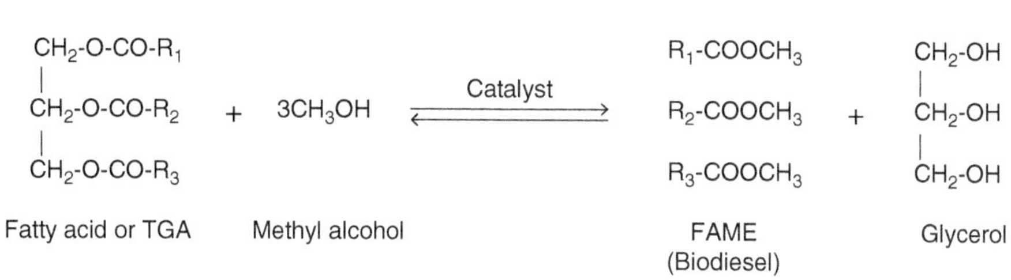

Figure 15.3 Transesterification of algal fatty acid (TGA) to glycerol and biodiesel.

High-density glycerol produced as a by-product can be routinely drawn out to maintain chemical equilibrium and be utilized in cosmetic and pharmaceutical industries.

Transesterification process is of two types, i.e. conventional extractive and direct or *in situ* transesterification. Conventional extractive transesterification is a multistep process involving controlled desiccation of biomass, fatty acid extraction, transesterification, and biodiesel refining. Despite its higher production cost and processing time, it yields a refined biofuel excellent for high-speed diesel engines [98]. Direct or *in situ* transesterification is a single-step method where untreated wet algal biomass is directly delivered into the reaction chamber for transesterification without extraction of fatty acids. It uses excess volumes of alcohol to tolerate water present in the biomass, facilitates fatty acid extraction, and serves as esterification reagent. This method has greater yield, low cost, and reduced waste compared to the conventional transesterification. Catalyst-free direct supercritical methyl alcohol transesterification of wet biomass can significantly improve the quality of biodiesel [99].

15.4.3 Biochemical Conversion

Biochemical conversion technologies of biofuel production rely on the presence of energy reserve carbohydrates in algal feedstock. These energy reserve polysaccharides differ from species to species and are different from structural polysaccharide of the cell wall. Their subcellular location and degree of polymerization vary with algal species. Most common carbohydrates stored in algal biomass are starch, floridean starch, chrysolaminarin, paramylon, and glycogen [25]. Microalgae have variable carbohydrate content like *Porphyridium cruentum* and *Chlamydomonas* have up to 57 and 17% carbohydrate reserves of total dry weight, respectively [42]. Algal polysaccharide content can be modified by altered growth conditions. Algal biomass can be directly used as feedstock in biochemical conversion pathways of biofuel generation without any pre-treatment. Theses pathways involve hydrolysis of energy reserve polysaccharide in algal biomass by fermenting bacteria into simple sugars through anaerobic digestion which can then be fermented into biofuels *viz.* biogas, bioalcohols, or biohydrogen.

Biogas is produced from delipidized, polysaccharide-rich wet algal biomass in consecutive stages of hydrolysis, fermentation (acidogenesis), acetogenesis, and methanogenesis. During hydrolysis, the wet biomass is broken into soluble sugars and amino acids by bacterial decomposition. In the next step, fermentable sugars are oxidized into alcohols, lactate, acetic acids, volatile fatty acids, and carbonic acid by acidogenic bacteria under anaerobic conditions. The fermented products are further oxidized into acetic acid substrates by acetogenesis. Methanogenic bacteria convert acetic acid into biogas (methane and CO_2) at final step of conversion (Figure 15.4).

The efficiency of anaerobic digestion is significantly impaired by C:N ratio of in feedstock, temperature, pH, and rate of feedstock input in the plant. Low C:N ratio is due to high protein content in feedstock that leads to ammonia accumulation that inhibits bacterial degradation reactions. Depleted content of lipid and protein in feedstock biomass have shown better yield of biogas [100]. Co-digestion of waste papers, use of halophilic bacteria, and microwave pretreatment of biomass can significantly improve the biogas production [101–103].

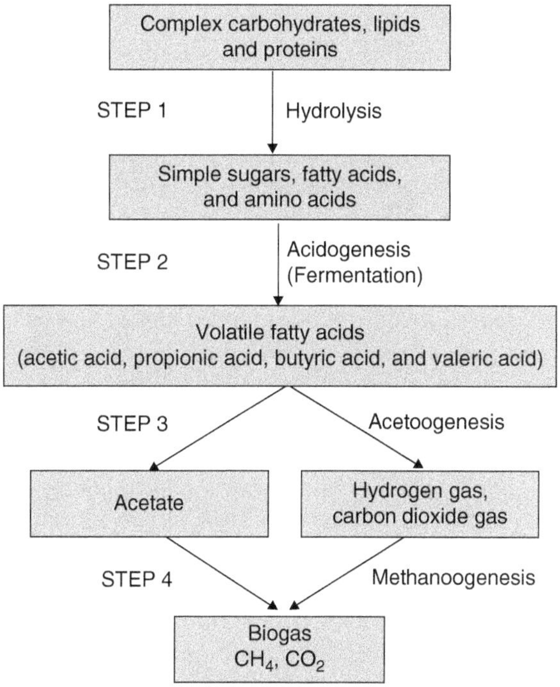

Figure 15.4 Biochemical conversion of algal biomass for biogas production.

Contrary to terrestrial plant biomass, algal biomass has low lignin and hemicellulose and high starch and cellulose content making it better choice for ethanol production. Algae like *Spirogyra, Chlorococcum vulgaris, Prymnesium parvum, Gracilaria, Gelidium amansii, Laminaria, Sargassum, Nizimuddinia zanardini, Ulva pertusa, Eucheuma cottonii, Alaria crassifolia,* etc. have been widely used for bioethanol synthesis [104, 105]. Saccharification of polysaccharide-rich algal biomass with hot water, acidic and enzymatic hydrolysis, and liquefaction is essential for efficient fermentation. Saccharification and fermentation can also be performed together by using amylase-producing strain of yeast to reduce the cost of bioethanol production. *Saccharomyces cerevisiae* and *Zymomonas mobilis* are widely used to perform fermentation under anaerobic conditions during bioethanol production. Since fermentation of mannitol takes place in the presence of oxygen, so the use of *S. cerevisiae* and *Z. mobilis* is not possible. A facultative anaerobic bacterium, *Zymobacter palmae*, is used for fermentation of mannitol-rich algal biomass. Major advantage of commercial bioethanol production is that it can be readily used as automobile fuel or blended with fossil fuel petroleum. Nonetheless, certain features of bioethanol like corrosiveness, low vapor pressure, flame luminosity, and energy density make it inferior fuel than diesel and petrol. Instead, butanol is considered as a better transportation fuel due to its higher energy density, and it has more potential for integration with petroleum fuel. Algae like *Ulva lactuca* and *Saccharina* (Phaeophyceae) provide promising biomass for butanol production [106].

Development of technologies for use of hydrogen gas as a fuel has leaped since last decade. Unlike fossil fuels that produce greenhouse and other toxic gases on combustion, hydrogen gas as fuel has unique features like high energy content, extremely low density, and production of water as a by-product on burning have made it the cleanest and most efficient fuel. Microalgae can synthesize biohydrogen during photosynthesis with high photon conversion efficiency. The ability of *Scenedesmus obliquus* to produce hydrogen gas through fermentative and photochemical processes [107] has opened the possibilities to use algae to assimilate the solar energy into clean fuel. Algae being oxygenic photosynthetic organism produce hydrogen through biochemical processes like biophotolysis and photofermentation. During the light reaction of photosynthesis, photolysis of water molecules generates hydrogen ions, electron, and molecular oxygen. Consequently, in the absence of oxygen, Fe-hydrogenase enzyme in microalgal cell converts hydrogen ions into biological hydrogen gas [108]. However, co-produced molecular oxygen tends to inhibit the activity of hydrogenase enzyme and impedes biohydrogen production. Such repression of Fe-hydrogenase activity by oxygen can be protected by exposing the algal culture to anaerobic conditions [109] which leads to a "two-step photosynthesis and biohydrogen production" system. First aerobic stage of such systems is composed of culturing of photosynthetically active algae, while during the second anaerobic stage, sulfur deficiency is stimulated in culture by biohydrogen production [110]. The sulfur deprivation in culture leads to gradual degradation of a sulfolipid in the thylakoid membrane exclusively associated with photosystem II and partially inactivates it, while photosystem I remains active. Thus, sulfur-deprived cells lower the oxygen production while steady respiratory oxygen consumption creates an anoxygenic environment in the algal cell. In this physiological condition, Calvin cycle stops the generation of $NADP^+$ and cell keeps releasing the electron to prevent oxidative damage. Some algae produce hydrogenase enzymes in such conditions to release the reductive pressure in the cell and produce hydrogen by accepting electron from ferredoxin [111]. *Chlamydomonas reinhardtii* and *Chlorella sorokiniana* are potential source of biohydrogen fuel production under a sulfur-deprived condition.

15.4.4 Photosynthetic Microbial Fuel Cell (MFC)

Energy demands have led to the development of alternative renewable energy sources in form of microbial fuel cell that can directly generate electricity in single or multiple steps [112, 113]. MFCs are bioelectrical equipment that directly harvest electrical energy generated during the anaerobic respiration of certain microbes grown on organic feedstocks. MFC is widely used as a bio-electrochemical system in wastewater management, sustainable electricity production, and regeneration of valuable reduced reaction end products in biochemical reactions.

In a hydrogen fuel cell, hydrogen gas is oxidized into protons and electrons at anode while oxygen is reduced to water at cathode in presence of catalyst. The structure of MFCs is comparable to hydrogen fuel cells in which protons flow from anode to cathode chamber through an electrically insulated proton-exchange membrane while electrons flow in a similar direction through an electric wire. In a MFC, anaerobic bacteria oxidize the organic feedstock in anode compartment and transfer the electrons to an electrode instead of an electron-acceptor molecule. The dissimilatory metal-reducing or anodophilic bacteria (*Geobacter metallireducens, Geothrix fermentans, Shewanella,* etc.), can make a biofilm on the surface of metal electrode (electron acceptor) in anode chamber and reduce it with the help of periplasmic c-type cytochrome protein as metal reductase [114–116]. Major constraint of typical MFC is the requirement of continuous aeration with oxygen as a potential terminal electron acceptor at cathode, which is an energetically intensive procedure in a real-world application.

The limitation of cathode aeration was eliminated by the assimilation of algal photosynthesis with MFCs, hence known as photo MFCs. In a photo MFC, bacterial respiration takes place in anaerobic anode chamber which is electrochemically coupled with an algal oxygenic photosynthesis cathode chamber as a potential terminal electron acceptor. Unicellular microalgae are widely used in photo MFCs for electricity generation. Thus, algal photo MFCs can efficiently treat wastewater, generate decent electricity, and fix CO_2 simultaneously [117].

Dominant anodophilic bacterial consortia for photo-MFCs are composed of *Shewanella oneidensis, Geobacter sulferreducens, Cytophaga xylanolytica, Cytophaga, Pseudomonas aeruginosa, Thiomonas perometabolis, Dechloromonas, Desulfovibrio,* etc. whereas algal species commonly used in cathode chamber are *Spirulina platensis, Anabaena cylindrical, Synechococcus leopoliensis, Scenedesmus obliquus, Chlorella vulgaris, Chlorella pyrenoidosa, Desmodesmus, Phormidium, Chlamydomonas reinhardtii, Laminaria saccharina,* etc. Photo-MFC with a combination of *Geobacter sulferreducens* suspension in acetate as anolyte and *Chlamydomonas reinhardtii* is shown to generate power output of $mW\,m^{-2}$ [118]. Thus, photo-MFCs propose a cost-effective and eco-friendly technology for electricity generation and wastewater management.

15.5 Algae in Additional Applications

15.5.1 Algae as Livestock Feed and Nutrition

Aquaculture, livestock animals, and poultry farm must be intensified in order to meet the growing needs of milk, meat, eggs, wool, and hides for increasing population. Productivity of reared animals greatly depends upon the quality of their feed so insufficient quantity or nutritionally incomplete agricultural products as feed lead to suboptimal production of meat, milk, and eggs. Algae serve as an excellent nutrition-rich alternative feed for a variety of animals. Algal supplemented feeds are more nutritive and rich in proteins, carbohydrates, PUFAs, carotenoids, essential vitamins, and mineral than conventional ones. These can boost immunity of ruminants and protect animals and fishes from diseases and lead to production of healthy animal products. At present, >30% of global algal production is used to feed reared animals, and >50% of global *Arthrospira platensis* production is used to prepare such feeds. Genetically divergent Australian sheep fed on *Arthospira*-fortified feed significantly gained body mass, growth, and body contour [119]. Algal feed increase ω-3 fatty acid content in milk without harming its lipid yield [120]. *Laminaria digitate* supplemented feed increased pig weight by 10% [121].

Chicken meat and eggs are among major sources of protein globally and algae serve as nutrition supplement for poultry. Algal biomass up to 10% of total feed weight is used as a partial replacement of traditional protein supplement in poultry feed. Chicken fed on algal biomass-supplemented feed exhibited bright yellow colour of egg yolk, skin, and shanks with 10% less cholesterol due to higher carotenoid content infeed [122]. Due to high carotenoids, use of excessive algal feeds for extended periods cause yellowing and change in texture of meat, so are not liked in some parts of the world, but it is not reported to cause toxicity in animal during feeding trails.

15.5.2 Algae as Feed in Aquaculture

Aquaculture, an important sector in animal husbandry, is a vital source of protein. Various microalgal species (*Chlorella, Arthrospira platensis, Pavlova, Tetraselmis, Isochrysis, Chaetoceros, Phaeodactylum, Thalassiosira, Skeletonema, Porphyra, Kappaphycus, Nannochloropsis,* etc.) are natural food for herbivores like rotifers, brine shrimps, copepods, and other crustacean that are directly or indirectly fed by fishes and secondary consumers in aquatic food chains. PUFAs (AA, EPA, and DHA) are essential for fecundity, egg quality, spawning, larvae hatching rates, growth and development, and pigmentation of fishes [123, 124]. Animals having limited ability of de novo synthesis of PUFA shave high demand of PUFA and its derivatives from fishes as food and nutraceuticals. The carotenoid-rich algal feeds that enhance the pigmentation of reared shrimps, salmonids, lobsters, red sea breams, etc., are an ideal alternative source of balanced nutrition for the blue revolution.

15.5.3 Algae as Bio-Fertilizer

Algal biomass of cyanobacteria and green algae is crucial in building organic component of soil on decomposition in an agro-ecosystem. Secretion of exopolysaccharides and other phytohormones promotes growth of crop, microflora, and fauna. Inoculation of fields with cyanobacteria and green algae augments microbial activities in soil and rhizosphere, increases plant growth promoters, *viz.*, vitamin B_{12}, indole-3-acetic acid, indole-3-propionic acid, or 3-methyl indole, and improves crop yield [125]. Inoculation of crop fields with heterocystous cyanobacteria (*Nostoc, Aulosira, Tolypothrix, Scytonema, Anabaena variabilis*, etc.) can fix atmospheric nitrogen, i.e. up to 40 kg of nitrogen per hectare and thereby significantly reduce the cost of crop production. Whereas non-heterocystous cyanophytes (*Plectonema boryanum, Oscillaoria, Lyngbya*) and green algae enrich soil by secreting insoluble salts. Thus, microalgae are used as biofertilizers and to maintain soil fertility.

15.6 Conclusion and Future Prospects

Algae have tremendous potential to produce a variety of high-value biomolecules. Microalgae offer a potential resource of renewable, clean fuels like biohydrogen, biogas, syngas, bioethanol, biodiesel, and photo-electricity to address current environmental imbalance and increased carbon foot printing due to fossil fuel combustion. Due to their remarkable ability to utilize carbon emitted from combustion as a source to recycle it in the form of biomass at an impressive rate algae helps to reduce global warming. Also, they have the potential to assimilate pollutants into biomass as nutrients and are greatly useful in wastewater management. Being a rich source of protein, carbohydrate, PUFAs, vitamins, and minerals, they help to meet global demands of healthy food. As per WHO recommendations on the intake of PUFA by infants and adults has increased its demand. However, the major PUFA demand is met with DHA obtained from fish oil of fishes feeding of microalgae producing it. Similarly, coloring agents, β-carotene, and astaxanthin are used as fish food and growing awareness of nutritional food will also drive their demand as health supplement molecules. At present, algal products and algae are gaining popularity and greater acceptability as a source of proteins, vitamins, minerals, polyphenols, and other value-added health supplements will improve their commercial production. Algae are rich source of pharmaceutical and cosmeceutical molecules and could drive their specific molecule-targeted commercial cultivation.

Although algae offer an array of value-added molecules, many challenges impede their exhaustive exploitation. A large gap exists between perceived commercial capabilities of various algal species and actual abilities of industrial processes. Evolution from niche market to the extraction of high-value algal products requires extensive research and improvement in methods to expand yields to gain multiple compounds from a single strategy. The extraction of bioactive algal compounds is greatly challenged by the cost of production, i.e. cultivation system, maintenance, limited productivity, and bio-refining techniques. Production and hyper-accumulation of single high-value molecule require identification and selection of appropriate algal strain, desired genetic engineering, and development of strain-specific culture cultivation strategies. Additionally, efforts are needed to understand the mechanisms of high-value molecule production and devise high-density culture production, dehydration, gentle methods of cell wall disruption, and suitable recovery procedures without altering its biochemical nature.

References

1 Lim, D.K. and Schenk, P.M. (2017). Microalgae selection and improvement as oil crops: GM vs non-GM strain engineering. *AIMS Bioengineering* 40 (1): 151–161.

2 Andersen, R.A. (1992). Diversity of eukaryotic algae. *Biodiversity and Conservation* 10 (4): 267–292.

3 Groendahl, S., Kahlert, M., and Fink, P. (2017). The best of both worlds: a combined approach for analyzing microalgal diversity via metabarcoding and morphologybased methods. *PLoS One* 12 (2): e0172808.

4 Ruggiero, M.A., Gordon, D.P., Orrell, T.M. et al. (2015). A higher level classification of all living organisms. *PLoS One* 10 (4): e0130114.

5 Round, F.E., Crawford, R.M., and Mann, D.G. (1990). The Diatoms: Biology and Morphology of the Genera, vol. ix, 747pp. Cambridge: Cambridge University Press.

6 Garcia, J.L., de Vicente, M., and Galan, B. (2017). Microalgae, old sustainable food and fashion nutraceuticals. *Microbial Biotechnology* 100 (5): 1017–1024.

7 Levasseur, W., Perré, P., and Pozzobon, V. (2020). A review of high value-added molecules production by microalgae in light of the classification. *Biotechnology Advances* 41: 107545.

8 Perin, G., Bellan, A., Bernardi, A. et al. (2019). The potential of quantitative models to improve microalgae photosynthetic efficiency. *Physiologia Plantarum* 1660 (1): 380–391.

9 Christaki, E., Florou-Paneri, P., and Bonos, E. (2011). Microalgae: a novel ingredient in nutrition. *International Journal of Food Sciences and Nutrition* 620 (8): 794–799.

10 Hamed, I. (2016). The evolution and versatility of microalgal biotechnology: a review. *Comprehensive Reviews in Food Science and Food Safety* 150 (6): 1104–1123.

11 Milledge, J.J. (2011). Commercial application of microalgae other than as biofuels: a brief review. *Reviews in Environmental Science and Bio/Technology* 100 (1): 31–41.

12 Ritchie, H. and Roser, M. (2017). CO_2 and Greenhouse Gas Emissions. OurWorld InData. org. https://ourworldindata.org/co2-and-other-greenhouse-gas-emissions (accessed 2019).

13 Skjanes, K., Lindblad, P., and Muller, J. (2007). BioCO_2 – a multidisciplinary, biological approach using solar energy to capture CO_2 while producing H_2 and high value products. *Biomolecular Engineering* 24: 405–413.

14 Tsai, D.D., Chen, P.H., and Ramaraj, R. (2017). The potential of carbon dioxide capture and sequestration with algae. *Ecological Engineering* 98: 17–23.

15 Sayre, R. (2010). Microalgae: the potential for carbon capture. *Bioscience* 60 (9): 722–727.

16 Herzog, H. and Golomb, D. (2004). Carbon capture and storage from fossil fuel use. *Encyclopedia of Energy* 1: 1–11.

17 Douskova, I., Doucha, J., Livansky, K. et al. (2009). Simultaneous flue gas bioremediation and reduction of microalgal biomass production costs. *Applied Microbiology and Biotechnology* 82: 179–185.

18 Keffer, J.E. and Kleinheinz, G.T. (2002). Use of *Chlorella vulgaris* for CO_2 mitigation in a photobioreactor. *Journal of Industrial Microbiology and Biotechnology* 29: 275–280.

19 Kumar, K. and Das, D. (2012). Growth characteristics of *Chlorella sorokiniana* in airlift and bubble column photobioreactors. *Bioresource Technology* 116: 307–313.

20 Tenenbaum, D.J. (2009). Biochar: carbon mitigation from the ground up. *Environmental Health Perspectives* 117: A70–A73.

21 Gifuni, I., Pollio, A., Safi, C. et al. (2019). Current bottlenecks and challenges of the microalgal biorefinery. *Trends in Biotechnology* 370 (3): 242–252.

22 Conde, E., Balboa, M., Parada, M., and Falque, E. (2013). Algal proteins, peptides and amino acids. In: Functional Ingredients from Algae for Foods and Nutraceuticals (ed. H. Dominguez), 135–180. Cambridge, UK: Woodhead Publishing Limited.

23 Becker, E.W. (2007). Micro-algae as a source of protein. *Biotechnology Advances* 250 (2): 207–210.

24 Khanra, S., Mondal, M., Halder, G. et al. (2018). Downstream processing of microalgae for pigments, protein and carbohydrate in industrial application: a review. *Food and BioproductsProcessing* 110: 60–84.

25 Bernaerts, T.M.M., Gheysen, L., Foubert, I. et al. (2019). The potential of microalgae and their biopolymers as structuring ingredients in food: a review. *Biotechnology Advances* 37 (8): 107419.

26 Kose, A., Ozen, M.O., Elibol, M., and Oncel, S.S. (2017). Investigation of in vitro digestibility of dietary microalga *Chlorella vulgaris* and cyanobacterium *Spirulina platensis* as a nutritional supplement. *3 Biotech* 70: 170.

27 Wild, K.J., Steingas, H., and Rodehutscord, M. (2018). Variability in nutrient composition and in vitro crude protein digestibility of 16 microalgae products. *Journal of Animal Physiology and Animal Nutrition* 1020 (5): 1306–1319.

28 Apone, F., Barbulova, A., and Colucci, M.G. (2019). Plant and microalgae derived peptides are advantageously employed as bioactive compounds in cosmetics. *Frontiers in Plant Science* 10: 756.

29 Yaakob, Z., Ali, E., Zainal, A. et al. (2014). An overview: biomolecules from microalgae for animal feed and aquaculture. *Journal of Biological Research (Thessaloniki)* 21: 6.

30 Spolaore, P., Joannis-Cassan, C., Duran, E., and Isambert, A. (2006). Commercial applications of microalgae. *Journal of Bioscience and Bioengineering* 101: 87–96.

31 Mourelle, M.L., Gomez, C., and Legido, J.L. (2017). The potential use of marine microalgae and cyanobacteria in cosmetics and thalassotherapy. *Cosmetics* 4: 46.

32 Galasso, C., Gentile, A., Orefice, I. et al. (2019). Microalgal derivatives as potential nutraceutical and food supplements for human health: a focus on cancer prevention and interception. *Nutrients* 110 (6): 1226.

33 Ko, S.-C., Kim, D., and Jeon, Y.-J. (2012). Protective effect of a novel antioxidative peptide purified from a marine *Chlorella ellipsoidea* protein against free radical-induced oxidative stress. *Food and Chemical Toxicology* 500 (7): 2294–2302.

34 Ko, S.-C., Kang, N., Kim, E.-A. et al. (2012). A novel angiotensin I-converting enzyme (ACE) inhibitory peptide from a marine *Chlorella ellipsoidea* and its antihypertensive effect in spontaneously hypertensive rats. *Process Biochemistry* 470 (12): 2005–2011.

35 Suetsuna, K. and Chen, J.R. (2001). Identification of antihypertensive peptides from peptic digest of two microalgae, *Chlorella vulgaris* and *Spirulina platensis*. *Marine Biotechnology* 30 (4): 305–309.

36 Darvish, M., Jalili, H., Ranaei-Siadat, S.-O., and Sedighi, M. (2018). Potential cytotoxic effects of peptide fractions from *Dunaliella salina* protein hydrolyzed by gastric proteases. *Journal of Aquatic Food Product Technology* 207 (2): 165–175.

37 Barkia, I., Ketata Bouaziz, H., Sellami Boudawara, T. et al. (2020). Acute oral toxicity study on Wistar rats fed microalgal protein hydrolysates from *Bellerochea malleus*. *Environmental Science and Pollution Research* 27 (16): 19087–19094.

38 Sahu, A., Pancha, I., Jain, D. et al. (2013). Fatty acids as biomarkers of microalgae. *Phytochemistry* 89: 53–58.

39 Raghukumar, S. (2008). Thraustochytrid marine protists: production of PUFAs and other emerging technologies. *Marine Biotechnology* 100 (6): 631–640.

40 Paes, C.R.P.S., Faria, G.R., Tinoco, N.A.B. et al. (2016). Growth, nutrient uptake and chemical composition of *Chlorella* sp. and *Nannochloropsis oculata* under nitrogen starvation. *Latin American Journal of Aquatic Research* 440 (2): 275–292.

41 Pignolet, O., Jubeau, S., Vaca-Garcia, C., and Michaud, P. (2013). Highly valuable microalgae: biochemical and topological aspects. *Journal of Industrial Microbiology and Biotechnology* 400 (8): 781–796.

42 Markou, G., Angelidaki, I., and Georgakakis, D. (2012). Microalgal carbohydrates: an overview of the factors influencing carbohydrates production, and of main bioconversion technologies for production of biofuels. *Applied Microbiology and Biotechnology* 960 (3): 631–645.

43 Suzuki, E. and Suzuki, R. (2013). Variation of storage polysaccharides in phototrophic microorganisms. *Journal of Applied Glycoscience* 60: 21–27.

44 Ruperez, P. and Saura-Calixto, F. (2001). Dietary fibre and physicochemical properties of edible Spanish seaweeds. *European Food Research and Technology* 212: 349–354.

45 Cantarel, B.L., Lombard, V., and Henrissat, B. (2012). Complex carbohydrate utilization by the healthy human microbiome. *PLoS One* 7: e28742.
46 Wells, M.M.L., Potin, P., Craigie, J.S. et al. (2016). Algae as nutritional and functional food sources: revisiting our understanding. *Journal of Applied Phycology* 29: 949–982.
47 Kulshreshtha, G., Rathgeber, B., Stratton, G. et al. (2014). Feed supplementation with red seaweeds, *Chondrus crispus* and *Sarcodiotheca gaudichaudii*, affects performance, egg quality, and gut microbiota of layer hens. *Poultry Science* 93: 2991–3001.
48 Liu, J.H., Kandasamy, S., Zhang, J.Z. et al. (2015). Prebiotic effects of dietsupplemented with the cultivated red seaweed *Chondrus crispus* or with fructo-oligo-saccharide on host immunity, colonic microbiota and gut microbial metabolites. *BMC Complementary and Alternative Medicine* 15: 279.
49 Gardeva, E., Toshkova, R., Minkova, K., and Gigova, L. (2009). Cancer protective action of polysaccharide, derived from red microalga porphyridium cruentum – a biological background. *Biotechnology and Biotechnological Equipment* 23: 783–787.
50 Bhattacharyya, S., Liu, H., Zhang, Z. et al. (2010). Carrageenan-induced innate immuneresponse is modified by enzymes that hydrolyze distinct galactosidic bonds. *Journal of Nutritional Biochemistry* 21: 906–913.
51 Bhattacharyya, S., O-Sullivan, I., Katyal, S. et al. (2012). Exposure to the common food additive carrageenan leads to glucose intolerance, insulin resistance and inhibition of insulin signaling in HepG2 cells and C57BL/6J mice. *Diabetologia* 55 (1): 194–203.
52 Rioux, L.-E., Turgeon, S.L., and Beaulieu, M. (2004). Characterization of polysaccharides extracted from brown seaweeds. *Carbohydrate Polymers* 69: 530–537.
53 Neyrinck, A.M., Mouson, A., and Delzenne, N.M. (2007). Dietary supplementation with laminaran, a fermentable marine ß(1–3) glucan, protects against hepatoxicity induced by LPS in rat by modulating immune response in the hepatic tissue. *International Immunopharmacology* 7: 1497–1506.
54 Fitton, J., Dell'Acqua, G., Gardiner, V.-A. et al. (2015). Topical benefits of two fucoidan-rich extracts from marine macroalgae. *Cosmetics* 2: 66–81.
55 Mishra, V.K., Bacheti, R.K., and Husen, A. (2011). Medicinal uses of chlorophyll: a critical overview. In: Chlorophyll: Structure, Function and Medicinal Uses (eds. H. Le and E. Salcedo), 177–196. Hauppauge, NY: Nova Science Publishers, Inc.
56 McQuistan, T.J., Simonich, M.T., Pratt, M.M. et al. (2012). Cancer chemoprevention by dietary chlorophylls: a 12,000-animal dose-dose matrix biomarker and tumor study. *Food and Chemical Toxicology* 50: 341–352.
57 Sher, I., Tzameret, A., Peri-Chen, S. et al. (2018). Synthetic 9-cis-beta-carotene inhibits photoreceptor degeneration in cultures of eye cups from rpe65rd12 mouse model of retinoid cycle defect. *Scientific Reports* 8: 6130.
58 Gammone, M.A., Riccioni, G., and D' Orazio, N. (2015). Carotenoids: potential allies of cardiovascular health? *Food & Nutrition Research* 59: 26762.
59 Stahl, W. and Sies, H. (2012). β-Carotene and other carotenoids in protection from sunlight. *American Journal of Clinical Nutrition* 96: 1179–1184.
60 Lee, R.E. (2018). Phycology. Cambridge University Press 978-1-107-55565-5.
61 Berthon, J.-Y., Nachat-Kappes, R., Bey, M. et al. (2017). Marine algae as attractive source to skin care. *Free Radical Research* 510 (6): 555–567.

62 Rammuni, M.N., Ariyadasa, T.U., Nimarshana, P.H.V., and Attalage, R.A. (2019). Comparative assessment on the extraction of carotenoids from microalgal sources: astaxanthin from *H. pluvialis* and β-carotene from *D. salina*. *Food Chemistry* 277: 128–134.

63 Rodrigues, D.B., Menezes, C.R., Mercadante, A.Z. et al. (2015). Bioactive pigments from microalgae *Phormidium autumnale*. *Food Research International* 77: 273–279.

64 Lorenz, R.T. and Cysewski, G.R. (2000). Commercial potential for *Haematococcus microalgae* as a natural source of astaxanthin. *Trends in Biotechnology* 18: 160–167.

65 Galasso, C., Orefice, I., Pellone, P. et al. (2018). On the neuroprotective role of astaxanthin: new perspectives? *Marine Drugs* 16: 247.

66 Park, J.S., Chyun, J.H., Kim, Y.K. et al. (2010). Astaxanthin decreased oxidative stress and inflammation and enhanced immune response in humans. *Nutrition & Metabolism (London)* 7: 18.

67 Yoshida, H., Yanai, H., Ito, K. et al. (2010). Administration of natural astaxanthin increases serum HDL-cholesterol andadiponectin in subjects with mild hyperlipidemia. *Atherosclerosis* 209 (2): 520–523.

68 Schalch, W. (1992). Carotenoids in the retina – a review of their possible role in preventing or limiting damage caused by light and oxygen. In: Free Radicals and Aging. EXS, vol. 62 (eds. I. Emerit and B. Chance), 280–298. Basel: Birkhauser.

69 Rasmussen, H.M. and Johnson, E.J. (2013). Nutrients for the aging eye. *Clinical Interventions in Aging* 8: 741–748.

70 Pasquet, V., Morisset, P., Ihammouine, S. et al. (2011). Antiproliferative activity of violaxanthin isolated from bioguided fractionation of *Dunaliella tertiolecta* extracts. *Marine Drugs* 9: 819–831.

71 Tarento, T.D.C., McClurea, D.D., Vasiljevskib, E. et al. (2018). Microalgae as a source of vitamin K1. *Algal Research* 36: 77–87.

72 Fabregas, J. and Herrero, C. (1990). Vitamin content of four marine microalgae. Potential use as source of vitamins in nutrition. *Journal of Industrial Microbiology* 5: 259–263.

73 Ortiz, J., Uquiche, E., Robert, P. et al. (2009). Functional and nutritional value of the Chilean seaweeds *Codium fragile, Gracilaria chilensisand,* and *Macrocystis pyrifera*. *European Journal of Lipid Science and Technology* 111: 320–327.

74 Manach, C., Scalbert, A., Morand, C. et al. (2004). Polyphenols: food sources and bioavailability. *American Journal of Clinical Nutrition* 79: 727–747.

75 Montero, L., del Pilar Sanchez-Camargo, A., Ibanez, E., and Gilbert-Lopez, B. (2018). Phenolic compounds from edible algae: bioactivity and health benefits. *Current Medicinal Chemistry* 250 (37): 4808–4826.

76 Jerez-Martel, I., García-Poza, S., Rodríguez-Martel, G. et al. (2017). Phenolic profile and antioxidant activity of crude extracts from microalgae and cyanobacteria strains. *Journal of Food Quality* 2017: 2924508.

77 Goiris, K., Muylaert, K., Fraeye, I. et al. (2012). Antioxidant potential of microalgae in relation to their phenolic and carotenoid content. *Journal of Applied Phycology* 240 (6): 1477–1486.

78 Yoon, N.-Y., Lee, S.-H., Wijesekara, I., and Kim, S.-K. (2011). in vitro and intracellular antioxidant activities of brown alga *Eisenia bicyclis*. *Fisheries and Aquatic Sciences* 14 (3): 179–185.

79 Kim, S.-K. and Kong, C.-S. (2010). Anti-adipogenic effect of dioxinodehydroeckol via AMPK activation in 3T3-L1 adipocytes. *Chemico-Biological Interactions* 186: 24–29.

80 Liu, M., Hansen, P.E., and Lin, X. (2011). Bromophenols in marine algae and their bioactivities. *Marine Drugs* 9 (7): 1273–1292. https://doi.org/10.3390/md9071273.

81 Luo, X., Su, P., and Zhang, W. (2015). Advances in microalgae-derived phytosterols for functional food and pharmaceutical applications. *Marine Drugs* 13: 4231–4254.

82 Lu, Y. and Xu, J. (2015). Phytohormones in microalgae: a new opportunity for microalgal biotechnology? *Trends in Plant Science* 20 (5): 273–282.

83 Fabregas, J. and Herrero, C. (1986). Marine microalgae as a potential source of minerals in fish diets. *Aquaculture* 51: 237–243.

84 Rupérez, P. (2002). Mineral content of edible marine seaweeds. *Food Chemistry* 79 (1): 23–26.

85 Circuncisão, A.R., Catarino, M.D., Cardoso, S.M., and Silva, A. (2018). Minerals from macroalgae origin: health benefits and risks for consumers. *Marine Drugs* 16: 400.

86 Nitschke, U. and Stengel, D.B. (2016). Quantification of iodine loss in edible Irish seaweeds during processing. *Journal of Applied Phycology* 28: 3527–3533. https://doi.org/10.1007/s10811-016-0868-6.

87 IEA (2019). World Energy Outlook 2019. Paris: IEA https://www.iea.org/reports/world-energy-outlook-2019.

88 Behera, S., Singh, R., Arora, R. et al. (2015). Scope of algae as third generation biofuels. *Frontiers in Bioengineering and Biotechnology* 90 (2): 1–13.

89 Milano, J., Ong, H.C., Masjuki, H.H. et al. (2016). Microalgae biofuels as an alternative to fossil fuel for power generation. *Renewable and Sustainable Energy Reviews* 58: 180–197.

90 Lee, O.K. and Lee, E.Y. (2016). Sustainable production of bioethanol from renewable brown algae biomass. *Biomass and Bioenergy* 92: 70–75.

91 Naik, S.N., Goud, V.V., Rout, P.K., and Dalai, A.K. (2010). Production of first and second generation biofuels: a comprehensive review. *Renewable and Sustainable Energy Reviews* 14: 578–597.

92 Adeniyi, O.M., Azimov, U., and Burluka, A. (2018). Algae biofuel: current status and future applications. *Renewable and Sustainable Energy Reviews* 90: 316–335.

93 Chiaramonti, D., Prussi, M., Bu, M. et al. (2017). Review and experimental study on pyrolysis and hydrothermal liquefaction of microalgae for biofuel production. *Applied Energy* 185: 963–972.

94 Raslavicius, L., Semenov, V.G., Chernova, N.I. et al. (2014). Producing transportation fuels from algae: in search of synergy. *Renewable and Sustainable Energy Reviews* 40: 133–142.

95 Pittman, J.K., Dean, A.P., and Osundeko, O. (2011). The potential of sustainable algal biofuel production using wastewater resources. *Bioresource Technology* 102: 17–25.

96 Bisen, P.S., Sanodiya, B.S., Thakur, G.S. et al. (2010). Biodiesel production with special emphasis on lipase-catalyzed transesterification. *Biotechnology Letters* 32: 1019–1030.

97 Al-lwayzy, S. and Yusaf, T. (2013). *Chlorella* protothecoides microalgae as an alternative fuel for tractor diesel engines. *Energies* 6: 766.

98 Salam, K.A., Velasquez-Orta, S.B., and Harvey, A.P. (2016). A sustainable integrated in situ transesterification of microalgae for biodiesel production and associated co-product-a review. *Renewable and Sustainable Energy Reviews* 65: 1179–1198.

99 Jazzar, S., Quesada-Medina, J., Olivares-Carrillo, P. et al. (2015). A whole biodiesel conversion process combining isolation, cultivation and in situ supercritical methanol transesterification of native microalgae. *Bioresource Technology* 190: 281–288.

100 Ramos-Suárez, J.L. and Carreras, N. (2014). Use of microalgae residues for biogas production. *Chemical Engineering Journal* 242: 86–95.

101 Passos, F., Solé, M., García, J., and Ferrer, I. (2013). Biogas production from microalgae grown in wastewater: effect of microwave pretreatment. *Applied Energy* 108: 168–175.

102 Brennan, L. and Owende, P. (2010). Biofuels from microalgae – a review of technologies for production, processing, and extractions of biofuels and co-products. *Renewable and Sustainable Energy Reviews* 14 (2): 557–577.

103 Marsolek, M.D., Kendall, E., Thompson, P.L., and Shuman, T.R. (2014). Thermal pretreatment of algae for anaerobic digestion. *Bioresource Technology* 151: 373–377.

104 Eshaq, F.S., Ali, M.N., and Mohd, M.K. (2011). Production of bioethanol from next generation feed-stock alga *Spirogyra* species. *International Journal of Engineering, Science and Technology* 3: 1749–1755.

105 Rajkumar, R., Yaakob, Z., and Takriff, M.S. (2014). Potential of the micro and macroalgae for biofuel production: a brief review. *BioResources* 9 (1): 1606–1633.

106 Potts, T., Du, J., Paul, M. et al. (2012). The production of butanol from Jamaica bay macro algae. *Environmental Progress & Sustainable Energy* 31: 29–36.

107 Gaffron, H. and Rubin, J. (1942). Fermentative and photochemical production of hydrogen in algae. *Journal of General Physiology* 26 (2): 219–240.

108 Cantrell, K.B., Ducey, T., Ro, K.S., and Hunt, P.G. (2008). Livestock waste-to-bioenergy generation opportunities. *Bioresource Technology* 99: 7941–7953.

109 Melis, A., Zhang, L., Forestier, M. et al. (2000). Sustained photobiological hydrogen gas production upon reversible inactivation of oxygen evolution in the green alga *Chlamydomonas reinhardtii*. *Plant Physiology* 122: 127–136.

110 Melis, A. and Happe, T. (2001). Hydrogen production. Green algae as a source of energy. *Plant Physiology* 127: 740–748.

111 Long, H., Chang, C.H., King, P.W. et al. (2008). Brownian dynamics and molecular dynamics study of the association between hydrogenase and ferredoxin from *Chlamydomonas reinhardtii*. *Biophysical Journal* 95: 3753–3766.

112 Angenent, L.T., Karim, K., Al-Dahhan, M.H. et al. (2004). Production of bioenergy and biochemicals from industrial and agricultural wastewater. *Trends in Biotechnology* 22: 477–485.

113 Mohan, S.V., Srikanth, S., Chiranjeevi, P. et al. (2014). Algal biocathode for in situ terminal electron acceptor (TEA) production: synergetic association of bacteria – microalgae metabolism for the functioning of biofuel cell. *Bioresource Technology* 166: 566–574.

114 Nevin, K.P. and Lovley, D.R. (2000). Lack of production of electronshuttling compounds or solubilization of Fe(III) during reduction of insoluble Fe(III) oxide by *Geobacter metallireducens*. *Applied and Environmental Microbiology* 66: 2248–2251.

115 Nevin, K.P. and Lovley, D.R. (2002). Mechanisms for accessing insoluble Fe(III) oxide during dissimilatory Fe(III) reduction by *Geothrix fermentans*. *Applied and Environmental Microbiology* 68: 2294–2299.

116 Rabaey, K., Lissens, G., Siciliano, S.D., and Verstraete, W. (2003). A microbial fuel cell capable of converting glucose to electricity at high rate and efficiency. *Biotechnology Letters* 25: 1531–1535.

117 Liu, T., Rao, L., Yuan, Y., and Zhuang, L. (2015). Bioelectricity generation in a microbial fuel cell with a self-sustainable photocathode. *The Scientific World Journal* 2015 Article ID 864568: 8. http://dx.doi.org/10.1155/2015/864568.

118 Nishio, K., Hashimoto, K., and Watanabe, K. (2013). Light/electricity conversion by defined cocultures of *Chlamydomonas* and *Geobacter*. *Journal of Bioscience and Bioengineering* 115: 412–417.

119 Holman, B., Kashani, A., and Malau-Aduli, A.E.O. (2012). Growth and body conformation responses of genetically divergent Australian sheep to Spirulina (*Arthrospira platensis*) supplementation. *American Journal of Experimental Agriculture* 2: 160–173.

120 Stamey, J.A., Shepherd, D.M., de Veth, M.J., and Corl, B.A. (2012). Use of algae or algal oil rich in *n*-3 fatty acids as a feed supplement for dairy cattle. *Journal of Dairy Science* 95: 5269–5275.

121 He, M.L., Hollwich, W., and Rambeck, W.A. (2002). Supplementation of algae to the diet of pigs: a new possibility to improve the iodine content in the meat. *Journal of Animal Physiology and Animal Nutrition* 86: 97–104.

122 Harun, R., Danquah, M.K., and Forde, G.M. (2010). Microalgal biomass as a fermentation feedstock for bioethanol production. *Journal of Chemical Technology and Biotechnology* 85: 199–203.

123 Bell, J.G., McEvoy, L.A., Estevez, A. et al. (2003). Optimising lipid nutrition in first-feeding flatfish larvae. *Aquaculture* 227: 211–220.

124 Lall, S.P. and Lewis-McCrea, L.M. (2007). Role of nutrients in skeletal metabolism and pathology in fish – an overview. *Aquaculture* 267: 3–19.

125 Guo, S., Wang, P., Wang, X. et al. (2020). Microalgae as biofertilizer in modern agriculture. In: Microalgae Biotechnology for Food, Health and High Value Products (eds. M. Alam, J.L. Xu and Z. Wang), 397–411. Singapore: Springer.

16

Plant Bioprospecting for Biopesticides and Bioinsecticides

Aradhana Lucky Hans and Sangeeta Saxena

Department of Biotechnology, Babasaheb Bhimrao Ambedkar University, Lucknow, India

16.1 Introduction

Agriculture holds an exceptional position in the growth of a country, especially developing ones. The major setback it faces is from the attack of various pests causing harm to crops. Loss due to harmful pest causes severe economic losses of agricultural crops and commodities. This threat to economic loss in the quality of agricultural crops has paved a way to major insights in the field of research and new discoveries of pest-controlling agents. Therefore, one of the crucial components of agricultural management is efficient management of these pests. They can be managed and controlled by the application of various pesticides in the different ways and at the different time points.

Over the years, chemical pesticides have made a great contribution to the fight against several pests and diseases. However, the repetitive and indiscriminate use of these pesticides has developed insecticide resistance in many major pests causing resurgence of minor pests, which in turn resulted in irreparable damage to crops. A new range of pests had even more and wider range of pesticide application, making the situation even more worse. Harmful effects, such as soil and water contamination and dramatic increase of the harmful residues, cause harm to both environment and human health. The use of chemical insecticides in crop pest control programs around the world had caused tremendous damage to the environment, pest resurgence, pest resistance to insecticides, and lethal effects on nontarget organisms [1].

It is evident that population rise has been a continuous process and to meet the needs, agriculture also has been pacing up at faster rate. The improvement in the yield and production of agriculture can be easily credited to the application of pesticides, therefore improving production in agriculture is most certain. The indiscriminate use of pesticides over the years, particularly in developing countries, has caused enough damage in various spheres. Overlooking the safety measures and recommended usage has led to many health-related problems. The risk is not only on humans but on the surrounding environment and the nontarget flora and fauna that are exposed to these harmful chemicals. The residues

Bioprospecting of Plant Biodiversity for Industrial Molecules, First Edition.
Edited by Santosh Kumar Upadhyay and Sudhir P. Singh.

linger in the air, water, and soil and persist for a long duration causing more damage to the environment. According to World Heath Organizations (WHO), about 25 million cases of acute occupational pesticide poisoning and 20 000 deaths occurred globally per year. This adverse effect of these pesticides is a matter of great concern, leading to awareness about safety of food and environment. A search for alternatives to these nonjudiciously used chemicals is catching up pace toward a more greener and cleaner alternative. In recent decades, there has been a heap change in the viewpoint, causing some changes in the use of conventional chemical pesticides [2]. Pertaining to the high risk with the use of chemical pesticides, integrated pest management (IPM) programs have brought out application of pesticides with more efficient and judicious manner. Eventually these changes have significantly reduced pesticide use and improved pest management practices. As a result, there is an increasing demand for less toxic or environmentally safe pesticides that would pose less risk to application as well as consumption. The need of the hour is to look for alternatives that are less adverse and possess lesser health risk and be more environment-friendly. Moving back from chemical toward natural resource would be more apt owing to immeasurable diversity of molecules present in the nature. To look for an alternative that would be ecologically better and environmentally safe, biopesticides seem to be the choicest option. Also, chemical pesticides are more expensive in comparison to biopesticides.

Plants are regarded as one of the most diverse and richest resources of various important bioactive natural products. Since ancient times, the use of plants by the natives of various parts of the world as pesticidal has been well utilized. Innumerable active compounds have been identified from plants and still the source has not exhausted its prospective. Hence, this interminable reservoir of bioactive compounds should be screened, searched, and secured for more applications. Biopesticides conceptualized natural enemy or predatory organism or their products that would include plants, microbial products, or by-products as they were capable of managing or reducing pests. There is an obvious urge for the impetus in developing biopesticides as they have the potential to protect plants from pest infestation. In search for novel bioactive substances from botanicals, screening of diverse plants can be of much utility. This process of discovering and further commercialization of new products based on biological resource is known as bioprospecting. Nature has been an unending reservoir wherein one can unleash plant compounds that can be helpful in controlling various pests. Biological resource comparatively has higher diversity than the chemical compounds, holding more potential for novel compounds for agricultural application [3]. In this chapter, the present scenario of plant-based biopesticides, advantages, and future prospects will be discussed.

16.2 Current Scenario in India

Due to its rich biodiversity, India offers plenty of scope in terms of sources for natural biological control organisms as well as natural plant-based pesticides. Traditional knowledge available with the highly diverse indigenous communities in India may provide valuable clues for developing newer and effective biopesticides. Owing to its biological background biopesticides are eco-friendly, easily biodegradable, and safer alternative to chemical pesticides. Although the production is not as par chemical pesticide, yet the demand is pacing up the production every passing year. Presently, biopesticides may represent approximately 4.2% of

Table 16.1 List of biopesticides from their resource plants and target pests.

Plant product used as biopesticide	Target pests	Plants
Limonene and Linalool	Fleas, aphids, and mites also kill fire ants, several types of flies, paper wasps, and house crickets	Citrus fruits
Neem/Azadirachtin	A variety of sucking and chewing insect	*Azadirachta indica*
Pyrethrum/Pyrethrins	Ants, aphids, roaches, fleas, flies, and ticks	*Chrysanthemum cineraria folium*
Rotenone	Leaf-feeding insects, such as aphids, certain beetles (asparagus beetle, bean leaf beetle, Colorado potato beetle, cucumber beetle, flea beetle, strawberry leaf beetle, and others) and caterpillars, as well as fleas and lice on animals	Genus Lonchocarpus majorly from *Lonchocarpus utilis* and *Lonchocarpus urucu*
Ryania	Caterpillars (European corn borer, corn earworm, and others) and thrips	*Ryana speciosa*
Sabadilla	Squash bugs, harlequin bugs, thrips, caterpillars, leaf hoppers, and stink bugs	South American lily *Schoenocaulon officinale*
Nicotine	Caterpillars chewing pests	*Nicotiana tabacum*
Acetogenins	Lepidopterans and Colorado potato beetle (*Leptinotarsa decemlineata*)	*Annona* spp., including *A. squamosa and A. muricate*
Capsaicinoids	Lepidopterans and Hemiptera insects	*Capsicum* spp.

the total pesticides market in India. There are more than 6000 species that are screened of which 2500 plant species were found to possess biologically active compounds against various kinds of pests. The important plant families containing the biopesticide plants are *Apocynaceae, Asteraceae, Euphorbiaceae, Fabaceae, Meliaceae (maximum), Myrtaceae, Ranunculaceae*, and *Rosaceae* [4]. The higher demand for organic, chemical-free food has augmented the demand of eco-friendly chemicals isolated from different parts of plants (Table. 16.1). They can be from leaves, barks, roots, fruits, flowers, seeds, or seed kernels. In nature, the higher plants have the ability to synthesize and produce various kinds of secondary metabolites that are naturally avoided by insects. Some are able to interfere in the insect life cycle, which are known as semiochemicals. Majorly four types of botanical products are used for insect control (neem, pyrethrum, rotenone, and essential oils); few others such as ryania, sabadilla, acetogenins, capsaicinoids, and nicotine have limited use. Additionally, plant oils and extracts also seem to have effect on insects that have even lesser use.

16.3 Plants-Based Active Compounds

16.3.1 Azadirachtin

The plant Neem or *Azadirachta indica* is the most extensively used plant for insect pest management. Several farmers in Indian villages collect the neem seeds to prepare crude seed kernel extract for pest control. Various reports highlight that neem-related products do not leave any toxic residues on application on plants. Neem oil is said to be effective against mites, soft-bodied insects, and phytopathogens. The compound Azadirachtin is said to have low toxicity toward mammals and pollinators [5] according to Environmental Protection Agency. This compound is isolated from the kernels and used in various bioformulations of biopesticides. Effect of azadirachtin on insects can be in two major ways. It blocks the synthesis and release of ecdysteroids (molting hormones), leading to incomplete ecdysis in young insects and sterility in adult female insects. The other effect seems to be its antifeedant property inhibiting pest attack. These excellent effects of neem on pest definitely make neem to the millennium paradigm toward the development of biopesticides. The reason is there is a plethora of neem-based commercial products Neemmark, Wellgrow, Azatin, Bio-neem, Nimbin, Neemark, to name a few, used effectively throughout the world.

16.3.2 Pyrethrins

Pyrethrum is obtained from dries flowers of *Chrysanthemum cineraria folium* and has been and is globally used as a biopesticide [6]. Pyrethrins are insecticidal chemicals extracted from the dried pyrethrum flower. The flower heads are processed into a powder to make dust. This dust can be used directly or infused into water to make a spray. Technical-grade pyrethrum, the resin used in formulating commercial insecticides, typically contains pyrethrin from a range of 20–25% [7]. Effect of pyrethrins on insects is characterized by rapid knockdown particularly in flying insects and hyperactivity and convulsions in other insects. The mode of action is quite similar to commercial chemical pesticides such as DDT (Dichloro-diphenyl-trichloroethane) and other organochloride-based insecticides, effecting mostly the neurotoxic action potential blocking voltage-gated sodium channels in nerve axons. Pyrethrins are less toxic and easily break down in sunlight as they are liable to the UV rays, making them safer alternatives to chemical pesticides.

16.3.3 Rotenone

Rotenone is prepared from the plant species belonging to the genus *Lonchocarpus* majorly from *Lonchocarpus utilis* and *Lonchocarpus urucu*. A cube resin, root extract is the major component consisting of active ingredients rotenone, deguelin, and tephrosin. Rotenone is also used as blends along with pyrethrins and approved as organic insecticides. Rotenone is an important insecticide extracted from various other leguminous plants. It is an effective insecticide, it blocks the electron transport chain and prevents energy production acting as mitochondrial toxin. To be effective it must be ingested. It is considered similar to DDT but is much lesser toxic in its formulated products. In insects, rotenone exerts its toxic effects primarily on nerve and muscle cells, causing rapid cessation of feeding. Death occurs several hours to a few days after exposure. The disruption energy metabolism and the subsequent loss of ATP result in a slowly developing toxicity, and the effects of all these compounds include inactivity, paralysis, and death.

16.3.4 Sabadilla

Sabadilla is derived from the seeds of South American lily *Schoenocaulon officinale*. They have active ingredients as celandine-type alkaloids and toxic to mammals, but in commercial preparations, the active ingredients is at very low concentration, diluting its toxic effect. These alkaloids are similar to pyrethrins and are used on citrus crops and avocado in organic farming.

The mode of action of these alkaloids affects nerve cell membrane action, causing loss of nerve action potential, causing paralysis and death [6].

16.3.5 Ryania

These botanical insecticides are derived from stem of *Ryana speciosa*. It contains an alkaloid ryanodine that interferes with calcium release in muscle tissue. It is a slow-acting abdominal toxin. The effect exerted by it is not rapid, rather delayed as insects do not immediately stop feeding after ingesting it. Although not much has been studied about its mode of action yet, it is said to be effective in hot weather. It is found to be toxic against citrus thrips and fruit moths [8].

16.3.6 Nicotine

Another alkaloid is nicotine, which is obtained from the leaves of tobacco plant or *Nicotiana tabacum* and few related species. Nicotine, along with nornicotine and anabasine, is a synaptic toxin that mimics the neurotransmitter acetylcholine. They cause similar effects as that of organophosphate and carbamate insecticides. It can be absorbed dermally and is toxic to humans; therefore, the use has seen a decline. It is a fast-acting killer for soft-bodied insects but not to all chewing pests. It mimics acetylcholine, neurotransmitter causing uncontrolled nerve firing. It is said that nicotine is fairly selective and affects only certain types of insects [9].

16.3.7 Acetogenins

Acetogenins have been traditionally prepared from *Annona* spp., including *A. squamosa* and *A. muricate*, which are used as botanical pesticides. These are source of fruit juice in some parts of Asia. The major acetogenins are obtained from the seeds of *A. squamosa* [8]. These compounds show toxicity as slow-acting gut poison, mostly effective against chewing pest, especially Lepidopteran pest and Colorado potato beetle (*Leptinotarsa decemlineata*). The mode of action of acetogenins is similar to rotenone. They work in blocking energy production in mitochondria.

16.3.8 Capsaicinoids

Peppers are very abundant in nature and found around easily. They are extremely pungent and hot in nature. The burning sensation in chilli owes to the presence of capsaicinoids found only in *Capsicum* spp. The main components of capsaicinoids are capsacin, dihydrocapsaicin, nordihydrocapsaicin, etc. Various reports have been indicating the broad-spectrum activity of capsaicin against insects such as *Myzus persicae, Bemicia tabaci, Sitophilus zeamais, Alfalfa weevil,* to name a few [10–11]. The mode of action is majorly by impairing the nervous system of insects or dysfunction of Na or K-gated channels.

16.3.9 Essential Oils

In addition to these, there are many plants that contain essential oil, which is a complex mixture of volatile compounds accumulated in seeds, flowers, and leaves. These are volatile in nature, complex secondary metabolites having a strong odor, and lipophilic in nature [12]. These compounds interfere with physiological, biochemical, metabolic functions of insects. Therefore, plant essential oil compounds are considered to be alternative insect-controlling compounds and act as biopesticides. They have a wide range of quality that is sufficient to manage insects and pests well. They have the ability to delay development activities such as adult emergence and fertility, deterrent effect on oviposition [13], arrestant and repellent action [14]. They have antifeedant and larvicidal activity [15].

The volatile components of essential oils are classified into four main groups: terpenes, benzene derivatives, hydrocarbons, and other miscellaneous compounds. Monoterpenoids are the most representative molecules consisting 90% of essential oils having diverse function. They have acyclic alcohols (linalool, citronellol, geraniol), cyclic alcohols (menthol, terpineol, isopulegol), bicyclic alcohols (borneol, verbenol), phenol (thymol, carvacrol), ketones (carvone, menthone, thujone), aldehydes (citronellal, citral), acids (chrysanthemic acid), and oxides (cineole) [16]. Linalool has been studied and found to have effect on the nervous system, affecting ion transport and release of acetylcholine esterase in insects [16]. The mode of action of most monoterpenes is by causing a drastic reduction in the number of intact mitochondria and Golgi bodies, impairing respiration and photosynthesis, and decreasing cell membrane permeability [17].

The essential oils of *Ocimum basilicum* contain active compounds juvocimenes, which are an analogue of juvenile hormones of insects. *Matricaria recutita* contains precocenes, which stimulate the production of juvenile hormone that suppresses the growth of insects during molting. Additionally, some aromatic plants such as coriander, marigold have strong odor, which act act insect repellents and act in managing the harmful pests. Therefore, an analytical approach in selecting right biomolecules for the production of biopesticides having broader range of activity would be a preferable alternative for eco-friendly pest management. From unravelling new and better molecules to utilizing them into useful application in the field is the actual need of the hour. Mere pilling up of numerous novel compounds would not be ideal, rather disseminating into biochemicals will be highly appreciated.

16.4 Advantages and Future Prospects of Bioinsecticides

Advantages attributed to biopesticide usage over chemically synthesized products are many (Figure 16.1). They are supposedly less toxic, more selective in combating unwanted biological targets, possess higher efficiency at lower concentrations, decompose more

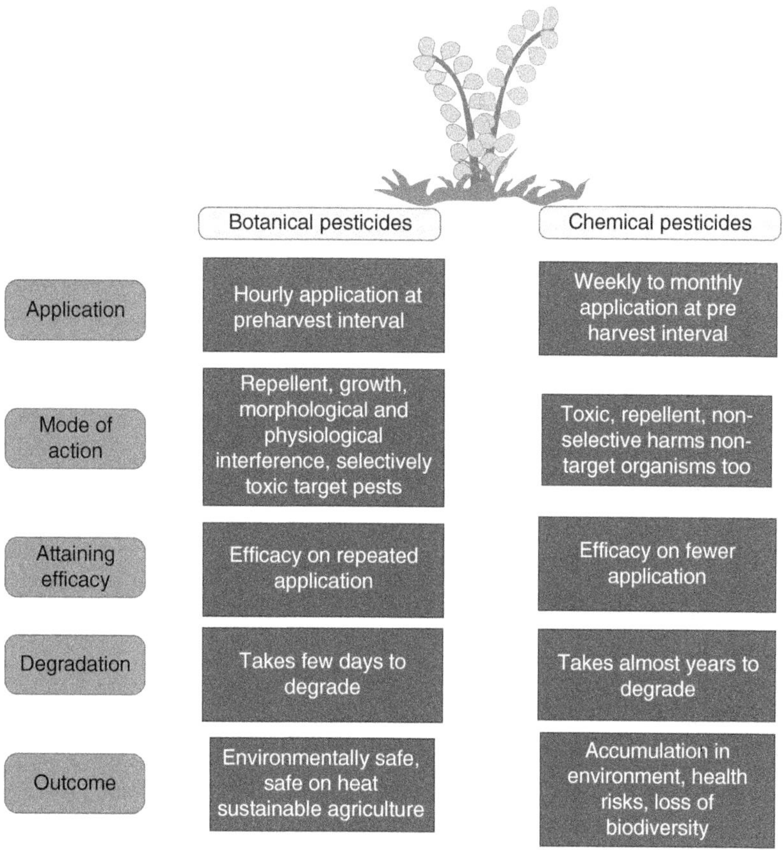

Figure 16.1 Advantages of botanical pesticides over chemical pesticides. *Source:* Adapted from Lengai et al. [18] Licence no. 4870610831544.

quickly, reducing adverse environmental effects. They can be used as an active component of IPM and reduce the use of conventional chemical pesticides while crop yields remain high. To use biopesticides effectively, however, one needs to know a great deal about managing pests.

In IPM-related programs, biopesticides are needed to be the preferred component in implementation of resistance management programs. The agrochemical industry is shifting from chemical to botanicals owing to its great implications. The demand for chemical-free food is increasing, and so is the awareness. The consumers are now more willing to pay beyond the regular prices for foods produced organically without the use of chemicals. Adaptation toward botanical pesticides from the more prominent chemical pesticides is gradual.

Although the scenario is fast changing, yet a lot has to be changed in the coming years to get a higher impact in modern agriculture. Resistance risk analysis is a key requirement, as efficacy demonstration of pesticide active substances and their formulations is mandatory. This kind of resistance risk assessment requires examination of the inherent risk (associated with the government authorities of the product and the pest) and the agronomic risk (influenced by the crop, the geographic area, and the use pattern). Therefore, resistance risk analysis is a complex issue, and the basis for risk assessment requires a clear willingness on the stakeholders to cooperate in order to maintain a sustainable, viable, and safe agricultural environment.

The actual benefits of biopesticides can be best realized in developing countries such as India, where farmers are not always able to afford chemical insecticides. The ancient and traditional use of plants and plant derivatives for protection of stored products is established well in our country. Although synthetic insecticides might be affordable to growers (e.g. through government subsidies), yet limited literacy and a lack of protective equipment result in thousands of accidental poisonings annually. The lack of judicious use of pesticides and the health and environmental risk it possesses can be overcome by the use of biopesticides. No security measure is taken care of while applying chemical pesticides; this kind of risk will be curbed by botanicals.

The area under organic cultivation (crops) in India is estimated to be around 100 000 hectare. Besides, there are lakhs of hectare of forest area being certified as organic. Furthermore, some states such as Uttaranchal and Sikkim have taken botanicals on a larger scale and declared their states as organic. The future is in need of making a shift from synthetic to natural source to minimize the adversity of chemicals. The nature has immense treasure instead of damaging approach; care should be taken to utilize and coexist benefitting the crop and its growers.

16.5 Conclusions

To acquire knowledge is good, but to apply the same is even better. The same stands true for biopesticides. Bioprospecting for novel molecules having varied properties should be carried out. But, perhaps it is time to converge the attention of the research community toward the development and application of known botanicals rather than screen more plants and isolate further novel bioactive substances that satisfy our curiosity but are unlikely to be of much utility [19].

The development of the biopesticide industries has to adopt a strategic, comprehensive and more forward, and applicable approach. With the growing population and the growing need of population, there will be greater need of crops and other products. The biopesticide industry has to pace up to stand par with the chemical industry. For this, the policymakers, government authorities, and the consumers demand has to work in sync. These together will boost up the importance and ease of trade will make it more available in the market for easy application. The increasing concern of consumers and government on food safety has led growers to explore new environmentally friendly methods to replace, or at least supplement, the current chemical-based practices. The use of biopesticides has emerged as promising alternative to chemical pesticides. Biopesticides have a precious role to play in the future of the IPM strategies. It is evident that the path would not be as easy although we have a wealth of knowledge and resource, but application still remains at a nascent stage.

The efficacy of biopesticides in managing insect pests is sure very effective, but the matter that needs to be pondered upon is their limited source. Though nature has immense treasure of these important botanicals, yet, the supply is limited, since a large number of source plants are needed for extraction of desired active compounds. And the supply in nature or natural habitat is marginal and would exhaust if used extensively. Therefore, commercial agriculture of such plants would raise income and keep the supply of required raw material at pace. To scale up the produce, landscapes that are not under direct cultivation can be utilized.

The process of extraction and processing of the biopesticides using appropriate methods should be worked upon also considering cost cutting during production. An efficient production with fluent distribution up to the smallhold farmers will see the real success of biopesticides. A better marketing strategy for biopesticides will also help in better circulation in the much upscale pesticide market. Also, a regular education or awareness program to farmers and supplier would boost up the botanicals immensely. All of such measures will surely help biopesticide to amplify its application in the market.

Apart from all these things, there are some areas that to my concern seem to be important to be pondered upon. Generally, discovering something hidden in nature and reporting seems to be satisfying, yet, how much is it valued or accepted is a critical point of discussion. A far more acceptable approach that also is rewarding is proving one's hypothesis; publication-wise too this seems better. Reporting a potent botanical compound and moving onto its application in the fields is a long process. Yet, there are many compounds successfully being applied in fields, such as Pyrethrin and Azadirachtin, etc. Therefore, the dawn of biopesticides is decent, but to see it a daylight, more effort with applicable focus is needed.

Acknowledgment

ALH and SS would like to acknowledge the funding from DST NPDF-SERB.

References

1 Gill, H.K. and Garg, H. (2014). Pesticide: environmental impacts and management strategies. *Pesticides-Toxic Aspects* 8: 187.
2 Pelaez, V. and Mizukawa, G. (2017). Diversification strategies in the pesticide industry: from seeds to biopesticides. *Ciência Rural* 47 (2): 1–7.
3 Nováková, J. and Farkašovský, M. (2013). Bioprospecting microbial metagenome for natural products. *Biologia* 68 (6): 1079–1086.
4 Das, S.K. (2014). Recent development and future of botanical pesticides in India. *Popular Kheti* 2 (2): 93–99.
5 Naumann, K. and Isman, M.B. (1996). Toxicity of a neem (*Azadirachta indica* A. Juss) insecticide to larval honey bees. *American Bee Journal* 136 (7): 518–520.
6 Oguh, C.E., Okpaka, C.O., Ubani, C.S. et al. (2019). Natural pesticides (biopesticides) and uses in pest management-a critical review. *Asian Journal of Biotechnology and Genetic Engineering* 2 (3): 1–18.
7 Casida, J.E. and Quistad, G.B. (1995). Pyrethrum flowers: production, chemistry, toxicology, and uses. In: International Symposium on Pyrethrum Flowers: Production, Chemistry, Toxicology and Uses, 356. New York: Oxford University Press.

8 Isman, M.B. (2006). Botanical insecticides, deterrents, and repellents in modern agriculture and an increasingly regulated world. *Annual Review of Entomology* 51: 45–66.

9 Liu, B., Chen, C., Wu, D., and Su, Q. (2008). Enantiomeric analysis of anatabine, nornicotine and anabasine in commercial tobacco by multi-dimensional gas chromatography and mass spectrometry. *Journal of Chromatography. B, Analytical Technologies in the Biomedical and Life Sciences* 865 (1-2): 13–17.

10 Li, B., Yang, M., Shi, R., and Ye, M. (2019). Insecticidal activity of natural capsaicinoids against several agricultural insects. *Natural Product Communications* 14 (7): 1934578X19862695.

11 Zhao, J.W., Zheng, Y., He, Y.X., and Weng, Q.Y. (2012). Biological activity of capsaicin and its effects on development and fecundity in *Bemisiatabaci* (Gennadius) (Hemiptera: Aleyrodidae). *Ying yong sheng tai xue bao* 23 (2): 531.

12 Bakkali, F., Averbeck, S., Averbeck, D., and Idaomar, M. (2008). Biological effects of essential oils–a review. *Food and Chemical Toxicology* 46 (2): 446–475.

13 Oyedele, A.O., Orafidiya, L.O., Lamikanra, A., and Olaifa, J.I. (2000). Volatility and mosquito repellency of Hemizygiawelwitschii Rolfe oil and its formulations. *International Journal of Tropical Insect Science* 20 (2): 123–128.

14 Landolt, P.J., Hofstetter, R.W., and Biddick, L.L. (1999). Plant essential oils as arrestants and repellents for neonate larvae of the codling moth (Lepidoptera: Tortricidae). *Environmental Entomology* 28 (6): 954–960.

15 Gbolade, A.A. (2001). Plant-derived insecticides in the control of malaria vector. *Journal of Tropical Medicinal Plants* 2 (1): 91–97.

16 Re, L., Barocci, S., Sonnino, S. et al. (2000). Linalool modifies the nicotinic receptor-ion channel kinetics at the mouse neuromuscular unction. *Pharmacological Research* 42: 177–181.

17 Tripathi, A.K., Upadhyay, S., Bhuiyan, M., and Bhattacharya, P.R. (2009). A review on prospects of essential oils as biopesticide in insect-pest management. *Journal of Pharmacognosy and Phytotherapy* 1 (5): 52–63.

18 Lengai, G.M., Muthomi, J.W., and Mbega, E.R. (2020). Phytochemical activity and role of botanical pesticides in pest management for sustainable agricultural crop production. *Scientific African* 7: e00239.

19 Isman, M.B. (2020). Botanical insecticides in the twenty-first century—fulfilling their promise? *Annual Review of Entomology* 65: 233–249.

17

Plant Biomass to Bioenergy

Mrinalini Srivastava and Debasis Chakrabarty

Molecular Biology and Biotechnology Division, Tissue Culture and Transformation Lab, CSIR- National Botanical Research Institute, Lucknow, India

17.1 Introduction

We are living in the age where continuous use of fossil fuels as energy resources enhances the pollution speedily and instigates serious health issues. As a result, there is high demand for alternative energy production ways. Whole scientific community across the world is trying to discover and promote renewable energies, which share the load and in some extent replace the conventional energy resources. Among alternative energy sources, plant biomass emerges as vital source of bioenergy and highly recommended by the European Union [1]. Renewable biomass sources can be converted to fuels and are a logical choice to replace oil [2]. From ancient times, biomass is considered as a main source of energy for humanity, and now there is a time for strong transition from old to modern source of energy [3].

Moreover, to overcome the crisis of environmental deterioration due to the impact of fossil fuels, maximum utilization of biofuels especially in transportation sector is obligatory. Plant-based biofuels are the most abundant source of renewable fuels, involved in the production of small (ethanol and butanol as gasoline additives) and long-chain hydrocarbons (for diesel additives or as jet fuels). Another advantage associated with biofuels is continuous cycling of carbon rather than being released in the environment as in case of fossil petroleum and natural gas [4].

Global warming is today's major problem, and CO_2 release can be minimized to a large extent with the use of biomass as an energy source instead of traditional one. Oil and natural gas are not infinite assets, definitely they all have end, and we cannot imagine the situation without these primary sources of fossil fuel until we have develop a backup plan. In these contexts, very high emphasis is needed to explore new sources of energy [5].

India generates approximately 450–500 million tons of biomass every year, and 32% of all the primary energy use in the country comes from biomass [6]. The composition of biomass has large diversity, for example, it may compose of cellulose, hemicellulose and

Bioprospecting of Plant Biodiversity for Industrial Molecules, First Edition.
Edited by Santosh Kumar Upadhyay and Sudhir P. Singh.

lignin; such variations are responsible for different chemical properties of biomass. Besides biofuel, biomass can also be utilized in the production of electricity and heat [7].

The Government of India has shown an enthusiastic interest toward renewable energy and aims to initiate a project regarding 175 GW renewable power installed capacity by the end of 2022. India is a country of huge biodiversity and considered as one of the biggest economies with a growing population. The demand for electricity is rapidly growing year by year due to more development in industrial sectors, increase in population, more urbanization, economic growth, and more electrification in rural areas. According to the Ministry of New and Renewable Energy ("MNRE"), more than 70% of the population of rural India directly or indirectly depends upon biomass-based energy, and it is very fortunate that bioenergy is a well-known form of energy in rural India due to easy availability of agricultural residues such as straw, etc. [8].

According to an estimate, 82% of total crude oil imports are needed to feed the domestic consumption demand; it exhibits how India is depended and affected by the unpredicted price hike in International market. Hence, it is very sensible and necessary to use and divert our potential in the production of bioenergy [8].

On the other side, continuous increase in world's population evokes us to explore new form of energy rather than depend on old sources and strengthen our country as "Aatmnirbhar Bharat." Data also suggested world population projected to increase further to nine billion by 2050 [9]. The MNRE ministry has started a project to reinforce the biomass-based industries up to year 2020 [10].

India generates 150 000 tons of municipal solid waste per day, use of this waste as biomass in power generation is the demand of time, and it can solve two major issues of the country: one is alternative power supply and other is waste management [11]. So, considering all these studies, we write this chapter and discussed about plant biomass, bioenergy, its conversion, and some other related major issues.

17.2 Plant Biomass

Plant biomass is considered as a renewable source of energy, basically it is an organic material that comes from plants. Biomass is used for various purposes as in electricity generation, in heat production, and in biofuel production. Biomass belongs to many categories such as wood, forest residue, garbage, and agricultural waste, etc.

It accommodates stored energy from the sun from the process known as photosynthesis, and the chemical energy is released as heat after burning of biomass. Biomass can be burned through two ways either directly (wood and wood processing wastes burned to produce heat and to generate electricity) or indirectly (agricultural crops and waste materials burned as a fuel); in latter case, it is converted into liquid biofuels [12].

Several types of biomass, such as corn, wood chips, and garbage, are used in the production of electricity. Biomass can be converted into liquid fuels called biofuels that can be helpful for cars, trucks, and tractors. Leftover food products such as vegetable oils can create biodiesel, while corn, sugarcane, and other plants can be fermented to produce ethanol [13].

Plant biomass is a feedstock to generate sustainable fuels and that fuels have potential to replace petroleum products. Biological fermentation of plant biomass is an assured method for liquid fuel production, but there are many challenges to overcome. Bioconversion of lignocellulose requires an appropriate deconstruction of the plant biomass and transformation of sugar into the desire fermentation products.

Thermal and chemical treatment of biomass and addition of enzymes for the hydrolysis raises the process economic costs. In this respect, CBP (consolidated bioprocessing) is emerging as promising transformative options, it combines all biological steps into one unit, and enzyme synthesis, hydrolysis, and fermentation steps are occurring in a single reactor. CBP is a cheaper and valuable option for cellulosic biofuel production. *Clostridium thermocellum* is well known for its role in degradation and utilization of cellulose, and it is also used as in CBP as biocatalyst for the synthesis of cellulosic ethanol [14].

Plants efficiently convert photons into electrons, in this regard new solar panels are developed with 30–36% efficiency. According to PV Magazine, current solar panels are more efficient in comparison to previous one. The electrons from the plants should be captured before its conversion into sugar. Once the electricity is captured from the plant, the electricity can be used for many purposes. Some electrical engineers at University of Washington designed a circuit that can convert the natural energy into valuable electricity [15]. Earlier in 2007, another Dutch research group at Wageningen University patented the process of collecting plant power; today, a Netherlands-based company named Plant-e holds this patent [16].

17.2.1 Types of Biomass (Source: [17])

- Wood from natural forests and woodlands
- Forest plants
- Forest residues
- Agricultural waste, for example, stover, straw, green agricultural wastes, and cane trash
- Agro-industrial wastes such as sugarcane bagasse and rice husk
- Industrial waste material, such as black liquor from paper manufacturing
- Sewage
- Food processing wastes
- Municipal solid wastes

17.3 Bioenergy

Bioenergy holds a remarkable position in the energy economy share about 9.5% of total primary energy supply and 70% of total renewable energy [18]. Usually bioenergy generates heat in buildings and industry, but it is expected that bioenergy would be involved in the 3% of electricity production and approximately 4% of transport energy till 2023 [19]. Extension of bioenergy relies on agricultural policies worldwide, there is a continuous increase observed in contribution of bioenergy of 40–55 EJ/year. A hike is estimated from 200 to 400 EJ during this century, and it makes biomass a significant option for energy supply rather than mineral oil [20]. There should be an awareness program regarding utility of biomass in power generation and in other sectors.

The concept of biomass energy is very simple and understandable. Plants take carbon dioxide from the air and perform photosynthesis; when plant dies, much of the carbon is again released into the atmosphere as carbon dioxide. According to Samuel Stevenson, a policy analyst at the Renewable Energy Association in London, "When we use biomass as an energy source, we are intercepting this carbon cycle, using that stored energy productively rather than it just being released into nature [21]."

Energy obtained from biomass has some advantages as it is free of fluctuation and does not require storage. Although biomass-generated energy has some challenges as biomass is not available throughout the year. So, there is a need to store biomass beyond the harvesting period [11].

Bioenergy accounts for 13–14% of the total energy consumption; presently among all renewable energy sources, bioenergy is the largest energy source and accounts for approximately second/third of the renewable energy mix. Bioenergy is a complex energy network, a range of steps, and various feedstock involved in the generation of energy from the biomass [22].

On the global platform, forestry sector is the biggest contributor to the bioenergy mix. Forestry products such as charcoal, pellets, fuelwood, and wood chips are responsible for more than 85% of biomass, which is used for energy purposes. Wood fuel (1.9 billion m^3) is the primary products, i.e. used for bioenergy production; mainly wood fuel is used for traditional cooking and heating in countries such as Asia and Africa [23].

17.4 Biomass Conversion into Bioenergy

There are some challenges to overcome when we want to shift from fossil fuel to biomass fuels. The heavy solid biomass has to be converted into liquid and gaseous fuels to make transportation convenient, and it can be done through two processes: one is biochemical and another one is thermochemical [24].

In regard to biomass conversion into bioenergy, combustion, pyrolysis, and anaerobic digestion are well-established technologies and already used at commercial level but liquefaction, gasification, hydrolysis, fermentation, and fractionation are in queue for future implementation [25]. Mainly three types of energy are generated through conversion of biomass, i.e. (i) thermal energy, (ii) electrical energy, (iii) transportation fuel energy.

Thermal energy is released by the transformation of woody biomass, it provides heat, which is needed for cooking. Steam generated in combustion can be used for domestic as well as industrial processes and can also be used in the production of electricity and donated as electrical energy. Synthesis gas and biogas forms in other conversion technologies such as pyrolysis, gasification, and anaerobic digestion are also suitable for electricity production. Transportation fuel is the energy obtained in self-propulsion motors from biofuels. These are also known as "second generation biofuels" and originated from lignocellulosic type of biomass. These biofuels are obtained through thermochemical and biochemical conversion processes. Cultivation of appropriate "energy crops" especially for bioenergy is actually the need of the hour [25].

Second-generation biofuels are different from first-generation biofuels in their origin. These are generated through lignocellulosic materials such as jatropha, cassava, switchgrass, wood, and straw and some other biomass residues. First-generation biofuels are that derived from edible food crops, i.e. sugarcane, wheat, barley, corn, potato, soybean, sunflower, and coconut [26].

17.4.1 Cogeneration

Cogeneration is a term used for simultaneous production of more than one form of energy via using one type of fuel. Cogeneration has more potential growth in comparison to generation of energy alone because it produces both heat and electricity.

17.5 The Concept of Biomass Energy (Source: [27])

Chemically biomass is composed of carbon and hydrogen elements, and in the transformation of biomass to bioenergy, the bound energy of these compounds is released. The energy can be release through two main kinds of conversion.

17.5.1 Thermochemical Conversion

This type of conversion is performed at high temperature; bond breaking and reforming of organic matter take place into biochar (solid), synthesis gas, and highly oxygenated bio-oil (liquid). Thermochemical conversion is divided into three subtypes. These are gasification, pyrolysis, and liquefaction. Mainly three factors decide which type of conversion is going to happen, i.e. nature, quantity of biomass, and the preferred type of energy. It was reported that thermal conversion has some advantages as it can produce energy from plastics, requires short processing time and reduced water usage, etc. Besides, thermochemical conversion is independent of environmental conditions [26].

17.5.1.1 Direct Combustion

It is the simplest and very economical; energy is released by burning the material through direct heat.

17.5.1.2 Pyrolysis

Pyrolysis is the temperature-based degradation of biomass by heat but in the absence of oxygen. Biomass is heated to a temperature between 400 and 750 °C, and end products are gas, fuel oil, and charcoal.

17.5.1.3 Gasification

Biomass is used in the production of methane through heating or anaerobic digestion. Syngas, a mixture of carbon monoxide and hydrogen, can be derived from biomass. Methane is the major constituent of natural gas, and decomposition of organic waste generates gas; this gas contains approximately 50% methane.

17.5.2 Biochemical Conversion

Biochemical conversion involves yeast or specialized bacteria to convert biomass or waste into beneficial energy. The subtypes are anaerobic digestion, alcoholic fermentation, and photobiological techniques, which produce different biofuels [26].

17.5.2.1 Anaerobic Digestion

Biomass such as manure, waterwaste (sewage), and food processing waste are mixed with water and added into a digester tank without air. This kind of digestion converts organic matter to a mixture of methane. Landfill gas is produced by anaerobic digestion of buried garbage in landfills.

17.5.2.2 Alcohol Fermentation

Feedstocks such as barley, wheat, potatoes, and waste paper, sawdust, and straw containing sugar, starch, or cellulose are used in fermentation. Alcohol is produced by converting starch into sugar, and then fermentation of sugar produces alcohol after that alcohol and water mixture are separated through distillation.

17.5.2.3 Hydrogen Production from Biomass

A serious issue is associated with biomass utilization, i.e. its low efficiency of utilizing biomass. In this regard, conversion of biomass into hydrogen is a very well approach, and hydrogen production plays a very significant role in the development of hydrogen economy.

Thermochemical process such as gasification or pyrolysis and biological process such as fermentation or biophotolysis can be practically applied to produce hydrogen [27].

17.6 Use of Biofuel in Transportation

Involvement of biofuel in transport is exceeded day by day and reached up to 6% on a yearly basis in 2019, and 3% annual production growth is expected over the next five years. There is a strong need of policymaking and remodeling in this sector to reduce the cost, and it will enhance the biofuel consumption and initiate its use in aviation and marine transport. In 2019, electricity generation from biomass increased by over 5%, and recent positive policy and market developments in emerging economies indicate an optimistic outlook for bioenergy, supporting its "on track" status [28].

17.7 Production of Biogas and Biomethane from Biomass

Biogas is a combination of methane, CO_2, and small quantities of other gases, and it can be used to generate power and fulfill heating and cooking demand. Biogas and biomethane can originate from a variety of feedstocks/biomass such as crop residues, animal manure, municipal solid waste, wastewater, but these two are dissimilar products and used in separate activities. Biomethane can be directly produced through gasification of forestry residues. Biogas provides a sustainable way to supply community energy demands, especially in area that requires more electricity or the area where power supply through national grids is not easily reachable. Biomethane can be prepared from biogas via removing the CO_2 and other impurities. Replacement of solid biomass with biogas for cooking purposes is improving health in developing countries. According to some reports, biogas provides a way toward clean cooking to 200 million people by 2040 [29].

17.8 Generation of Biofuel

As present Indian government is giving more emphasis to make in India program, biofuel from biomass is gelling very well in it, and it also gives right direction toward some more government schemes such as Swachh Bharat Abhiyan, skill development, doubling of farmer's income, employment generation, etc.

Recently, some national policies on biofuels are finalized and approved by the Union Cabinet; hopefully these policies will provide a massive force to the small growing bioenergy sector in India. A policy about usage and production of ethanol from damaged farm products and food grains has cleared the way to utilize the agricultural waste for the generation of bio-power. This policy also has provision to convert waste such as plastic as well as municipal solid waste to fuel. It will certainly help India to achieve the target of 10 GW of biomass power by 2022. The policies provide a practical and acknowledgeable support for the bioenergy sector in India [11].

Production of plant-based fuels definitely decreases our dependence on fossil fuels. Besides, many other advantages are also associated with flora fuels as these fuels release very less quantity of greenhouse gases such as carbon dioxide, etc. Biofuels are sustainable, and energy companies mix them with gasoline. In contrast to coal, oil, or natural gases, biofuels are renewable at some levels.

Biofuels are divided into two major headings, bioalcohol and biodiesel. The bioalcohol as ethanol is made from corn and some other plants because of breakdown of starch via yeast and bacteria. On the other hand, biodiesel is manufactured in refineries by using the oil crops such as soybeans. Vegetable oils are treated with alcohol for conversion into biodiesel [30]. According to the study, any plant material can be converted to biofuel, as waste products such as sawdust corn stalks, and corn kernels. Research is continuously going on to explore plants that can produce fuel with very less requirement of marginal land and little or no irrigation or no need of fertilizer. Some invasive species can also fulfill the need as feedstock for biofuel plants [31].

17.8.1 Bioethanol

Bioethanol can be produced by two ways: one is through sugar fermentation process and another one is by chemical process through the reaction of ethylene with steam. Corn, wheat, maize, willow, waste straw, and popular trees, reed canary grass, sawdust, cord grasses, artichoke, *Jerusalem sorghum* plants, and *Miscanthus* are mainly use for ethanol production. Ethanol is biodegradable, causes lesser toxicity and pollution, and it is a high octane fuel. Bioethanol has huge importance over conventional fuels. This fuel is manufactured from the renewable sources such as plants, by creating low pollution, it will be helpful in improving the air quality. Rural economy also gets benefited due to more need of necessary crops for bioethanol production.

Plant cell wall is made up of cellulose, hemicellulose, and lignin, these are complicated mixture of carbohydrate and create problem during burning of biomass. Ethanol is produced through the hydrolysis and sugar fermentation of biomass. Biomass is treated with acids or enzymes, and it degrades the polymers into monomers. The cellulose and the hemicellulose are broken down into sugar and fermented into ethanol. The lignin is usually used as a fuel for the ethanol production [30].

17.8.2 Biodiesel

Biodiesel is manufactured through chemical reaction by the reaction of vegetable oil with an alcohol such as methanol or ethanol. The chemical reaction that produces biodiesel is called "transesterification." Oils and fats are categorized into the ester family. Biodiesel can be made from any usual source such as canola oil, soybean oil, pork lard, beef tallow, and also from some exotic oils, for example, walnut oil or avocado oil. Although waste/used oil can also be used to make biodiesel, but due to the presence of some contaminants, these oils should be filtered before use in reaction.

Ethanol is quite expensive; that's why methanol is the ultimate choice for biodiesel production although it is very toxic in comparison to ethanol. Quality masseurs for biodiesel are determined by the American Society for Testing and Materials (ASTM) specification D6751. Fuel testing and quality check are expensive, but there is no alternative, and it is the most reliable method to supply the best-quality product to consumers [31].

A very interesting study was conducted by Mao et al. [32]; they make bibliometric analysis by collecting data from 9514 studies with the help of searching key words "Biomass energy" and "Environment." Data was collected from the year 1998 to 2017, and many fascinating topics on biomass utilization and bioenergy production were covered in this analytical paper. List of some important biofuel crops and next generation biofuel crops are given in the Tables 17.1 and 17.2 [34] and [35].

17.9 Advanced Technologies in the Area of Bioenergy

Various research proposals are funded by the several agencies in public as well as in private sectors to promote, establish, and advance renewable energy, especially bioenergy. Among them, some of the advanced biomass-related technologies are named here. Many of them are already used at commercial level and some are ready to be commercialized [33].

- Advanced plantation systems for some unexplored or newly explored plants.
- Development of advanced biorefinery system for uninterrupted production of products such as biodiesel.
- Modification in anaerobic treatment methods to induce higher destruction of pathogens in wastewaters to achieve more biogas yields along with high production rates.
- Development of close-coupled biomass gasification–combustion systems to generate hot water and steam for the utilization in commercial buildings and schools.
- Strengthen the biomass gasification processes to meet the high-efficiency production of medium-energy content fuel gas and power.

- Use of genetically engineered and efficient microbes, which can convert pentose and hexose sugars from cellulosic biomass and further convert it into ethanol through fermentation.
- Use of zero-emissions waste biomass-combustion systems for joint association of disposal-energy recovery and recycling.
- Selection of catalysts for thermochemical conversion especially for gasification of biomass to produce gases at higher yield.
- In case of pyrolysis process, emphasis is given to short residence time to produce chemicals and liquid fuels from appropriate biomass.

17.10 Conclusion

The present chapter aims to collect the information about plant biomass, bioenergy, and its conversion. We have concluded from this chapter how biomass-based energy is needed at present time, it can deal with several issues at a single time as pollution, waste management, oil crisis, etc. It can also generate employment and strengthen the economy as well. Biomass holds significant potential to overcome the current energy demand and have strong ability to compensate the excessive dependence on fossil fuels. Due to continuous depletion in fossil fuel resources, biomass-based energy along with other biotechnological approaches would become focus of future research.

Table 17.1 Important biofuel crops.

S. no	Plant	Uses
1	Corn	Corn is the major crop in the world, used for ethanol-based biofuels production.
2	Rapeseed/ Canola	It's an important biodiesel crop, Rapeseed oil has been used in cooking. Canola gains special importance due to presence of low erucic acid content, and it makes it healthier for animals as well as humans to eat.
3	Sugarcane	Sugarcane is also used in bioethanol production, and it is six times cheaper than corn. It is exclusively used in Brazil.
4	Palm oil	Palm oil is obtained from the fruit of palm trees, and it is among the energy-efficient biodiesel fuels. Palm-oil-based biodiesel causes lesser pollution in comparison to gasoline. Palm oil is helpful in boosting the economies of Malaysia and Indonesia.
5	Jatropha	This crop is a master player in the biofuel market. Presently, India is the world's largest *Jatropha* producer. The main benefit associated with this crop is requirement of very normal agricultural land, and in this way it will be helpful for rural farmers. *Jatropha* plants can also live on land that suffered from the drought and pests for approximately up to 50 years. Oil is extracted from the seeds although seed cases and vegetable matter are also good source of biomass fuel.
6	Soybeans	Soybeans exclusively used for biodiesel production in the United States. Motor vehicles, buses, and heavy equipment can run on pure soybean-based biodiesel or, blending is also done with more traditional diesel fuels. According to The National Academy of Sciences, soybean diesel yields more energy than corn ethanol.

Source: Ref. [34].

Table 17.2 Next-generation biofuel crops.

S. no.	Name of plant	Uses
1	Hemp	Hemp plant can produce oil equivalent to soybean.
2	Switchgrass	This plant has potential to replace corn as a feedstock for ethanol production.
3	Carrizo Cane	It is native in Europe and utilized at the commercial scale in Europe and produces almost highest biomass per acre than any other plant.
4	Algae	Algae produce up to 200 times more oil per acre in comparison to soybean. Several world largest companies have invested hundreds of millions of dollars for scaling up the algal fuel production.

Source: Ref. [35].

Acknowledgment

Chapter manuscript no. CSIR-NBRI_MS/2021/01/04.

References

1 Tursi, A. (2019). A review on biomass: importance, chemistry, classification, and conversion. *Biofuel Research Journal* 22: 962–979.

2 Mohan, D., Pittman, C.U., and Steele, P.H. (2006). Pyrolysis of wood/biomass for bio-oil: a critical review. *Energy & Fuels* 20: 848–889.

3 Rosillo-Calle, F. (2016). A review of biomass energy-shortcomings and concerns. *Journal of Chemical Technology & Biotechnology* 91: 1933–1945.

4 Hood, E.E. (2016). Plant-based biofuels [version 1; peer review: 2 approved]. *F1000Research* 5 (F1000FacultyRev): 185. https://doi.org/10.12688/f1000research.7418.1.

5 https://escientificpublishers.com/environment-and-development-bioenergy-for-future-ACCE-01-0003.

6 www.eai.in › ref › bio › bio.

7 Freiberg, A., Scharfe, J., Murta, V.C., and Seidler, A. (2018). The use of biomass for electricity generation: a scoping review of health effects on humans in residential and occupational settings. *International Journal of Environmental Research and Public Health* 15: 354. https://doi.org/10.3390/ijerph15020354.

8 https://www.roedl.com/insights/renewable-energy/2020-02/market-overview-bioenergy-india.

9 Perea-Moreno, M.A., Samerón-Manzano, E., and Perea-Moreno, A.J. (2019). Biomass as renewable energy: worldwide research trends. *Sustainability* 11: 863. https://doi.org/10.3390/su11030863.

10 https://energy.economictimes.indiatimes.com/news/renewable/india-can-generate-18000-megawatt-renewable-energy-using-biomass-power-minister-r-k-singh/72435991.

11 https://yourstory.com/2018/09/india-biomass-power-generation.

12 https://www.eia.gov/energyexplained/biomass/.

13 https://archive.epa.gov/climatechange/kids/solutions/technologies/biomass.html.
14 Brown, S.D., Sander, K.B., Wu, C.W., and Guss, A.M. (2015). *Clostridium thermocellum*: engineered for the production of bioethanol. In: Direct Microbial Conversion of Biomass to Advanced Biofuels (ed. M.E. Himmel). Elsevier https://doi.org/10.1016/B978-0-444-59592-8.00016-6.
15 https://www.earth.com/news/plants-generate-electricity/.
16 https://www.smithsonianmag.com/innovation/turning-energy-plants-produce-usable-electricity-180955110/
17 https://www.bioenergyconsult.com/tag/types-of-biomass/.
18 International Energy Agency, 2017.
19 Reid, W.V., Ali, M.K., and Field, C.B. (2019). The future of bioenergy. *Global Change Biology* 26: 1–13. https://doi.org/10.1111/gcb.14883.
20 https://www.ieabioenergy.com/wp-content/uploads/2013/10/Potential-Contribution-of-Bioenergy-to-the-Worlds-Future-Energy-Demand.pdf.
21 https://physicsworld.com/a/biomass-energy-green-or-dirty/.
22 https://worldbioenergy.org/uploads/191129%20WBA%20GBS%202019_LQ.pdf.
23 Basu, P. (2010). Biomass gasification and pyrolysis practical design. 1–25. doi: https://doi.org/10.1016/B978-0-12-374988-8.00001-5. ISBN no- 978-0-12-374988-8
24 Görgens, J.F., Carrier, M., and García-Aparicio, M.P. (2014). Biomass conversion to bioenergy products. In: Bioenergy from Wood: Sustainable Production in the Tropics, Managing Forest Ecosystems 26 (ed. T. Seifert). Dordrecht: Springer Science C Business Media https://doi.org/10.1007/978-94-007-7448-3.
25 Lee, S.Y., Sankaran, R., Chew, K.W. et al. (2019). Waste to bioenergy: a review on the recent conversion technologies. *BMC Energy* 1: 4. https://doi.org/10.1186/s42500-019-0004-7.
26 https://ypte.org.uk/factsheets/renewable-energy-biomass-energy/the-theory-of-biomass-energy.
27 Ni, M., Leung, D.Y.C., Leung, M.K.H., and Sumathy, K. (2006). An overview of hydrogen production from biomass. *Fuel Processing Technology* 87: 461–472.
28 https://www.iea.org/fuels-and-technologies/bioenergy.
29 https://www.iea.org/reports/outlook-for-biogas-and-biomethane-prospects-for-organic-growth.
30 http://www.esru.strath.ac.uk/EandE/Web_sites/0203/biofuels/what_bioethanol.htm#:~:text=Bioethanol%20fuel%20is%20mainly%20produced,from%20fuel%20or%20energy%20crops.&text=Ethanol%20burns%20to%20produce%20carbon%20dioxide%20and%20water.
31 https://farm-energy.extension.org/biodiesel-production-principles-and-processes/.
32 Mao, G., Huang, N., Chen, L., and Wang, H. (2018). Research on biomass energy and environment from the past to the future: a bibliometric analysis. *Science of the Total Environment* 635: 1081–1090.
33 Sharma, S., Meena, R., Sharma, A., and Goyal, P.K. (2014). Biomass conversion technologies for renewable energy and fuels: a review note. *Journal of Mechanical and Civil Engineering* 11: 1–8.
34 https://interestingengineering.com/seven-biofuel-crops-use-fuel-production.
35 https://www.smithsonianmag.com/innovation/next-generation-biofuels-could-come-from-these-five-crops-180965099/.

18

Bioenergy Crops as an Alternate Energy Resource

Garima Pathak[1] *and Shivanand Suresh Dudhagi*[2]

[1] *B.D. College – A Constituent Unit of Patiliputra University, Patna, India*
[2] *CSIR-National Botanical Research Institute, Lucknow, India*

18.1 Introduction

A significant challenge in the twenty-first century is to provide adequate energy to the rising world population, which needs more and more energy per person. Today, this energy supply is based predominantly on fossil energy carriers which adversely affect the climate and degrade the natural resources by emitting greenhouses gases, carbon dioxide, and nitrogen oxide. Nuclear fission-generated electricity requires huge infrastructure and has adverse effects on the environment and human health [1]. The use of fossil fuels is linked to long-term environmental impacts, which can contribute to land degradation and fertile soil desertification [1]. For instance, coal emits greenhouse gasses such as carbon dioxide, particulate soot, and sulfur-containing compounds which lead to soil acidification. Now concern is rising about how such rising demands can continue to be met by finite and slowly depleting resources. There is definitely a need for renewable energy sources. Several countries changed their energy fulfillment goals from non-renewables to renewables. There are only few energy sources which are renewable and having lesser impact on the environmental. One such potential alternative with long-term positive future results is the use of "bioenergy crops" for generating energy [2]. Energy from bioenergy crops is drawn from plant and animal biomass [3]. Bioenergy crops reduce carbon dioxide, reduce greenhouse gas emissions, increase soil carbon, reduce soil erosion, increase transpiration, and provide heat and electricity [4–6]. The bioenergy crops also help in remediation of heavy metals from the contaminated soil [7]. Large-scale bioenergy cultivation could also have a positive impact on the wildlife. Bioenergy concept is very popular mango scientific community because of its renewability and eco-friendly nature. However, the worldwide market has more conventional use of bioenergy crops as food, which raises food safety issues for energy use. Bioenergy plants also compete with food crops for agricultural land, water resources, and nutrient needs. Another negative impact linked to the use of bioenergy

Bioprospecting of Plant Biodiversity for Industrial Molecules, First Edition.
Edited by Santosh Kumar Upadhyay and Sudhir P. Singh.

crops involves the destruction of wildlife habitats and increased dispersion of invasive plant species [8]. Different types of bioenergy crops and their features are described in this chapter.

18.2 Classification of Bioenergy Crops

The concept of "traditional biofuel" was introduced to overcome the environmental and associated issues. Traditional biofuels were derived from cultivated vegetables. Their use in bioenergy is debatable because of issues relating to food security. Bioenergy crops are screened based on specific characteristics such as oil yields, oil quality, and mitigation of global climate change. Traditional bioenergy crops could improve the production of food and fodder, with the additional advantage of mitigating global climate change [9]. Bioenergy crops are mainly classified into five groups namely, first-generation, second-generation, and third-generation, energy crops dedicated, and halophytes (Figure 18.1), and components of various types of bioenergy crops are listed in Table 18.1.

18.2.1 First-Generation Bioenergy Crops

Biofuel generation programs have been initiated with the first-generation bioenergy crops (FGECs). These crops are also a common local or global food source. FGECs such as sweet sorghum, corn, sugarcane, oil palm, and rapeseed were initially used for biofuel preparation [10]. However, bioenergy crops of first generation have limited capacity to replace petrol oil products due to higher production costs [11–14]. Such limitations were overcome by the use of lingo-cellulosic materials from crop residues in the fuel extraction phase in the second-generation bioenergy processing model [15]. Some common FGECs are discussed below.

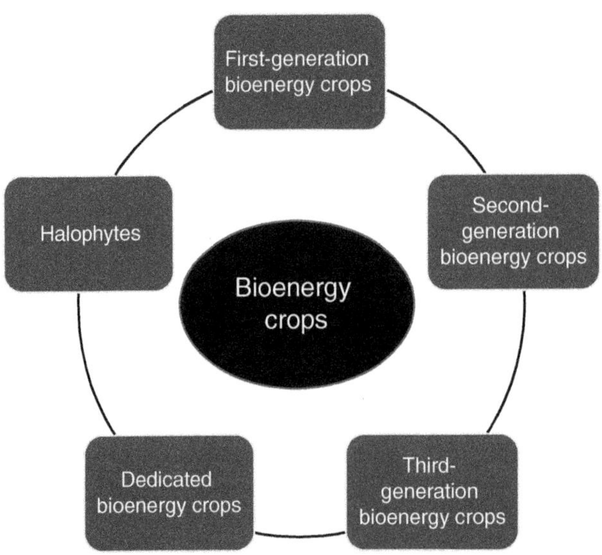

Figure 18.1 Types of bioenergy crops.

Table 18.1 Constituents of various types of bioenergy crops.

First-generation bioenergy crop	Sweet sorghum, corn, sugarcane, oilseed rape, linseed, field mustard, hemp, sunflower, safflower, castor oil, olive, palm, coconut, and groundnut
Second-generation bioenergy crop	Switchgrass, reed canary grass, alfalfa, Napier grass, and Bermuda grass
Third-generation bioenergy crop	Boreal plants, crassulacean acid metabolism (CAM) plants, eucalyptus, and microalgae
Dedicated bioenergy crop	Cellulosic plants (eucalyptus, poplar, willow, birch, etc.), perennial grasses (giant reed, reed canary grass, switchgrass, elephant grass, etc.), non-edible oil crops (castorbean, physic nut, oil radish, pongamia, etc.), and oil plants (*Jatropha curcas, Pistacia chinensis, Sapium sebiferum* and *Vernicia fordii*)
Halophytes	*Acacia, Eucalyptus, Casuarina, Melaleuca, Prosopis, Rhizophora*, and *Tamarix*

18.2.1.1 Sugarcane

Sugarcane (*Saccharum officinarum* L.) is the largest sugar-producing plant perennial plant that grows year-round and adapted to warm temperate or tropical climates. Therefore, sugarcane feedstock remains available year-round at comparatively lower costs compared to other bioenergy crops [3]. Sugarcane is primarily produced to obtain sugar. The sugarcane juice contains a high proportion of sucrose, which is a substratum for biofuel. Several breeding projects are underway to boost the germplasm of the sugarcane to increase the production of sucrose and cellulosic biomass. Commercial bioethanol is derived from the molasses, a sugar industry by-product.

18.2.1.2 Corn

Corn (*Zea mays*) is an effective crop of feed due to high grain yield and better accumulation of starch in grains [16]. Corn has high content of volatile compounds and simple processing method makes it a preferred crop for bioconversion. Corn is used in the manufacture of ethanol in the United States and elsewhere. The key drawback of maize feedstock, however, is its predominant use in many countries as a staple food. The use of corn in the production of bioenergy fuel could increase food prices worldwide, contributing to hunger and famine. In order to overcome this issue, sweet corn variety was produced in the corn kernel endosperm by spontaneous recessive mutations in genes that regulate conversion of sugars to starch. Using dual-purpose and photosynthetically efficient sweet corn hybrids could benefit farmers by contributing to energy generation without affecting the environment and food supply [17, 18].

18.2.1.3 Sweet Sorghum

There are several varieties of sweet sorghum (*Sorghum bicolour* L.) which have high content of sugar. Sweet sorghum accumulates a significant quantity of fermented sugar in stems to produce greater biomass. The plant needs less fertilizer and therefore is easily grown on marginal lands. Agronomic characteristics of sorghum include high resistance to drought and C4 photosynthesis. Limited research attempts have been made to classify the genetic and molecular features of sorghum compared to crops such as maize and sugarcane. Sweet sorghum is a used as model bioenergy crop for understanding the complex genomes of other bioenergy crops such as corn, sugarcane, miscanthus, and switchgrass [19]. Sweet sorghum comprises high amounts of sugar in stems, and hence higher activity of sugar metabolic enzymes detected during stem development [20]. The sorghum crop has good nitrogen use efficiency and accumulates higher amounts of sugar in the stem during drought [21, 22]. Crops of sorghum and sweet sorghum could be cross-breed to improve crop productivity, and the desirable attributes could be detected by genetic mapping [23, 24].

18.2.1.4 Oil Crops

Oil crops consists of oilseed rape, linseed, field mustard, hemp, sunflower, safflower, castor oil, olive, palm, coconut, and groundnuts. Vegetable oils could be refined to generate biofuels for transport or used directly as fuel for heating [25].

18.2.2 Second-Generation Bioenergy Crops

Second-generation bioenergy crops (SGECs) include perennial forage crops (switchgrass, reed canary grass, alfalfa, Napier grass, and Bermuda grass) [26, 27]. The second-generation bioenergy generation takes the more pragmatic approach of using crop remains as feedstock. SGECs are more efficient than FGECs, as SGECs utilized cellulosic biomass for biofuel generation and the biofuel is non-oxygenated and in pure form [27]. SGECs reduce many environmental problems and require very less cost of production of biofuels. Biofuels from SGECs are thermo-chemically or biochemically derived from lingo-cellulosic crop wastes [14, 28]. Second-generation biofuel's main components are annual grain crops and annual biomass crops [4]. The SGECs require less processing and generate high energy with reduced emissions of greenhouse gases compared to FGEC. Growing SGECs generate important biomass for bioenergy generation [29]. The sugarcane industry has an enormous potential as a SGEC because the remains of sugarcane stalks (bagasse) are currently being burned in sugarcane factories to produce steam and electricity. Bagasse is mainly composed of cellulose, which mainly contained a linear chain of thousands of $\beta(1 \rightarrow 4)$-linked D-glucose units. Upon bacterial fermentation, bagasse releases cellulose residues that could be used to produce bioenergy using the latest technologies [30]. However, the second-generation ethanol produced from sugarcane has not yet been commercialized due to the lower rate of bagasse to sugar conversion. But certain countries like Brazil have met their energy requirements for bioethanol produced from sugarcane. The following are some of the principal bioenergy crops of the second generation.

18.2.2.1 Switchgrass

Switchgrass (*Panicum virgatum* L.) is a warm-season, perennial, C4 grass which is grown on marginal and erosive lands. This crop needs less nutrients and water for growth, making it an environmentally friendly crop for large-scale production of biofuels [31, 32]. Switchgrass, however, has a slow setting time that takes about two years for complete encroachment [31]. This plant has not grabbed much attention from researchers, particularly in the plant breeding sector [33]. Consequently, the germplasm of most switchgrass cultivars is not far from the native genomes. Depending on genetic composition, few types of switchgrass from natural populations are non-differentiable. Switchgrass, therefore, holds immense potential for genetic advancement in effective development of biomass.

18.2.2.2 Miscanthus

The genus *Miscanthus* contains 14–20 species of tall, perennial grasses, native to Asia, grown as ornamental plants [34]. The morphology of the plant constrains its use as crop forage. The plant is the main feedstock of herbaceous biomass in Europe. The *Miscanthus* plant performs C4 photosynthesis, has a high fixation rate of carbon dioxide, and requires less water and nitrogen than the C3 plants [35]. This grass is considered a dedicated energy crop due to its rapid growth, disease resistance, high productivity, and 10–15 years of comparatively longer plant life [36]. *Miscanthus* biomass yield was 33% higher than switchgrass [37]. One prime example of the genus *Miscanthus* is *M. Giantiseus* L. This needs 87% less land to generate equivalent biomass to prairie species [34]. The disadvantage of growing *Miscanthus* crops includes longer propagation duration of two to three years for rhizome cuttings, excessive irrigation, and energy consumption during greenhouse propagation.

18.2.2.3 Alfalfa

Alfalfa (*Medicago sativa* L.) is North America's oldest forage crop [38]. The *Alfalfa* stems are fibrous and burnt for electricity production in the gasification cycle. The leaves are enriched with protein [39]. The plant is a feedstock for the production of biofuels and also an excellent feed for animals [40]. *Alfalfa* has higher concentrations of polysaccharide and lignin in stem cell walls which lead to higher yields of stem dry matter and theoretical yields of ethanol [41].

18.2.2.4 Reed Canary Grass

Reed canary grass (*Phalaris arundinacea* L.) is a North American C3 grass. It is tall-growing perennial grass that is efficient in the recycling of internal nitrogen from shoots to roots. Several characteristics of canary reed grass are common to switchgrass such as slow growth and low yield. It is a native species of wetlands [42]. The grass yields relatively higher biomass and therefore could produce enough biofuel [43].

18.2.2.5 Other Plants

Because of the related benefits, several other plants often contribute to bioenergy. For example, a tall, perennial, and tropical grass, called Napier grass (*Pennisetum purpureum* Schumach) is preferred bioenergy crop due to the ease of establishment, persistent, and drought tolerance ability. The grass is full of taste and nutritious [44]. Napier grass's ability as a bioenergy crop was recognized by its low-lignin content and higher biomass yield per acre [[45]]. The Napier grass biomass has higher levels of volatile matter, carbon content, and lower levels of ash, nitrogen, and sulfur [46]. Reportedly, Napier grass's simultaneous saccharification and fermentation (SSF) yielded 74.1% ethanol. Bermuda grass (*Cynodon dactylon* L.) is an another plant used in bioenergy. It is a highly diverse perennial grass, short-lived, and mostly used as a warm-season forage. Because of its innovative existence and resistance to salinity, Bermuda grass acts as a soil binder in riverbanks or sea shore sand dams. It is a precious crop in irrigated lands [47]. Other possible perennial grass feedstocks are Eastern gamagrass (*Tripsacum dactyloides* L.) and prairie cordgrass (*Spartina pectinata* Link) [48].

18.2.3 Third-Generation Bioenergy Crops

Third-generation bioenergy crops (TGECs) include boreal plant, CAM plants, eucalyptus, and microalgas. CAM and boreal plants are the feedstock for direct cellulosic biomass fermentation [49, 50]. Thermo-conversion process is used to generate bioenergy from eucalyptus [14, 51]. Some species of microalgae could be potential feedstock for biodiesel production. The success of TGECs as a reliable source of biofuel depends on the efficient metabolism of cellulolytic bacteria during the fuel conversion process. Cellulose is broken down into water and carbon dioxide in the aerobic system. However, cellulose degrades to CH_4 and H_2 in anaerobic systems. Newer methodologies such as genomics, biodiversity studies, system biology, and metabolic engineering improve biofuel yields. TGECs are being introduced to develop a renewable and non-polluting energy source that could reduce global climate change [52–54].

18.2.3.1 Boreal Plants

Perennial grasses such as *Phleum pratense* and *Phalaris arundinacea* are examples of boreal plant species. Under boreal conditions, perennial grass is a major producer of herbaceous biomass. Boreal plants could be easily cultivated, harvested, stored, and used for the production of CH_4. Plants are tolerant of most phyto-pathogenic diseases, such as drought and frost. Boreal plants can withstand cold winters and grow on low-nutrition soils [55]. Few boreal plants such as *Ananas comosus*, *Opuntia ficus-indica*, *Agave sisalana*, and *Agave tequilana* are commonly used for bioenergy production [56].

18.2.3.2 Crassulacean Acid Metabolism (CAM) Plants

Plants with CAM are well adapted to photosynthesis. These plants have potential to absorb carbon dioxide at night. In arid habitats, CAM plants improve water efficiency and carbon assimilation. CAM plants are drought tolerant and used as bioenergy crops [57]. The CAM plants have three to six times higher water use efficiency than the C3 and C4 plants. CAM plants such as cardoon are multifunctional bioenergy crops. These plants are used for the production of solid and liquid biofuels [58, 59].

18.2.3.3 Eucalyptus

Eucalyptus (*Eucalyptus* sp.) is a native Australian plant. The plant grows faster with indefinite growth and has a large genetic resource base. The plant is resistant to drought, fire, insects, acid soils, low fertile soils, and other harsh conditions. *Eucalyptus* is cultivated in tropical countries due to faster growth and higher yields ($70\,m^3\,ha^{-1}\,year^{-1}$). The rotation period of the plant is as short as five years. Only four species and their hybrids (*E. grandis, E. urophylla, E. camaldulensis*, and *E. globulus*) contribute 80% of lantations worldwide. Of these four species, *E. globulus* is a widely adapted plant that is used in breeding programs due to faster growth. Eucalyptus oil extracted by thermo-conversion from plant parts holds enormous potential in the production of biofuel and bioenergy [14, 60].

18.2.3.4 Agave

Agave (*Agave* sp.) is a monocot plant native to hot, arid regions of Mexico. A plant species, *Agave tequilana*, is used for the production of tequila. *Agave* nectar is used as an alternative sugar for cooking. The plant grows in arid regions and has thick, fleshy leaves that end at a sharp point. *Agave* uses the CAM path for photosynthesis. It opens up stomata for CO_2 uptake during the night, causing less water loss during transpiration. The plant is used to make alcoholic beverages, sweeteners, and fibers. *Agave* is preferred feedstock for biofuels as it has minimal water requirements, is easily grown in wastelands, and does not compete with food crop feedstocks [61].

18.2.3.5 Microalgae

Microalgae are an important feedstock for the production of biodiesel, bioethanol, biomethane, and bio-hydrogen [62]. Photosynthetically, they are more efficient than terrestrial plants. Microalgae reduce greenhouse gas emissions by absorbing carbon dioxide released by plants. They produce vast biomass in a short period of time through efficient photosynthesis [50]. Microalgae minimize atmospheric carbon dioxide by sequestrating it. Compared to conventional biofuel-producing crops, microalgae biofuels have less impact on the environment and food supply in the world [49, 50, 63]. Microalgae have a very high potential to mitigate global climate change [49] as they have an efficient conversion rate of photons to photosynthates. In addition, they could be harvested all year round [64]. Microalgae provide nontoxic and biodegradable biofuels. Several programs are underway to improve the rate of biofuel production by enhancing the efficiency of strains through genetic engineering. Compared to other bioenergy crops, microalgae-derived fuel is considered greener due to higher conversion rates to biofuels.

18.2.4 Dedicated Bioenergy Crops

Examples of dedicated energy crops are perennial herbaceous and woody plant species. To produce biomass, they require lesser biological, chemical, or physical treatments. Such crops are considered environmentally friendly and would help mitigate global climate change [28, 65]. Such crops may solve numerous environmental problems by reducing salinity, carbon sequestration, enriching biodiversity, and improving the quality of soil and water [53, 66]. Cellulosic plants (Eucalyptus, Poplar, Willow, Birch, etc.), perennial grasses (giant reed, canary reed grass, switchgrass, elephant grass, etc.), non-edible oil crops (castor bean, physic nut, oil radish, pongamia, etc.), and oil plants (*Jatropha curcas, Pistacia chinensis, Sapium sebiferum* and *Vernicia fordii*) are the devoted bioenergy crops. These crops have a shorter life cycle and may therefore be harvested many times a year with long harvesting periods [67, 68]. Short rotation coppice (SRC) is among the most promising bioenergy-dedicated crop [69]. Countries such as Sweden and the United Kingdom are leaders in the comprehensive cultivation of dedicated bioenergy crops [70].

18.2.5 Halophytes

Halophytes are unique plants growing in sandy, marshy, and semi-deserted soils. They also inhabit coastal area, mangrove swamps, and estuaries [71]. The plants grow and propagate better at higher concentrations of salt [72]. Halophytes help to sequestrate carbon and rehabilitate degraded land, stabilizing habitats by providing the ecological niches required to mitigate climate change. In addition, they protect the associated flora and fauna from the environment and pathogens [73]. Frost-sensitive eucalyptus species and the frost-tolerant populus species are the most likely to survive under saline conditions [60]. Halophytes are easily established in salt-degraded soils and could also phyto-remediate soils contaminated with heavy metals [74, 75]. Dicot halophytes have been shown to be more tolerant to saline than monocots [76]. Halophytes may be used for food, medicine, and ornamental landscaping purposes. They also protect the environment by promoting wildlife [75, 77]. Halophytes of the genera *Acacia, Eucalyptus, Casuarina, Melaleuca, Prosopis, Rhizophora*, and *Tamarix* are commonly used in the production of biofuels. Perennial halophyte (*Kosteletzkya pentacarpos*) seeds have been shown to be used for the production of biodiesel [78]. Halophytes have a high rate of biofuel conversion efficiency because of high percentage of secondary metabolites [79].

18.3 Characteristics of Bioenergy Crops

Bioenergy crops can protect the environment in many ways [80]. Because of their perennial nature, they are resistant to disease and pest [55]. Bioenergy plans have advanced phenotypic, architectural, biochemical, and physiological characteristics that are desirable for biofuel production. In addition, cultivars of bioenergy crops are resistant to biotic and abiotic stresses that grows faster than other crops. Furthermore, crops with bioenergy need fewer biological, chemical, or physical pretreatment, thereby reducing the costs involved in processing biomass. New high-yielding energy crop varieties need to be introduced to meet energy needs that could be accomplished through large screening of productive botanical plants around the globe.

18.3.1 Physiological and Ecological Traits

Bioenergy plants store thermo-chemical and solar energy in a number of biochemical forms. These plants need a variety of physiological and ecological features to maximize radiation absorption, water efficiency, nutrient utilization, and environmental sustainability [31, 67]. These physiological features include efficient nutrient cycling, low nutrient requirement, carbon sequestration, low plant group competition, long canopy duration, efficient C4 or CAM photosynthetic pathway, and effective light capture. All these physiological features help plants to increase over-ground biomass during the growing season [66, 81]. Eco-physiological features of perennial short rotation coppice and lignocellulose grass germplasm show great diversity [12, 82]. Bioenergy crops have vegetative storage organs to store food reserves for longer periods of time. Vegetative storage structures are reported to reduce environmental stress and minimize metabolic loss [14]. The ratio of carbon and nitrogen is the deciding factor in generating bioenergy from plant biomass. Higher C:N ratio of bioenergy crops yields more bioenergy in the form of methane [83].

18.3.2 Agronomic and Metabolic Traits

Bioenergy crops need low energy for establishment, are well suited to marginal lands, and have higher biomass content. Such plants are reducing global warming and minimizing the effects of global climate change. The bioenergy crop should carry features of long canopy length, perennial production, sterility, less dry matter to reproductive structures, and less moisture content at harvest, according to agronomic characteristics. *Miscanthus* spp., a perennial C4 grass, retains most of these agronomic characteristics [81, 84, 85]. The metabolic architecture of the dedicated energy crop reduces "plant-to-plant" and "weed" competition. The plant metabolic improvement also decreases the interception of radiation, increases the efficiency of water usage, and accelerates drying on the field. These plants are simple, dense with upright stem branching, and waterlogging-resistant.

18.3.3 Biochemical Composition and Caloric Content

The biochemical composition of carbohydrates, proteins, lipids, and organic acids varies between plants. Their use in the bioenergy field is based on biochemical composition being unique. Bioenergy crops are a good source of energy, keep low cost of production, and reduce greenhouse gas emissions [86]. In terms of calorific value, the plant bioenergy is calculated, which is characterized as the expression of released heat value and energy content during the burning of material in air. In terms of calorific value, each type of bioenergy plant has its own merits and demerits. For instance, more energy is obtained from poplar plants than switchgrass and canary grass, whereas canary grass emits more greenhouse gasses compared to switchgrass and hybrid poplar [87, 88]. Plant growth energy and crop suitability issues are critical and related to bioenergy and food production [10]. Improving the biochemical composition and structure of bioenergy crops increases its caloric value, generating higher energy per ton of biomass [89]. Accumulated plant biomass is not proportional to the energy absorbed during photosynthesis, as the magnitude of the accumulated chemical forms varies in their energy densities. This difference depends on the species and the plant's stage of development. Carbohydrate generation is a valuable feature in bioenergy crops. Carbon hydrates are used for the production of biofuel in the fermentation process. Cellulose crops have more potential in the generation of bioenergy since their degradation releases a large number of glucose units. Higher yields of biofuel from cellulosic crops correspond to lower greenhouse gas emissions per hectare and per unit of biofuel produced compared to FGECs [12].

18.4 Genetic Improvement of Bioenergy Crops

Plants are commonly grown for food and feed. Traditional genetic modifications in breeding techniques have helped to develop plant varieties with the desired morphological, phenotypic, and biochemical characteristics [90, 91]. The main focus of these efforts is on improving crop productivity and quality. In addition, food crops could be modified for bioenergy generation by genotype alteration to produce more starch and a higher C:N ratio. Such modification could alter the pathway of lignin biosynthesis for better pre-processing via cellulose and cellulose expression. Bioenergy crop characteristics can be improved by identifying natural variations and genetic alterations in the production of transgenic plants [92, 93]. Genetically engineered bioenergy crops have greater adaptability to unfavorable conditions and higher growth rates and caloric value. A high degree of similarity is present between the grass genomes and poplar species. The transfer of gene function in such species into more genetically recalcitrant grass plants, such as switchgrass, *Miscanthus*, and short rotation coppice, could be encouraged. Because of easy propagation and faster growth in short rotation coppice cycles with lesser fertilizer requirement, *Willow* was established as a promising biomass crop. Willow plants need to be kept free of pests and diseases for better yield [94].

18.5 Environmental Impacts of Bioenergy Crops

Bioenergy crops provide numerous environmental and human benefits. The positive environmental impacts of bioenergy crops can be measured through the review of the sustainability indicators [95], risk–vulnerability–reliability assessment [96] and absolute impact measurement or percentage change with baseline comparison [97, 98]. Different environmental impacts of the production of bioenergy crops are shown in Figure 18.2 and well described below.

18.5.1 Soil Quality

Prevalent cropping systems and crop characteristics affect soil quality by affecting nutrient supply, availability of organic matter, soil structure, and pH. *Miscanthus*, switch-grass, and other fiber crops, for example, are mild to nutrient requirements, while giant reeds and

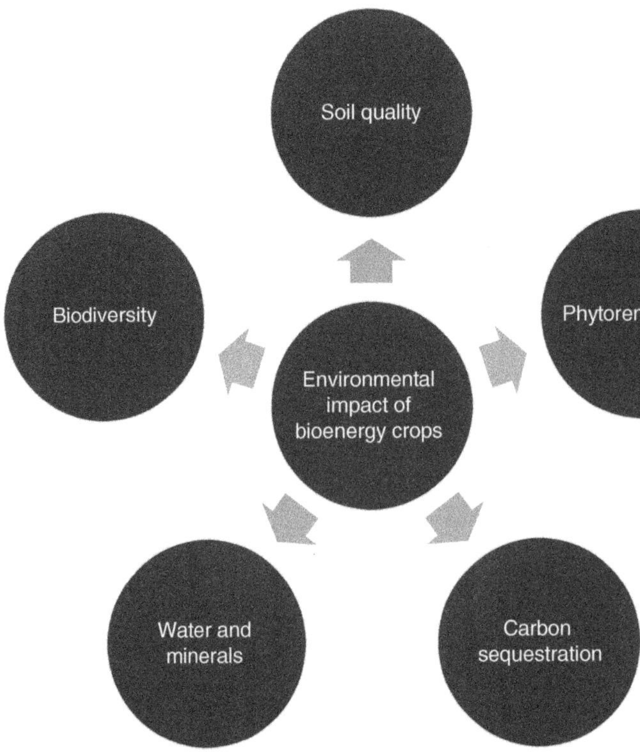

Figure 18.2 Impacts of bioenergy crops on environment.

cardoons are heavily depleted. Supplementation of soil with proper nutrients is necessary to ensure soil quality. In addition, nutrient supplementation needs to be carefully adjusted with concentration. For example, sweet sorghum and potato crops require a relatively lower concentration of phosphorus. Moderate concentrations of nitrogen and potassium are needed by crops to prevent plant malnutrition. Lack of adequate nutrition decreases plant biomass, and nutrient deficiency in the form of outward symptoms is evident. Sunflower, giant reed, and cardoon exhibit deeper nitrogen deficits. Giant reed, cardoon, sugar beet, sweet sorghum, canary grass, and wheat also reveal high deficiencies in potassium [99].

18.5.2 Water and Minerals

Cultivation of bioenergy crops could be water demanding to the point of compromising the availability of natural water resources. The water requirements of the crop should therefore be taken into account before planting bioenergy crops. The scarcity of water could hinder the successful establishment of bioenergy crops as a biofuel resource. Careful selection of bioenergy crops with tolerance to water stress is required in arid and semi-arid regions. Some deep-rooted bioenergy crops are drought-tolerant and capable of efficient carbon sequestration. However, such crops alter the dynamics of water and nutrients in soils to negatively impact biodiversity [53]. Corn, sugarcane, and oil palm crops required plenty of water for cultivation therefore well suited for growing in high-rainfall tropical areas [57]. Sugar beet, hemp, and potato also significantly affect water supplies [99]. Still, *Miscanthus*, and *Eucalyptus* plants have a lower cumulative effect on water supplies. Bioenergy crops have been known to influence minerals in the soil. The sorghum plant accumulates, for example, Pb, Ni, and Cu in roots and shoots. Applying phosphorus and potassium to bioenergy crop fields greatly decreases the loss of soil mineral content. Perennial crops are less demanding for macronutrients, and their pattern of use of nutrients does not vary significantly from annual crops. *Eucalyptus* and *Willow* plants impact mineral resources at lower levels, while sweet sorghum and potato face higher risks of depletion of nutrients [99].

18.5.3 Carbon Sequestration

Carbon sequestration requires extracting CO_2 from the environment through plant mediation. Bioenergy crops minimize atmospheric CO_2 by means of a large accumulation of biomass. The use of perennial crops could improve soil quality by increasing carbon sequestration through high production of biomass and deep-rooted systems [100]. Bioenergy crops will now be used to sequester atmospheric CO_2 and increase the productivity of biomass for generating bioenergy [101].

18.5.4 Phytoremediation

Phytoremediation means usage of plants for remediating polluted soil, sediments, and groundwater by eliminating or degrading pollutants [102]. This technology is innovative and cost-effective and maintains long-term applicability [103]. Bioenergy plant may extract heavy metals from the soil to improve the quality of soil. The approach has an added benefit of treating polluted site without digging [104, 105]. Phyto-stabilization and phyto-extraction include the main phytoremediation methods used to remediate heavy metal-contaminated property. Phyto-stabilization requires the use of growing root plants, which reduces the bioavailability of stabilized metals in the substrate [106]. Phyto-extraction involves the use of plants with the ability to accumulate heavy metals from soils, sediments, and water. This way of handling metal-polluted land seems economically feasible [107]. Phytoremediation is common to many plant genera. Effective phytoremediation, however, requires the selection of appropriate plants. Plant selection depends on its availability, adaptation to specific climate conditions, ability to extract heavy metals, rate of production of biomass, and economic value [103]. A study on *Sorghum bicolor* for the phytoremediation of heavy metals has shown that the plant is efficient in the absorption of metals due to high biomass. The plant accumulates a high concentration of metal in its shoots. *Sorghum* plants were able to absorb metals such as Ni, Pb, and Zn efficiently [108]. A major source of water pollution in agricultural land is widespread and unrestricted use of fertilizer in the field. High levels of nitrate fertilizer are used in fields to increase crop yields. The use of high nitrate fertilizers creates pollution of the surface and groundwater by nitrates. Few bioenergy plants have the potential to remove soil or water contaminants. *Poplar* plant is known to accumulate high nitrate levels from water streams draining from agricultural lands [109]. This plant extracts nitrate from water bodies and thereby decreases its concentration in polluted water [110]. *Poplar* is well suited for growing in nitrate-rich soil through high- and low-affinity nitrate transporter proteins [111]. *Miscanthus* crops are also used for phytoremediation [112, 113]. In phytoremediation, the crop is favored because of its perennial existence, high productivity, better growth rate, effective sequestration of CO_2, higher efficiency in water use, and the ability to protect soil erosion. However, the use of *Miscanthus* has been associated with the disadvantage of lower numbers of viable seeds for oil extraction [113, 114], making it unsuitable for biofuel extraction.

18.5.5 Biodiversity

Biodiversity defines the variety of species that exist on earth. This increases the efficiency of the habitats, where each species contributes in its own way. Biodiversity preservation is thus necessary for a healthy ecosystem. The biodiversity of nature is diminished by many environmental factors, among which land conversions, deforestation, and grassland conversions contribute to great duration. Most such factors linked to the environment could be regulated by growing crops with bioenergy. Bioenergy crops protect biodiversity by reducing emissions of greenhouse gases and mitigating global climate change [80]. Furthermore, the blossoming period of biodiversity and other crops also increases the abundance and diversity of birds or insects, particularly in the fields of sunflower [99, 115]. However, growing annual crops reduces biodiversity due to short soil impacts and high growth requirements. The development of biofuel-based lignocellulose systems that use a range of feedstocks could increase the diversity of agricultural landscapes and increase arthropod-mediated ecosystem services [116]. For example, perennial grasses with a high content of lignocellulose reduce soil tillage and agrochemical usage, yield high above and below ground biomass, favor soil microfauna, and provide shelter for invertebrates and birds [80, 117]. In contrast to perennial grasses, willow and poplar plants support more biodiversity due to longer life cycles and habitat development for birds, vertebrates, and flora. The overall effect of these crops on biodiversity, however, may be marginal or not even positive [118–120]. Bioenergy plants such as eucalyptus do not support biodiversity because of a more vigorous cultivation management.

18.6 Conclusion and Future Prospect

Plants expand by absorbing released CO_2 during the combustion of biomass. No net CO_2 is produced by the use of crop biomass for energy generation, as the amount emitted during use has been previously fixed during plant development. Using bioenergy crops for generating energy may help to make use of this alternative source of renewable energy. Commercial development of bioenergy fuels could reduce our reliance on fossil fuel transport using existing engine technology. Feedstock for bioenergy crops (cellulose or sugar, starch plants) may play a major role in ethanol and biodiesel generation to improve the rural economy, provide greater energy efficiency, and use environmentally degraded lands in a sustainable way. A large-scale plantation of bioenergy crops may control these environmental factors. Since bioenergy crops may alter the soil's water and nutrient dynamics, their pattern of water use should also be taken into account before field planting. A suitable bioenergy crop should be recommended according to the land type.

References

1 Gresshoff, P.M., Rangan, L., Indrasumunar, A., and Scott, P.T. (2017). A new bioenergy crop based on oil-rich seeds from the legume tree Pongamia pinnata. *Energy and Emission Control Technologies* 5: 19–26.

2 Karp, A. and Shield, I. (2008). Bioenergy from plants and the sustainable yield challenge. *New Phytologist* 179 (1): 15–32.

3 Taylor, G. (2008). Bioenergy for heat and electricity in the UK: a research atlas and roadmap. *Energy Policy* 36 (12): 4383–4389.

4 Yuan, J.S., Tiller, K.H., Al-Ahmad, H. et al. (2008). Plants to power: bioenergy to fuel the future. *Trends in Plant Science* 13 (8): 421–429.

5 Adler, P.R., Grosso, S.J., and Parton, W.J. (2007). Life-cycle assessment of net greenhouse-gas flux for bioenergy cropping systems. *Ecological Applications* 17 (3): 675–691.

6 Wang, S., Hastings, A., and Smith, P. (2012). An optimization model for energy crop supply. *GCB Bioenergy* 4 (1): 88–95.

7 Kim, H.K., Parajuli, P.B., and To, S.F. (2013). Assessing impacts of bioenergy crops and climate change on hydrometeorology in the Yazoo River Basin, Mississippi. *Agricultural and Forest Meteorology* 169: 61–73.

8 Barbosa, B., Boléo, S., Sidella, S. et al. (2015). Phytoremediation of heavy metal-contaminated soils using the perennial energy crops *Miscanthus* spp. and Arundo donax L. *BioEnergy Research* 8 (4): 1500–1511.

9 Dipti, P. (2013). Bioenergy crops an alternative energy. *International Journal of Environmental Engineering and Management* 4: 265–272.

10 Singh, P.P. (2008). Exploring biodiversity and climate change benefits of community-based forest management. *Global Environmental Change* 18 (3): 468–478.

11 Lobell, D.B., Burke, M.B., Tebaldi, C. et al. (2008). Prioritizing climate change adaptation needs for food security in 2030. *Science* 319 (5863): 607–610.

12 Chhetri, A.B., Tango, M.S., Budge, S.M. et al. (2008). Non-edible plant oils as new sources for biodiesel production. *International Journal of Molecular Sciences* 9 (2): 169–180.

13 Carroll, A. and Somerville, C. (2009). Cellulosic biofuels. *Annual Review of Plant Biology* 60: 165–182.

14 Lorenz, A.J., Coors, J.G., De Leon, N. et al. (2009). Characterization, genetic variation, and combining ability of maize traits relevant to the production of cellulosic ethanol. *Crop Science* 49 (1): 85–98.

15 Wang, Y. and Yan, L. (2008). CFD studies on biomass thermochemical conversion. *International Journal of Molecular Sciences* 9: 1108–1130. https://doi.org/10.3390/ijms9061108.

16 Eisenbies, M.H., Vance, E.D., Aust, W.M., and Seiler, J.R. (2009). Intensive utilization of harvest residues in southern pine plantations: quantities available and implications for nutrient budgets and sustainable site productivity. *BioEnergy Research* 2 (3): 90–98.

17 Mabee, W.E., McFarlane, P.N., and Saddler, J.N. (2011). Biomass availability for lignocellulosic ethanol production. *Biomass and Bioenergy* 35 (11): 4519–4529.

18 Takamizawa, K.A., Anderson, W.I., and Singh, H.P. (2010). Ethanol from lignocellulosic crops. In: *Industrial Crops and Uses*, vol. 25 (ed. H.P. Singh), 104–139. Oxfordshire: CABI.

19 Zhao, R., Wu, X.I., Bean, S.C., and Wang, D.O. (2010). Ethanol from grain crops. In: *Industrial Crops and Uses*, vol. 25 (ed. H.P. Singh), 84–101. CABI International.

20 Paterson, A.H., Bowers, J.E., Bruggmann, R. et al. (2009). The Sorghum bicolor genome and the diversification of grasses. *Nature* 457 (7229): 551–556.

21 Qazi, H.A., Paranjpe, S., and Bhargava, S. (2012). Stem sugar accumulation in sweet sorghum–activity and expression of sucrose metabolizing enzymes and sucrose transporters. *Journal of Plant Physiology* 169 (6): 605–613.

22 Thomas, H. and Howarth, C.J. (2000). Five ways to stay green. *Journal of Experimental Botany* 51 (suppl_1): 329–337.

23 Harris, K., Subudhi, P.K., Borrell, A. et al. (2007). Sorghum stay-green QTL individually reduce post-flowering drought-induced leaf senescence. *Journal of Experimental Botany* 58 (2): 327–338.

24 Okada, M., Lanzatella, C., Saha, M.C. et al. (2010). Complete switchgrass genetic maps reveal subgenome collinearity, preferential pairing and multilocus interactions. *Genetics* 185 (3): 745–760.

25 Swaminathan, K., Alabady, M.S., Varala, K. et al. (2010). Genomic and small RNA sequencing of Miscanthus× giganteus shows the utility of sorghum as a reference genome sequence for Andropogoneae grasses. *Genome Biology* 11 (2): R12.

26 Sims, R.E., Hastings, A., Schlamadinger, B. et al. (2006). Energy crops: current status and future prospects. *Global Change Biology* 12 (11): 2054–2076.

27 Sanderson, M.A. and Adler, P.R. (2008). Perennial forages as second generation bioenergy crops. *International Journal of Molecular Sciences* 9 (5): 768–788.

28 Oliver, R.J., Finch, J.W., and Taylor, G. (2009). Second generation bioenergy crops and climate change: a review of the effects of elevated atmospheric CO2 and drought on water use and the implications for yield. *GCB Bioenergy* 1 (2): 97–114.

29 Petersen, J.E. (2008). Energy production with agricultural biomass: environmental implications and analytical challenges. *European Review of Agricultural Economics* 35 (3): 385–408.

30 Kotchoni, S.O. and Gachomo, E.W. (2008). Biofuels production: a promising alternative energy for environmental cleanup and fuelling through renewable resources. *Journal of Biological Sciences* 8: 693–701.

31 Waclawovsky, A.J., Sato, P.M., Lembke, C.G. et al. (2010). Sugarcane for bioenergy production: an assessment of yield and regulation of sucrose content. *Plant Biotechnology Journal* 8 (3): 263–276.

32 McLaughlin, S.B., Kiniry, J.R., Taliaferro, C.M., and Ugarte, D.D. (2006). Projecting yield and utilization potential of switchgrass as an energy crop. *Advances in Agronomy* 90: 267–297.

33 Vogel, K.P. and Mitchell, R.B. (2008). Heterosis in switchgrass: biomass yield in swards. *Crop Science* 48 (6): 2159–2164.

34 Bouton, J.H. (2007). Molecular breeding of switchgrass for use as a biofuel crop. *Current Opinion in Genetics & Development* 17 (6): 553–558.

35 Heaton, E.A., Dohleman, F.G., Miguez, A.F. et al. (2010). Miscanthus: a promising biomass crop. In: *Advances in Botanical Research*, vol. 56 (eds. J.-C. Kader and M. Delseny), 75–137. Academic Press.

36 Villaverde, J.J., Li, J., Ek, M. et al. (2009). Native lignin structure of Miscanthus x giganteus and its changes during acetic and formic acid fractionation. *Journal of Agricultural and Food Chemistry* 57 (14): 6262–6270.

37 Villaverde, J.J., Ligero, P., and Vega, A.D. (2010). Miscanthus x giganteus as a source of biobased products through organosolv fractionation: a mini review. *The Open Agriculture Journal* 4 (1).

38 Heaton, E. and Voigt, T. (2004). Long SP. A quantitative review comparing the yields of two candidate C4 perennial biomass crops in relation to nitrogen, temperature and water. *Biomass and Bioenergy* 27 (1): 21–30.

39 Russelle, M.P. (2001). Alfalfa: After an 8,000-year journey, the "Queen of Forages" stands poised to enjoy renewed popularity. *American Scientist* 89 (3): 252–261.

40 Lamb, J.F., Sheaffer, C.C., and Samac, D.A. (2003). Population density and harvest maturity effects on leaf and stem yield in alfalfa. *Agronomy Journal* 95 (3): 635–641.

41 Delong, M.M., Swanberg, D.R., and Oelke, E.A. (1995). *Sustainable Biomass Energy Production and Rural Economic Development Using Alfalfa as Feedstock*. Golden, CO: National Renewable Energy Lab.

42 Lamb, J.F., Jung, H.J., Sheaffer, C.C., and Samac, D.A. (2007). Alfalfa leaf protein and stem cell wall polysaccharide yields under hay and biomass management systems. *Crop Science* 47 (4): 1407–1415.

43 Merigliano, M.F. and Lesica, P. (1998). The native status of reed canarygrass (Phalaris arundinacea L.) in the inland Northwest, USA. *Natural Areas Journal* 18 (3): 223–230.

44 Tahir, M.H., Casler, M.D., Moore, K.J., and Brummer, E.C. (2011). Biomass yield and quality of reed canarygrass under five harvest management systems for bioenergy production. *BioEnergy Research* 4 (2): 111–119.

45 Yasuda, M., Nagai, H., Takeo, K. et al. (2014). Bio-ethanol production through simultaneous saccharification and co-fermentation (SSCF) of a low-moisture anhydrous ammonia (LMAA)-pretreated napiegrass (Pennisetum purpureum Schumach). *SpringerPlus* 3 (1): 333.

46 Schmer, M.R., Vogel, K.P., Mitchell, R.B., and Perrin, R.K. (2008). Net energy of cellulosic ethanol from switchgrass. *Proceedings of the National Academy of Sciences of the United States of America* 105 (2): 464–469.
47 Mohammed, I.Y., Abakr, Y.A., Kazi, F.K. et al. (2015). Comprehensive characterization of Napier grass as a feedstock for thermochemical conversion. *Energies* 8 (5): 3403–3417.
48 Index, F.G. (2011). *A Searchable Catalogue of Grass and Forage Legumes*. Rome: FAO.
49 Springer, T.L. and Dewald, C.L. (2004). Eastern gamagrass and other Tripsacum species. In: *Warm-Season (C4) Grasses* (eds. L.E. Moser, B.L. Burson and L.E. Sollenberger), 955–973. Madison, WI: American Society of Agronomy.
50 Patil, V., Tran, K.Q., and Giselrød, H.R. (2008). Towards sustainable production of biofuels from microalgae. *International Journal of Molecular Sciences* 9 (7): 1188–1195.
51 Schenk, P.M., Thomas-Hall, S.R., Stephens, E. et al. (2008). Second generation biofuels: high-efficiency microalgae for biodiesel production. *Bioenergy Research* 1 (1): 20–43.
52 Carere, C.R., Sparling, R., Cicek, N., and Levin, D.B. (2008). Third generation biofuels via direct cellulose fermentation. *International Journal of Molecular Sciences* 9 (7): 1342–1360.
53 Bush, D.R. and Leach, J.E. (2007). Translational genomics for bioenergy production: there's room for more than one model. *The Plant Cell* 19 (10): 2971–2973.
54 Ehrlich, P.R. and Pringle, R.M. (2008). Where does biodiversity go from here? A grim business-as-usual forecast and a hopeful portfolio of partial solutions. *Proceedings of the National Academy of Sciences of the United States of America* 105 (Supplement 1): 11579–11586.
55 Rubin, E.M. (2008). Genomics of cellulosic biofuels. *Nature* 454 (7206): 841–845.
56 Finckh, M.R. (2008). Integration of breeding and technology into diversification strategies for disease control in modern agriculture. In: *Sustainable Disease Management in a European Context* (eds. D.B. Collinge, L. Munk and B.M. Cooke), 399–409. Dordrecht: Springer.
57 Lehtomäki, A., Viinikainen, T.A., and Rintala, J.A. (2008). Screening boreal energy crops and crop residues for methane biofuel production. *Biomass and Bioenergy* 32 (6): 541–550.
58 De Fraiture, C., Giordano, M., and Liao, Y. (2008). Biofuels and implications for agricultural water use: blue impacts of green energy. *Water Policy* 10 (S1): 67–81.
59 Grammelis, P., Malliopoulou, A., Basinas, P., and Danalatos, N.G. (2008). Cultivation and characterization of *Cynara cardunculus* for solid biofuels production in the Mediterranean region. *International Journal of Molecular Sciences* 9 (7): 1241–1258.
60 Borland, A.M., Griffiths, H., Hartwell, J., and Smith, J.A. (2009). Exploiting the potential of plants with crassulacean acid metabolism for bioenergy production on marginal lands. *Journal of Experimental Botany* 60 (10): 2879–2896.
61 Rockwood, D.L., Rudie, A.W., Ralph, S.A. et al. (2008). Energy product options for Eucalyptus species grown as short rotation woody crops. *International Journal of Molecular Sciences* 9 (8): 1361–1378.
62 Escamilla-Treviño, L.L. (2012). Potential of plants from the genus Agave as bioenergy crops. *Bioenergy Research* 5 (1): 1–9.
63 Ahmad, A.L., Yasin, N.M., Derek, C.J., and Lim, J.K. (2011). Microalgae as a sustainable energy source for biodiesel production: a review. *Renewable and Sustainable Energy Reviews* 15 (1): 584–593.
64 Tilman, D., Socolow, R., Foley, J.A. et al. (2009). Beneficial biofuels—the food, energy, and environment trilemma. *Science* 325 (5938): 270–271.

65 Williams, C., Black, I., Biswas, T., and Heading, S. (2007). Pathways to prosperity: second generation biomass crops for biofuels using saline lands and wastewater. *Agricultural Sciences* 20 (3): 28.

66 Taherzadeh, M.J. and Karimi, K. (2008). Pretreatment of lignocellulosic wastes to improve ethanol and biogas production: a review. *International Journal of Molecular Sciences* 9 (9): 1621–1651.

67 Lal, R. Carbon sequestration. *Philosophical Transactions of the Royal Society, B: Biological Sciences* 363: 815–830.

68 Boe, A. and Lee, D. (2007). Genetic variation for biomass production in prairie cordgrass and switchgrass. *Crop Science* 47 (3): 929–934.

69 Ranade, S.A., Srivastava, A.P., Rana, T.S. et al. (2008). Easy assessment of diversity in *Jatropha curcas* L. plants using two single-primer amplification reaction (SPAR) methods. *Biomass and Bioenergy* 32 (6): 533–540.

70 Rae, A.M., Street, N.R., Robinson, K.M. et al. (2009). Five QTL hotspots for yield in short rotation coppice bioenergy poplar: the poplar biomass loci. *BMC Plant Biology* 9 (1): 23.

71 Mola-Yudego, B. and González-Olabarria, J.R. (2010). Mapping the expansion and distribution of willow plantations for bioenergy in Sweden: lessons to be learned about the spread of energy crops. *Biomass and Bioenergy* 34 (4): 442–448.

72 Glenn, E.P., Brown, J.J., and Blumwald, E. (1999). Salt tolerance and crop potential of halophytes. *Critical Reviews in Plant Sciences* 18 (2): 227–255.

73 Ventura, Y., Eshel, A., Pasternak, D., and Sagi, M. (2015). The development of halophyte-based agriculture: past and present. *Annals of Botany* 115 (3): 529–540.

74 Jaradat, A.A. (2010). Genetic resources of energy crops: biological systems to combat climate change. *Australian Journal of Crop Science* 4 (5): 309–323.

75 Hasanuzzaman, M., Nahar, K., Alam, M. et al. (2014). Potential use of halophytes to remediate saline soils. *BioMed Research International* 2014: 589341.

76 Panta, S., Flowers, T., Lane, P. et al. (2014). Halophyte agriculture: success stories. *Environmental and Experimental Botany* 107: 71–83.

77 Flowers, T.J. and Colmer, T.D. (2008). Salinity tolerance in halophytes. *The New Phytologist* 1: 945–963.

78 Cassaniti, C., Romano, D., Hop, M.E., and Flowers, T.J. (2013). Growing floricultural crops with brackish water. *Environmental and Experimental Botany* 92: 165–175.

79 Moser, B.R., Dien, B.S., Seliskar, D.M., and Gallagher, J.L. (2013). Seashore mallow (*Kosteletzkya pentacarpos*) as a salt-tolerant feedstock for production of biodiesel and ethanol. *Renewable Energy* 50: 833–839.

80 Hastilestari, B.R., Mudersbach, M., Tomala, F. et al. (2013). *Euphorbia tirucalli* L.–comprehensive characterization of a drought tolerant plant with a potential as biofuel source. *PLoS One* 8 (5): e63501.

81 Boehmel, C., Lewandowski, I., and Claupein, W. (2008). Comparing annual and perennial energy cropping systems with different management intensities. *Agricultural Systems* 96 (1-3): 224–236.

82 Jakob, K., Zhou, F., and Paterson, A.H. (2011). Genetic improvement of C4 grasses as cellulosic biofuel feedstocks. In: *Biofuels* (eds. D. Tomes, P. Lakshmanan and D. Songstad), 113–138. New York, NY: Springer.

83 Tharakan, P.J., Robison, D.J., Abrahamson, L.P., and Nowak, C.A. (2001). Multivariate approach for integrated evaluation of clonal biomass production potential. *Biomass and Bioenergy* 21 (4): 237–247.

84 Long, S.P., ZHU, X.G., Naidu, S.L., and Ort, D.R. (2006). Can improvement in photosynthesis increase crop yields? *Plant, Cell & Environment* 29 (3): 315–330.

85 Lewandowski, I., Clifton-Brown, J.C., Scurlock, J.M., and Huisman, W. (2000). Miscanthus: European experience with a novel energy crop. *Biomass and Bioenergy* 19 (4): 209–227.

86 Leakey, A.D. (2009). Rising atmospheric carbon dioxide concentration and the future of C4 crops for food and fuel. *Proceedings of the Royal Society B: Biological Sciences* 276 (1666): 2333–2343.

87 Monti, A., Di Virgilio, N., and Venturi, G. (2008). Mineral composition and ash content of six major energy crops. *Biomass and Bioenergy* 32 (3): 216–223.

88 Ferré, C., Leip, A., Matteucci, G. et al. (2005). Impact of 40 years poplar cultivation on soil carbon stocks and greenhouse gas fluxes. *Biogeosciences Discussions* 2 (4): 897–931. https://doi.org/10.5194/bgd-2-897-2005.

89 Boe, A. and Beck, D.L. (2008). Yield components of biomass in switchgrass. *Crop Science* 48 (4): 1306–1311.

90 Sticklen, M. (2006). Plant genetic engineering to improve biomass characteristics for biofuels. *Current Opinion in Biotechnology* 17 (3): 315–319.

91 Lee, M. (1998). Genome projects and gene pools: new germplasm for plant breeding? *Proceedings of the National Academy of Sciences of the United States of America* 95 (5): 2001–2004.

92 Baenziger, P.S., Russell, W.K., Graef, G.L., and Campbell, B.T. (2006). Improving lives: 50 years of crop breeding, genetics, and cytology (C-1). *Crop Science* 46 (5): 2230–2244.

93 Gressel, J. (2008). Transgenics are imperative for biofuel crops. *Plant Science* 174 (3): 246–263.

94 Ortiz, R. (2008). Crop genetic engineering under global climate change. *Annals of Arid Zone* 47 (3): 343.

95 Karp, A., Hanley, S.J., Trybush, S.O. et al. (2011). Genetic improvement of willow for bioenergy and biofuels free access. *Journal of Integrative Plant Biology* 53 (2): 151–165.

96 McBride, A.C., Dale, V.H., Baskaran, L.M. et al. (2011). Indicators to support environmental sustainability of bioenergy systems. *Ecological Indicators* 11 (5): 1277–1289.

97 Hoque, Y.M., Raj, C., Hantush, M.M. et al. (2014). How do land-use and climate change affect watershed health? A scenario-based analysis. *Water Quality Exposure and Health* 6 (1-2): 19–33.

98 Feng, Q., Chaubey, I., Her, Y.G. et al. (2015). Hydrologic and water quality impacts and biomass production potential on marginal land. *Environmental Modelling and Software* 72: 230–238.

99 Cibin, R., Trybula, E., Chaubey, I. et al. (2016). Watershed-scale impacts of bioenergy crops on hydrology and water quality using improved SWAT model. *GCB Bioenergy* 8 (4): 837–848.

100 S.M Boléo. Environmental impact assessment of energy crops cultivation in the Mediterranean Europe. https://run.unl.pt/handle/10362/7434.

101 Ma, Z., Wood, C.W., and Bransby, D.I. (2000). Impacts of soil management on root characteristics of switchgrass. *Biomass and Bioenergy* 18 (2): 105–112.

102 Lemus, R. and Lal, R. (2005). Bioenergy crops and carbon sequestration. *Critical Reviews in Plant Sciences* 24 (1): 1–21.

103 Fritioff, Å. and Greger, M. (2003). Aquatic and terrestrial plant species with potential to remove heavy metals from stormwater. *International Journal of Phytoremediation* 5 (3): 211–224.

104 P.W. Team. Phytoremediation decision tree. http://citeseerx.ist.psu.edu/viewdoc/download?doi=10.1.1.122.5430&rep=rep1&type=pdf

105 Vaněk, T., Podlipna, R., and Soudek, P. (2010). General factors influencing application of phytotechnology techniques. In: *Application of Phytotechnologies for Cleanup of Industrial, Agricultural, and Wastewater Contamination* (eds. P. .A. Kulakow and V. .V. Pidlisnyuk), 1–13. Dordrecht: Springer.

106 Zhu, K., Chen, H., and Nan, Z. (2010). Phytoremediation of loess soil contaminated by organic compounds. In: *Application of Phytotechnologies for Cleanup of Industrial, Agricultural, and Wastewater Contamination* (eds. P. .A. Kulakow and V. .V. Pidlisnyuk), 159–176. Dordrecht: Springer.

107 Salt, D.E., Blaylock, M., Kumar, N.P. et al. (1995). Phytoremediation: a novel strategy for the removal of toxic metals from the environment using plants. *Bio/Technology* 13 (5): 468–474.

108 Oh, K. (2015). Phytoremediation potential of sorghum as a biofuel crop and the enhancement effects with microbe inoculation in heavy metal contaminated soil. *Journal of Biosciences and Medicines* 3 (06): 9.

109 Al Chami, Z., Amer, N., Al Bitar, L., and Cavoski, I. (2015). Potential use of Sorghum bicolor and *Carthamus tinctorius* in phytoremediation of nickel, lead and zinc. *International Journal of Environmental Science and Technology* 12 (12): 3957–3970.

110 Rennenberg, H., Wildhagen, H., and Ehlting, B. (2010). Nitrogen nutrition of poplar trees. *Plant Biology* 12 (2): 275–291.

111 O'Neill, G.J. and Gordon, A.M. (1994). The nitrogen filtering capability of Carolina poplar in an artificial riparian zone. *Journal of Environmental Quality* 23 (6): 1218–1223.

112 Bai, H., Euring, D., Volmer, K. et al. (2013). The nitrate transporter (NRT) gene family in poplar. *PLoS One* 8 (8): e72126.

113 Xie, X., Zhou, F., Zhao, Y., and Lu, X. (2008). A summary of ecological and energy-producing effects of perennial energy grasses. *Acta Ecologica Sinica* 28 (5): 2329–2342.

114 Masarovičová, E., Kráľová, K., and Peško, M. (2009). Energetic plants–cost and benefit. *Ecological Chemistry and Engineering* 16 (3): 263–276.

115 Miller, C.A. and Gage, C.L. (2011). Potential adverse environmental impacts of greenhouse gas mitigation strategies. In: *Global Climate Change-The Technology Challenge* (ed. F. Princiotta), 377–415. Dordrecht: Springer.

116 Jones, G.A. and Sieving, K.E. (2006). Intercropping sunflower in organic vegetables to augment bird predators of arthropods. *Agriculture, Ecosystems & Environment* 117 (2-3): 171–177.

117 Landis, D.A., Gardiner, M.M., van der Werf, W., and Swinton, S.M. (2008). Increasing corn for biofuel production reduces biocontrol services in agricultural landscapes.

Proceedings of the National Academy of Sciences of the United States of America 105 (51): 20552–20557.

118 Börjesson, P. (1999). Environmental effects of energy crop cultivation in Sweden—II: economic valuation. *Biomass and Bioenergy* 16 (2): 155–170.

119 Berg, Å. (2002). Breeding birds in short-rotation coppices on farmland in central Sweden—the importance of Salix height and adjacent habitats. *Agriculture, Ecosystems and Environment* 90 (3): 265–276.

120 Paine, L.K., Peterson, T.L., Undersander, D.J. et al. (1996). Some ecological and socio-economic considerations for biomass energy crop production. *Biomass and Bioenergy* 10 (4): 231–242.

19

Marine Bioprospecting

Seaweeds for Industrial Molecules

Achintya Kumar Dolui[1,2]

[1] *Department of Lipid Science, CSIR-Central Food Technological Research Institute, Mysuru, Karnataka, India*
[2] *Academy of Scientific and Innovative Research, Ghaziabad, Uttar Pradesh, India*

19.1 Introduction

Seaweed is another term for marine algae, which is benthic in nature (since they reside on the bottom of water bodies). They do not feature typical plant-like structures like root, stem, and leaves and belong to the division of Thallophyta in the plant kingdom [1]. Seaweeds generally lack a common ancestor and form a polyphyletic group as a result of their convergent evolution. They usually refer to unicellular to multicellular and mostly macroscopic forms of algae. Seaweeds are classified into three different groups, i.e. green algae (phylum: Chlorophyta, classes: Bryopsidophyceae, Chlorophyceae, Dasycladophyceae, Prasinophyceae, and Ulvophyceae, about 1200 species), brown algae (phylum: Ochrophyta, Classes: Phaeophyceae, about 1750 species), and red algae (phylum: Rhodophyta, about 6000 species). This classification is based on the types of pigment (composition) they possess, characteristics of their cell walls, and the type of reserve polysaccharides [2, 3]. Sometimes tuft forming blue green algae (Cyanobacteria) is also lumped together with the seaweeds. The presence of seawater (or at least brackish water) is an absolute requirement for the seaweed to be physiologically functional. Besides, the presence of sufficient light is needed to make the seaweeds photosynthetically active. From the ecological point of view, another prerequisite for seaweeds is a robust attachment point. Thus, most of the seaweeds occupy the littoral zone, preferably or within that zone on rocky shores than on sand or shingles. Some species of algae are living deepest into the sea, where accessibility of sunlight is a major challenge. Seaweed like *Sagassum* is a free-floating planktonic seaweeds (brown algae) thanks to their gas-filled sacs, which keep them afloat in an acceptable depth. However, the free-floating seaweeds are subjected to major challenges in the form of changing temperature, salinity, and periodic drying [2]. Considering the plethora of benefits that seaweeds offer to humankind, the cultivation of seaweeds on a commercial scale has been put into practice for biomass production. Unlike terrestrial agriculture, there is no requirement for freshwater as well as arable land for its cultivation. Besides, it does not require the application of fertilizer in most cases. From the industrial perspective,

Bioprospecting of Plant Biodiversity for Industrial Molecules, First Edition.
Edited by Santosh Kumar Upadhyay and Sudhir P. Singh.

seaweed cultivation is an ever-expanding industry that is being exploited for the isolation of pharmaceuticals, extraction of functional food ingredients, and the production of biofuels [4]. Globally, seaweeds cultivation is a lucrative industry which is dominated mainly by Asian countries like China (47.9%) Indonesia (38.7%), the Philippines (4.7%), the Republic of Korea (4.5%), Japan (1.3%), and Malaysia (0.7%) [5]. Seaweed is considered as superfoods and called as sea vegetables, since it is one of the nutrient-dense food in the plant kingdom. Edible seaweeds are also a rich source of macronutrients and micronutrients, beyond the source of essential nutrients, and also a reservoir of various health-promoting bioactive components. Even though seaweeds make a huge contribution to the traditional human diets in oriental countries, its inclusion into the Western diet was limited only to the coastal communities. However, recently it has gained popularity in the Western population due to the impetus from the health, food, and nutraceutical industry [6]. Since seaweeds are loaded with several health-enhancing elements such as antioxidants, phycocolloids, carotenoids, soluble dietary fibers, polyunsaturated fatty acids (PUFAs), phycobilins, polysaccharides, sterols, tocopherols, terpenes, and phycocyanins, it has huge potential for the health supplement markets in the form of value-added products or novel foods. Seaweeds can alleviate lifestyle-related disorders, such as hyperglycemia, hypercholesterolemia, and hyperlipidemia [7]. Marine algae are also a rich source of potential bioactive components of industrial interest. The list includes isolated polysaccharides (e.g. alginate, fucoidan), proteins (e.g. phycobiliproteins), polyphenols (e.g. phlorotannins), carotenoids (e.g. fucoxanthin), and *n*-3 long-chain PUFAs (e.g. eicosapentaenoic acid) [8]. Moreover, seaweeds produce a lot of secondary metabolites with substantial industrial and therapeutic potential. These secondary metabolites exhibit potent biological activities such as antimicrobial, antitumor, antidiabetic, anticoagulant, antioxidant, anti-inflammatory, antiviral, antimalarial, anti-tubercular, anti-aging, antifouling, and antiprotozoal [9–11]. For optimum exploitation of their full potential for human health and nutrition, there should be well-established cultivation (aquaculture) practices along with good extraction protocol to enrich and isolate the target bioactive compounds specifically. Also, the efficacy of the target molecule for pharmaceutical and industrial applications should be validated with a proper clinical assay, including a safety profile. Recent advancements in "omics" techniques (genomics, metagenomics, and proteomics), along with molecular biology approaches such as combinatorial biosynthesis, synthetic biology, selection methods, expression systems, and bioinformatics platform have also played a significant role and contributed substantially toward the discovery of new drug leads with pharmaceutical significance from seaweeds [12, 13]. In this book chapter, I have discussed the various applications of seaweeds in human health and nutrition (functional foods and nutraceuticals), industrial therapeutics and pharmaceutical application, cosmeceuticals, and therapy to lifestyle-related disorders. I have also highlighted various extraction techniques usually adopted for the isolation of seaweed compounds. Finally, the market potential of seaweeds is also briefly discussed.

19.2 Seaweeds as Nutraceuticals and Functional Foods

Functional foods are a new category of food products that offer health-promoting extra-nutritional benefits such as lowering the level of plasma cholesterol, antioxidant activities, body weight regulation, and improved gut health. Since seaweeds are heterogeneous in their composition, they have the potential to be used as functional foods. They have a unique profile of bioavailable carbohydrates, protein content, soluble dietary fiber, and higher levels of PUFA [14]. For example, phytosterol extracted from marine algae, especially from the brown algae, finds wider application as nutraceuticals since they are capable of lowering the cholesterol thereby reducing the risk of cardiovascular disease [15]. In addition, there are pieces of evidence which provide the experimental proof of the potential of seven phytosterols namely fucosterol, saringosterol, 24-hydroperoxy-24-vinyl-cholesterol, 29-hydroperoxy-stigmasta-5, 24(28)-dien-3β-ol, 24-methylene-cholesterol, 24-keto-cholesterol, and 5α, 8α-epidioxyergosta-6, 22-dien-3β-ol from *Sargassum fusiforme* being cholesterol-lowering agents in cell line study. The authors concluded that saringosterol was more potent than other phytosterols in stimulating LXRα which in turn lower the level of cholesterol [16]. In another study, it was reported that the lipid extracts of *Nostoc commune* Vaucher seaweed lowered the expression level of genes that are involved in cholesterol metabolism. The observed hypocholesterolemic effect was attributed to the campesterol, sitosterol, and clionasterol of this alga [17]. However, increased phytosterol consumption is reported to increase serum phytosterol levels. Thus, the consumption of phytosterol should be avoided by phytosterolemic patients [18]. The United States has authorized the use of oil, which is obtained from *Schizochytium* sp. as food ingredients since it is very high in docosahexaenoic acid (22:6 ω–3), including very high levels of squalene and phytosterols. It also has threefold less cholesterol than fish oil [14]. Although the lipid content in seaweeds is low, they have a unique fatty acid profile, which is different from terrestrial plants. The predominant fatty acid present in seaweeds is ω–3 fatty acid [19]. The fatty acids usually have a linear chain with an even number of carbon atoms and one or two double bonds. Red algae are rich in eicosapentaenoic acid (20:5 ω–3) arachidonic acid (20:4 ω–6), whereas brown alga such as Wakame has a higher level of palmitic acids (16:0) and oleic acids (18:1 ω–9). Green algae like *U. pertusa* have ample of hexadecatetraenoic acid (16:4 ω–3), and oleic and palmitic acids. Octadecatetraenoic acid (18:4 ω–3) is the predominant fatty acid in *Laminaria* sp. and *U. pinnatifida* seaweeds, whereas the hexadecatetraenoic acid (16:4 ω–3) is plentiful in *Ulva* sp. [20, 21]. Seaweeds are an excellent source of dietary fibers. The predominant soluble fibers from seaweeds include alginate, carrageenan, and agar. Besides, minor polysaccharides which feature fucoidans, xylans, and ulvans are also found [22]. Algal species, such as *Hypnea* spp., *Ulva lactuca*, possess more soluble fiber than terrestrial plants on a dry weight basis [23]. Alginate is one such polysaccharide which constitutes 14%–40% of the dry mass of *Laminaria* sp. There are several advantages associated with the consumption of alginate-rich algae. It helps in decreased cholesterol uptake, alters the colonial bacterial profile favorably, and absorbs toxin from the body [24]. Since alginates are excellent metal-chelating agents, they are considered valuable nutraceuticals as they can scavenge toxic and harmful elements in the human gut. In the scavenging process, they may also reduce the concentration of essential elements that are di- or polyvalent. Further, algal polysaccharides act as satiety inducer and therefore are used as weight control management agents [25]. Soluble fibers offer other health benefits such as improved gut health through enhanced water-binding capacity as well as optimum nutrient absorption by decreasing the digestive transit time [26]. Higher intake of dietary fiber is associated with a lower risk of lifestyle-related health complications such as cancers, diabetes, and heart disease [27]. Virtually, all seaweeds are an excellent source of macro elements such as Ca, Mg, Na, P, and K, as well as minor elements like Zn, I, and Mn, which are required for growth, metabolism, and development [26]. It is reported that phosphorus and calcium content of seaweeds surpasses that of potatoes, carrots, apples, and oranges. Green and brown algae are a rich source of vitamin C. It varies anything between 50 and 300 mg/100 g dry matter and comparable to that of parsley and peppers. The vitamin C content of red algae varies from 10 to 80 mg/100 g of dry matter. *Monostroma undulatum,* green algae, which are found in the coastal region of Southern Argentina coast, are very rich in vitamin C, which is 159–455 mg/100 g of dry matter [14]. Seaweeds are also considered as an alternative and viable protein source for humans. Spirulina is one such alga that is considered a superfood by the World Health Organization (WHO) and is consumed widely because it is a rich source of protein. It offers several health benefits, such as renal protection, anti-hyperlipidemia, anti-hypertension, and anti-hyperglycemic [28]. Moreover, it is also a rich source of hypocholesterolemic γ-linoleic acid (GLA), B-vitamins, and free-radical scavenging phycobiliproteins [29]. Since spirulina is nutrient-rich and contains 180% more calcium than milk, 670% more protein than tofu, 3100% more β-carotene than carrots, and 5100% more iron than spinach, it has been sent to international space stations by NASA [30]. As far as industrial applications concerned, lectins and phycobiliproteins are two bioactive proteins which have been utilized by industry. Many patents have also been filed regarding the health-enhancing bioactivities of phycobiliproteins as nutraceutical [31]. Some seaweeds (kelp, wakame nori, and Kombu) from japan is very high in iodine content. Thus, dietary supplements containing iodine from kelp seaweed is very popular [32].

19.3 Seaweeds in the Alleviation of Lifestyle Disorders

There is literature that evidences the usage of fucoxanthin as potential food supplements for managing obesity and diabetes in *in vitro;* studies as well as in animal models. For instance, fucoxanthin is reported to inhibit α-amylase and α-glucosidase digestive enzymes. As a result of this, lipid metabolism is modulated favorably by the downregulation of lipid synthesis and upregulation of lipid hydrolysis [33]. In another trend, fucoxanthin supplementation in animal study was correlated with a reduction of blood glucose and plasma level as well as improvement of plasma lipid profile, which eventually reduce the risk of insulin resistance [34]. Fucoxanthin was also found to have an antiobesogenic effect by reducing visceral fat and BMI in human clinical trials spanning over four weeks [35]. Consumption of seaweeds and seaweed isolates rich in carotenoids and alginates was found to exhibit a positive effect on satiety, appetite, blood glucose, and cholesterol level [36]. Supplementation of diets with *Undaria pinnatifida* and *Sacchariza polyschides* seaweeds in the human clinical trial had favorable glucose levels, reduced level of serum triglyceride, and increased concentration of high-density lipoprotein in type 2 diabetes persons [37]. Phlorotannins, fucoxanthin, polyphenolics, and polysaccharides have been identified as the main seaweed components behind the mitigation of diabetes and its related health complications. There are several pharmaceuticals such as Captopril, Eplerenone, and the angiotensin-I converting enzyme (ACE) inhibitor available in the market for the treatment of high blood pressure and heart disease. However, these commercial drugs have some reported side effects, such as impaired renal function, dry cough, and extremely low blood pressure [38]. Thus, a seaweed-derived functional molecule can be an alternative to these commercial drugs. In epidemiological studies, it was found that an inverse relationship exists between daily consumption of seaweeds and lower risk of hypertension and cardiovascular disease [39]. Seaweed-derived bioactive peptides have the potential to bind to the active site of the ACE and mitigate the high blood pressure. For instance, in an *in vitro;* study, *Gracilariopsis lemaneiformis* peptides had potent ACE inhibitory activity [40].

19.4 Anti-Inflammatory Activity of Seaweeds

There are reports of seaweeds (marine algae) harboring compounds which are potent anti-inflammatory. For example, methanolic extracts from the green seaweed *Ulva conglobata* has been recognized as an anti-inflammatory in the cell (neuronal HT22 cells and microglial BV2) line study [41]. Among the many algal bioactive compounds, carbohydrates are reported to possess anti-inflammatory properties. For example, oligosaccharides derived from alginate are shown to inhibit neuroinflammation. Similarly, laminarin, a polysaccharide found in *Laminaria* species, was found to regulate neuroinflammation. A sulfated polysaccharide fraction derived from the red seaweed *Gracilaria cornea* was demonstrated to possess anti-inflammatory attributes and modulate the acute inflammatory process by histamine inhibition, vascular permeability, and neutrophil migration [42]. In addition to polysaccharides, lipids derived from algae also have potent anti-inflammatory properties. PUFA, from microalgae, has multiple health benefits and reduces the risk of diseases such as diabetes, arthritis, and obesity, mostly in the context of neuroinflammation. Sterols (cholesterol in red algae, fucosterol in brown algae and isofucosterol, and clionasterol in green algae) are another class of lipids that are proposed to alleviate neuroinflammation, since they can cross the blood–brain barrier. However, there is limited literature regarding the neuroprotective ability of algal sterols [43]. Lectins represent a group of glycoproteins. They are functional biopeptides which are found in marine algae and possess anti-inflammatory activity. Further, phenolic and polyphenolic compounds such as phlorotannins from marine algae have also gained much attention for their anti-inflammatory actions [44]. Carotenoids and terpenoids are an important class of isoprenoids that belong to marine algae. Fucoxanthin is present in brown algae, β-carotene is predominantly present in green microalgae. Carotenoids, in particular, fucoxanthin, has exhibited anti-inflammation and anti-oxidative damage activity [41]. Brown seaweed, such as *Sargassum*, is reported to have anti-inflammatory activity. Supplementation of diets with whole seaweed powder (*Sargassum hystrix*) reported improving stress-induced liver inflammation conditions in Wistar rats [45].

19.5 Seaweed Is a Source of Anticoagulant Agent

In the biomedical industry, heparin is the only predominant drug which is being used extensively in the treatment of thromboembolic disorders. However, there are some long-term side effects such as thrombocytopenia, a hemorrhagic effect associated with heparin.

This necessitates a hunt for an alternative source for the antithrombotic agent [46]. There are reports of seaweeds polysaccharides being recognized for anticoagulant activity as their usage does not involve any potential contamination from harmful prions or viruses associated with commercial heparins [47]. Seaweeds-derived polysaccharides do not pose any ill effects toward cellular metabolism, and drugs derived from algal extracts are very much affordable. From a pharmacological point of view, the mode of actions of these algal polysaccharides is dependent on the extent and position of the sulfate group present in the polysaccharides, including their molecular weight. This property of algal polysaccharides indicates that the length, shape, and density of negative charge of the molecule are the prerequisite for the desirable anticoagulant activity. Literature evidence that phlorotannins and fucoidans, sulfated polysaccharides from brown algae, have been identified as anticoagulant agents [48]. Further, carrageenans from red algae and ulvans in green algae also have an anticoagulant property that is proven in *in vitro;* model. Faggio et al. [47] evaluated the anticoagulant activity of sulfated polysaccharides from *Ulva fasciata* (Chlorophyta) and *Agardhiella subulata* (Rhodophyta) on human blood. Both the agents were shown to be effective in prolonging coagulation time in PT and APTT assay [47]. In another study, two sulfated polysaccharides, namely MP and SP, were extracted from *Enteromorpha linza*, purified by ion-exchange and size exclusion chromatography, and characterized by fourier-transform infrared analysis. These two molecules demonstrated potent anticoagulant activity in *in vitro;* system [49]. However, for the exploitation of the sulfated polysaccharides, their efficacy, as well as safety profile, should be evaluated extensively and exclusively in *in vivo;* model.

19.6 Anticancer Property of Seaweed

Cancer is one of the major causes of death around the world, and researchers are continuously striving toward developing a new strategy or drugs to manage this pandemic in a better way. Epidemiological studies support the health benefit of consumption of plant and algal-derived foods with the reduction of risk of cancer. There are reports of certain algal compounds being useful in the treatment of cancer because of the antioxidant property of their compounds [50]. *Bryopsis* sp. is a green marine alga which holds a promising future in the anticancer treatment since it contains compounds such as depsipeptides kahalalide A and F that are reported to be potent drugs for tumors, AIDS, and lung cancers [51]. However, kahalalide A could not advance beyond the phase II trial because of its short shelf life and lack of efficacy. Besides, it did not exhibit the expected response in patients. Anyways, it is high cytotoxic agents that pave the way for the development of many synthetic analogs of kahalalide A [52]. Bis-indolic amides such as chondriamide A and B are derived from the red alga *Chondria* sp., and they have cytotoxicity. These two are potent anticancer drug candidates/leads as they prevent the proliferation of human nasopharyngeal and colorectal cancer cells [53]. Algal polysaccharides such as fucoidans, laminarans, alginic acids, carrageenans, along with other modified polysaccharides have also attracted attention over the year for their capacity to ameliorate various forms of human cancer [54]. In addition to polysaccharide, terpene is another group of bioactive compounds, in particular from *Chlorella sorokiniana* and *Chaetoceros calcitrans* extracts, which have an encouraging anticancer profile in comparison with commercially available drugs [55]. The anticancer properties of different terpene compounds such as ursane-type triterpenoids, lupane-type triterpenoids as well as known diterpenoids extracted from *Acanthopanax trifoliatus* were evaluated in the cell model. These compounds exhibited strong to moderate anticancer activity against SF-268, MCF-7, HepG2, and NCIH460 cancer cells [56]. Further, brominated cyclic diterpenes from *Sphaerococcus coronopifolius*, a red alga, (*Rhodophyta Phylum)* has antitumor potential and cytotoxic activity against malignant cell lines which are understudied, since the mechanism behind the observed effects are still unknown [52]. Liu et al. [57] reported that bis (2, 3-dibromo-4, 5-dihydroxybenzyl) ether, a bromophenol compound from algae, is capable of inducing apoptosis of K562 cells by arresting cell cycle at S phase[57]. The compounds such as lauren diterpenol, thyrsiferol, and caulerpin have been reported to be involved in antitumor activity since they inhibit the transcription factor HIF-1 [58]. Lee et al. [59] reported that phlorofucofuroeckol-A, which has been isolated from brown alga *Eisenia bicyclis*, to be a potent antitumor compound as it inhibits Aldo-keto reductase family 1 B10 AKR1B10, which is a therapeutic cancer target [59]. In another instance, the compound sulfoquinovosyl diacylglycerol (SQDG), from red algae, exhibited remarkable inhibition of telomerase enzyme, which is involved in the uncontrolled division of malignant cells [60]. In a recent study, ethanolic extracts of *Chaetomorpha* sp., green algae, were used for the extraction of a novel anticancer molecule and tested in cell culture (MDA-MB-231 breast cancer cell lines). It was found that the presence of anticancer property of this alga was primarily due to the antitumor chemicals like terpinol, oximes, and dichloracetic acid (DCA). In addition, these compounds also contain calcium, silicon, along with other essential metals which make these molecules a novel target for anticancer drug as well as a nutritional supplement [61]. Brentuximab vedotin (D) (Adcetris®) is a drug of cyanobacteria origin which has got Food and Drug Administration (FDA) approval for the treatment of Hodgkin's lymphoma. There are other drug candidates, mostly algal secondary metabolites, which are in phases of the clinical trial or in the pipeline to be approved by the FDA [52]. Sometimes, structural characteristics are one of the parameters which have to be considered when developing a therapeutic from marine source, since in many cases the putative drug from different sources exhibits varied response *in vivo;* [52]. For example, fucoidans showed promising activity on clinical trials but exhibited varied responses when administered intraperitoneal, oral, or intravenous [62]. Algal-derived compounds are often administered along with therapeutic molecules as co-adjuvant to improve the efficacy of the drugs. For example, when $HepG_2$ cells were pre-treated with fucoxanthin, a well-known pigment from brown algae, the efficiency of cisplastin (well-known therapeutic used for chemotherapy) was improved [63]. In another study, λ-carrageenan was conjugated with fluorouracil (5-FU), marketed as adrucil. As a result of this, the antitumor activity was enhanced as well as the by-standing effect of immunocompetence damage caused by 5-FU was fine-tuned or mitigated [64]. Most of the identified compounds, such as polysaccharides, polyphenols, pigments, alkaloids, and terpenes, exert their inhibitory effects by mediating specific cellular processes [52]. For example, laminarin, a storage polysaccharide from brown algae, is reported to induce apoptosis on HT-29 colon cancer cells by arresting the cells in G_2/M phase of the cell cycle [65].

19.7 Seaweeds as Antiviral Drugs and Mosquitocides

Seaweed-derived compound has also been used for the treatment of various vector-borne diseases as well as mosquitocides. Among the mosquito-borne viral disease, dengue is one of the diseases which outbreaks each year, in particular in India, and kills many people. Female mosquitos that belong to *Aedes aegypti* and *Aedes albopictus* sp. are the common carrier of this viral fever. In addition, it also transmits related viruses like Zika and Chikungunya [66]. The multiplication of the chikungunya virus starts with its attachment to the host cell surface. Thus, inhibiting the binding of the virus will be a vital strategy to manage the virus. Since the seaweeds polysaccharides can modify their cell surface properties, the use of algal polysaccharides is touted as an effective approach to prevent this deadly viral disease. To this end, sulfated polysaccharides are being exploited for the development of possible antiviral agents. For example, ulvan is one such compound that contains 11% of sulfate and 6% of uronic acids in its structure. It is extracted from *Caulerpa cupressoidesis, a* green alga found in Brazil. Its antiviral potential was tested against DENV-1 in cell line study by Rodrigues et al. [67]. Results are very promising with a satisfactory selective index (>714) and no cytotoxicity (CC50 > 1000 $\mu g\,ml^{-1}$). Further, the analysis of sulfated polysaccharides by infrared revealed a distinct pattern of molecular weight (8–100 kDa). In support of their observed results, the authors pointed out that sulfate residue on C_6 galactose residues was critical for the observed inhibitory effect. Further, they mentioned the extent of sulfation (at least by 1.8 margins than the content of uronic acid) as well as they (<100 kDa) also have a secondary role in the antiviral mechanism of ulvan [67]. Fucoidan is another sulfated polysaccharide extracted from the brown algae *Cladosiphon okamuranus*. It has sulfated fucose and glucuronic acid in a typical ratio of 6.1:1.0:2.9 (fucose, glucuronic acid, and sulfate, respectively) [68]. This compound inhibited the activity of DENV-2 virus by 20% in cell line study (BHK-21) with $LC_{50}$of 4.7 $\mu g\,ml^{-1}$. This compound also exhibited antiviral activity in a dose-dependent manner [69]. When desulfated, the efficiency of inhibition was reduced markedly (reduced to 1%), whereas fucoidan derivative wherein glucuronic acid was replaced by glucose had no inhibition on DENV-2 virus. It was revealed that both glucuronic acid and sulfated fucose residues are crucial for the interaction with the viral glycoprotein present in the envelope of DENV-2. Among the different serotypes (DENV1-, DENV-2, DENV3, and DENV4) of the dengue virus, DENV-2 was found to be more vulnerable to fucoidan. DENV-2 and DENV3 had moderate susceptibility, whereas the DENV1 strain showed no vulnerability to fucoidan. To explain this phenomenon, the authors analyzed the nucleotide sequence of the four serotypes. It was revealed that amino acid residue located at 295 and 310 in their protein sequence plays a vital role in the recognition of sulfated glycosaminoglycans of fucoidan. Later, the antiviral activity of carrageenan and agaran (DL galactans) from *Gymnogongrus torulosus* (Rhodophyta) against DENV-2 was demonstrated in *in vitro;* cell line study. These novel series of DL galactans showed remarkable inhibitory activity with no cytotoxicity effects [70]. There are also a few studies reported regarding the inhibitory effects of seaweeds extracts against ZIKV virus [71, 72]. For instance, extracts from marine algae such as *Osmundaria obtusiloba, Kappaphycus alvarezii,* and *Caulerpa racemosa* from Brazilian coast were successfully tested against ZIKV. Inhibition of ZIKV replication was more than 90% in a dose-dependent with LC_{50} value ranging from 1.38, 1.98, and 1.82 $\mu g\,ml$. Among the three algae, *Caulerpa racemosa* was found to be more potent than other algae [72].

SQDG, (1, 2-di-*O*-acyl-3-*O*-(6-deoxy-6-sulfo-α-D-glucopyranosyl)-*sn*-glycerol), a sulfolipid, also exhibited antiviral activity against HSV-1 virus [71]. The mosquitocidal activity of seaweed-derived compounds has also been shown against mosquito, the carrier of most viral disease. Yu et al. [73] reported that halogenated sesquiterpene (−)-elatol derived from *Laurencia dendroidea*, a red algae typically found in Brazil, exhibited potent larvicidal effects when tested against *Aedes aegypti*. The mortality rate was demonstrated to be very high (>91% at 50 ppm). LC_{50} value was calculated to be 10.7 ppm, which was promising [73]. In the same study, certain fatty acids such as palmitoleic, myristic, lauric, and capric acids isolated from *Cladophora glomerata* were evaluated against *Aedes triseriatus* with LC_{50}value varying from 3 to 14 ppm. In another study, Salvador-Neto et al. [74] shown the herbicidal effect of halogenated sesquiterpene, (+)-obtusol, and elatol derived from *Laurencia dendroidea*. The mortality rate of elatol compound was found to be 30% at 10 ppm against *A. aegypti* within 24 hour, whereas the same concentration (10 ppm) of obtusol exhibited more potent larvicidal activity (with a mortality rate of 90%). It was also revealed that the obtusol exhibited dose-dependent larvicidal activity with LC_{50} of 3.5 ppm. On further probing, it was found that larvae that were subjected to larvicide had their intestinal epithelium damaged. In their chemical structure, these two compounds differ by one double bond and by an additional bromine atom [74]. This might explain the differences in their activity profile.

19.8 Use of Seaweeds in the Cosmeceutical Industry

Recently, microalgae occupied a permanent place in the cosmeceutical industry and consolidated its presence in the markets. Cosmeceuticals are basically sold out as cosmetics that contain biologically active ingredients. They are essential ingredients in a wide variety of skincare formulation, mainly topical applications, and their intended use is limited to as a beautifying agent for the enhancement of skin tone and color. For example, Arthrospira- and Chlorella-derived extracts containing bioactive compounds are incorporated in face and skincare products [75]. There are reports of anti-aging, skin-whitening, and anti-pigmentation agents that are derived from marine algae [76]. Laminaria, Fucus, and Chondrus are the algae that are predominantly being exploited in the cosmetics thanks to their ability to rehydrate and nourish the skin [77]. Laminaran, a polysaccharide extracted from brown algae *Laminaria*, is used as anticellulite cosmetics products due to its broad range of bioactive properties [78]. Cellulite is basically not a health issue but a cosmetic issue and can be alleviated by routine skincare regiments to improve the visual appearance of skin. Fucoidan has found application in the anti-aging formulation as it enhances hydration and elasticity of cells by stimulating the production of heparin-growth factor (HGF), which in turn helps in the growth of cells and tissues [79]. Sunscreen and skin-whitening products is one segment that is steadily expanding because it plays a protective role from the sunlight, sunburn, tanned skin, and pigmentation. Whitonyl®, a skin-whitening product, commercialized by Silab is oligosaccharides derived from *Palmaria palmata* (red alga, Rhodophyta). It alleviates the symptoms, such as deep wrinkles, loss of elasticity, sun freckles, or brown spots, which are caused due to chronic exposure to the sun. These symptoms are localized in the face, arms, and hands in women who are exposed to sun and are accentuated with age and with the extension of exposure [80]. Among polysaccharides, alginic acid is the predominant polysaccharide in several species of brown algae. Alginic acid is reported to inhibit scar formation and helps in the wound-healing process. Alginate, along with collagen, is used in the clinical industry for repairing tissues [81]. One of the important characteristics of alginate is the hydrogel formation capacity in the presence of several cations. Thus, alginates are produced by the conversion of insoluble alginic acid into soluble alginates in conjugation with salts of sodium or potassium [82]. At low pH, they form gel-like structures. Alginates are used in a wide variety of gelling agents in drugs and cosmeceuticals as emulsion stabilizers, protective colloids, as thickeners. They are also used as a lotion, hand jellies, ointment bases, pomades, and hair products. They are also used as skin products such as facial cream or beauty masks [83]. Carrageenans are another class of polysaccharides that are found in carrageenophytes (*Betaphycus gelatinum, Chondrus crispus, Eucheuma denticulatum, Gigartina skottsbergii, Kappaphycus alvarezii, Hypnea musciformis, Mastocarpus stellatus, Mazzaella laminaroides,* and *Sarcothalia crispata*), red algae. There are three major types of carrageenans, namely kappa (κ), iota (ι), and lambda (λ). Iota and kappa carrageenans possess gelling property, whereas lambda carrageenan is preferred as a thickening/viscosifier [84]. Carrageenans find application in a wide array of personal hygiene and grooming products such as toothpaste, hair conditioners, lotions, medicines, shampoos, deodorants, sunray filters, shaving creams, foams, and sprays. As much as 20% of carrageenans are exploited in cosmetology and pharmacy [82]. There is also evidence of seaweeds extract being useful in maintaining a slim figure through gene regulation and protein expression [85].

19.9 Use of Seaweed as Contraceptive Agents

Three different varieties of red algae from coastal waters in Srilanka were evaluated for post-coital contraceptive agents. Methanol and methylene chloride (1:1) extracts of two varieties namely *Gelidiella acerosa* and *Gracilaria corticata* showed potent post-coital contraceptive activity in female rats without showing any by-standing effects [86]. In another study, Bhakuni et al. [87] demonstrated the anti-implantation activity of ethanol extracts of *G. edulis* and *G. corticata* algae in the mouse model [87]. There is also a report of 100% inhibition of sperm motility by ethanolic extracts from *G. edulis,* which disrupts plasma membrane in a similar fashion as spermicidal compounds [88]. *Halimeda gracilis,* Indian seaweed, exhibited 100% inhibition of human spermatozoa at a dose of $10\,mg\,ml^{-1}$ in just 20 seconds. EC_{50}value was calculated to be $2.05\,mg\,ml^{-1}$ in 20 seconds. It was revealed that the plasma membrane of sperm was damaged due to the exposure to extracts of *Halimeda gracilis.* Phytochemical analysis of the extracts of this seaweed showed the presence of secondary metabolites such as alkaloids and flavonoids and protein and sugar [89].

During the course of evolution, marine organisms acquired exceptional metabolic capacity by virtue of their production of an array of secondary metabolites, which are quite specific and potent in their actions. These secondary metabolites are produced by marine organisms in defense of predation, in competition for space and food as well as to maintain an ecological relationship in the marine ecosystem [90]. Here, in the table mentioned below (Table 19.1), I have highlighted the different bioactive molecules present in the seaweeds along with their biological actions and applications.

Table 19.1 Bioactive compounds from various seaweeds and their potential actions/applications.

Species	Extracts/compound	Application	References
Rhodophyta (red algae)			
Gelidiella acerosa, Gracilaria corticata	Methanolic extracts	Contraceptive agents	[86]
G. edulis, G. edulis	Ethanolic extracts	Male contraceptive	[88]
Sphaerococcus coronopifolius	Brominated cyclic diterpenes	Antitumor, anti-bacterial	[52, 91]
Porphyra yezoensis	SQDG	Antitumor	[60]
Laurencia dendroidea	Halogenated sesquiterpene	Lavicidal	[73]
Chondrus crispus, Eucheuma denticulatum, Kappaphycus alvarezii	Carrageenan, source of zinc	Hydrogel, cosmoceuticals, lubricant, anticoagulant, nutraceuticals	[84, 92, 93]
Agardhiella subulata	Sulfated polysaccharide	Anticoagulant	[47]
Palmaria palmata	Extracts, source of iron, and iodine	Skin-lightening and skin-whitening agent, anti-pigmentation activity, health supplements (nutraceuticals)	[82, 94]
Gelidium sp., *Gracilaria* sp.	Agar	Adhesives, suppositories, capsules, textile printing/dyeing	[94]
Porphyra sp.	Porphyrans	Hypolipidemic, anticancer, anti-inflammatory	[95]
Chlorophyta (green algae)			
Ulva conglobata and *Ulva reticulata, Ulva fasciata*	Sulfated polysaccharide	Anticoagulant	[47, 96, 97]
Ulva conglobata	Methanolic extracts	Anti-inflamatory	[41]
Bryopsis sp.	Depsipeptides kahalalide A and F	Anticancer	[51]
Chaetomorpha sp.	Terpinol, oximes and DCA	Anticancer	[61]
Caulerpa cupressoidesis	Ulvan	Anti-viral	[67]
Dunaliella, Chlorella, Chlamydomonas, Arthrospira, Sargassum, Spirulina, Gracilaria, Prymnesium parvum, Euglena gracilis and *Scenedesmus*	Bio-ethanol, β-carotene, biogas	Biofuel/biodiesel, dye, vitamin C supplement	[98, 99]
Haematococcus pluvialis,	Astaxanthin	Antioxidant	[100]

(*Continued*)

Table 19.1 (Continued)

Species	Extracts/compound	Application	References
Ochrophyta (brown algae)			
Myagropsis myagroides	Fucoxanthin	Anti-inflamatory, antioxidative	[101]
Laminaria sp. (*Laminaria digitata, Laminaria hyperborean)*	Laminaran	Cosmetics, anticellulite	[78]
Fucus vesiculosus, Eisenia bicyclis, Ecklonia cava	Phlorotanins, Dieckol, Eckol, Fucosterol, and 8,8′-bieckol	Cosmetics, Alzheimer's disease, MMP inhibition, antidiabetic	[102, 103, 104, 105]
Brown algae sp.	Alginates	Tissue engineering	[65]
Ectocarpus siliculosus	Fucostatin	Anticancer	[106]
Ecklonia cava, Eisenia arborea, Ecklonia stolinifera, Eisenia bicyclis	Polyphenol	Antioxidant, anti-inflammatory, antidiabetic, antitumor, antihypertensive, anti-allergic	[107]

19.10 Extraction of Active Ingredients from Seaweed

The selection of the extraction method is the most critical step, which ultimately influences the extraction yield of the target molecule. There is no single and well-established protocol that ensures maximum recovery and compatible with most of the molecule with industrial importance. Seaweeds extraction usually involves many steps, and during the process of isolation, the seaweeds are often subjected to treatments such as exposure to solvents, high temperature, extreme pH, and extended period of extraction. For each target compound, the extraction method should be optimized and scaled up to a commercial scale. The end products should be enriched in the active ingredients and devoids of any impurities to maximize its efficiency. Here, we have discussed the different extraction methods usually adopted for the extraction of bioactives from seaweeds.

19.10.1 Supercritical Fluid Extraction (SFE)

The supercritical fluid extraction (SFE) is an extraction technique which is adopted to extract valuable component (target molecule) from a sample matrix using a suitable solvent. Although a solid sample is the preferred matrix for this extraction process, liquid samples are also amenable to this process. The solvent which acts as extractant has a major role for the selective extraction of the molecule of interest. For example, phlorotannins such as phlorethols, fucols, or fucophlorethols are selectively extracted from *Cystoseira abies-marina* algae by pure ethanol at 100 °C in subcritical state. Their chemical characterization was confirmed by two-dimensional liquid chromatography (LC × LC-MS/MS) method [108]. Polyphenols of *Cystoseira tamariscifolia* extracted with 100% methanol and chloroform exhibited more cytotoxic activity than water extract in the cell (leukemia, HL-60, PC3, and THP-1 cells) line study with methanol extracts being seventeen times more potent [109]. Nowadays, green solvents are preferred over organic solvents for safe and environmentally friendly extraction of bioactive molecules from seaweeds. For example, 2-methyltetrahydrofuran (MTHF) is one such green solvent that has been utilized for the successful extraction of carotenoids from *Chlorella vulgaris* [110]. It is obtained from renewable and biodegradable sources (lignocellulosic biomass), which can be recycled. It was also reported that ethanol and MTHF at 1:1 ratio and at 110 °C for 30 minutes yielded satisfactory carotenoids and can be an alternative to *n*-hexane. There are pieces of evidence that suggest that a pretreatment method in the form of drying the raw material is often helpful to recover maximum target compounds since it will enhance the direct contact between the sample and the solvent. Moreover, the drying mode also influences the recovery of carotenoids. Maximum carotenoids were obtained by SFE method in the optimized condition at 60 °C and 20–40 MPa with 23% water content of raw material from *Dunaliella salina* [111]. More sophisticated techniques which allow integration of extraction, identification as well as characterization of compounds into a single platform are also emerging. These platforms not only simplify the process of bioprospecting but also avoid possible damage to the target molecule by minimizing the entire extraction process. For instance, Abrahamsson et al. [112] developed an SFE-UV/Vis-ELSD equipment/platform. They demonstrated its efficiency in the detection of ergosterol, chlorophyll A, carotenoids, and total lipids from the algal extract obtained by SFE [112].

19.10.2 Ultrasound-Assisted Extraction (UAE)

Ultrasound-assisted extraction (UAE) of marine bioactive is another strategy that offers several advantages such as simplicity, lower solvent consumption, and less extraction time. Further, it can be operated at lower temperatures, which avoids the degradation of heat-sensitive compounds. Moreover, it is affordable techniques that can be easily adapted to any industrial scale [113]. Polysaccharides extracted from brown *Ascophyllum nodosum* and *Laminaria hyperborea* algae by UAE method had better Laminaran yield than conventional liquid–solid extraction and better antioxidant and antimicrobial profile [114]. UAE has been exploited for the isolation of pigments such as fucoxanthin from brown marine algae. Recovery of fucoxanthin to the tune of to 0.197 g/100 dry sample was achieved using 70% ethanol with high antioxidant activity [115]. UAE has also been utilized for the isolation of protein from seaweeds. For instance, the extraction of protein from *Palmaria palmate* was achieved by applying ultrasounds. Further, the crude protein isolates were treated with papain to obtain hydrolyzed bioactive peptides. The recovered peptides had anti-atherosclerosis activity and reduced blood pressure [116]. This report evidenced that the combination of ultrasounds with proteolytic enzymes fine-tuned the overall recovery of bioactive peptides, which ultimately is reflected in their activity profile.

19.10.3 Microwave-Assisted Extraction (MAE)

Microwave-assisted extraction (MAE) is an extraction technique that relies on microwave energy to heat the solvents and increase the mass transfer of the solute embedded in the sample into the solvents. The advantages of this method over soxhlet are manifold since it drastically reduces the extraction time to 30–60 minutes. It is applied in the extraction of organic compounds from plant and marine sources with better yield. However, the efficiency of MAE can be optimized by manipulating several factors, such as the selection of solvent, microwave power, temperature, and the solvent-to-solid ratio [117]. For example, microwave power of 800 W and temperature of 40 °C for a period of 60 minutes was found to be critical for optimal recovery (2.2 and 37.7 μg ml^{-1}, respectively) of carotenoids and chlorophylls from *Ulva flexuosa* [118]. The MAE has been successfully applied for obtaining phenolics from brown seaweed *Lessonia trabeculate*. With an extraction time of 15 minutes, a satisfactory yield of polyphenols to the tune of 74.13 GAE mg/100 g dry seaweed was achieved compared with the conventional extraction method, which yields 49.80 GAE mg/100 g over a period of four hours. The extracts also had inhibition activity on pancreatic lipase, α-glucosidase, α amylase, and tyrosinase [119]. In addition, MAE of fucoxanthin from *Undaria pinnatifida* was performed at 60 °C with a solid-to solvent ratio of 1:15 (g ml^{-1}) for 10 minutes. With a microwave power of 300 W, an optimal amount of fucoxanthin (109.3 mg/100 g dry weight) was achieved [120]. However, it is not compatible with the thermostable compound and sometimes cannot sustain the accuracy and activity of the target molecule.

19.10.4 Enzyme-Assisted Extraction (EAE) and EMEA

Enzyme-assisted extraction (EAE) of bioactive compounds from marine algae is another promising strategy, since hydrolytic actions of enzymes render the cell disrupted and allow enrichment of the bioactive compounds in the fractionations. Several brown algae species such as *Sargassum angustifolium*, *Sargassum boveanum*, *P. gymnospora*, *C. cervicornis*, *Colpomenia sinuosa*, *Feldmannia irregularis*, and *Iyengaria stellate* were subjected to carbohydrase and proteases treatment for recovery of polyphenols, polysaccharides, and proteins [121]. In another study, the extraction of ulvans from *Ulva armoricana* was enhanced upon treatment with endo-protease. Further, the antiviral activity of the extracted ulvans was established in this study [122]. It can be pointed out that the EAE is the method of choice for the extraction of food grade and food compatible compounds for the purpose of development of functional foods or nutraceuticals. It will increase the recovery rate and enhance the biological attributes of the novel compounds. For example, EAE of *Sargassum muticum* exhibited a better prebiotic and antioxidant potential than fructooligosaccharides. In a similar way, EAE of *Osmundea pinnatifida* showed more radical scavenging activity, and *Codium tomentosum* extracts exhibited potent α-glucosidase inhibition, which is a great attribute for an antidiabetic compound [123]. For better results, enzymatic extraction is coupled with MAE. Charoensiddhi et al. [124] applied enzymatic and microwave-assisted enzymatic extraction (MAEE) for higher recovery of phlorotannins as well as antioxidant compounds from *Ecklonia radiata*. They reported that a short period of extraction of 30 minutes with MAEE yielded more phlorotanins than the enzymatic treatment alone at 24 hours. MAEE enables high-performance recovery of the bioactive compound since the microwave radiation, as well as hydrolytic actions of enzymes, act in synergy to disintegrate the cell wall structure, which makes way for the release of the target compounds [124].

19.11 Market Potential of Seaweeds

According to the recent report released by Grand View Research, Inc., the global market volume of commercial seaweeds is predicted to reach USD 11.9 billion by 2027 and will grow at a compound annual growth rate (CAGR) of 9.1%. In the Asian subcontinent, the market is expected to witness a substantial growth at CAGR of 9.3% from 2020 to 2027, whereas, in North America, the same market is estimated to witness a CAGR of 8.8% over the same forecast period [125]. The main driving force that fuels this persistent growth of global seaweeds market comprises of various factors such as diversified applications, acknowledgment of the health benefits of seaweeds, the rapid increase in awareness among the consumers, and aggressive promotions by manufacturers as well as increasing penetration levels in both established and emerging markets. The leading players in the seaweeds market are always trying to expand their business either by collaborating with or acquiring small-scale business partners in order to increase their product portfolio. Some of the giant companies in the seaweeds market are Cargill, Inc., Roullier Group, E.I. DuPont Nemours and Company, Biostadt India Ltd., and Compo GmbH and Co. Moreover, the environment-friendly regulation process is expected to further augment the expansion of this market. The market segment includes upstream producers, vendors, end users, several stakeholders, and government organizations [125]. The market for the product derived from seaweeds is segmented into green, red, and brown seaweeds. Among these, red and brown seaweed are largely exploited in the food and pharma industry. These two industries are predicted to capture more than 70% market size of commercial seaweed market by 2025 [126]. Based on formulation, the commercial seaweed market is bifurcated into liquid, powdered, and flakes. For example, the seaweed hydrocolloid industry, which is segmented into agar, alginate, and carrageenan, is steadily growing at the rate of 2–3% per year. The Asia Pacific region is the predominant market known for the raw material and manufacturing process of this particular industry [127].

19.12 Conclusion

Seaweeds are boon to humankind as the potential benefits of seaweed consumption are multilayered. Seaweeds offer an economical and sustainable source of health-enhancing functional compounds in human nutrition and commercial value-added products. They are also an incredible source of active ingredients and are exploited to full potential in nutraceuticals, pharmaceuticals, cosmeceuticals, consumer products, and industrial therapeutics. Seaweeds are also a suitable and viable alternative source of drugs for several life-threatening diseases (cancer and viral) and lifestyle-related diseases (cardiovascular, diabetes, and hypertension).

References

1 Kolanjinath, K., Ganesh, P., and Sanraj, P. (2014). Pharmacological importance of seaweeds: a review. *World Journal of Fish and Marine Sciences* 6: 1–15.

2 Kılınç, B., Cirik, S., Turan, G. et al. (2013). Seaweeds for food and industrial applications. In: *Food Industry* (eds. B. Kılınç, S. Cirik, G. Turan, et al.), 735–748. InTech publishers.

3 Anantharaman, P. (2002). Manual on identification of seaweed. All India coordinate project on survey and inventorization of coastal and marine biodiversity. *Journal of the Marine Biological Association of India* 29: 1–9.

4 Hasselström, L., Visch, W., Gröndahl, F. et al. (2018). The impact of seaweed cultivation on ecosystem services – a case study from the west coast of Sweden. *Marine Pollution Bulletin* 133: 53–64.

5 FAO (2020). *The State of World Fisheries and Aquaculture*. Rome: FAO.
6 Bouga, M. and Combet, E. (2015). Emergence of seaweed and seaweed-containing foods in the UK: focus on labeling, iodine content, toxicity and nutrition. *Food* 4: 240–253.
7 Collins, K.G., Fitzgerald, G.F., Stanton, C., and Ross, R.P. (2016). Looking beyond the terrestrial: the potential of seaweed derived bioactives to treat non-communicable diseases. *Marine Drugs* 14: 60.
8 Cherry, P., O'Hara, C., Magee, P.J. et al. (2019). Risks and benefits of consuming edible seaweeds. *Nutrition Reviews* 77: 307–329.
9 Blunt, J.W., Copp, B.R., Keyzers, R.A. et al. (2013). Marine natural products. *Natural Product Reports* 30: 237–323.
10 Mayer, A., Rodríguez, A., Taglialatela-Scafati, O., and Fusetani, N. (2013). Marine pharmacology in 2009–2011: marine compounds with antibacterial, antidiabetic, antifungal, anti-inflammatory, antiprotozoal, antituberculosis, and antiviral activities; affecting the immune and nervous systems, and other miscellaneous mechanisms of action. *Marine Drugs* 11: 2510–2573.
11 Agrawal, S., Adholeya, A., Barrow, C.J., and Deshmukh, S.K. (2018). Marine fungi: an untapped bioresource for future cosmeceuticals. *Phytochemistry Letters* 23: 15–20.
12 Molinski, T.F., Dalisay, D.S., Lievens, S.L., and Saludes, J.P. (2009). Drug development from marine natural products. *Nature Reviews Drug Discovery* 8: 69–85.
13 Bucar, F., Wube, A., and Schmid, M. (2013). Natural product isolation – how to get from biological material to pure compounds. *Natural Product Reports* 30: 525–545.
14 Bocanegra, A., Bastida, S., Benedí, J. et al. (2009). Characteristics and nutritional and cardiovascular-health properties of seaweeds. *Journal of Medicinal Food* 12: 236–258.
15 Ibañez, E., Herrero, M., Mendiola, J.A., and Castro-Puyana, M. (2012). Extraction and characterization of bioactive compounds with health benefits from marine resources: macro and micro algae, cyanobacteria, andinvertebrates. In: *Marine Bioactive Compounds* (ed. M. Hayes), 55–98. Boston, MA: Springer.
16 Chen, Z., Liu, J., Fu, Z. et al. (2014). 24(*S*)-saringosterol from edible marine seaweed *Sargassum fusiforme* is a novel selective LXRβ agonist. *Journal of Agricultural and Food Chemistry* 62: 6130–6137.
17 Rasmussen, H.E., Blobaum, K.R., Park, Y.K. et al. (2008). Lipid extract of *Nostoc commune* var. sphaeroides Kützing, a blue-green alga, inhibits the activation of sterol regulatory element binding proteins in HepG2 cells. *The Journal of Nutrition* 138: 476–481.
18 John, S., Sorokin, A.V., and Thompson, P.D. (2007). Phytosterols and vascular disease. *Current Opinion in Lipidology* 18: 35–40.
19 Jeong, B.Y., Cho, D.M., Moon, S.K., and Pyeum, J.H. (1993). Quality factors and functional components in the edible seaweeds. I. Distribution on *n*-3 fatty acids in 10 species of seaweeds by their habitats. *Journal of the Korean Society of Food Science and Nutrition* 22: 612–628.
20 Jiménez-Escrig, A. and Goñi, I. (1999). Evaluaciónnutricional y efectosfisiológicos de macroalgas marinas comestibles. *Nutrition Transition: A Review of Latin American Profile* 49: 114–120.
21 Dembitsky, V.M., Pechenkina-Shubina, E., and Rozantsvet, O.A. (1991). Glycolipids and fatty acids of some seaweeds and marine grasses from the black sea. *Phytochemistry* 30: 2279–2283.

22 Burtin, P. (2003). Nutritional value of seaweeds. *Electronic Journal of Environment Agriculture Food Chemistry* 2: 498–503.

23 Wong, K.H.P. and Cheung, C.K. (2000). Nutritional evaluation of some subtropical red and green seaweeds part I: proximate composition, amino acid profiles and some physico-chemical properties. *Food Chemistry* 71: 475–482.

24 Brownlee, I.A., Allen, A., Pearson, J.P. et al. (2005). Alginate as a source of dietary fiber. *Critical Reviews in Food Science and Nutrition* 45: 497–510.

25 Yavorska, N. (2012). Sodium alginate – a potential tool for weight management: effect on subjective appetite, food intake, and glycemic and insulin regulation. *Journal of Undergraduate Life Science* 6: 66–69.

26 Matanjun, P., Mohamed, S., Mustapha, N.M., and Muhammad, K. (2009). Nutrient content of tropical edible seaweeds, *Eucheuma cottonii, Caulerpa lentillifera* and *Sargassum polycystum*. *Journal of Applied Phycology* 21: 1–6.

27 Eyre, H., Kahn, R., and Robertson, R.M. (2004). Preventing cancer, cardiovascular disease, and diabetes: a common agenda. *Stroke* 35: 1999–2010.

28 Spolaore, P., Joannis-Cassan, C., Duran, E., and Isambert, A. (2006). Commercial applications of microalgae. *Journal of Bioscience and Bioengineering* 101: 87–96.

29 Sajilata, M., Singhal, R., and Kamat, M. (2008). Fractionation of lipids and purification of γ-linolenic acid (GLA) from *Spirulina platensis*. *Food Chemistry* 109: 580–586.

30 Capelli, B. and Cysewski, G.R. (2019). Potential health benefits of *Spirulina* microalgae. *Nutrafoods* 9: 19–26.

31 Sekar, S. and Chandramohan, M. (2008). Phycobiliproteins as a commodity: trends in applied research, patents and commercialization. *Journal of Applied Phycology* 20: 113–136.

32 Aquaron, R., Delange, F., Marchal, P. et al. (2002). Bioavailability of seaweed iodine in human beings. *Cellular and Molecular Biology (Noisy-le-Grand, France)* 48: 563–569.

33 Harris, R.B.S. (2014). Direct and indirect effects of leptin on adipocyte metabolism. *BiochimicaetBiophysica Acta (BBA) – Molecular Basis of Disease* 1842: 414–423.

34 Maeda, H., Tsukui, T., Sashima, T. et al. (2018). Seaweed carotenoid, fucoxanthin, as a multi-functional nutrient. *Asia Pacific Clinical Nutrition* 1: 196–199.

35 Hitoe, S.S.H. (2017). Seaweed fucoxanthin supplementation improves obesity parameters in mild obese Japanese subjects. *Functional Foods in Health and Disease* 7: 246–262.

36 Brown, E.M., Allsopp, P.J., Magee, P.J. et al. (2014). Seaweed and human health. *Nutrition Reviews* 72: 205–216.

37 Kim, M.S., Kim, J.Y., Choi, W.H., and Lee, S.S. (2008). Effects of seaweed supplementation on blood glucose concentration, lipid profile, and antioxidant enzyme activities in patients with type 2 diabetes mellitus. *Nutrition Research and Practice* 2: 62–67.

38 Shannon, E. and Abu-Ghannam, N. (2019). Seaweeds as nutraceuticals for health and nutrition. *Phycologia* 58: 563–577.

39 Cornish, M.L., Critchley, A.T., and Mouritsen, O.G. (2015). A role for dietary macroalgae in the amelioration of certain risk factors associated with cardiovascular disease. *Phycologia* 54: 649–666.

40 Cao, D., Lv, X., Xu, X. et al. (2017). Purification and identification of a novel ACE inhibitory peptide from marine alga *Gracilariopsis lemaneiformis* protein hydrolysate. *European Food Research and Technology* 243: 1829–1837.

41 Jin, D.Q., Lim, C.S., Sung, J.Y. et al. (2006). *Ulva conglobata*, a marine algae, has neuroprotective and anti-inflammatory effects in murine hippocampal and microglial cells. *Neuroscience Letters* 402: 154–158.

42 Coura, C.O., Souza, R.B., Rodrigues, J.A.G. et al. (2015). Mechanisms involved in the anti-inflammatory action of a polysulfated fraction from *Gracilaria cornea* in rats. *PLoS One* 10: e0119319.

43 Barbalace, M.C., Malaguti, M., Giusti, L. et al. (2019). Anti-inflammatory activities of marine algae in neurodegenerative diseases. *International Journal of Molecular Sciences* 20: 3061.

44 Montero, L., delPilar Sánchez-Camargo, A., Ibáñez, E., and Gilbert-López, B. (2018). Phenolic compounds from edible algae: bioactivity and health benefits. *Current Medicinal Chemistry* 25: 4808–4826.

45 Husni, A., Lailatussifa, R., and Isnansetyo, A. (2019). *Sargassum hystrix* as a source of functional food to improve blood biochemistry profiles of rats under stress. *Preventive Nutrition and Food Science* 24: 150–158.

46 Jiao, G., Yu, G., Zhang, J., and Ewart, H.S. (2011). Chemical structures and bioactivities of sulfated polysaccharides from marine algae. *Marine Drugs* 9: 196–223.

47 Faggio, C., Pagano, M., Dottore, A. et al. (2015). Evaluation of anticoagulant activity of two algal polysaccharides. *Natural Product Research* 30: 1934–1937.

48 Liu, X., Wang, S., Cao, S. et al. (2018). Structural characteristics and anticoagulant property *in vitro;* and *in vivo;* of a seaweed sulfated rhamnan. *Marine Drugs* 16: 243.

49 Qi, X., Mao, W., Chen, Y. et al. (2012). Chemical characteristics and anticoagulant activities of two sulfated polysaccharides from *Enteromorpha linza* (Chlorophyta). *Journal of Ocean University of China* 12: 175–182.

50 Gomez-Zavaglia, A., Prieto Lage, M.A., Jimenez-Lopez, C. et al. (2019). The potential of seaweeds as a source of functional ingredients of prebiotic and antioxidant value. *Antioxidants* 8: 406.

51 Smit, A.J. (2004). Medicinal and pharmaceutical uses of seaweed natural products: a review. *Journal of Applied Phycology* 16: 245–262.

52 Alves, C., Silva, J., Pinteus, S. et al. (2018). From marine origin to therapeutics: the antitumor potential of marine algae-derived compounds. *Frontiers in Pharmacology* 9: 777.

53 Palermo, J.A., Flower, P.B., and Seldes, A.M. (1992). Chondriamides A and B, new indolic metabolites from the red alga *Chondria* sp. *Tetrahedron Letters* 33: 3097–3100.

54 Fedorov, S.N., Ermakova, S.P., Zvyagintseva, T.N., and Stonik, V.A. (2013). Anticancer and cancer preventive properties of marine polysaccharides: some results and prospects. *Marine Drugs* 11: 4876–4901.

55 Martínez Andrade, K.A., Lauritano, C., Romano, G., and Ianora, A. (2018). Marine microalgae with anti-cancer properties. *Marine Drugs* 16: 165.

56 Li, D.L., Zheng, X., Chen, Y.C. et al. (2015). Terpenoid composition and the anti-cancer activity of *Acanthopanax trifoliatus*. *Archives of Pharmacal Research* 39: 51–58.

57 Liu, M., Zhang, W., Wei, J. et al. (2012). Marine bromophenolbis (2, 3-dibromo-4,5-dihydroxybenzyl) ether, induces mitochondrial apoptosis in K562 cells and inhibits topoisomerase I *in vitro;*. *Toxicological Letters* 211: 126–134.

58 Ke, Q. and Costa, M. (2006). Hypoxia-inducible factor-1 (HIF-1). *Molecular Pharmacology* 70: 1469–1480.

59 Lee, J., Kim, S., Jung, W.S. et al. (2012). Phlorofucofuroeckol-A, a potent inhibitor of aldo-keto reductase family 1 member B10, from the edible brown alga *Eisenia bicyclis*. *Journal of Korean Society for Applied Biological Chemistry* 55: 721–727.

60 Eitsuka, T., Nakagawa, K., Igarashi, M., and Miyazawa, T. (2004). Telomerase inhibition by sulfoquinovosyldiacylglycerol from edible purple laver (*Porphyra yezoensis*). *Cancer Letters* 212: 15–20.

61 Haq, S.H., Al-Ruwaished, G., and Al-Mutlaq, M.A. (2019). Antioxidant, anticancer activity and phytochemical analysis of green algae, *Chaetomorpha* collected from the Arabian Gulf. *Scientific Repeorts* 9: 18906.

62 Kwak, J.Y. (2014). Fucoidan as a marine anti-cancer agent in preclinical development. *Marine Drugs* 12: 851–870.

63 Yang, L., Wang, P., Wang, H. et al. (2013). Fucoidan derived from *Undaria pinnatifida* induces apoptosis in human hepatocellular carcinoma SMMC-7721 cells via the ROS-mediated mitochondrial pathway. *Marine Drugs* 11: 1961–1976.

64 Zhou, G., Sheng, W., Yao, W., and Wang, C. (2006). Effect of low molecular λ-carrageenan from *Chondrus ocellatus* on anti-tumor H-22 activity of 5-Fu. *Pharmacological Research* 53: 129–134.

65 Park, H., Kim, I., Kim, J., and Nam, T. (2013). Induction of apoptosis and the regulation of ErbB signaling by laminarin in HT-29 human colon cancer cells. *International Journal of Molecular Medicine* 32: 291–295.

66 Freile-Pelegrín, Y. and Tasdemir, D. (2019). Seaweeds to the rescue of forgotten diseases: a review. *Botanica Marina* 62: 211–226.

67 Rodrigues, J.A.G., Eloy, Y.R.G., Vanderlei, E.D.S.O. et al. (2017). An anti-dengue and anti-herpetic polysulfated fraction isolated from the coenocytic green seaweed *Caulerpa cupressoides* inhibits thrombin generation *in vitro;*. *Acta Scientiarum. Biological Sciences* 39: 149–159.

68 Nagaoka, M., Shibata, H., Kimura-Takagi, I. et al. (1999). Structural study of fucoidan from Cladosiphon okamuranus tokida. *Glycoconjugate Journal* 16: 19–26.

69 Hidari, K.I.P.J., Takahashi, N., Arihara, M. et al. (2008). Structure and anti-dengue virus activity of sulfated polysaccharide from a marine alga. *Biochemical and Biophysical Research Communications* 376: 91–95.

70 Estevez, J.M., Ciancia, M., and Cerezo, A.S. (2001). dl-Galactan hybrids and agarans from gametophytes of the red seaweed *Gymnogongrus torulosus*. *Carbohydrate Research* 331: 27–41.

71 De Souza, L.M., Sassaki, G.L., Romanos, M.T.V., and Barreto-Bergter, E. (2012). Structural characterization and anti-hsv-1 and hsv-2 activity of glycolipids from the marine algae *Osmundaria obtusiloba* isolated from southeastern brazilian coast. *Marine Drugs* 10: 918–931.

72 Cirne-Santos, C.C., Teixeira, V.L., Castello-Branco, L.R. et al. (2006). Inhibition of HIV-1 replication in human primary cells by a *dolabellane diterpene* isolated from the marine algae *Dictyota pfaffii*. *Planta Medica* 72: 295–299.

73 Yu, K.X., Jantan, I., Ahmad, R., and Wong, C.L. (2014). The major bioactive components of seaweeds and their mosquitocidal potential. *Parasitology Research* 113: 3121–3141.

74 Salvador-Neto, O., Azevedo Gomes, S., RibeiroSoares, A. et al. (2016). Larvicidal potential of the halogenated sesquiterpene (+)-obtusol, isolated from the alga *Laurencia dendroidea*

J. Agardh (Ceramiales: Rhodomelaceae), against the dengue vector mosquito *Aedesaegypti* (Linnaeus) (Diptera: Culicidae). *Marine Drugs* 14: 20.
75 Kim, S.H., Shin, C., Min, S.K. et al. (2012). *in vitro;* evaluation of the effects of electrospun PCL nanofiber mats containing the microalgae *Spirulina* (*Arthrospira*) extract on primary astrocytes. *Colloids and Surfaces. B, Biointerfaces* 90: 113–118.
76 Wang, H.D., Chen, C.C., Huynh, P., and Chang, J.S. (2015). Exploring the potential of using algae in cosmetics. *Bioresource Technology* 184: 355–362.
77 De Roeck-Holtzhauer, Y. (1991). Uses of seaweeds in cosmetics. In: *Seaweed Resources in Europe: Uses and Potential* (eds. M. Guiry and G. Blunden), 83–94. Chichester: Wiley.
78 Fabrowska, J., Łęska, B., Schroeder, G. et al. (2015). Biomass and extracts of algae as material for cosmetics. In: *Marine Algae Extracts* (eds. S.-K. Kim and K. Chojnacka), 681–706. Weinheim: Wiley-VCH, Verlag GmbH & Co. KGaA.
79 Fitton, J.H., Dell'Acqua, G., Gardiner, V.A. et al. (2015). Topical benefits of two fucoidan-rich extracts from marine microalgae. *Cosmetics* 2: 66–81.
80 Silab. WHITONYL® and the Complexion Becomes Porcelain, 2018. https://www.silab.fr/produit-55-whitonyl_usa.html (accessed August 2020).
81 Park, D.H., Choi, W.S., Yoon, S.H. et al. (2007). A developmental study of artificial skin using the alginate dermal substrate. *Key Engineering Materials* 342-343: 125–128.
82 Pereira, L. (2018). Seaweeds as source of bioactive substances and skin care therapy – cosmeceuticals, algotheraphy, and thalassotherapy. *Cosmetics* 5: 68.
83 Mafinowska, P. (2011). Algae extracts as active cosmetic ingredients. *ZeszyNaukowe* 212: 123–129.
84 Rinaudo, M. (2007). Seaweed polysaccharides. In: *Comprehensive Glycoscience*, Analysis of Glycans; Polysaccharide Functional, Properties, vol. 2 (ed. J.P. Kamerling), 691–735. Oxford: Elsevier.
85 Pereira, L. and Correia, F. (2015). *AlgasMarinhas da Costa Portuguesa—Ecologia, Biodiversidade e Utilizações*, 341. Paris: Nota de RodapéEditores. ISBN: 978-989-20-5754-5.
86 Ratnasooriya, W.D., Premakumara, G.A.S., and Tillekeratne, L.M.V. (1994). Post-coital contraceptive activity of crude extracts of Sri Lankan marine red algae. *Contraception* 50: 291–299.
87 Bhakuni, D.S., Dhawan, B.N., Garg, H.S. et al. (1992). Bioactivity of marine organisms: part VI-screening of some marine flora from Indian coasts. *Indian Journal of Experimental Biology* 30: 512–517.
88 Babuselvam, M. and Ravikumar, S. (1993). *Screening of Male Anti-Fertility Compounds from Marine Seaweed Macro Algae*, 1–14. Rajakkamangalam: Division of Marine Microbiology and Medicine, ManonmaniamSundaranar University.
89 Prakash, S., Ravikumar, S., Reddy, K.V.R., and Kannapiran, E. (2013). Spermicidal activity of Indian seaweeds: an *in vitro;* study. *Andrologia* 46: 408–416.
90 Martins, A., Vieira, H., Gaspar, H., and Santos, S. (2014). Marketed marine natural products in the pharmaceutical and cosmoceutical industries: tips for success. *Marine Drugs* 12: 1066–1101.
91 Etahiri, S., Bultel-Poncé, V., Caux, C., and Guyot, M. (2001). New bromoditerpenes from the red alga *Sphaerococcus coronopifolius*. *Journal of Natural Products* 64: 1024–1027.
92 Mišurcová, L., Machů, L., and Orsavová, J. (2011). Seaweed minerals as nutraceuticals. *Advances in Food and Nutrition Research* 64: 371–390.

93 Shanmugam, M. and Mody, K.H. (2020). Heparinoid-active sulphated polysaccharides from marine algae as potential blood anticoagulant agents. *Current Science* 79: 1672–1683.

94 Cotas, J., Leandro, A., Pacheco, D. et al. (2020). A comprehensive review of the nutraceutical and therapeutic applications of red seaweeds (Rhodophyta). *Lifestyles* 10: 19.

95 Jiang, Z., Hama, Y., Yamaguchi, K., and Oda, T. (2012). Inhibitory effect of sulphated polysaccharide porphyran on nitric oxide production in lipopolysaccharide-stimulated RAW264.7 macrophages. *The Journal of Biochemistry* 151: 65–74.

96 Mao, W., Zang, X., Li, Y., and Zhang, H. (2006). Sulfated polysaccharides from marine green algae *Ulva conglobata* and their anticoagulant activity. *Journal of Applied Phycology* 18: 9–14.

97 Minh Thu, Q.T. (2018). Effect of sulfation on the structure and anticoagulant activity of ulvan extracted from green seaweed *Ulva reticulata*. *Vietnam Journal of Science and Technology* 54: 373.

98 Sharma, P. and Sharma, N. (2017). Industrial and biotechnological applications of algae: a review. *Journal of Advances in Plant Biology* 1: 01–25.

99 Dixon, C. and Wilken, L.R. (2018). Green microalgae biomolecule separations and recovery. *Bioresources and Bioprocessing* 5: 14.

100 Panis, G. and Carreon, J.R. (2016). Commercial astaxanthin production derived by green alga *Haematococcus pluvialis*: a microalgae process model and a techno-economic assessment all through production line. *Algal Research* 18: 175–190.

101 Heo, S.J., Yoon, W.J., Kim, K.N. et al. (2010). Evaluation of anti-inflammatory effect of fucoxanthin isolated from brown algae in lipopolysaccharide-stimulated RAW 264.7 macrophages. *Food and Chemical Toxicology* 48: 2045–2051.

102 Liu, H. and Gu, L. (2012). Phlorotannins from brown algae (*Fucus vesiculosus*) inhibited the formation of advanced glycation end products by scavenging reactive carbonyls. *Journal of Agriculture and Food Chemistry* 60: 1326–1334.

103 Ryu, B.M., Li, Y., Qian, Z.J. et al. (2009). Differentiation of human osteosarcoma cells by isolated phlorotannins is subtly linked to COX-2, iNOS, MMPs, and MAPK signaling: implication for chronic articular disease. *Chemico-Biological Interactions* 179: 192–201.

104 Jung, H.A., Jin, S.E., Ahn, B.R. et al. (2013). Anti-inflammatory activity of edible brown alga *Eisenia bicyclis* and its constituents fucosterol and phlorotannins in LPS-stimulated RAW264.7 macrophages. *Food Chemistry and chemical Toxicology* 59: 199–206.

105 Lee, S., Youn, K., Kim, D.H. et al. (2019). Anti-neuroinflammatory property of phlorotannins from *Ecklonia cava* on Aβ25-35-induced damage in PC12 cells. *Marine Drugs* 17: 7.

106 Mikami, K. and Hosokawa, M. (2013). Biosynthetic pathway and health benefits of fucoxanthin, an algae-specific xanthophyll in brown seaweeds. *International Journal of Molecular Sciences* 14: 13763–13781.

107 Thomas, N.V. and Kim, S.K. (2011). Potential pharmacological applications of polyphenolic derivatives from marine brown algae. *Environmental Toxicology and Pharmacology* 32: 325–335.

108 Sánchez-Camargo, A.P., Montero, L., Cifuentes, A. et al. (2016). Application of Hansen solubility approach for the subcritical and supercritical selective extraction of phlorotannins from *Cystoseiraabies-marina*. *RSC Advances* 6: 94884–94895.

109 Mansur, A.A., Brown, M.T., and Billington, R.A. (2020). The cytotoxic activity of extracts of the brown alga *Cystoseira tamariscifolia* (Hudson) Papenfuss, against cancer cell lines changes seasonally. *Journal of Applied Phycology* 32: 2419–2429.

110 Damergi, E., Schwitzguébel, J.P., Refardt, D. et al. (2017). Extraction of carotenoids from *Chlorella vulgaris* using green solvents and syngas production from residual biomass. *Algal Research* 25: 488–495.

111 Mouahid, A., Crampon, C., Toudji, S.A.A., and Badens, E. (2016). Effects of high water content and drying pre-treatment on supercritical CO_2 extraction from *Dunaliella salina* microalgae: experiments and modelling. *The Journal of Supercritical Fluids* 116: 271–280.

112 Abrahamsson, V., Jumaah, F., and Turner, C. (2018). Continuous multicomponent quantification during supercritical fluid extraction applied to microalgae using in-line UV/Vis absorption spectroscopy and on-line evaporative light scattering detection. *The Journal of Supercritical Fluids* 131: 157–165.

113 Kadam, S.U., Tiwari, B.K., and O'Donnell, C.P. (2013). Application of novel extraction technologies for extraction of bioactives from marine algae. *Journal of Agricultural and Food Chemistry* 61: 4667–4675.

114 Kadam, S.U., O'Donnell, C.P., Rai, D.K. et al. (2015). Laminarin from Irish Brown seaweeds *Ascophyllum nodosum* and *Laminaria hyperborea*: ultrasound assisted extraction. *Characterization and Bioactivity Marine Drugs* 13: 4270–4280.

115 Dang, T.T., Van Vuong, Q., Schreider, M.J. et al. (2017). Optimisation of ultrasound-assisted extraction conditions for phenolic content andantioxidant activities of the alga *Hormosirabanksii* using response surface methodology. *Journal of Applied Phycology* 29: 3161–3173.

116 Fitzgerald, C., Gallagher, E., O'Connor, P. et al. (2013). Development of a seaweed derived platelet activating factor acetylhydrolase (PAF-AH) inhibitory hydrolysate, synthesis of inhibitory peptides and assessment of their toxicity using the Zebrafish larvae assay. *Peptides* 50: 119–124.

117 Gullón, B., Gagaoua, M., Barba, F.J. et al. (2020). Seaweeds as promising resource of bioactive compounds: overview of novel extraction strategies and design of tailored meat products. *Trends in Food Science & Technology* 100: 1–18.

118 Fabrowska, J., Messyasz, B., Szyling, J. et al. (2017). Isolation of chlorophylls and carotenoids from freshwater algae using different extraction methods. *Phycological Research* 66: 52–57.

119 Yuan, Y., Zhang, J., Fan, J. et al. (2018). Microwave assisted extraction of phenolic compounds from four economic brown macro algae species and evaluation of their antioxidant activities and inhibitory effects on α-amylase, α-glucosidase, pancreatic lipase and tyrosinase. *Food Research International* 113: 288–297.

120 Xiao, X., Si, X., Yuan, Z. et al. (2012). Isolation of fucoxanthin from edible brown algae by microwave-assisted extraction coupled with high-speed countercurrent chromatography. *Journal of Separation Science* 35: 2313–2317.

121 Habeebullah, S.F.K., Alagarsamy, S., Sattari, Z. et al. (2019). Enzyme-assisted extraction of bioactive compounds from brown seaweeds and characterization. *Journal of Applied Phycology* 32: 615–629.

122 Hardouin, K., Bedoux, G., Burlot, A.S. et al. (2016). Enzyme-assisted extraction (EAE) for the production of antiviral and antioxidant extracts from the green seaweed *Ulva armoricana* (Ulvales, Ulvophyceae). *Algal Research* 16: 233–239.

123 Rodrigues, D., Sousa, S., Silva, A. et al. (2015). Impact of enzyme- and ultrasound-assisted extraction methods on biological properties of red, brown and green seaweeds from the central west coast of Portugal. *Journal of Agricultural and Food Chemistry* 63: 3177–3188.

124 Charoensiddhi, S., Franco, C., Su, P., and Zhang, W. (2014). Improved antioxidant activities of brown seaweed *Ecklonia radiata* extracts prepared by microwave-assisted enzymatic extraction. *Journal of Applied Phycology* 27: 2049–2058.

125 Commercial seaweeds market size worth $ 11.9 billion by 2027. www.grandviewresearch.com (accessed August 2020).

126 Commercial Seaweed Market Growing at 7.5% CAGR to be Worth over USD 92 Billion by 2025: Global Market Insights, Inc. www.gminsights.com (accessed August 2020).

127 Porse, H. and Rudolph, B. (2017). The seaweed hydrocolloid industry: 2016 updates, requirements, and outlook. *Journal of Appllied Phycology* 29: 2187–2200.

20

Bioprospection of Orchids and Appraisal of Their Therapeutic Indications

Devina Ghai[1], Jagdeep Verma[2], Arshpreet Kaur[1], Kranti Thakur[3], Sandip V. Pawar[4], and Jaspreet K. Sembi[1]

[1]*Department of Botany, Panjab University, Chandigarh, UT, India*
[2]*Department of Botany, Government College, Rajgarh, Himachal Pradesh, India*
[3]*Department of Botany, Shoolini Institute of Life Sciences and Business Management (SILB), Solan, Himachal Pradesh, India*
[4]*University Institute of Pharmaceutical Sciences, Panjab University, Chandigarh, UT, India*

20.1 Introduction

Bioprospecting or biodiversity prospecting refers to the systematic search for biochemical and genetic information in nature in order to develop commercially important products for pharmaceutical, agricultural, cosmetic, and other applications [1]. It is a purposeful exploration, extraction, and evaluation of wild biological materials to develop products that are valuable for the mankind in various important forms such as pharmaceuticals, agrochemicals, nutritional supplements, cosmetics, flavorings, fragrances, biological controls, industrial enzymes, molecular probes etc. According to Shaw [2], global biodiversity centers can act as potential bioprospecting regions because of their unique bioresource richness.

A wide range of established industries (such as pharmaceuticals, manufacturing, and agriculture) as well as a wide range of comparatively newer industries (such as aquaculture, bioremediation, biomining, biomimetic engineering, and nanotechnology) are actively engaged in bioprospection [3]. As the bioprospecting researches are fueled by technological developments, these remain confined to industries of the developed nations. It has been seen in majority of cases that the industries exploit traditional knowledge accumulated over centuries in developing and underdeveloped nations and give nothing or bare minimum in return. The rights linked with revenue generation (manufacturing, refinement, selling, etc.) remain reserved with the industry that produces the final product or service produced at the end of the discovery and development chain. The controversy surrounding the involvement of some multinational companies in developing weight loss products based on the Hoodia cactus (*Hoodia gordonii*) used by the San tribe of the Kalahari Desert for centuries as an appetite suppressant is an example of such problems [4]. Another example is that of turmeric (powdered dry rhizome of *Curcuma longa*), a herb used in a variety of culinary and therapeutic formulations since ages in India, where a patent granted

Bioprospecting of Plant Biodiversity for Industrial Molecules, First Edition.
Edited by Santosh Kumar Upadhyay and Sudhir P. Singh.

to researchers at University of Mississippi Medical Center for wound healing property of turmeric was revoked by United States Patent and Trademark Office (USPTO) based on the evidences produced by Council of Scientific and Industrial Research (CSIR), India [5], for protection of traditional knowledge. Such a gap between traditional knowledge and product development (leading to revenue generation) usually results in triggering debates over indigenous claims to intellectual and cultural property and importance of ownership rights. Convention on Biological Diversity (CBD), a multilateral treaty, attempts to bridge these gaps by recognizing the sovereign rights of each country to control the access to the biodiversity existing within its borders and ensuring benefit sharing and technology transfer in exchange for access to bioresources [6].

Plants and their products have been utilized since long for quenching a vast variety of human needs including food, clothing, shelter, and medicine. A large number of important drugs have plant origin and many others are semisynthetic derivatives of such naturally occurring products,and were discovered by using traditional knowledge of several ethnic communities across the globe. The drugs such as atropine (*Atropa belladonna*), cocaine (*Erythroxylum coca*), codeine and morphine (*Papaver somniferum*), colchicine (*Colchicum autumnale*), digitoxin (*Digitalis purpurea*), ephedrine (*Ephedra sinica*), curcumin (*Curcuma longa*), hyoscyamine (*Hyoscyamus niger*), methyl salicylate (*Gaultheria procumbens*), podophyllotoxin (*Podophyllum peltatum*), prostratin (*Homalanthus nutans*), quinine (*Cinchona robusta*), reserpine (*Rauvolfia serpentina*), scopolamine (*Datura* spp.), taxol (*Taxus brevifolia*), theobromine (*Theobroma cacao*), vincristine (*Catharanthus roseus*), etc. are in use throughout the world as therapeutics. Researchers have also stressed the importance of marine fungi in producing several bioactive natural products with potential curative applications [7]. Microbial endophytes have yielded antibiotics, antivirals, anticancer, and antidiabetic agents, antioxidants, immunosuppressant, and insecticidal compounds [8, 9]. Microbes inhabiting varied habitats possess excellent bioprospecting potential in the discovery of plant growth promoting substances and novel enzymes [10, 11]. Rapid sequencing and analysis of bacterial and fungal genomes have led to the discovery of certain gene clusters in these organisms that potentially govern the biosynthesis of novel biologically active compounds in them [12].

20.2 Orchids as a Bioprospecting Resource

Orchids represent one of the largest and highly evolved angiosperm families, the Orchidaceae, and comprise plants possessing strikingly beautiful flowers of incredible shapes, colors, and size range. Though popular as affluent ornamentals, orchids were first discovered for their therapeutic properties. These plants find mention in various ancient scriptures as curatives and still find a place of pride in traditional medicine throughout the world [13–17]. According to Nugraha et al. [18], the vascular epiphytes including orchids are important sources of therapeutic agents with diverse biological activities. Being one of the largest plant families with nearly 28 500 species on record [19], Orchidaceae offers a vast scope for varied researches including bioprospection. Their ability to have distributed across almost every corner of the earth with variously adapted habits such as terrestrial, epiphytic, lithophytic, and even subterranean makes these plants a valuable source of a wide range phytochemicals. Their inherently slow-growing nature and unique associations with symbiotic fungal partners for germination and growth in nature add to the uniqueness of their biosynthetic pathways. The obligation to an insect for pollination ensures the production of varied compounds to offer interesting rewards for the insect pollinator and add on to the diverse array of phytochemicals being produced by the plant [20, 21]. All this leads to the evolution of a variety of new compounds through activation of newer metabolic pathways and thus providing tremendous resource for bioprospecting for therapeutic indications.

20.3 Orchids as Curatives in Traditional India

In India, orchids find utility in local systems of medicine since the Vedic period [22–24]. The earliest reference to Indian orchids can be seen in "Charaka Samhita" wherein the medicinal importance of some orchidaceous taxa is reported [25]. However, the first scientific account of these plants appeared in Hortus Malabaricus, a 12-volume work by Van Rheede published during 1678–1693, wherein the medicinal properties of some orchids including *Acampe praemorsa, Bulbophyllum sterile, Cleisostoma tenuifolium, Cymbidium aloifolium, Dendrobium ovatum, Eulophia epidendraea, E. graminea, Liparis odorata, Pholidota imbricata, Rhynchostylis retusa, Rhytionanthos rheedei, Seidenfia rheedii, Taprobanea spathulata*, etc. from peninsular region were provided [26]. In ancient Indian literature, there are references to a group of eight plants, popularly known as "*Astavarga*," which were used for preparation of a number of rejuvenating herbal formulations, out of these, four are orchids, i.e. Jivak (*Malaxis muscifera*), Rishbhaka (*Crepidium acuminatum*), Riddhi (*Habenaria intermedia*) and Vriddhi (*Platanthera edgeworthii*) [27, 28]. Astavarga is an important constituent of "*Chyawanprash*," the popular immune booster Ayurvedic formulation in India [29]. Some other botanicals including Ban-alu (*Gastrodia falconeri*), Jewanti (*Dendrobium* spp., *Flickingeria macraei*), Salam/Salampanja/Hathpanja (*Dactylorhiza* spp., *Eulophia* spp.), Shwethuli (*Zeuxine strateumatica*), Salabmisri (*Eulophia dabia*), and Rasna (*Acampe papillosa, Vanda tessellata*) are used as aphrodisiacs, blood purifiers, and general restorative tonics [30]. Rahamtulla et al. [31] provided ethnobotanical aspects of 25 orchid species from the Darjeeling Himalaya (India). Over the years, there are several reports on orchids curing a variety of human ailments [30, 32–43].

20.4 Therapeutics Indications of Orchids in Asian Region

It is believed that the Chinese were first to document their medicinal uses. *Bletilla striata* and one *Dendrobium* species were described in "Materia Medica" of Shen-nung during twenty-eighth century BCE. Shi-Hu, Tian-Ma, and Bai-Ji are three orchid-based therapeutic formulations used in China [14]. The first one (*Shi-Hu*) is derived from *Dendrobium nobile* and allied species and is valued as an important tonic because of its effectiveness in lung, kidney, and stomach diseases, hyperglycemia, and diabetes [44–46]. "*Tian-Ma*" is prepared from tubers of an achlorophyllous (mycoheterotrophic) orchid named *Gastrodia elata* and finds use to treat headaches, migraine, epilepsy, high blood pressure, rheumatism, fever, and nervous problems; till date, more than 50 chemical substances have been isolated from this species [17, 45, 47]. Tubers of *Bletilla striata* are used for the preparation of third drug called "*Bai-Ji*," which is used for curing tuberculosis and gastric ulcers, boils, malaria, hemorrhage, inflammation, and malignant swellings [47]. Preparations from *Dactylorhiza hatagirea* (dbang lag) provide sustenance for Tibetan monks (yogins) practicing in remote caves [17]. Some orchids including *Acriopsis javanica, Corymborchis longiflora, Epidendrum bifidum, Eria pannea, Grammatophyllum scriptum, Nervilia aragoana, Tropidia curculigoides*, etc. are used in some regions of Malaysia and Indonesia to cure a number of problems such as postdelivery sickness in women, malaria, scabies and skin lesions, intestinal tapeworms, and skin sores [48–50]. Orchids are famous by the name of "*Sunakhari*," "*Sungava*," "*Chandigava*" in Nepal, and about 60 species find use as local medicines in addition to energizers and aphrodisiacs [51, 52]. Vaidya [53] presented information on medicinal properties of 130 Nepalese orchid species. Yonzone et al. [54] reported 74 orchid species used for various medicinal purposes in the Himalayan region. Different species of *Dendrobium, Vanda, Cymbidium*, and many other genera are used in Indonesia, Malaysia, Singapore, Vietnam, Sri Lanka, Thailand, Myanmar, Taiwan, Korea, and Japan as tonic and to cure a variety of disorders [55–57]. Chauhan et al. [58] mentioned three orchid species (*Dactylorhiza hatagirea, Eulophia nuda, Vanda tessellata*) in the list of plants used for improving virility. *Flickingeria nodosa, Dendrobium aqueum*, and *Pholidota pallida* have shown the antioxidant activity [59–61]. Minh et al. [62, 63] suggested the possible use of root extracts of *Phalaenopsis* hybrids as a potential source of natural antioxidants. Eating tubers of *Geodorum* sp. is linked with promoting longevity in Myanmar [17]. Hossain et al. [64] screened the bioactive phytochemicals in three epiphytic orchids (*Luisia zeylanica, Rhynchostylis retusa, Papilionanthe teres*) of Bangladesh.

20.5 Evidences of Medicinal Uses of Orchids in Ethnic African Groups

Africans use traditional medicine as an important part of their cultural beliefs. Orchids form an important part of their traditional medicine system because of their therapeutic properties. People of Zulu community use an infusion of *Ansellia humilis* and *Habenaria foliosa* as emetic. Some species of *Eulophia* are considered to prevent miscarriage and relieve pain. *Ansellia gigantea* and *A. humilis* were used by Zulus as an aphrodisiac and as antidote to bad dreams [65, 66]. Species such as *Angraecum augustipetallum, Bulbophyllum falcatum, B. maximum, B. pumilum, B. melinostachyum, Cyrtorchis arcuata, Diaphananthe bidens, Eulophia cucullata, Graphorkis laurida, Habenaria procera, Liparis nervosa, Polystachya cultriformis*, etc. are still used for treatment of diabetes, skin infections, epilepsy and fertility problems, arthritis, tuberculosis, and gastritis, paste of *Ansellia africana* pseudobulbs is used as a contraceptive [67, 68]. According to Hutchings et al. [69] and Chinsamy et al. [66], around 50 orchid species are informally traded and used in South African traditional medicine especially by the Zulu community. In South Africa, "*Iphamba*" refers to a group of 12 different orchids, namely *Cyrtorchis arcuata, Diaphananthe millarii, D. xanthopollinia, Eulophia ensata, E. ovalis, E. leontoglossa, Microcoelia exilis, Mystacidium capense, M. venosum, Polystachya transvaalensis, Tridactyle bicaudata*, and *T. tridentate* [66]. Tubers of some *Disa, Habenaria*, and *Satyrium* species are used as food in Malawi [70], whereas the cooked root tubers of *Eulophia cucullata* are used as a poultice [71]. Infusion prepared from the roots of *Disa aconitoides* is administered as an emetic for women (to promote conception) and that prepared from the tubers of *Eulophia clavicornis* and *E. tenella* to treat infertility [71, 72]. *Ansellia africana*, an African endemic, popularly known as leopard orchid has been reported to be a species with high medicinal potential and reserve of many important biomolecules [73]. Root infusion of *A. africana* is also given to children for treatment of cough [71]. Chinsamy et al. [74] tested the anti-inflammatory, antioxidant, anticholinesterase activity, and mutagenicity of seven orchid species that are most commonly traded in herbal markets of South Africa and found that root extract of *A. africana* to be the most effective. Vanilla flavor is the most important commercial produce of orchids in present time. Besides this, species of *Vanilla* (mainly *Vanilla planifolia*) possess medicinal and aphrodisiac properties and cultural reliance [17, 75, 76].

20.6 Orchids as a Source of Restoratives in Europe

In Europe, orchids are thought to have many healing properties besides the ability to enhance virility and potency. During the first century, Dioscorides in his book "De Materia Medica" described nearly 500 medicinal plants including two terrestrial orchids. This book highlighted the effectiveness of orchids as a determinant of sex of the offspring and promoted the "Doctrine of Signatures" [14]. Some species of *Ophrys, Orchis, Serapias*, and *Dactylorhiza* were used against alcoholic gastritis, and tuberous orchids were expected to increase fertility in men [77]. An orchid-based nutritious drink, "Salep," was sold at street stalls in London [78], and its popularity declined only after introduction of tea and coffee [17]. Salep was prepared from the tubers of around 35 orchids especially those belonging to genus *Orchis, Dactylorhiza, Neotinea, Ophrys, Himantoglossum, Serapias, Steveniella*, and *Anacamptis* [79, 80] and is rich in mucilage, sugar, starch, nitrogenous substance, and traces of volatile oil and is considered as a valued tonic and aphrodisiac [48]. According to Ghorbani et al. [81], wild orchids were traditionally collected for "Salep" in Iran for their use in traditional medicine and ice-cream industry. Some species of *Dactylorhiza* are still promoted as aphrodisiacs in many Asian countries, especially in Himalayan region.

20.7 Remedial Uses of Orchids in American and Australian Cultures

Vanilla, the spice orchid used since ancient times to flavor cocoa, finds mention in an Aztec herbal of 1552 [14, 17] and was described to treat fevers, impotency, and rheumatism. There are reports of *Arethusa bulbosa* roots being used to relieve toothache and hot juice of the roasted fruits of *Bulbophyllum vaginatum* to treat earache [82]. Orchids such as *Goodyera pubescens* (against mad dog bites), *Bromheadia finlaysoniana* (pain reliever), *Spathoglottis plicata* (treatment of joint pains), *Cypripedium* spp. (as sedative, to treat anxiety, fever, headache, nervous tension, etc.) are widely used in America [48, 83, 84]. Different species of *Cypripedium* were used in North America by different ethnic groups for its sedative and antispasmodic properties [84]. *C. parviflorum* is an important orchid employed to treat hysteria and disorders of the nervous system by North American Indians [17]. In

Mexico, people of Mixtec and Triqui ethnicities use many native plants including orchids as food and medicine [85]. According to Garcia et al. [86], the epiphytic orchids are known by the name "*ita ndeka*" in the Mixtec language, and *Prosthechea karwinskii* (endemic to Southern Mexico) are used to cure coughs, wounds and burns, diabetes, and miscarriage. Orchids such as *Bletia purpurea, Calanthe calanthoides, Catasetum integerrimum, Cyrtopodium macrobulbon, Laelia autumnalis, L. speciosa, Myrmecophila tibicinis*, and *Rhyncholaelia digbyana* find mention in Mexican traditional medicine to heal wounds, burns, and respiratory problems [13, 66, 87]. Asseleih et al. [88] listed 12 orchid species (*Epidendrum chlorocorymbos, Habenaria floribunda, Isochillus latibracteatus, I. major, Mormodes maculata, Oestlundia luteorosea, Oncidium ascendens, Scaphyglottis fasciculata, Sobralia macrantha, Spiranthes eriophora, Stanhopea oculata, Vanilla planifolia*), which are used for their ethnobotanical and pharmacological properties in Veracruz state of Mexico.

Australians use the pseudobulbs of *Cymbidium* sp. for treatment of dysentery. *Dendrobium teratifolium* and *D. discolor* were used to relieve pain and control ringworm [89]. Interestingly, the seed capsules of *Selenipedium chica* were occasionally used as a substitute for *Vanilla*. *Liparis reflexa* is considered harmful for human health because of its toxic properties [89]. Tubers and bulbs of some orchids (*Gastrodia sesamoides, Dendrobium speciosum, Caladenia* spp.) are reported as emergency bush foods [78]. According to Teoh [17], at least 20 Australian species are edible, and only a few *Cymbidium* and *Dendrobium* species are used as medicine.

20.8 Scientific Appraisal of Therapeutic Indications of Orchids

As evident from the preceding text, orchids are the backbone of the traditional medicine systems across the globe. With the advent of medical science and technology, efforts have been made to tap the immense potential as indicated by the crude ethnomedicinal evidences, to develop products with sound scientific scaffolding for therapeutic use. This role of orchids in traditional literature as restoratives and therapeutics has been the basis of modern research to evaluate their potential as anticancer, antioxidant, antimicrobial, immunomodulatory, antidiabetic, and anti-inflammatory agents (Figure 20.1; Table 20.1). The following sections discuss in detail the various therapeutic indications in orchids:

20.8.1 Orchids as Potent Anticancer Agents

Cancer is a leading cause of death worldwide. Although several synthetic drugs are available for its treatment, none of them is completely effective and possesses many side effects too. On the other hand, plant-based anticancer drugs have been proved to be effective and safe for cancer treatment to some extent. Several reports elucidating the anticancer properties of the compounds extracted from orchids have been documented. The isolated bibenzyl compounds from *Dendrobium officinale* and *D. findlayanum* displayed high cytotoxicity against the growth of Hela human cervical cancer cell line [122, 145]. Moscatilin, a bibenzyl derivative obtained from *D. loddigesii* showed an inhibitory effect toward human melanoma, esophageal cancer, and breast cancer [130–132]. Similarly, moscatilin isolated from

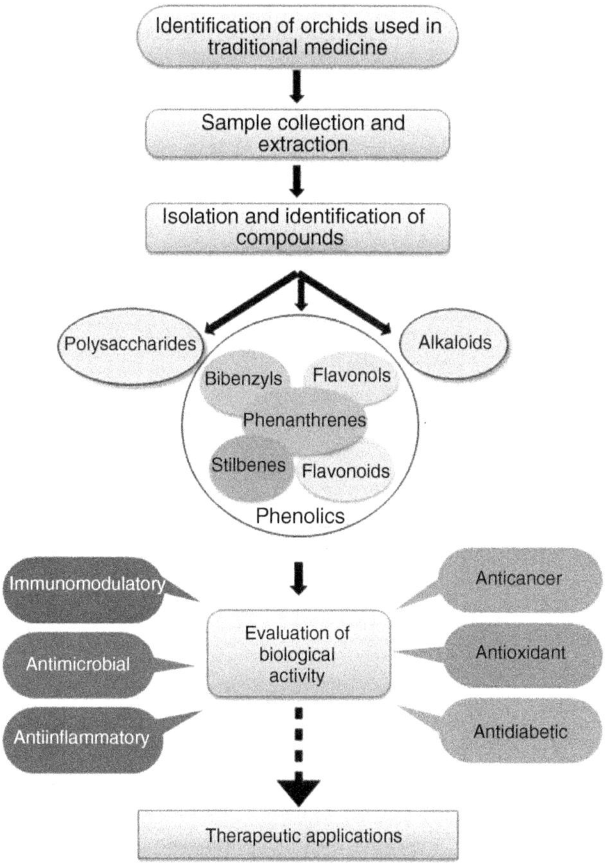

Figure 20.1 An outline of bioprospecting in orchids and its therapeutic applications.

Table 20.1 Bioactive compounds derived from different orchid species.

Plant name	Compound/Class	Activity	Reference
Anoectochilus formosanus	Type II arabino-galactan (Polysaccharide)	Immunomodulatory activity	[90]
Anoectochilus roxburghii	ARPP80 (Polysaccharide)	Antioxidant activity	[91]
Bletilla ochracea	Blestriarene A, blestriarene B, blestriarene C (Phenathrenes)	Antibacterial activity	[92]
	4-Methoxyphenanthrene-2,7-diol (Phenanthrene)	Anti-inflammatory activity	[93]
Bletilla striata	BSP (Polysaccharide)	Healing oral ulcers	[94]
	Coelonin (Dihydrophenanthrene)	Anti-inflammatory activity	[95]

(*Continued*)

Table 20.1 (Continued)

Plant name	Compound/Class	Activity	Reference
	2,7-Dihydroxy-3,4-dimethoxyphenanthrene, 2,7-dihydroxy-4-methoxy-9,10-dihydrophenanthrene, 2,7-dihydroxy-3,4-dimethoxy-9,10-dihydrophenanthrene, shanciol C, shanciol F, shanciol D blestriarene A (Stilbenoids)	Antibacterial activity	[96]
	Bletistrin F, bletistrin G, bletistrin J, bulbocol, shanciguol and shancigusin B (Bibenzyl derivatives)	Antibacterial activity	[97]
	4,7,7′-trimethoxy-9′,10′-dihydro(1,3′-biphenanthrene)-2,2′,5′-triol, 4,7,4′-trimethoxy-9′,10′-dihydro(1,1′-biphenanthrene)-2,2′,7′-triol; 4,7,3′,5′-tetramethoxy-9′,10′-dihydro(1,1′-biphenanthrene)-2,2′,7′-triol, 4,8,4′,8′-tetramethoxy(1,1′-biphenanthrene)-2,7,2′,7′-tetrol, blestriarene C [4,4′-dimethoxy(1,1′-biphenanthrene)-2,7,2′,7′-tetrol] (Biphenanthrenes)	Antibacterial activity	[98]
	pFSP (Polysaccharides)	Antioxidant activity	[99]
	BSP (Polysaccharides)	Antioxidant activity	[100]
	BSP-1 (Polysaccharide)	Immunomodulatory activity	[101]
	BSPF2 (Polysaccharide)	Immunomodulatory activity	[102]
	Dihydropinosylvi, batatasin III, 3′-hydroxy-2-(4-hydroxybenzyl)-3,5-dimethoxy-bibenzyl, gymconopin D and 5-[2-(3-methoxyphenyl)ethyl]-1,3-benzenediol (Stilbenes)	Anti-neuroinflammatory activity	[103]
	Phochinenin K, bleformin F and 4,8,4′,8′-tetramethoxy-(1,1′-biphenanthrene)-2,7,2′,7′-tetrol	Anti-inflammatory activity	[104]
Bulbophyllum retusiusculum	Retusiusines B (Phenylpropanoid)	Antifungal activity	[105]
Cremastra appendiculata	Coelonin, orchinol (Phenanthrenes)	Antioxidant activity	[106]
Cymbidium finlaysonianum	Cymbinodin-A (Phenathrenequinone)	Anticancer activity	[107]
Dendrobium aphyllum	DAP (Polysaccharide)	Immunomodulatory activity	[108]
	Aphyllone B (Bibenzyl derivative)	Antioxidant activity	[109]
Dendrobium aurantiacum var. *denneanum*	Moscatilin (Bibenzyl derivative)	Anticancer activity	[110]

Table 20.1 (Continued)

Plant name	Compound/Class	Activity	Reference
Dendrobium christyanum	*n*-Docosyl 4-hydroxy-trans-cinnamate (Ester of cinnamic acid)	Antidiabetic property	[111]
Dendrobium chrysotoxum	Erianin (Bibenzyl)	Inhibits diabetic retinopathy	[112]
Dendrobium crepidatum	(+)-Homocrepidine A (Indolizidine)	Anti-inflammatory activity	[113]
	(+)-Dendrocrepidamine A, dendrocrepidamine B, (+)-homocrepidine A (Octahydroindolizine-type alkaloids)	Anti-inflammatory activity	[114]
	Dendrocrepine (Indolizidine alkaloid)	Antidiabetic property	[115]
Dendrobium denneanum	2,5-Dihydroxy-4-methoxy-phenanthrene 2-*O*-β-D-glucopyranoside (Phenanthrene glycosides), 5-methoxy-2,4,7,9*S*-tetrahydroxy-9,10-dihydrophenanthrene [9,10-dihydrophenanthrenes]	Anti-inflammatory activity	[116]
Dendrobium devonianum	5-Hydroxy-3-methoxy-flavone-7-*O*-(β-D-apiosyl-(1-6))-β-D-glucoside (Flavonol glycoside)	Antidiabetic property	[117]
	DvP-1 (Polysaccharide)	Immunomodulatory activity	[118]
	DDP (Polysaccharide)	Immunomodulatory activity	[119]
Dendrobium draconis	Gigantol (Bibenzyl compound)	Anticancer activity	[120]
Dendrobium falconeri	Dendrofalconerol A (Bibenzyl)	Anticancer activity	[121]
Dendrobium findlayanum	4,4′-Dihydroxy-3,3′,5-trimethoxy bibenzyl (Bibenzyl)	Anticancer activity	[122]
Dendrobium formosum	Confusarin (Phenanthrene), 5-methoxy-7-hydroxy-9,10-dihydro-1,4-phenanthrenequinone (Phenanthrenequinone)	Antidiabetic property	[123]
Dendrobium huoshanense	DHP1A (Polysaccharide)	Hepatoprotective activity	[124]
	GXG (galactoxyloglucan) (Polysaccharide)	Protection of intestine	[125]
	GXG (galactoxyloglucan) (Polysaccharide)	Antidiabetic	[126]
	DHP-4A (Polysaccharide)	Immunomodulatory activity	[127]
Dendrobium infundibulum	Dendrosinen B (Bibenzyl derivative)	Inhibit pancreatic lipase	[128]
Dendrobium loddigesii	Loddigesiinols G–J (Polyphenols)	Antidiabetic	[129]
	Moscatilin (Bibenzyl derivative)	Anticancer activity	[130–132]

(*Continued*)

Table 20.1 (Continued)

Plant name	Compound/Class	Activity	Reference
Dendrobium nobile	Moscatilin (Bibenzyl)	Protect retinal cells from hypoxia/ischemia	[133]
	3′,4-Dihydroxy-3,5′-dimethoxybibenzyl, 3-hydroxy-5-methoxybibenzyl, batatasin III, tristin, 3,3′,5-trihydroxybibenzyl (Bibenzyl)	Antifungal activity	[134]
	Nudol (Phenanthrene)	Anticancer activity	[135]
	Dendronbibisline A, Dendronbibisline B (dihydrophenanthrofurans)	Anticancer activity	[136]
	Dendronbibisline C Dendronbibisline D (*Bisbibenzyl* derivative)		
Dendrobium officinale	3,4-Dihydroxy-4′,5-dimethoxybibenzyl (Bibenzyl)	Antidiabetic property	[137]
	DOP (Polysaccharide)	Prevent lung inflammation	[138]
	DOPS (Polysaccharide)	Protective role in liver injury in acute colitis	[139]
	DOP (Polysaccharide)	Treating constipation	[140]
	Dendronan (Polysaccharide)	Maintaining colonic health	[141]
	DOP (Polysaccharide)	Gastroprotective activity	[142]
	DOPS (Polysaccharide)	Neuroprotective activity	[125]
	DOPW-1 and DOPW2 (Polysaccharide)	Immunomodulatory activity	[143]
	DOP-1-1 (Polysaccharide)	Immunomodulatory activity	[144]
	4,4′-Dihydroxy-3,5-dimethoxy bibenzyl (Bibenzyl)	Anticancer activity	[145]
	Rutin (Flavonoid)	Antioxidant activity	[146]
Dendrobium palpebre	Dendroflorin (Phenolic)	Antioxidant activity	[147]
Dendrobium parishii	Dendroparishiol (Bibenzyl-dihydrophenanthrene derivative)	Anti-inflammatory activity Antioxidant activity	[148]
Dendrobium plicatile	4′-Dihydroxy-3′,5-dimethoxybibenzyl (Bibenzyl)	Anticancer activity	[149]
Dendrobium scabrilingue	Dendroscabrol B (Bisbibenzyl) RF-3192C (Dinaphthalenone)	Antidiabetic property	[150]
Dendrobium tortile	Dendrofalconerol A (Bisbibenzyl)	Antidiabetic property	[151]

Table 20.1 (Continued)

Plant name	Compound/Class	Activity	Reference
Dendrobium tosaense	DTP (Polysaccharide)	Immunomodulatory activity	[152]
Eulophia macrobulbon	4-Methoxy-9,10-dihydro-2,7-phenanthrenediol, 4-methoxy-2,7-phenanthrenediol, 1,5-dimethoxy-2,7-phenanthrenediol, 1,5,7-trimethoxy-2,6-phenanthrenediol, 1-(4-hydroxybenzyl)-4,8-dimethoxy-2,7-phenanthrenediol (Phenanthrene)	Anti-inflammatory activity	[153]
	4-Methoxy-2,7-phenanthrenediol	Antioxidant activity	
Gastrodia elata	GPs (Polysaccharides)	Immunomodulatory activity	[154]
	Bis(4-hydroxybenzyl)ether mono-β-L galactopyranoside	Antioxidant activity	[155]
	Gastrodin (4-hydroxyapatite-4-hydroxyapatite-glucoside)	Antioxidant activity	[156]
	Gastrodinol (tetra-*p*-cresol substituted cyclopenta [a] naphthalene derivative)	Antibacterial activity	[157]
Gavilea lutea	Gavilein (Bibenzyl derivative)	Antifungal activity	[158]
Liparis regnieri	Erianthridin, gigantol, hircinol, nudol, coelonin, moscatin (Phenanthrene derivatives)	Antibacterial activity	[159]
Paphiopedilum callosum	3′-Hydroxy-2,6,5′-trimethoxystilbene, 3′-hydroxy-2,5′-dimethoxystilbene, 2,3′-dihydroxy-5′-methoxystilbene (Stilbenes), galangin (Flavonoid)	Anticancer activity	[160]
Paphiopedilum godefroyae	5,6-Dimethoxy-2-(3-hydroxy-5-methoxyphenyl)benzofuran (Stilbenes)	Anticancer activity	[161]
Pleione bulbocodioide	2,5,2′,5′-Tetrahydroxy-3-methoxybibenzyl and 2,5,2′,3′-tetrahydroxy-3-methoxybibenzyl (Bibenzyl)	Anti-neuroinflammatory	[162]
Spiranthes sinensis	Spiranthesphenanthrene A (Phenanthrene)	Anticancer activity	[163]
Vanda teres	Eucomicacid (Auxin) and vandateroside II (Glycopyranosyloxybenzyleucomate)	Antiaging activity	[164]

D. aurantiacum has also depicted its potential in the treatment of pancreatic cancer [110]. Likewise, the studies on *D. draconis* and *D. falconeri* also yielded bibenzyl compounds, gigantol and dendrofalconerol A, which were cytotoxic against lung cancer cells [120, 121]. A phenanthrene derivative nudol, isolated from *D. nobile*, played a critical role in the inhibition of osteosarcoma (U2OS) cell growth [135]. Spiranthes phenanthrene A from *Spiranthes sinensis* inhibited the growth and migration of melanoma (B16–F10) cells [163]. Furthermore, the phenolic compounds isolated from *Cymbidium finlaysonianum*, *Paphiopedilum callosum*, and *P. godefroyae* were found to be cytotoxic against human small cell lung cancer (NCI-H187) cell lines [107, 160, 161]. So, in a nutshell, it can be said that the metabolites of orchids have the potential to be used in cancer treatment after more extensive investigations.

20.8.2 Immunomodulatory Activity in Orchids

The immune-enhancing potential of plant-derived compounds leads toward maintaining a disease-free state in human beings. Both in vivo and in vitro; testing methods have been employed for evaluating the immunomodulatory effects of various orchids. The polysaccharides obtained from many species belonging to the genus *Dendrobium* have exhibited noteworthy immunomodulatory activity due to their ability to promote the proliferation of macrophages and lymphocytes, increase in the phagocytosis activity and upregulation of nitric oxide and cytokine production, etc. These activities have been testified in the polysaccharides of *Dendrobium aphyllum*, *D. devonianum*, *D. huoshanense*, *D. officinale* and *D. tosaense* [108, 118, 119, 127, 143, 144, 152]. Type II arabino-galactan, a polysaccharide purified from *Anoectochilus formosanus*, stimulated the maturation of dendritic cells, which play a crucial role in the induction of immune responses against pathogens [90]. Further, the polysaccharides from *Bletilla striata* and *Gastrodia elata* improved the spleen and thymus indices, which attributed toward their immune-enhancing potential [101, 154].

20.8.3 Orchids and Their Antioxidant Potential

As free radicals are the causal factors leading to many diseases such as cancer and cardiovascular diseases. Therefore, the compounds having antioxidant activity can be utilized for the treatment of such grave diseases. Many investigations have been undertaken so far to evaluate the antioxidant potential of various orchids using both in vitro; and in vivo; methods. The polysaccharides extracted from *Anoectochilus roxburghii* and *Bletilla striata* displayed remarkable antioxidant effects [91, 99, 100]. Several phenolic compounds also possessed antioxidant potential. For instance, Rutin (flavonoid), Aphyllone B (bibenzyl derivative), and 4-methoxy-2,7-phenanthrenediol isolated from *Dendrobium officinale, D. aphyllum*, and *Eulophia macrobulbon*, respectively, showed significant DPPH (2,2-Diphenyl-1-picrylhydrazyl) radical scavenging activity, which is a standard test to evaluate the antioxidant potential [109, 146, 153]. Moreover, phenanthrenes such as coelonin and orchinol derived from *Cremastra appendiculata* showed efficiency in both DPPH and ABTS (2,2′-azino-bis(3-ethylbenzothiazoline-6-sulfonic acid)) radical scavenging activities [106]. Furthermore, dendroflorin and dendroparishiol isolated from *D. palpebre* and *D. parishii* exhibited antioxidant activity by decreasing reactive oxygen species (ROS) in H_2O_2-treated cells and increasing the expression of antioxidant enzymes such as superoxide dismutase (SOD), glutathione peroxidase (GPX), and catalase (CAT) [147, 148]. The enhancement of the activity of these enzymes has also been reported in a similar study conducted in *Gastrodia elata* [155].

20.8.4 Antimicrobial Studies in Orchids

The antimicrobial effects of the compounds extracted from orchids have been studied against a diverse species of bacteria and fungi. The stilbenoids and the bibenzyl derivatives isolated from the tubers of *Bletilla striata* showed significant antibacterial activity against

Staphylococcus aureus [96, 97]. In the earlier researches on *Bletilla striata* and *Bletilla ochracea*, the derived biphenanthrenes and phenanthrenes have also been reported to possess inhibitory effects against gram-positive bacteria such as *Staphylococcus aureus, Staphylococcus epidermis*, and *Bacillus subtilis* [92, 98]. In the same manner, the phenanthrene derivatives (erianthridin, gigantol, hircinol, nudol, coelonin, and moscatin) in *Liparis regnieri* displayed antibacterial activity against *Streptococcus agalactiae* and *Bacillus subtilis* [159]. Furthermore, gavilein and retusiusines B isolated from *Gavilea lutea* and *Bulbophyllum retusiusculum*, respectively, exhibited noteworthy antifungal activity against *Candida albicans* [105, 158]. Additionally, broad-spectrum antifungal activity has been shown by bibenzyl derivatives of *Dendrobium nobile* against *Alternaria brassicicola, Phytophthora parasitica* var. *nicotianae, Colletotrichum capsici, Bipolaris oryzae, Diaporthe medusae nitschke, Ceratocystis paradoxa moreau, Exserohilum turcicum, Pestallozzia theae*, and *Alternaria citri* [134].

20.8.5 Orchids and Anti-inflammatory Activity

Lipopolysaccharide induces the production of nitric oxide synthase in macrophages, which upon stimulation produces different inflammatory factors such as tumor necrosis factor-α, interleukin-1β, etc. Different studies have been conducted to decrease inflammation with the help of compounds derived from different orchid species by inhibiting nitric oxide production. The stilbenes (dihydropinosylvi, batatasin III, 3′-hydroxy-2-(4-hydroxybenzyl)-3,5-dimethoxy-bibenzyl, gymconopin D, and 5-[2-(3-methoxyphenyl)ethyl]-1,3-benzenediol) isolated from *Blettia striata* were reported to have anti-neuroinflammatory activity [103]. Zhou et al. [104] studied the effect of phochinenin K, bleformin F, 4,8,4′,8′-tetramethoxy-(1,1′-biphenanthrene)-2,7,2′,7′-tetrol on reducing nitric oxide in lipopolysaccharide-induced BV-2 microglial cells. Bibenzyl derivative isolated from pseudobulbs of *Pleione bulbocodioides* also exhibited anti-neuroinflammatory potential [162]. 2,5-Dihydroxy-4-methoxy-phenanthrene 2-*O*-β-D-glucopyranoside and 5-methoxy-2,4,7,9*S*-tetrahydroxy-9,10-dihydrophenanthrene of *Dendrobium denneanum* exert anti-inflammatory effects by inhibiting MAPKs and nuclear factor κB pathways [116]. An indolizidine, (+)-homocrepidine A, isolated from stems of *D. crepidatum*, inhibited accumulation of nitric oxide and reduced inflammatory responses [113]. Similarly, a bibenzyl-dihydrophenanthrene derivative of *D. parishii* and phenanthrene derivatives of *Eulophia macrobulbon* could also invoke anti-inflammatory responses [148, 153]. A dihydrophenanthrene, coelonin, identified from ethanolic *B. striata* extract, and 4-methoxyphenanthrene-2,7-diol from *B. ochracea* have remarkable anti-inflammatory properties [93, 95].

20.8.6 Antidiabetic Prospects in Orchids

Diabetes is a disease with defective glucose metabolism caused by reduced insulin uptake or inefficient insulin production. Unmanaged diabetes could lead to several health-related problems. Several studies have been conducted to evaluate the orchid polysaccharides for their hypoglycemic potential. Noteworthy among these is a comparative study on various species of *Dendrobium* [165]. GXG (galactoxyloglucan), a purified polysaccharide of *D. huoshanense*, prevents hyperglycemia in type 2 diabetes in mice by improving insulin sensitivity [126]. Loddigesiinols G–J extracted from stems of *D. loddigesii* and 3,4-Dihydroxy-4′,5-dimethoxybibenzyl from *D. officinale* were used in treating type 2 diabetes [129, 137]. 5-Hydroxy-3-methoxy-flavone-7-*O*-(β-D-apiosyl-(1-6))-β-D-glucoside, a flavanol derivative of *D. devonianum*, inhibited the production of α-glucosidase, thereby reducing the blood sugar levels [117]. *D. chrysotoxum*–derived erianin could prevent diabetic retinopathy [112]. A phenol from *D. christyanum* (*n*-Docosyl 4-hydroxy-trans-cinnamate) and a bisbenzyl (dendrofalconerol A) from *D. tortile* exhibited high α-glucosidase inhibition showing their antidiabetic potential [111, 151]. Similarly, confusarin and 5-methoxy-7-hydroxy-9,10-dihydro-1,4-phenanthrenequinone of *D. formosum* also showed antidiabetic activity [123]. These compounds can form the basis of future development of plant based antidiabetic drugs.

20.8.7 Other Analeptic Properties in Orchids

There are several records of orchid-derived compounds exhibiting medicinal properties. *Dendrobium officinale* polysaccharides could be suggested for treating chronic obstructive pulmonary disease (COPD), liver injury in acute colitis, and constipation [138–140]. Dendronan, a polysaccharide from *D. officinale*, was reported to be playing role in maintenance of colonic health in rat [141]. Moscatilin, a bibenzyl derivative from *D. nobile*, was able to protect retinal cells from hypoxia/ischemia [133]. *Bletilla striata* polysaccharide composed of mannose and glucose plays a role in healing oral ulcers [94]. Various polysaccharides extracted from *Dendrobium* were reported to have multiple protective roles such as DHP1A has hepatoprotective potential by decreasing the expression of several inflammatory responses, DOP provided gastroprotective effect, and GXG (galactoxyloglucan) enhanced mucosal lining in intestinal areas [124, 142, 166]. Dendrosinen B, a bibenzyl derivative of *D. infundibulum,* was found to inhibit pancreatic lipase [128]. Polysaccharides from *D. officinale* were proved to be protective against neurodegenerative disease [125].

Also, orchids are being used as cosmetic agents. A detailed review by Hadi et al. [167] emphasized upon the importance of antiaging properties of *Vanda* extract. Similarly, stimulation of cytochrome c oxidase in *Vanda teres* by eucomic acid and vandateroside II provided antiaging effects [164].

20.9 Conclusions

Orchids are an immense source of active phytochemicals due to their varied life modes and physiology and their wide adaptive potential. The traditional literature is dotted with reports of their use as curatives and restoratives. However, to establish these plants as therapeutics and to exploit their vast reserve of phytochemicals and produce pharmaceutically active dosage forms, systemized observational studies need to be undertaken to create scientific evidences and link the ethnobotanical knowledge to state-of-the-art research and development in order to facilitate new drugs discovery.

Acknowledgments

JKS is grateful to CSIR for research support No. 38(1443)/17/EMR-II dated 25/5/17. SVP acknowledges financial support from the UGC-Start up grant No.F.4-5(65-FRP) (lv-cycle)/201 7 (BSR) under UGC Faculty Recharge Programme (GOI), New Delhi. DG is grateful for Senior Research Fellowship (File No. 09/135(0809)/2018-EMR-I) by CSIR. AK is grateful for DST INSPIRE (Provisional offer: DST/INSPIRE/03/2019/000643). The authors are also grateful to Department of Science and Technology, Government of India, for partial financial support under Promotion of University Research and Scientific Excellence (PURSE) grant scheme.

References

1 UNDP. Bioprospecting, 2016. http://www.undp.org/content/sdfinance/en/home/solutions/bioprospecting.html.
2 Shaw, J. (2017). The role of biodiversity centres in bioprospecting: a case study from Sarawak. In: *Bioprospecting*, 295–298. Cham: Springer.
3 Beattie, A.J., Hay, M., Magnusson, B. et al. (2011). Ecology and bioprospecting. *Austral Ecology* 36 (3): 341–356.
4 Gebru, A. (2017). Intellectual property and bioprospecting: a model legal framework. *North Carolina Journal of Law and Technology* 19 (2): 257–336.
5 Jayaraman, K.S. (1997). US patent office withdraws patent on Indian herb. *Nature* 389 (6646): 6.
6 Artuso, A. (2002). Bioprospecting, benefit sharing, and biotechnological capacity building. *World Development* 30 (8): 1355–1368.
7 Gomes, A.R., Duarte, A.C., and Rocha-Santos, T.A.P. (2016). Analytical techniques for discovery of bioactive compounds from marine fungi. In: *Fungal Metabolites*, Reference Series in Phytochemistry (eds. J.M. Mérillon and K. Ramawat), 415–434. Cham: Springer.
8 Strobel, G. and Daisy, B. (2003). Bioprospecting for microbial endophytes and their natural products. *Microbiology and Molecular Biology Reviews* 67 (4): 491–502.
9 Toghueo, R.M.K. (2020). Bioprospecting endophytic fungi from *Fusarium* genus as sources of bioactive metabolites. *Mycology* 11 (1): 1–21.
10 Pandey, A. and Yarzábal, L.A. (2019). Bioprospecting cold-adapted plant growth promoting microorganisms from mountain environments. *Applied Microbiology and Biotechnology* 103 (2): 643–657.
11 Thakur, K., Chownk, M., Kumar, V. et al. (2020). Bioprospecting potential of microbial communities in solid waste landfills for novel enzymes through metagenomic approach. *World Journal of Microbiology and Biotechnology* 36 (3): 1–15.
12 Zotchev, S.B., Sekurova, O.N., and Katz, L. (2012). Genome-based bioprospecting of microbes for new therapeutics. *Current Opinion in Biotechnology* 23 (6): 941–947.
13 Lawler, L.J. (1984). Ethnobotany of the Orchidaceae – a manual. In: *Orchid Biology Reviews and Perspectives*, vol. III (ed. J. Arditti), 27–149. Ithaca, NY: Cornell University Press.
14 Hossain, M.M. (2011). Therapeutic orchids: traditional uses and recent advances – an overview. *Fitoterapia* 82 (2): 102–140.

15 Verma, J. (2014). Orchids as nutraceuticals. *ZOOS'PRINT* 29 (9): 16.

16 Sut, S., Maggi, F., and Dall'Acqua, S. (2017). Bioactive secondary metabolites from orchids (Orchidaceae). *Chemistry and Biodiversity* 14 (11): e1700172.

17 Teoh, E.S. (2019). *Orchids as Aphrodisiac, Medicine or Food*. Springer.

18 Nugraha, A.S., Triatmoko, B., Wangchuk, P., and Keller, P.A. (2020). Vascular epiphytic medicinal plants as sources of therapeutic agents: their ethnopharmacological uses, chemical composition, and biological activities. *Biomolecules* 10 (2): 181.

19 Govaerts, R., Bernet, P., Kratochvil, K. et al. (2020). *World Checklist of Orchidaceae*. Kew: Facilitated by the Royal Botanic Gardens http://wcsp.science.kew.org/ (accessed 16 July 2020).

20 Pansarin, E.R. and Maciel, A.A. (2017). Evolution of pollination systems involving edible trichomes in orchids. *AoB Plants* 9 (4): plx033.

21 Stipkova, Z., Tsiftsis, S., and Kindlmann, P. (2020). Pollination mechanisms are driving orchid distribution in space. *Scientific Reports* 10 (1): 1–13.

22 Kaushik, P. (1983). *Ecological and Anatomical Marvels of the Himalayan Orchids*, vol. 8. New Delhi: Today and Tomorrow's Printers and Publishers.

23 Handa, S.S. (1986). Orchids for drugs and chemicals. *Biology, Conservation, and Culture of Orchids*: 89–100.

24 Behera, D., Rath, C.C., and Mohapatra, U. (2013). Medicinal orchids in India and their conservation: a review. *Floriculture and Ornamental Biotechnology* 7 (1): 53–59.

25 Singh, D.K. (2001). Orchid diversity in India: an overview. In: *Orchids: Science and Commerce* (eds. P. Pathak, R.N. Sehgal, N. Shekhar, et al.), 35–65. Dehradun, India: Bishen Singh Mahendra Pal Singh.

26 Van Rheede, H.A. (1678–1693). *Hortus Malabaricus*, vol. 12. Amsterdam: V. S. Joannis and D. V. Joannis.

27 Kant, R., Verma, J., and Thakur, K. (2012). Distribution pattern, survival threats and conservation of 'Astavarga' orchids in Himachal Pradesh, Northwest Himalaya. *Plant Archives* 12 (1): 165–168.

28 Suyal, R., Bhatt, D., Rawal, R.S., and Tewari, L.M. (2019). Status of two threatened astavarga herbs, *Polygonatum cirrhifolium* and *Malaxis muscifera*, in West Himalaya: conservation implications. *Proceedings of the National Academy of Sciences, India Section B: Biological Sciences* 90 (3): 695–704.

29 Sharma, R., Martins, N., Kuca, K. et al. (2019). Chyawanprash: a traditional Indian bioactive health supplement. *Biomolecules* 9 (5): 161.

30 Vij, S.P., Verma, J., and Kumar, S.C. (2013). *Orchids of Himachal Pradesh*. Dehradun: Bishen Singh Mahendra Pal Singh.

31 Rahamtulla, M., Pradhan, U.C., Roy, A.K. et al. (2020). Ethnobotanical aspects of some orchids from Darjeeling Himalaya (India). In: *Orchid Biology: Recent Trends and Challenges* (eds. S. Khasim, S. Hegde, M. González-Arnao and K. Thammasiri), 451–472. Singapore: Springer.

32 Chauhan, N.S. (1990). Medicinal orchids of Himachal Pradesh. *Journal of the Orchid Society of India* 4 (1-2): 99–105.

33 Kumar, P.K.S., Subramoniam, A., and Pushpangadan, P. (2000). Aphrodisiac activity of *Vanda tessellata* (Roxb.) Hook. ex Don extract in male mice. *Indian Journal of Pharamacology* 32 (5): 300–304.

34 Reddy, K.N., Reddy, C.S., and Jadhav, S.N. (2005). Ethnobotany of certain orchids of Eastern Ghats of Andhra Pradesh. *The Indian Forester* 131 (1): 90–96.

35 Medhi, R.P. and Chakrabarti, S. (2009). Traditional knowledge of NE people on conservation of wild orchids. *Indian Journal of Traditional Knowledge* 8: 11–16.

36 Pathak, P., Chattopadhyay, A., Vij, S.P. et al. (2010). An update on the medicinal orchids of Himachal Pradesh with brief notes on their habit, distribution and flowering period. *Journal of Non Timber Forest Products* 17 (3): 365–372.

37 Bhatt, D., Sharma, P., Sharma, L., and Joshi, G.C. (2012). Folk herbal remedies for skin care in Kumaun Himalaya. *Journal of Nin-Timber Forest Products* 19: 309–312.

38 Nongdam, P. (2014). Ethno-medicinal uses of some orchids of Nagaland, North-east India. *Research Journal of Medicinal Plants* 8 (3): 126–139.

39 De, L.C., Rao, A.N., Rajeevan, P.K. et al. (2015). Medicinal and aromatic orchids – an overview. *International Journal of Current Research* 7 (9): 19931–19935.

40 Khajuria, A.K., Kumar, G., and Bisht, N.S. (2017). Diversity with ethnomedicinal notes on Orchids: a case study of Nagdev forest range, Pauri Garhwal, Uttarakhand, India. *Journal of Medicinal Plants Studies* 5 (1): 171–174.

41 Kumar, A., Samant, S.S., Tewari, L.M., and Paul, S. (2018). Diversity, distribution, indigenous uses, and status of orchids in Kalatop – Khajjiar Wildlife Sanctuary, Chamba District, Himachal Pradesh. *Journal of Orchid Society of India* 32: 93–98.

42 Chamoli, K.P. and Sharan, H. (2019). Ethno-medicinal properties of *Dactylorhiza hatagirea* in higher Himalayan villages of Rudraprayag district of Uttarakhand. *Journal of Mountain Research* 14 (2): 85–88.

43 Pal, R., Meena, N.K., Dayamma, M., and Singh, D.R. (2020). Ethnobotany and recent advances in Indian medicinal orchids. In: *Orchids Phytochemistry, Biology and Horticulture*, Reference Series in Phytochemistry (eds. J.M. Merillon and H. Kodja), 1–27. Cham: Springer.

44 Bulpitt, C.J., Li, Y., Bulpitt, P.F., and Wang, J.J. (2007). The use of orchids in Chinese medicine. *Journal of the Royal Society of Medicine* 100 (12): 558–563.

45 Teoh, E.S. (2016). *Medicinal Orchids of Asia*. Springer.

46 Leon, C. and Lin, Y.L. (2017). *Chinese Medicinal Plants, Herbal Drugs and Substitutes: An Identification Guide*. Kew Publishing.

47 Kong, J.M., Goh, N.K., Chia, L.S., and Chia, T.F. (2003). Recent advances in traditional plant drugs and orchids. *Acta Pharmacologica Sinica* 24 (1): 7–21.

48 Duggal, S.C. (1971). Orchids in human affairs (a review). *Quarterly Journal of Crude Drug Research* 11 (2): 1727–1734.

49 Sulistiarini, D. (2003). Grammatophyllum scriptum Bl. In: *Plant Resources of South-East Asia No 12(3): Medicinal and Poisonous Plants 3* (eds. R.H.M.J. Lemmens and N. Bunyapraphatsara), 222. Leiden: Backhuys.

50 Garvita, R.V. and Wawangningrum, H. (2020). Potential of terrestrial orchid as a medicine and in vitro; multiplication. *Prosiding Seminar Nasional Masyarakat Biodiversitas Indonesia* 6: 512–519.

51 Marasini, R. and Joshi, S. (2012). Antibacterial and antifungal activity of medicinal orchids growing in Nepal. *Journal of Nepal Chemical Society* 29: 104–109.

52 Subedi, A., Kunwar, B., Choi, Y. et al. (2013). Collection and trade of wild-harvested orchids in Nepal. *Journal of Ethnobiology and Ethnomedicine* 9 (1): 64.

53 Vaidya, B.N. (2019). Nepal: a global hotspot for medicinal orchids. In: *Medicinal Plants* (eds. N. Joshee, S. Dhekney and P. Parajuli), 35–80. Cham: Springer.

54 Yonzone, R., Lama, D., and Bhujel, R.B. (2011). Medicinal orchids of the Himalayan region. *Pleione* 5 (2): 265–273.

55 Basu, K., Dasgupta, B., Bhattacharya, S. et al. (1971). Anti-inflammatory principles of *Vanda roxburghii. Current Science* 40: 80–86.

56 Pant, B. (2013). Medicinal orchids and their uses: tissue culture a potential alternative for conservation. *African Journal of Plant Science* 7 (10): 448–467.

57 Chowdhury, M.A., Rahman, M.M., Chowdhury, M.R.H. et al. (2014). Antinociceptive and cytotoxic activities of an epiphytic medicinal orchid: *Vanda tessellata* Roxb. *BMC Complementary and Alternative Medicine* 14 (1): 464.

58 Chauhan, N.S., Sharma, V., Dixit, V.K., and Thakur, M. (2014). A review on plants used for improvement of sexual performance and virility. *BioMed Research International* 2014: 868062.

59 Chhajed, M.R., Tomar, G.S., Gautam, S.P., and Hariharan, A.G. (2008). Phytochemical investigation and evaluation of in vitro; free radical scavenging activity of *Flickingeria nodosa* Lindl. *Indian Journal of Pharmaceutical Education and Research* 42 (4): 377–380.

60 Mukherjee, S., Phatak, D., Parikh, J. et al. (2012). Antiglycation and antioxidant activity of a rare medicinal orchid *Dendrobium aqueum* Lindl. *Medicinal Chemistry & Drug Discovery* 2 (2): 17–29.

61 Nagananda, G.S., Ashwini, P., Jaykumar, V.K., and Rajath, S. (2014). Phytochemical evaluation and *in vitro;* free radical scavenging activity of cold and hot successive pseudobulb extracts of medicinally important orchid *Pholidota pallida* Lindl. *Advances in Bioresearch* 5: 100–105.

62 Minh, T.N., Khang, D.T., Tuyen, P.T. et al. (2016). Phenolic compounds and antioxidant activity of *Phalaenopsis* orchid hybrids. *Antioxidants* 5 (3): 31.

63 Minh, T.N., Tuyen, P.T., Khang, D.T. et al. (2017). Potential use of plant waste from the moth orchid (*Phalaenopsis* Sogo Yukidian "V3") as an antioxidant source. *Foods* 6 (10): 85.

64 Hossain, M.M., Akter, S., and Uddin, S.B. (2020). Screening of bioactive phytochemicals in some indigenous epiphytic orchids of Bangladesh. In: *Orchid Biology: Recent Trends and Challenges* (eds. S. Khasim, S. Hegde, M. González-Arnao and K. Thammasiri), 481–506. Singapore: Springer Nature.

65 Morris, B. (2003). Children of the wind-orchids as medicine in Malawi. *Orchid Review* 111: 271–277.

66 Chinsamy, M., Finnie, J.F., and Van Staden, J. (2011). The ethnobotany of South African medicinal orchids. *South African Journal of Botany* 77 (1): 2–9.

67 Berliocchi, L. and Griffiths, M. (2000). *The Orchid in Lore and Legend*. Portland, Oregon: Timber Press.

68 Fonge, B.A., Essomo, S.E., Bechem, T.E. et al. (2019). Market trends and ethnobotany of orchids of Mount Cameroon. *Journal of Ethnobiology and Ethnomedicine* 15 (1): 29.

69 Hutchings, A., Scott, A.H., Lewis, G., and Cunningham, A.B. (1996). *Zulu Medicinal Plants: An Inventory*. Pietermaritzburg: University of Natal Press.

70 Kasulo, V., Mwabumba, L., and Cry, M. (2009). A review of edible orchids in Malawi. *Journal of Horticulture and Forestry* 1 (7): 133–139.

71 Watt, J.M. and Breyer-Brandwijk, M.G. (1962). *The Medicinal and Poisonous Plants of Southern and Eastern Africa*, 2e. Edinburgh: E&S Livingstone.

72 Hulme, M.M. (1954). *Wild Flowers of Natal*. Pietermaritzburg: Schuter and Shooter.

73 Bhattacharyya, P. and Van Staden, J. (2016). *Ansellia africana* (Leopard orchid): a medicinal orchid species with untapped reserves of important biomolecules – a mini review. *South African Journal of Botany* 106: 181–185.

74 Chinsamy, M., Finnie, J.F., and Van Staden, J. (2014). Anti-inflammatory, antioxidant, anti-cholinesterase activity and mutagenicity of South African medicinal orchids. *South African Journal of Botany* 91: 88–98.

75 Lubinsky, P., Bory, S., Hernández, J.H. et al. (2008). Origins and dispersal of cultivated vanilla (*Vanilla planifolia* Jacks. [Orchidaceae]). *Economic Botany* 62 (2): 127–138.

76 Randriamiharisoa, M.N., Kuhlman, A.R., Jeannoda, V. et al. (2015). Medicinal plants sold in the markets of Antananarivo, Madagascar. *Journal of Ethnobiology and Ethnomedicine* 11 (1): 60.

77 Parkinson, J. (1640). *Theatrum Botanical: The Theatre of Plants*. London: Thomas Cotes.

78 Bulpitt, C.J. (2005). The uses and misuses of orchids in medicine. *QJM* 98 (9): 625–631.

79 Sezik, E. (2002). Turkish orchids and salep. *Acta Pharmaceutica Turcica* 44: 151–157.

80 Ghorbani, A., Gravendeel, B., Selliah, S. et al. (2017). DNA barcoding of tuberous Orchidoideae: a resource for identification of orchids used in Salep. *Molecular Ecology Resources* 17 (2): 342–352.

81 Ghorbani, A., Gravendeel, B., Naghibi, F., and de Boer, H. (2014). Wild orchid tuber collection in Iran: a wake-up call for conservation. *Biodiversity and Conservation* 23 (11): 2749–2760.

82 Castle, L. (1886). *Orchids: Their Structure, History and Culture (Illustrated)*. London: Fleet Street.

83 Moerman, D.E. (1986). *Medicinal Plants of the Native Americans*, vol. 2. University of Michigan Museum.

84 Wilson, M.F. (2007). *Medicinal Plant Fact Sheet: Cypripedium: Lady's Slipper Orchids*. Virginia: *Arlington*.

85 Solano Gómez, R., Cruz-Lustre, G., Martinez-Feria, F., and Lagunez-Rivera, L. (2010). Plantas utilizadas en la celebracion de la Semana Santa en Zaachila, Oaxaca, Mexico. *Polibotanica* 29: 263–279.

86 Garcia, G.C., Gomez, R.S., and Rivera, L.L. (2014). Documentation of the medicinal knowledge of *Prosthechea karwinskii* in a Mixtec community in Mexico. *Revista Brasileira de Farmacognosia* 24 (2): 153–158.

87 Hartmann, W. (1972). Las orquideas en la medicina y otros usos practicos. *Orquidea (Mexico City)* 2: 70–71.

88 Asseleih, L.M.C., García, R.A.M., and Cruz, J.Y.S.R. (2015). Ethnobotany, pharmacology and chemistry of medicinal orchids from Veracruz. *Journal of Agriclutural Science and Technology* A, 5: 745–754.

89 Lawler, L.J. and Slaytor, M. (1970). Uses of Australian orchids by aborigines and early settlers. *The Medical Journal of Australia* 2 (26): 1259–1261.

90 Lai, C.Y., Yang, L.C., and Lin, W.C. (2015). Type II arabinogalactan from *Anoectochilus formosanus* induced dendritic cell maturation through TLR2 and TLR4. *Phytomedicine* 22 (14): 1207–1214.

91 Zeng, B., Su, M., Chen, Q. et al. (2016). Antioxidant and hepatoprotective activities of polysaccharides from *Anoectochilus roxburghii*. *Carbohydrate Polymers* 153: 391–398.

92 Yang, X., Tang, C., Zhao, P. et al. (2012). Antimicrobial constituents from the tubers of *Bletilla ochracea*. *Planta Medica* 78 (6): 606–610.

93 Li, J.Y., Kuang, M.T., Yang, L. et al. (2018). Stilbenes with anti-inflammatory and cytotoxic activity from the rhizomes of *Bletilla ochracea* Schltr. *Fitoterapia* 127: 74–80.

94 Liao, Z., Zeng, R., Hu, L. et al. (2019). Polysaccharides from tubers of *Bletilla striata*: physicochemical characterization, formulation of buccoadhesive wafers and preliminary study on treating oral ulcer. *International Journal of Biological Macromolecules* 122: 1035–1045.

95 Jiang, F., Li, M., Wang, H. et al. (2019). Coelonin, an anti-inflammation active component of *Bletilla striata* and its potential mechanism. *International Journal of Molecular Sciences* 20: 4422.

96 Jiang, S., Chen, C.F., Ma, X.P. et al. (2019). Antibacterial stilbenes from the tubers of *Bletilla striata*. *Fitoterapia* 138: 104350.

97 Jiang, S., Wan, K., Lou, H.Y. et al. (2019). Antibacterial bibenzyl derivatives from the tubers of *Bletilla striata*. *Phytochemistry* 162: 216–223.

98 Qian, C.D., Jiang, F.S., Yu, H.S. et al. (2015). Antibacterial biphenanthrenes from the fibrous roots of *Bletilla striata*. *Journal of Natural Products* 78 (4): 939–943.

99 Chen, Z., Zhao, Y., Zhang, M. et al. (2020). Structural characterization and antioxidant activity of a new polysaccharide from *Bletilla striata* fibrous roots. *Carbohydrate Polymers* 227: 115362.

100 Qu, Y., Li, C., Zhang, C. et al. (2016). Optimization of infrared-assisted extraction of *Bletilla striata* polysaccharides based on response surface methodology and their antioxidant activities. *Carbohydrate Polymers* 148: 345–353.

101 Wang, Y., Han, S., Li, R. et al. (2019). Structural characterization and immunological activity of polysaccharides from the tuber of *Bletilla striata*. *International Journal of Biological Macromolecules* 122: 628–635.

102 Peng, Q., Li, M., Xue, F., and Liu, H. (2014). Structure and immunobiological activity of a new polysaccharide from *Bletilla striata*. *Carbohydrate Polymers* 107: 119–123.

103 Zhou, D., Chang, W., Liu, B. et al. (2020). Stilbenes from the tubers of *Bletilla striata* with potential anti-neuroinflammatory activity. *Bioorganic Chemistry* 97: 103715.

104 Zhou, D., Chen, G., Ma, Y.P. et al. (2019). Isolation, structural elucidation, optical resolution, and antineuroinflammatory activity of phenanthrene and 9, 10-Dihydrophenanthrene derivatives from *Bletilla striata*. *Journal of Natural Products* 82: 2238–2245.

105 Fang, Y.S., Yang, M.H., Cai, L. et al. (2018). New phenylpropanoids from *Bulbophyllum retusiusculum*. *Archives of Pharmacal Research* 41 (11): 1074–1081.

106 Tu, Y., Huang, J., and Li, F. (2018). Anticholinesterase, antioxidant, and beta-amyloid aggregation inhibitory constituents from *Cremastra appendiculata*. *Medicinal Chemistry Research* 27: 857–863.

107 Lertnitikul, N., Pattamadilok, C., Chansriniyom, C., and Suttisri, R. (2020). A new dihydrophenanthrene from *Cymbidium finlaysonianum* and structure revision of cymbinodin-A. *Journal of Asian Natural Products Research* 22 (1): 83–90.

108 Liu, H., Ma, J., Gong, F. et al. (2017). Structural characterisation and immunomodulatory effects of polysaccharides isolated from *Dendrobium aphyllum*. *International Journal of Food Science and Technology* 53 (5): 1185–1194.

109 Yang, D., Liu, L.Y., Cheng, Z.Q. et al. (2015). Five new phenolic compounds from *Dendrobium aphyllum*. *Fitoterapia* 100: 11–18.

110 Zhang, L., Fang, Y., Xu, X.F., and Jin, D.Y. (2017). Moscatilin induces apoptosis of pancreatic cancer cells via reactive oxygen species and the JNK/SAPK pathway. *Molecular Medicine Reports* 15 (3): 1195–1203.

111 San, H.T., Boonsnongcheep, P., Putalun, W. et al. (2020). α-Glucosidase inhibitory and glucose uptake stimulatory effects of phenolic compounds from *Dendrobium christyanum*. *Natural Product Communications* 15: 1934578X20913453.

112 Yu, Z., Zhang, T., Gong, C. et al. (2016). Erianin inhibits high glucose-induced retinal angiogenesis via blocking ERK1/2-regulated HIF-1α-VEGF/VEGFR2 signaling pathway. *Scientific Reports* 6: 34306.

113 Hu, Y., Zhang, C., Zhao, X. et al. (2016). (±)-Homocrepidine A, a pair of anti-inflammatory enantiomeric octahydroindolizine alkaloid dimers from *Dendrobium crepidatum*. *Journal of Natural Products* 79: 252–256.

114 Hu, Y., Yang, H., Ding, X. et al. (2020). Anti-inflammatory octahydroindolizine alkaloid enantiomers from *Dendrobium crepidatum*. *Bioorganic Chemistry* 100: 103809.

115 Xu, X., Chen, X., Yang, R. et al. (2020). Crepidtumines A and B, two novel indolizidine alkaloids from *Dendrobium crepidatum*. *Scientific Reports* 10 (1): 1–8.

116 Lin, Y., Wang, F., Yang, L.J. et al. (2013). Anti-inflammatory phenanthrene derivatives from stems of *Dendrobium denneanum*. *Phytochemistry* 95: 242–251.

117 Sun, J., Zhang, F., Yang, M. et al. (2014). Isolation of α-glucosidase inhibitors including a new flavonol glycoside from *Dendrobium devonianum*. *Natural Product Research* 28: 1900–1905.

118 Wu, Y.G., Wang, K.W., Zhao, Z.R. et al. (2019). A novel polysaccharide from *Dendrobium devonianum* serves as a TLR4 agonist for activating macrophages. *International Journal of Biological Macromolecules* 133: 564–574.

119 Deng, Y., Li, M., Chen, L.X. et al. (2018). Chemical characterization and immunomodulatory activity of acetylated polysaccharides from *Dendrobium devonianum*. *Carbohydrate Polymers* 180: 238–245.

120 Bhummaphan, N. and Chanvorachote, P. (2015). Gigantol suppresses cancer stem cell-like phenotypes in lung cancer cells. *Evidence-based Complementary and Alternative Medicine* 2015: 836564.

121 Pengpaeng, P., Sritularak, B., and Chanvorachote, P. (2015). Dendrofalconerol a suppresses migrating cancer cells via EMT and integrin proteins. *Anticancer Research* 35 (1): 201–206.

122 Liu, G.Y., Tan, L., Cheng, L. et al. (2020). Dendrobine-type alkaloids and bibenzyl derivatives from *Dendrobium findlayanum*. *Fitoterapia* 142: 104497.

123 Inthongkaew, P., Chatsumpun, N., Supasuteekul, C. et al. (2017). α-Glucosidase and pancreatic lipase inhibitory activities and glucose uptake stimulatory effect of phenolic compounds from *Dendrobium formosum*. *Revista Brasileira de Farmacognosia* 27: 480–487.

124 Tian, C.C., Zha, X.Q., and Luo, J.P. (2015). A polysaccharide from *Dendrobium huoshanense* prevents hepatic inflammatory response caused by carbon tetrachloride. *Biotechnology and Biotechnological Equipment* 29: 132–138.

125 Liang, J., Wu, Y., Yuan, H. et al. (2019). *Dendrobium officinale* polysaccharides attenuate learning and memory disabilities via anti-oxidant and anti-inflammatory actions. *International Journal of Biological Macromolecules* 126: 414–426.

126 Wang, H.Y., Li, Q.M., Yu, N.J. et al. (2019). *Dendrobium huoshanense* polysaccharide regulates hepatic glucose homeostasis and pancreatic β-cell function in type 2 diabetic mice. *Carbohydrate Polymers* 211: 39–48.

127 Li, F., Cui, S.H., Zha, X.Q. et al. (2015). Structure and bioactivity of a polysaccharide extracted from protocorm-like bodies of *Dendrobium huoshanense*. *International Journal of Biological Macromolecules* 72: 664–672.

128 Ranong, S.N., Likhitwitayawuid, K., Mekboonsonglarp, W., and Sritularak, B. (2019). New dihydrophenanthrenes from *Dendrobium infundibulum*. *Natural Product Research* 33: 420–426.

129 Lu, Y., Kuang, M., Hu, G.P. et al. (2014). Loddigesiinols G-J: α-glucosidase inhibitors from *Dendrobium loddigesii*. *Molecules* 19: 8544–8555.

130 Cardile, V., Avola, R., Graziano, A.C., and Russo, A. (2020). Moscatilin, a bibenzyl derivative from the orchid *Dendrobium loddigesii*, induces apoptosis in melanoma cells. *Chemico-Biological Interactions* 323: 109075.

131 Chen, W.K., Chen, C.A., Chi, C.W. et al. (2019). Moscatilin inhibits growth of human esophageal cancer xenograft and sensitizes cancer cells to radiotherapy. *Journal of Clinical Medicine* 8 (2): 187.

132 Pai, H.C., Chang, L.H., Peng, C.Y. et al. (2013). Moscatilin inhibits migration and metastasis of human breast cancer MDA-MB-231 cells through inhibition of Akt and Twist signaling pathway. *Journal of Molecular Medicine* 91 (3): 347–356.

133 Chao, W.H., Lai, M.Y., Pan, H.T. et al. (2018). *Dendrobium nobile* Lindley and its bibenzyl component moscatilin are able to protect retinal cells from ischemia/hypoxia by dowregulating placental growth factor and upregulating Norrie disease protein. *BMC Complementary and Alternative Medicine* 18: 193.

134 Zhou, X.M., Zheng, C.J., Gan, L.S. et al. (2016). Bioactive phenanthrene and bibenzyl derivatives from the stems of *Dendrobium nobile*. *Journal of Natural Products* 79 (7): 1791–1797.

135 Zhang, Y., Zhang, Q., Xin, W. et al. (2019). Nudol, a phenanthrene derivative from *Dendrobium nobile*, induces cell cycle arrest and apoptosis and inhibits migration in osteosarcoma cells. *Drug Design, Development and Therapy* 13: 2591–2601.

136 Cheng, L., Guo, D.L., Zhang, M.S. et al. (2020). Dihydrophenanthrofurans and bisbibenzyl derivatives from the stems of *Dendrobium nobile*. *Fitoterapia* 143: 104586.

137 Chu, C., Li, T., Pedersen, H.A. et al. (2019). Antidiabetic constituents of *Dendrobium officinale* as determined by high-resolution profiling of radical scavenging and α-glucosidase and α-amylase inhibition combined with HPLC-PDA-HRMS-SPE-NMR analysis. *Phytochemistry Letters* 31: 47–52.

138 Song, T.H., Chen, X.X., Tang, S.C.W. et al. (2016). *Dendrobium officinale* polysaccharides ameliorated pulmonary function while inhibiting mucin-5AC and stimulating aquaporin-5 expression. *Journal of Functional Foods* 21: 359–371.

139 Liang, J., Chen, S., Hu, Y. et al. (2018). Protective roles and mechanisms of *Dendrobium officinale* polysaccharides on secondary liver injury in acute colitis. *International Journal of Biological Macromolecules* 107: 2201–2210.

140 Luo, D., Qu, C., Lin, G. et al. (2017). Character and laxative activity of polysaccharides isolated from *Dendrobium officinale*. *Journal of Functional Foods* 34: 106–117.

141 Zhang, G.Y., Nie, S.P., Huang, X.J. et al. (2016). Study on *Dendrobium officinale* O-Acetyl-glucomannan (Dendronan). 7. Improving effects on colonic health of mice. *Journal of Agricultural and Food Chemistry* 64: 2485–2491.

142 Zeng, Q., Ko, C.H., Siu, W.S. et al. (2017). Polysaccharides of *Dendrobium officinale* Kimura & Migo protect gastric mucosal cell against oxidative damage-induced apoptosis in vitro; and in vivo;. *Journal of Ethnopharmacology* 208: 214–224.

143 Tao, S., Lei, Z., Huang, K. et al. (2019). Structural characterization and immunomodulatory activity of two novel polysaccharides derived from the stem of *Dendrobium officinale* Kimura et Migo. *Journal of Functional Foods* 57: 121–134.

144 He, T.B., Huang, Y.P., Yang, L. et al. (2016). Structural characterization and immunomodulating activity of polysaccharide from *Dendrobium officinale. International Journal of Biological Macromolecules* 83: 34–41.

145 Ren, G., Deng, W.Z., Xie, Y.F. et al. (2020). Bibenzyl derivatives from leaves of *Dendrobium officinale. Natural Product Communications* 15 (2): 1–5.

146 Zhang, Y., Zhang, L., Liu, J. et al. (2017). *Dendrobium officinale* leaves as a new antioxidant source. *Journal of Functional Foods* 37: 400–415.

147 Kyokong, N., Muangnoi, C., Thaweesest, W. et al. (2019). A new phenanthrene dimer from *Dendrobium palpebrae. Journal of Asian Natural Products Research* 21 (4): 391–397.

148 Kongkatitham, V., Muangnoi, C., Kyokong, N. et al. (2018). Anti-oxidant and anti-inflammatory effects of new bibenzyl derivatives from *Dendrobium parishii* in hydrogen peroxide and lipopolysaccharide treated RAW264.7 cells. *Phytochemistry Letters* 24: 31–38.

149 Chen, D.N., Wang, Y.Y., Liu, W.J. et al. (2020). Stilbenoids from aerial parts of *Dendrobium plicatile. Natural Product Research* 34 (3): 323–328.

150 Sarakulwattana, C., Mekboonsonglarp, W., Likhitwitayawuid, K. et al. (2020). New bisbibenzyl and phenanthrene derivatives from *Dendrobium scabrilingue* and their α-glucosidase inhibitory activity. *Natural Product Research* 34 (12): 1694–1701.

151 Limpanit, R., Chuanasa, T., Likhitwitayawuid, K. et al. (2016). a-Glucosidase inhibitors from *Dendrobium tortile. Records of Natural Products* 10: 609–616.

152 Yang, L.C., Lu, T.J., Hsieh, C.C., and Lin, W.C. (2014). Characterization and immunomodulatory activity of polysaccharides derived from *Dendrobium tosaense. Carbohydrate Polymers* 111: 856–863.

153 Schuster, R., Zeindl, L., Holzer, W. et al. (2017). *Eulophia macrobulbon* – an orchid with significant anti-inflammatory and antioxidant effect and anticancerogenic potential exerted by its root extract. *Phytomedicine* 24: 157–165.

154 Bao, Q., Qian, L., Gong, C., and Shen, X. (2017). Immune-enhancing activity of polysaccharides from *Gastrodia elata. Journal of Food Processing and Preservation* 41 (4): e13016.

155 Farooq, U., Pan, Y., Lin, Y. et al. (2019). Structure characterization and action mechanism of an antiaging new compound from *Gastrodia elata* Blume. *Oxidative Medicine and Cellular Longevity* 2019: 5459862.

156 Jiang, T., Chu, J., Chen, H. et al. (2020). Gastrodin inhibits H_2O_2-induced ferroptosis through its antioxidative effect in rat glioma cell line C6. *Biological and Pharmaceutical Bulletin* 43 (3): 480–487.

157 Yang, L., Jiang, R., Li, H.H. et al. (2020). Three new compounds from the flower branch of *Gastrodia elata* Blume and anti-microbial activity. *RSC Advances* 10 (25): 14644–14649.

158 Cretton, S., Oyarzún, A., Righi, D. et al. (2018). A new antifungal and antiprotozoal bibenzyl derivative from *Gavilea lutea. Natural Product Research* 32 (6): 695–701.

159 Ren, J., Qian, X.P., Guo, Y.G. et al. (2016). Two new phenanthrene glycosides from *Liparis regnieri* finet and their antibacterial activities. *Phytochemistry Letters* 18: 64–67.

160 Nwe, S.Y., Tungphatthong, C., Laorpaksa, A. et al. (2020). Bioassay-guided isolation of topoisomerase I poison from *Paphiopedilum callosum* (Rchb.f.) stein. *Records of Natural Products* 14 (2): 89–97.

161 Lertnitikul, N., Jittham, P., Khankhampoch, L. et al. (2016). Cytotoxic stilbenes from the roots of *Paphiopedilum godefroyae*. *Journal of Asian Natural Products Research* 18 (12): 1143–1150.

162 Li, Y., Zhang, F., Wu, Z.H. et al. (2015). Nitrogen-containing bibenzyls from *Pleione bulbocodioides*: absolute configurations and biological activities. *Fitoterapia* 102: 120–126.

163 Liu, L., Yin, Q.M., Yan, X. et al. (2019). Bioactivity-guided isolation of cytotoxic phenanthrenes from *Spiranthes sinensis*. *Journal of Agricultural and Food Chemistry* 67 (26): 7274–7280.

164 Simmler, C., Antheaume, C., André, P. et al. (2011). Glucosyloxybenzyleucomate derivatives from *Vanda teres* stimulate HaCaT cytochrome c oxidase. *Journal of Natural Products* 74: 949–955.

165 Pan, L.H., Li, X.F., Wang, M.N. et al. (2014). Comparison of hypoglycemic and antioxidative effects of polysaccharides from four different *Dendrobium* species. *International Journal of Biological Macromolecules* 64: 420–427.

166 Xie, S.Z., Liu, B., Ye, H.Y. et al. (2019). *Dendrobium huoshanense* polysaccharide regionally regulates intestinal mucosal barrier function and intestinal microbiota in mice. *Carbohydrate Polymers* 206: 149–162.

167 Hadi, H., Razali, S.N.S., and Awadh, A.I. (2015). A comprehensive review of the cosmeceutical benefits of *Vanda* species (Orchidaceae). *Natural Product Communications* 10: 1483–1488.

Index

Bioprospecting of Plant Biodiversity for Industrial Molecules, First Edition.
Edited by Santosh Kumar Upadhyay and Sudhir P. Singh.

e

f

g